SMECTIC C

Point group C_{2h}

SMECTIC C*

Point group C_2

z

x

y

Rotation 180° around $\hat{y}$

$(P_x, P_y, P_z) \rightarrow (-P_x, P_y, -P_z)$ $(P_x, P_y, P_z) \rightarrow (-P_x, P_y, -P_z)$

$\Downarrow$ $\Downarrow$

$\vec{P} = (0, P_y, 0)$ $\vec{P} = (0, P_y, 0)$

Reflexion in the xz plane

$(0, P_y, 0) \rightarrow (0, -P_y, 0)$ not allowed due to chirality

$\Downarrow$ $\Downarrow$

$\vec{P} = (0, 0, 0)$ $\vec{P} = (0, P_y, 0)$

Ferroelectricity not allowed Ferroelectricity allowed

The Physics of Ferroelectric and Antiferroelectric Liquid Crystals

I. MUŠEVIČ
R. BLINC
B. ŽEKŠ
J. STEFAN INSTITUTE, LJUBLJANA
UNIVERSITY OF LJUBLJANA
SLOVENIA

THE PHYSICS OF FERROELECTRIC AND ANTIFERROELECTRIC LIQUID CRYSTALS

World Scientific
Singapore • New Jersey • London • Hong Kong

Published by

World Scientific Publishing Co. Pte. Ltd.
P O Box 128, Farrer Road, Singapore 912805
USA office: Suite 1B, 1060 Main Street, River Edge, NJ 07661
UK office: 57 Shelton Street, Covent Garden, London WC2H 9HE

British Library Cataloguing-in-Publication Data
A catalogue record for this book is available from the British Library.

THE PHYSICS OF FERROELECTRIC AND ANTIFERROELECTRIC LIQUID CRYSTALS

Fig. 1.3.6 is reprinted with permission from "Spontaneous formation of macroscopic chiral domains in a fluid smectic phase of achiral molecules", by Link, D.R., Natale, G., Shao, R., Maclennan, J.E., Clark, N.A., Körblova, E., Walba, D., *Science* Vol. 278, pp. 1924–1927(1997). Copyright © 1997 American Association for the Advancement of Science.

Fig. 9.2.5 is reprinted with permission from "Smectic liquid crystal monolayers on graphite observed by STM", by Smith, D.P.E., Horber, H., Gerber, Ch., Binnig, G., *Science* Vol. 245, pp. 43–45(1989). Copyright © 1989 American Association for the Advancement of Science.

Fig. 9.2.6 is reprinted with permission from "Near atomic resolution imaging of ferroelectric liquid crystal molecules on graphite by STM", by Walba, D.M., Stevens, F., Parks, D.C., Clark, N.A., Wand, M.D., *Science* Vol. 267, pp. 1144–1147(1995). Copyright © 1995 American Association for the Advancement of Science.

ISBN 981-02-0325-X

Printed in Singapore by World Scientific Printers

To Maja, Sašo and Nataša for their patience and understanding.

Igor

Contents

Chapter 8 Antiferroelectricity in Liquid Crystals

Chapter 9 STM and AFM in Complex Liquids

Chapter 10 Electrooptic Effects in Ferro- and Antiferroelectric Phases and Their Applications

Appendix A Landau Theory of Second Order Phase Transitions

Preface

When ferroelectricity was first reported in liquid crystals by R.B.Meyer and co-workers in 1975, the idea of having a spontaneous electric polarization in the liquid crystalline state was revolutionary. In those days, it was difficult for liquid crystal community to accept this idea, as scientists have been looking for ferroelectric liquids for a long time with no success. Whereas ferroelectricity in liquid crystals was first considered only as a beautiful and extraordinary natural phenomenon, the technological importance of these materials was realized after the discovery of fast electrooptic switching in ferroelectric liquid crystals by N.A.Clark and S.T.Lagerwall in 1980. This discovery triggered a long and fruitful period of research in the field of ferroelectric liquid crystals that culminated in discovery of antiferroelectricity and intermediate ferrielectric phases in liquid crystals by Chandani et al. in 1989. Again, this discovery was accepted with great reluctance by the liquid crystal community, as seems to be the case for outstanding achievements in general. However, if we look back and consider the development of this field in the past, we see that our understanding of the physics of these materials is now consistent. This seemed to us to be the right moment to write this book.

The book presents mainly the results of the work that has been done by us and our colleagues from the J.Stefan Institute, Ljubljana and the University of Ljubljana. We were lucky that a coherent group of theoreticians and experimentalists was formed, that were working on different aspects of liquid crystals from the very beginning of this field. We would in particular like to acknowledge the contributions of Martin Čopič, Adrijan Levstik, Cene Filipič, Mojca Čepič, Boštjan Zalar, Janez Pirš, Silva Pirš, Miha Škarabot, Irena

Drevenšek, Slobodan Žumer, Marija Vilfan and others. We would also like to thank Theo Rasing for his cooperation in high magnetic field experiments.

We have written this book in a way that it can be easily followed by both graduate students of physics, as well as industrial scientists working in application of liquid crystals. Individual chapters present a coherent subject, with not much overlap between each other. The first chapter is introductory and gives an overview of the phenomena that are discussed in detail in other chapters. Some chapters, like for example 3.4., 3.5., 7.8, 8.8 and 10.6. are very specialized and readers who do not want to get bored by mathematics or the specialized subject are advised to skip them at the first reading. Finally, we give a long list of references that is nevertheless far from complete. It would take too much space to include all relevant references in the field and we are sorry that we had to select them according to the contents of the book.

Ljubljana,
December 31, 1999

Igor Muševič
Robert Blinc
Boštjan Žekš

Chapter 1

Ferroelectricity and Antiferroelectricity in Liquid Crystals

1.1. Symmetry and Ferroelectricity in Liquid Crystals

Robert Meyer and co-workers discovered ferroelectricity in liquid crystals in 1975. The paper entitled "Ferroelectric Liquid Crystals" appeared in Le Journal de Physique Lettres, Vol.36, page L-69 and was written by R.B.Meyer, L.Liebert, L.Strzelecki and P.Keller. In the paper, Meyer and his coworkers reported a large linear electrooptic effect in a new chiral smectic material, p-decyloxybenzylidene p'-amino 2-methyl butyl cinnamate (DOBAMBC), shown in Fig.1.1.1.

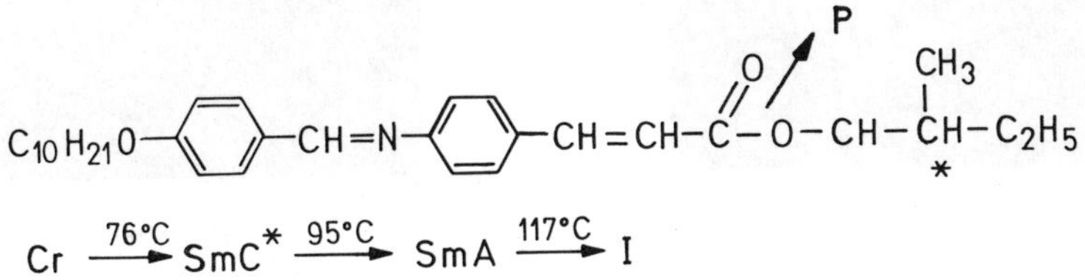

$$C_{10}H_{21}O-C_6H_4-CH=N-C_6H_4-CH=CH-C(=O)-O-CH_2-\overset{*}{C}H(CH_3)-C_2H_5$$

$$\mathrm{Cr} \xrightarrow{76^\circ C} \mathrm{SmC}^* \xrightarrow{95^\circ C} \mathrm{SmA} \xrightarrow{117^\circ C} \mathrm{I}$$

Fig.1.1.1. Molecular structure of the first ferroelectric liquid crystal p-decyloxybenzylidene p'-amino 2-methyl butyl cinnamate (DOBAMBC), used in the experiments of Meyer et al. (1975).

Meyer and his coworkers were investigating the effect of a DC electric field, applied perpendicularly to the helical axis, on the shape of the conoscopic figure in homeotropically aligned samples. In a certain temperature range, they observed a very large linear electrooptic effect, presented in Figs.1.1.2.: For a given polarity of the field, the originally uniaxial conoscopic figure tilted as a whole in a plane perpendicular to the direction of the applied field. At large fields, the conoscopic image became biaxial and saturated. At saturation, the tilt of the conoscopic image roughly corresponded to the molecular tilt. By reversing the electric field, the direction of the tilt of the conoscopic image would reverse, which was a direct evidence of a very large, linear electrooptic effect.

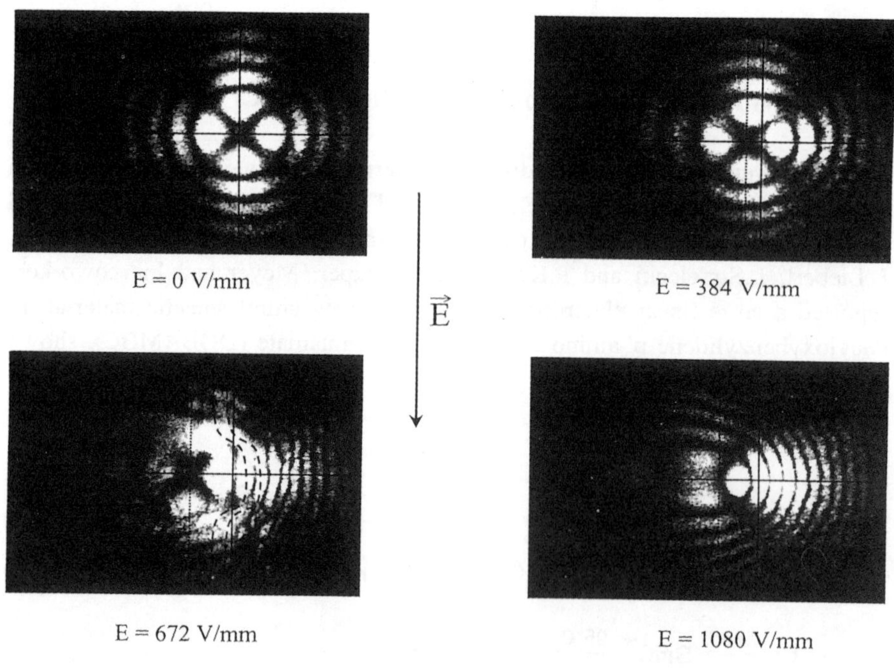

Fig.1.1.2. The conoscopic figure of a ferroelectric smectic liquid crystal in a transverse DC electric field. At small fields, the direction of the optical axis tilts in a plane perpendicular to the electric field. At higher fields, the image becomes biaxial and saturates above the critical field. After Conradi et al., 1999. For details on conoscopic measurements, see Note 1.1.4.

At that time, the only known linear electrooptic effect in liquid crystals was the flexoelectric effect in nematic liquid crystals, which was also investigated by R.B.Meyer (1969). In fact, as noted by Meyer himself (1977), the flexoelectric effect in nematics and the related work of F.C.Frank and W.L.McMillan has led him to the conclusion that by an "***unusual combination of symmetry properties, smectic C liquid crystals composed of chiral molecules ought to be ferroelectric***" (R.B.Meyer, 1977). This resulted in a straightforward explanation of the large linear electrooptic effect observed in DOBAMBC: *The chiral* $smectic-C^*$ *liquid crystal is a ferroelectric liquid crystal.* Meyer's explanation was as follows: In chiral tilted smectic phases, molecules are arranged in liquid-like smectic layers. Each smectic layer can posses an in-plane spontaneous polarization. By symmetry, the in-plane polarization is allowed in a direction perpendicular to the molecular tilt, as shown in Fig.1.1.3.

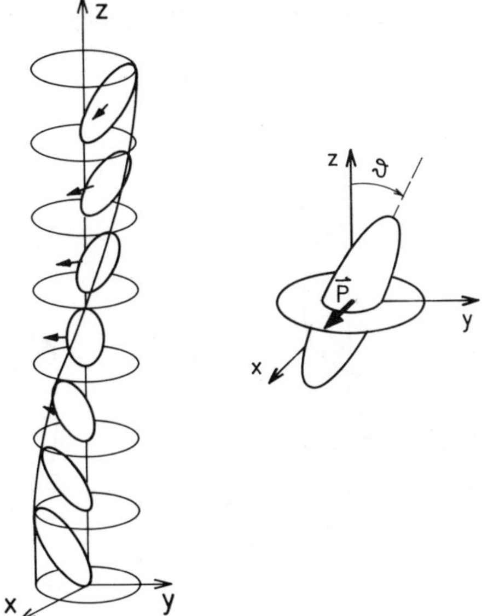

Fig.1.1.3. The structure of the chiral ferroelectric $smectic-C^*$ phase, as proposed by R.B.Meyer et al. (1975). The small arrow indicates the spontaneous polarization in each layer. The picture on the right shows the direction of the tilt and the in-plane direction of spontaneous polarization.

There is therefore a linear coupling between the electric polarization and the tilt, which is allowed under the symmetry group of the $smectic - C^*$ phase. The effect of the coupling on the dielectric properties of the homogeneous ferroelectric $smectic - \overline{C}^*$ was first discussed by Blinc (1975). Due to the chirality of the molecules, the director rotates as we move along the normal to the smectic layers, and the chiral tilted smectic phase is helicoidally modulated. On the macroscopic scale, this helical modulation results in a uniaxial conoscopic figure because of the averaging of the refractive index.

When an external DC field is applied in a direction parallel to the smectic layers, the linear coupling between the electric field and the spontaneous polarization distorts the modulation, as shown in Fig.1.1.4. More and more molecules are aligned in a plane perpendicular to the field direction, until at a critical electric field the structure unwinds and we obtain a homogeneous, uniformly polarized tilted smectic phase. The corresponding conoscopic image shifts during the unwinding process and becomes biaxial, as expected for biaxial and homogeneously tilted smectic above the critical electric field.

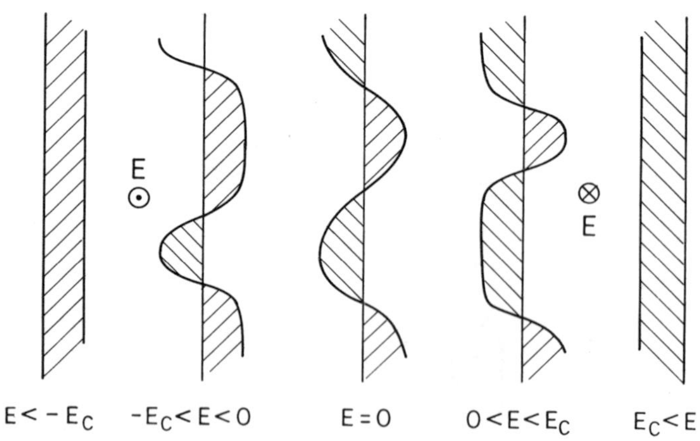

Fig.1.1.4. Response of the helix of a ferroelectric $smectic - C^*$ phase to an external electric field, as proposed by R.B.Meyer and co-workers (1977).

Whereas the existence of long range polar order in solid ferroelectric crystals was known from the beginning of this century, the idea of the existence of long range polar order in liquids was revolutionary at the time of Meyer's experiment. However, it became very soon clear that Meyer's explanation was correct, as it was followed by similar reports by other groups that performed complementary and independent experiments on DOBAMBC and other chiral and polar smectics.

The discovery of ferroelectricity in liquid crystals was not accidental. Instead, it was a result of a brilliant conjecture, which was based on purely symmetry arguments (R.B.Meyer, 1977). Let us first consider the symmetry arguments, which allow for the existence of ferroelectricity in chiral $smectic-C^*$ phase. The polarization

$$\vec{P} = (P_x, P_y, P_z) \tag{1.1.1}$$

is a polar vector. For spontaneous polarization to exist in a given phase we must demand that the vector $\vec{P}$ is invariant under symmetry operations that are allowed in the structure of question. Let us now see if this requirement explains the fact that the $smectic-C^*$ phase is ferroelectric, whereas the achiral $smectic-C$ phase is not. In order to avoid the complication that arises from the helical structure of the $smectic-C^*$ phase, let us consider the case of a thin film of a few smectic layers only.

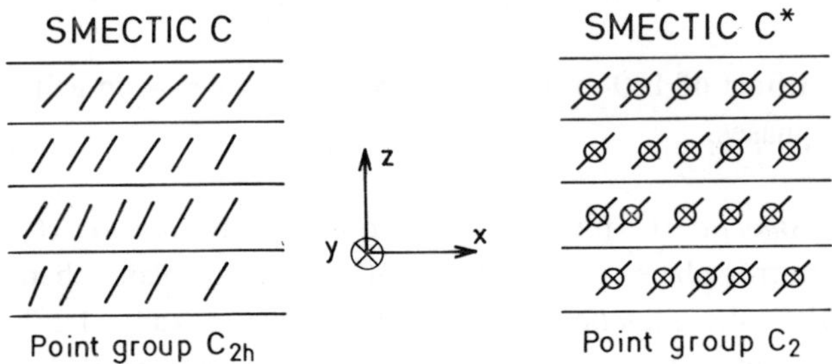

Fig.1.1.5. The structure and the point group of chiral and achiral polar tilted smectics. The cross on the right picture represents the direction of polarization, which is in smectic layers, pointing into the plane of the paper.

Such a system can be considered to be homogeneous as the precession of the tilt and polarization is slow on the molecular scale, $d/p \approx 10^{-3}$, where d is the thickness of the smectic layers and p is the period of the helix. This changes the symmetry of the system globally and not locally.

The x, y, z coordinate system and the arrangement of the molecules in the achiral $smectic-C$ and chiral $smectic-C^*$ phases are shown in Fig.1.1.5. In both the achiral $smectic-C$ (point group symmetry C_{2h}) and chiral $smectic-C^*$ phases (point group symmetry C_2), a 180° rotation around the y-axis is an allowed symmetry operation (see Table 1.1.1). Under this transformation, the polarization vector transforms according to $P=(P_x, P_y, P_z) \Rightarrow (-P_x, P_y, -P_z)$. Since these two expressions must be identical, we can only allow the possibility of a spontaneous polarization in the direction of the two-fold axis: $P=(0, P_y, 0)$.

SmC	SmC*
C_{2y}	C_{2y}
$(P_x, P_y, P_z) \Rightarrow (-P_x, P_y, -P_z)$	$(P_x, P_y, P_z) \Rightarrow (-P_x, P_y, -P_z)$
$\Downarrow$	$\Downarrow$
$\vec{P}=(0, P_y, 0)$	$\vec{P}=(0, P_y, 0)$

Table 1.1.1. Effect of 180° rotation on polarization vector in achiral and chiral tilted smectic phases.

In the achiral $smectic-C$ phase (but not in the chiral $smectic-C^*$ phase), the xz plane is a mirror plane. The existence of this mirror plane changes $(0, P_y, 0)$ into $(0, -P_y, 0)$ in the $smectic-C$ phase, as shown in Table 1.1.2. Since these two expressions must be identical, $\bar{P}$ must be zero and ferroelectricity is not allowed in the achiral $smectic-C$ phase. On the other hand, in the chiral $smectic-C^*$ phase the molecules are not any more mirror images of themselves. Thus, the xz plane is not any longer a mirror plane and ferroelectricity is permitted by symmetry. The spontaneous polarization in the $smectic-C^*$ phase points in the direction of the y-axis, i.e. perpendicular to the plane of the tilt. For a given direction of the tilt, there are two possible directions of the spontaneous polarization, depending on the molecular structure, as shown in Fig.1.1.6. By convention, the spontaneous polarization is considered negative for the direction shown in Fig.1.1.6a and is positive for the opposite direction, shown in Fig.1.1.6b.

SmC	SmC*
m_{xz} $(0, P_y, 0) \Rightarrow (0, -P_y, 0)$ $\Downarrow$ $\vec{P} = (0,0,0)$	Reflection not allowed due to chirality $\Downarrow$ $\vec{P} = (0, P_y, 0)$

Table 1.1.2. Effect of reflection in the *xz* plane.

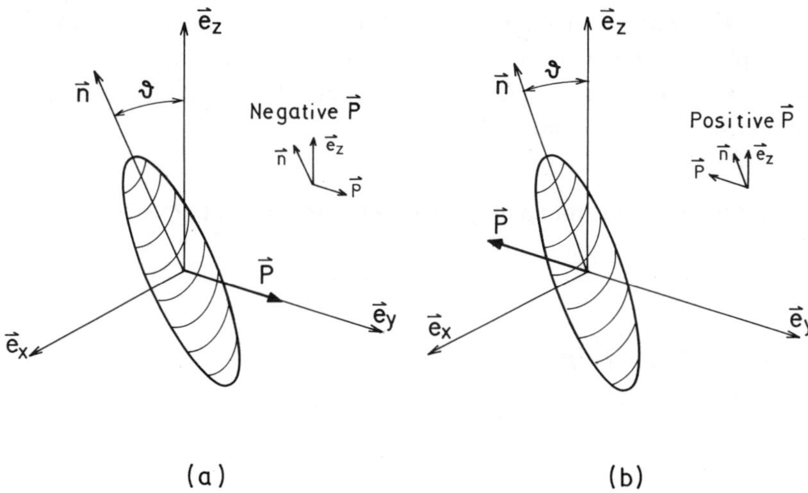

Fig.1.1.6. By convention, the direction of the spontaneous polarization in *smectic* $-C^*$ phase is negative for the molecule shown in (a) and is positive for the molecule, shown in (b) (Hall et al., 1997). The direction of polarization is determined by the chemical structure of the molecules.

The existence or non-existence of ferroelectricity in any liquid crystalline or solid phase can be deduced from the Neumann's principle which states that ***"The symmetry elements of any physical property of a medium must include the symmetry elements of the point group of the medium"***. This means that the symmetry of any physical property can be higher, but not lower than the point group symmetry. Of the 32 crystal point groups, 22 contain too many symmetry elements to allow for the existence of a spontaneous polarization, i.e., to allow for ferroelectricity. For the same symmetry reasons as in $smectic-C^*$ phase, ferroelectricity is allowed in the chiral tilted $smectic-I^*$ and $smectic-F^*$ phases. Here, the molecular arrangement is similar to the $smectic-C^*$ phase, except for the additional bond orientational order and the hexatic packing.

It should be stressed that the $smectic-C^*$ phase is an improper ferroelectric phase. In contrast to proper ferroelectrics, ferroelectricity in the $smectic-C^*$ phase is due to the linear coupling of the electric polarization with the tilt angle. The tilt angle is the primary order parameter that drives the transition and the polarization is a secondary order parameter.

Following similar symmetry arguments, it is straightforward to see that ferroelectricity is not allowed in the *isotropic (I*)*, *nematic (N*)* and $smectic-A^*$ phases, although they are formed by chiral molecules. The isotropic phase with chiral (optically active) molecules has the point symmetry of the SO(3) group, i.e. no reflections are allowed. However, rotation around any axis is a symmetry operation:

(i) A 180° rotation around the *y*-axis yields $P_x = 0$ and $P_z = 0$

$$\vec{P} = \begin{pmatrix} P_x \\ P_y \\ P_z \end{pmatrix} \rightarrow \begin{pmatrix} -P_x \\ P_y \\ -P_z \end{pmatrix} \Rightarrow P_x = 0, P_z = 0, \vec{P} = \begin{pmatrix} 0 \\ P_y \\ 0 \end{pmatrix}$$

(ii) A rotation of 90° around the *z*-axis transforms P_y into P_x, so that likewise it has to vanish in view of *(i)*

$$\begin{pmatrix} 0 \\ P_y \\ 0 \end{pmatrix} \rightarrow \begin{pmatrix} P_x \\ 0 \\ 0 \end{pmatrix} \Rightarrow \vec{P} \equiv 0$$

An isotropic liquid can be thus optically active but not ferroelectric. This is because the polarization $\vec{P}$ is a polar vector, whereas the optical activity coefficient is a pseudoscalar.

The point group of the chiral nematic and the $smectic - A^*$ phase consisting of chiral molecules, is D_∞. The same symmetry arguments can be applied as in the isotropic phase, since the 180° rotation around the *y*-axis and 90° rotation around the *z*-axis are allowed symmetry operations. No spontaneous polarization can therefore occur.

Let us now turn to ***theoretical predictions about hypothetical, not yet observed phases***, exhibiting a transverse or longitudinal spontaneous polarization. In all of our symmetry considerations we have implicitly assumed that the $+\vec{n}$ and $-\vec{n}$ molecular directions are fully equivalent, or, in other words, this symmetry is not spontaneously broken in the phases that we have considered. However, in principle, there is no physical reason, why this symmetry could not be spontaneously broken in systems, where the constituent molecules are of low enough symmetry, or where the phase is more structured. By some microscopic mechanism, this could results in a ferroelectric phase, which is achiral, yet ferroelectric (Tournilhac et al., 1992). In this case, a thermodynamic stability analysis of the phase in question has to be performed. This analysis gives the answer if the ordered-ferroelectric phase can in fact exist.

A very interesting idea about the origin of a new type of polar order in tilted smectics has been given by Photinos and Samulski (1995), see also Tredgold, (1990). They have considered possible polar ordering of achiral "*S*"-like molecules, like those, shown in Fig.1.1.7. As one can see, these molecules are clearly not chiral. However, due to their specific, "bent" shape, the orientation of the molecule on the right figure is energetically different from the orientation, shown on the left figure. This means that one of the conformations is energetically preferable, which can lead to the transverse polar order of non-polar molecules. The question is, however, if this type of spontaneous polar ordering can be observed experimentally.

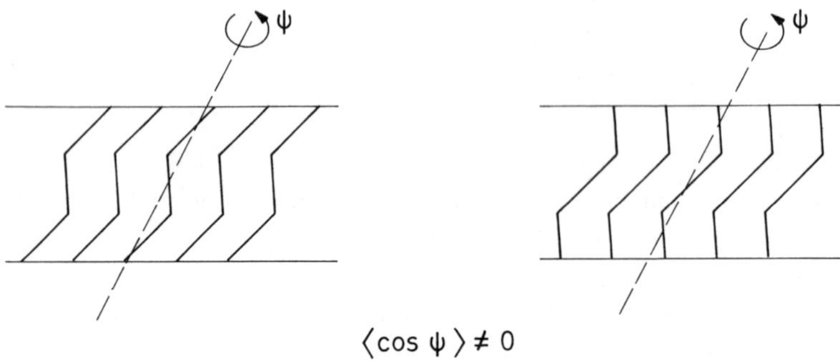

Fig.1.1.7. Two different orientations of non-chiral and non-polar molecules are energetically different due, for example, to their interaction with molecules i neighboring layers (Photinos and Samulski, 1995).

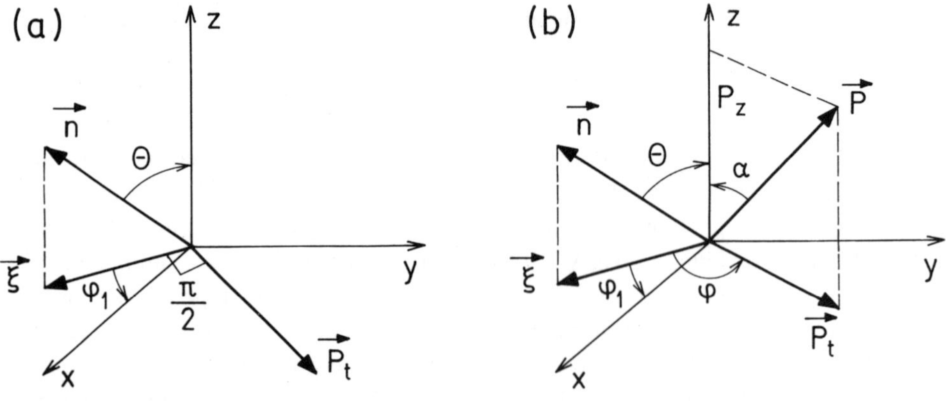

Fig.1.1.8. (a) In the traditional ferroelectric $smectic-C^*$ phase, the polarization is perpendicular to $\vec{n}$ and the normal to the smectic layers, which is due to the $+\vec{n} \rightarrow -\vec{n}$ symmetry. (b) This symmetry can be spontaneously broken in the $smectic-C_p^*$ phase, as predicted by Toledano and Neto. The phase is of triclinic C_1 symmetry. After P.Toledano and A.M.Figueiredo Neto (1997).

The existence of different ferroelectric phases was analyzed theoretically by P.Toledano and A.M.Figueiredo Neto (1997). They have shown that besides the well-known chiral $smectic-C^*$ phase of monoclinic (C_2) symmetry, a triclinic chiral phase of C_1 symmetry can be thermodynamically stable. In this phase, the traditional $+\vec{n} \to -\vec{n}$ symmetry is spontaneously broken resulting in a component of spontaneous polarization along the smectic layer normal, as shown in Fig.1.1.8. In other words, we have a spontaneous longitudinal (along $\vec{n}$) component of polarization in the bulk chiral smectic phase, and the phase is denoted as $smectic-C_p^*$. Toledano and Neto have predicted that this phase could occur just below the ordinary ferroelectric $smectic-C^*$ phase. It is interesting to note that the stability of such a phase was analyzed in an early work of Michelson, Benguigui and Cabib (1977b). They have predicted that such a triclinic ferroelectric smectic phase could be thermodynamically stable. At the time of writing, there is no experimental evidence for the existence of such a phase in bulk samples.

Finally, we should comment a question of the existence of ferroelectricity in the nematic-like phase, or, in other words, is it possible to have a orientationaly ordered (but positionally disordered, i.e. nematic-like) ferroelectric phase? The existence of such a polar order in an orientationaly ordered phase obviously requires the spontaneous breaking of the $+\vec{n} \to -\vec{n}$ symmetry and was the subject of many theoretical considerations in the past. So far, these considerations are inconclusive. For example, N.Boccara found that the spontaneous breaking of the $+\vec{n} \to -\vec{n}$ symmetry in ordinary nematic phases leads to a ***polar nematic state, which is thermodynamically unstable*** (see de Gennes 1974). Michelson, Benguigui and Cabib (1977b) have come to the same result by simply considering the dipolar electrostatic energy and entropy within the mean-field approximation. This was followed by the work of Alexander, Hornreich and Shtrikman (1981), who found that both positionally ordered and disordered nematic ferroelectric phases are stable for a certain set of parameters of the Landau free energy expansion. Recently, density-functional theories (Groh and Dietrich, 1994), and Monte-Carlo simulations (Ayton and Patey, 1996) have shown that it is possible to have a ferroelectric nematic-like ordered phase, where the $+\vec{n} \to -\vec{n}$ symmetry has been spontaneously broken. However, this state of matter was found in model systems only, like the Stockmayer liquid and discotic nematics with a specific shape of the molecules. So far there is no experimental evidence for that.

Note 1.1.1. Chirality in liquid crystals:
A given object, e.g. a liquid crystalline phase or a molecule is chiral if it is different from its mirror image. For example, hands are chiral objects: If one looks at the right hand in a mirror, one sees a left hand, hence the expression left- and right-handed. An example of a chiral molecule is given in Fig.1.1.1., where the chiral part of the molecule is an asymmetric carbon indicated by *. An example of a chiral arrangement of atoms in molecules is shown on the figure below (after Hall et al. 1997). Here, the atoms or groups labeled by A, B, C and D, can be arranged in two different ways to produce two optical isomers or *enantiomers*, labeled (R), i.e. rectus and (S), i.e. sinister, respectively.

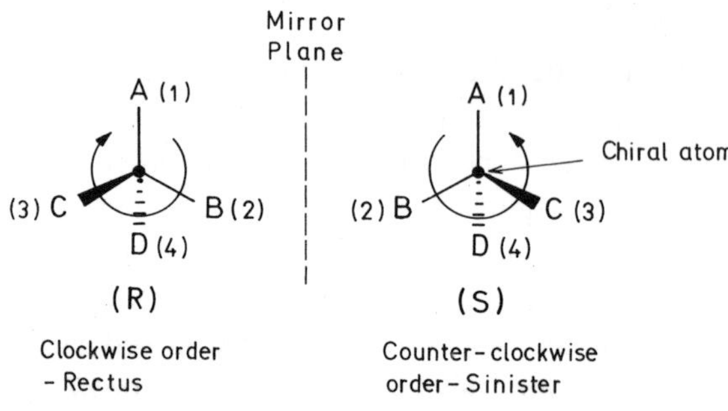

Note 1.1.2. Polarity in liquid crystals:
A ferroelectric chiral smectic liquid crystalline phase can be obtained only if the constituent molecules possess a transverse dipole moment. This dipole moment of the molecule originates from the asymmetric distribution of electron density in the vicinity of various sites of the molecules. Permanent dipole moments therefore only occur in asymmetric molecules, where various bond or group dipole moments contribute to the total dipole moment of the molecule. The dipole moments of some selected molecules, bonds and groups are given in the table below (after Israelachvili, 1995).

Dipole moments of molecules, bonds and groups in Debye units ($1D=3.336\times10^{-30}$ A·s·m)					
		Molecules			
CO_2	0	NaCl	8.5	Acetone	2.9
CO	0.11	CsCl	10.4	Aniline	1.5
H_2O	1.85	Hexanol	1.7	Benzene	0
		Bond moments			
$C—H^+$	0.4	C—C	0	$C^+—Cl$	1.5-1.7
$N—H^+$	1.31	C=C	0	$C^+—O$	2.3-2.7
$O—H^+$	1.51	$C^+—O$	0.74	$N^+—O$	2.0
		Group moments			
$C—{}^+OH$	1.65	$C—{}^+CH_3$	0.4	$C—{}^+COOH$	1.7
$C—{}^+NH_2$	1.2-1.5	$C^+—NO_2$	3.1-3.8	$C—{}^+OCH_3$	1.3

Note 1.1.3. Spontaneous polarization in ferroelectric liquid crystals:

The spontaneous polarization of a typical ferroelectric liquid crystals, such as DOBAMBC and DOBA-1-MPC is several nano-Coulombs per cm^2 (1 Coulomb=A·s) and is temperature dependent, as shown in the figure below (after Z.Kutnjak, 1991).

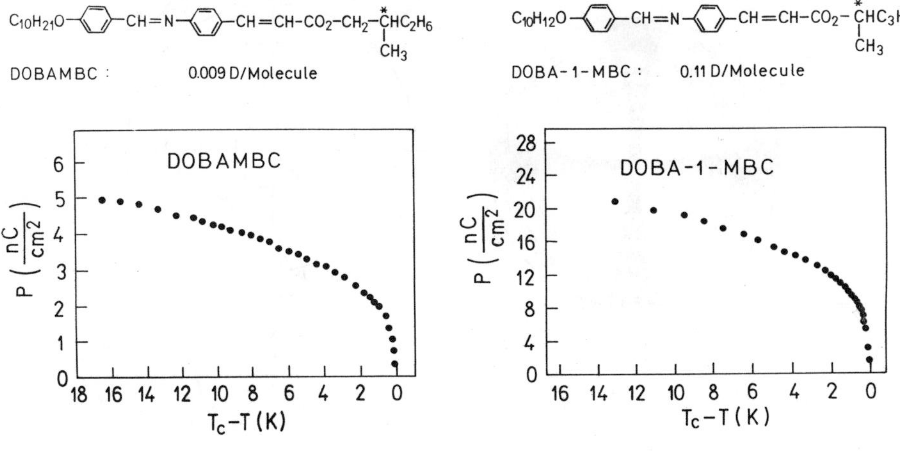

The spontaneous polarization in ferroelectric liquid crystals is typically 100 to 1000 times smaller than in solid ferroelectrics. For example, the spontaneous polarization of TGS (triglycine sulfate) is 2.8μC/cm^2 and is thousand times larger than in DOBAMBC. There are some ferroelectric liquid crystals with such a large polarization. However, in view of their very large viscosity they are not interesting for application (Collings and Hird, 1997).

The spontaneous polarization in ferroelectric phases is only a fraction (i.e. several %) of what we would expect by simply adding the dipole moments of all molecules in the sample. This indicates that the averaging due to fast thermal molecular motion is very efficient in decreasing the spontaneous polarization of the ferroelectric liquid crystalline phase. In particular, biasing of the molecular rotation around the long molecular axis is important and can be determined with NMR, as shown in Chapter 7. The spontaneous polarization is strongly affected by the position of the chiral center with respect to the molecular core and is higher when the chiral center is closer to the core. This can be seen from the above figure, which compares spontaneous polarizations in DOBAMBC and DOBA-1-MBC. Note that in DOBA-1-MBC the chiral center is closer to the core, resulting in a larger polarization.

Note 1.1.4. Conoscopic observations:

Conoscopic observations can be performed either using an optical microscope or by projecting a convergent laser beam onto a sample and observing the light polarization pattern on a screen behind the sample (Hartshorne and Stuart, 1964).

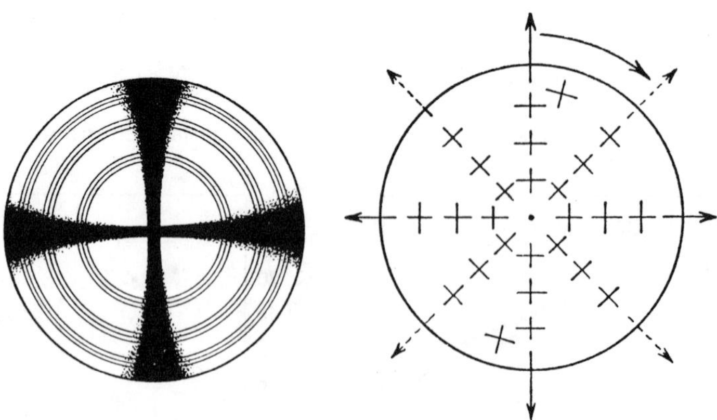

When using a polarizing microscope, a Bertrand lens has to be inserted and one observes the conoscopic figure of the sample in highly convergent illuminating light. For a uniaxial crystal with the optical axis parallel to the optical axis of the microscope, one observes in a monochromatic light a typical interference pattern, shown on the previous page. This "basal" uniaxial interference figure consists of concentric dark rings and the "Maltese cross", oriented at 45 degrees with respect to the directions of the polarizer and the analyser, which are crossed. The interference rings appear because of the destructive interference of ordinary and extraordinary rays. As the light travels at an angle with respect to the optical axis, the phase difference between the two eigenpolarizations depends on the angle of propagation. When the phase difference is half-wave, a black ring appears, because of the uniaxiality of the pattern. The Maltese cross appears because of the polarization properties of light, indicated on the previous figure. The small lines represent the directions of the oscillation of the electric field in a convergent light cone. There are two angles, where this direction is crossed with respect to the analyser, hence the dark cross appears along these two directions.

1.2. Antiferroelectricity in Liquid Crystals

Whereas the first antiferroelectric liquid crystal was synthesized by Levelut et al. in 1983, antiferroelectricity in smectic liquid crystals was first recognized in the conoscopic experiments of Chandani and co-workers in 1989. Surprisingly, Chandani et al. have used the same optical conoscopic method, which led fourteen years before R.B.Meyer to the first observation of ferroelectricity in liquid crystals. They have performed a conoscopic experiment on a newly synthesized compound with a very large spontaneous polarization 4-(1-methyl-heptyloxycarbonyl-phenyl) 4'-octylbiphenyl-4-carboxylate (MHPOBC). Under the application of a strong transverse electric field, they have observed in a certain temperature range a strange sequence of conoscopic figures, which are shown in Fig.1.2.1.

In contrast to the conoscopic figures of the ferroelectric phase, the conoscopic figure of an antiferroelectric phase of a liquid crystal does not shift as a whole under the application of a transverse electric field. Instead, the image remains fully symmetric and just splits in two parts with no tilting. This is a clear indication of the absence of a linear ferroelectric coupling between the field and the polarization and consequently of the absence of spontaneous polarization in this phase. The symmetric splitting of the conoscopic image however indicates the presence of quadratic dielectric coupling. Since this coupling is symmetric

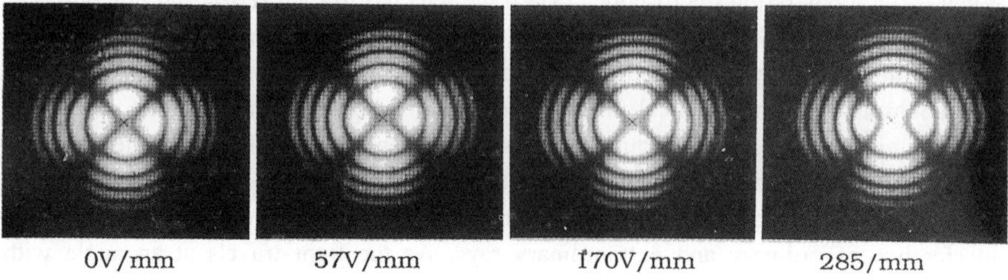

Fig.1.2.1. The electric field dependence of the conoscopic images in the antiferroelectric phase of MHPBC (after Okabe et al., 1992). Note the symmetric splitting of the "two eyes" of the conoscopic image at high fields, far right figure. For higher fields, the antiferroelectric structure unwinds and we obtain a uniformly polarized ferroelectric state.

with respect to the $+\vec{E}$ and $-\vec{E}$ directions of the external field, it preserves the original mirror symmetry of the conoscopic figure.

This conoscopic observation has led to the following conjecture on the molecular ordering in the so-called antiferroelectric $smectic-C_A^*$ phase, which is shown in Fig.1.2.2. The antiferroelectric liquid crystalline $smectic-C_A^*$ phase is characterized by an alternation of the tilt direction of the average molecular orientation and the direction of the in-plane spontaneous polarization $P(\vec{r})$ by nearly 180° on going from one smectic layer to another (Fig.1.2.2a). Two neighboring layers thus form an antiferroelectric unit cell with two antiparallel electric dipoles and a zero value of the equilibrium electric polarization $\vec{P}_\circ(\vec{r}) = \vec{P}_i + \vec{P}_{i+1} = 0$. Because of chirality, the directions of the spontaneous tilt and the in-plane polarization slowly precess around the layer normal as one moves along the direction perpendicular to the smectic plane. This causes a small deviation from the 180° alternation in the tilt between two consecutive layers and the formation of a modulated, helicoidal structure. Since the basic structural unit of the $smectic-C_A^*$ phase is formed of two neighboring layers, we have here in fact a double-twisted helicoidal structure, formed by two identical ferroelectric $smectic-C^*$ helices gearing into each other as shown in Fig.1.2.2b. The periodicity of this helical modulation is of the order of the wavelength of the visible light and is in general incommensurate to the basic antiferroelectric unit cell of the *smectic* $-C_A^*$ phase.

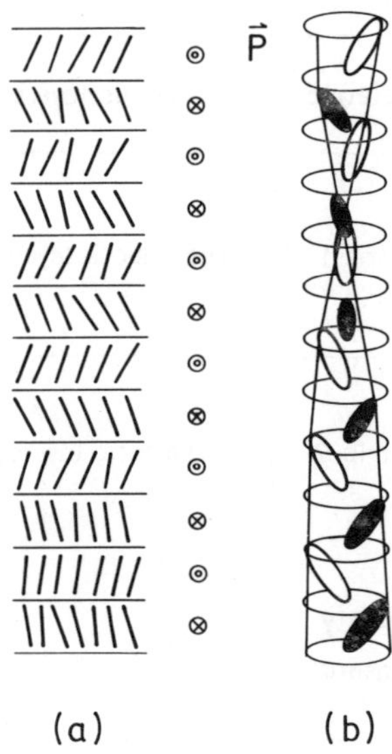

Fig.1.2.2. (a) The alternating-tilt model of the homogeneous antiferroelectric *smectic* $-C_A^*$ phase of liquid crystals. (b) The antiferroelectric double helix.

At the time of these observations, the general opinion was not very favorable towards the interpretation of Chandani's experiments with the alternating tilt model. It seemed nearly impossible to have tilt alternation on a molecular scale, because as one expects intuitively, this would be energetically very costly. In fact, it turned out later, that this intuitive guessing was wrong. Following simple arguments, it can be even shown that the alternating tilt structure in fact lowers the energy of van der Waals attractive forces between pairs of molecules in adjacent layers. Soon after this conjecture on the alternating tilt structure, several experiments on freely suspended antiferroelectric films (Galerne and Liebert, 1990; Bahr and Fliegner, 1993) clearly confirmed the alternating tilt structure of the antiferroelectric *smectic* $-C_A^*$.

Symbol	Type	Unit cell	Top view of a unit cell
smectic - C^*	ferroelectric	one smectic layer	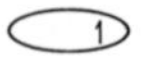 1
smectic - C_A^*	antiferroeletric	two smectic layers	 2 1
smectic - C_{FI1}^* ($\equiv$ smectic - C_γ^*)	ferrielectric 1	three smectic layers	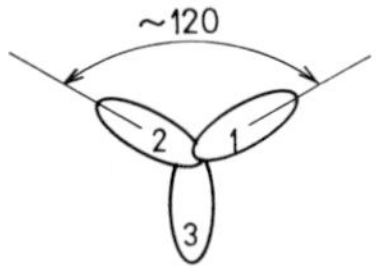 ~120 2 1 3
smectic - C_{FI2}^*	ferrielectric 2	four smectic layers	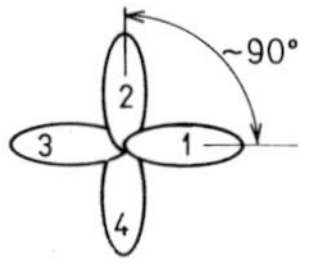 ~90° 2 3 1 4
smectic - C_α^*	presumably incommensurate	none	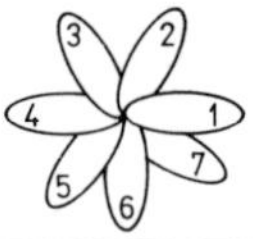 3 2 4 1 7 5 6

Fig.1.2.3. The list of experimentally confirmed tilted smectic phases of polar chiral liquid crystal materials. After Mach et al., 1998.

The final proof of the antiferroelectric alternating tilt ordering was given by the resonant X-ray experiment, performed by Mach et al. (1998). Normally, X-rays cannot determine the spatial orientation of the molecular tilt, as the X-ray scattering cross-section depends only on the modulation of the mass density, which is a scalar. However, by fine tuning of the X-ray energy, one can come closer to the absorption band of some atoms that form the liquid crystalline molecules. In this case, the resonant X-ray cross-section becomes sensitive to the direction of the molecular axis and can therefore discriminate between different molecular directions on the nanometer scale. This has allowed for the first unambiguous determination of the nearly antiparallel orientation of molecular dipoles in the antiferroelectric phase. It has also brought new discoveries about the structure of the so-called ferrielectric phases. Mach's experiment has clearly shown that not only the alternating tilt (herringbone) phase is stable, but that ferrielectric phases with a periodicity of three and four smectic layers exist. In all these structures, the magnitude of the tilt angle is constant throughout the sample. Figure 1.2.3. presents the list of the experimentally confirmed tilted smectic phases, as deduced from the resonant X-ray experiment of Mach et al. (1998).

1.3. Chiral Phases of Achiral Molecules

Whereas molecular chirality results in a macroscopic spontaneous polarization and leads to the formation of ferroelectric and antiferroelectric phases in tilted smectics, it was reported that the opposite may happen as well: tilted smectic phases of achiral but polar molecules exhibit a spontaneous formation of polarized macroscopic chiral domains (Niori et al., 1996; D.R.Link et al., 1997; Macdonald et al., 1998). This phenomenon can be observed in phases formed by the so-called "bow" (banana) shaped molecules, which are illustrated in Fig.1.3.1.

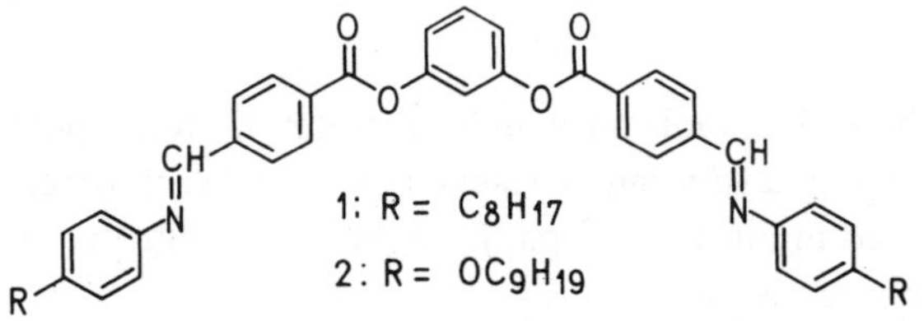

1: R = C_8H_{17}

2: R = OC_9H_{19}

Fig.1.3.1. Typical structure of a bow-shaped (banana) molecule. Note that the molecule is achiral and possesses C=O groups with a large electric dipole moment. After Link et al. (1997).

The unusual properties of this new class of liquid crystalline materials have been first reported by Niori et al.(1996) and later by Sekine et al. (1997) and Heppke et al. (1997). These new materials did not mix with any of the "standard" smectic materials, which was attributed to their particular shape. For this reason, phases that are formed by bow-shaped molecules were denoted by B_1, B_2, B_3, ...B_7. Further on, the structure of some of these phases was identified in a free standing-film experiment of Link et al. (1997), who has also proposed a new classification, based on synclinic and anticlinic ordering of the tilt. In this short Section we shall briefly discuss possible structures that could appear in these materials. A comprehensive review of the physical properties of phases formed by bow-shaped molecules can be found in a review by Pelzl et al.(1999).

Fig.1.3.2. illustrates the possible orientational and positional ordering of bow-shaped molecules, when cooled from their isotropic phase. Three different scenarios can be foreseen:

(i) A nematic phase of bow-shaped molecules is formed, similar to the nematic phase of rod-like molecules.

(ii) Molecules spontaneously order in non-polar, smectic-A-like layers. When stacked on top of each-other, these layers can form the usual non-polar $smectic-A$ phase.

(iii) More likely, the molecules will spontaneously form a polar, nontilted smectic-A layer. This is due to strongly anisotropic interactions because of the very specific shape of the molecules. This is a proper ferroelectric phase.

Experiments show that molecules indeed prefer to form polarized, nontilted smectic layers. These layers can be stacked on top of each-other in two different ways, as illustrated in Fig.1.3.3. First, it might be possible that we obtain a polar $smectic-A$ phase, which is in the Boulder nomenclature designated $smectic-AP_F$. Here A assigns that the molecules are on the average oriented perpendicular to the smectic layers. With bow-shaped molecules this means that the "string" of the bow is a director, which is oriented perpendicular to the layers and P_F assigns ferroelectric ordering of molecular dipoles, i.e. "arrows" of the bows. A second possibility, which is energetically more probable, is antiferroelectric ordering of the dipoles. In this case, we obtain the $smectic-AP_A$ phase, shown on the right side of Fig.1.3.3.

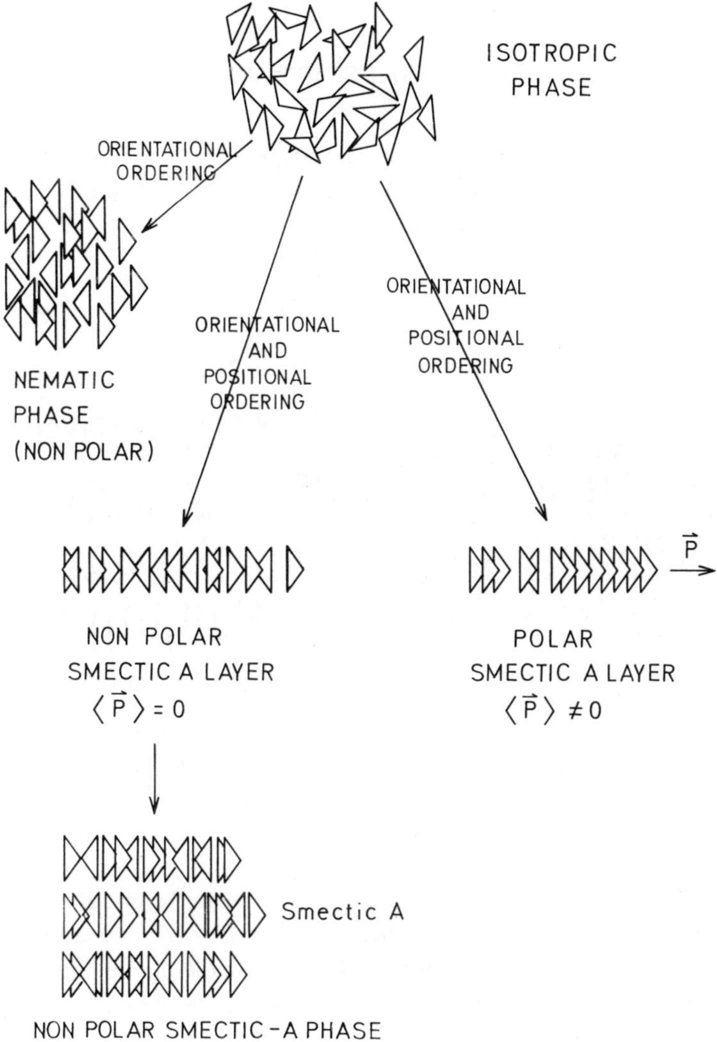

Fig.1.3.2. Possible scenarios of the spontaneous breaking of the orientational and translational symmetry of the isotropic phase of bow-shaped molecules. The existence of a standard nematic phase is very unlikely due to the specific shape of molecules, that prefer polar packing within a smectic layer.

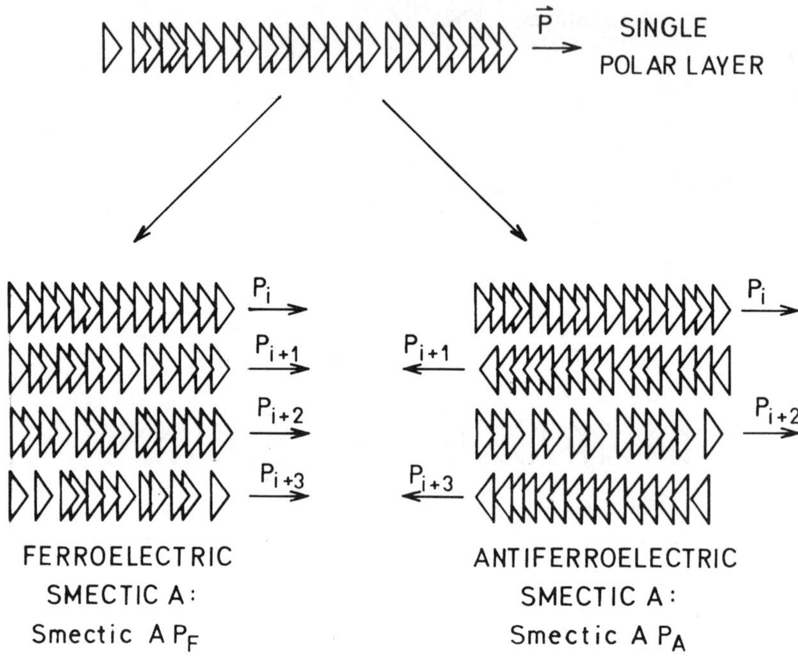

Fig.1.3.3. Stacking of individual polar smectic layers, formed by bow-shaped molecules. Either a ferroelectric $smectic-AP_F$ or antiferroelectric $smectic-AP_A$ phase can be formed. Due to dipole-dipole interactions, the antiferroelectric $smectic-AP_A$ phase is energetically preferred.

Due to specific shape of bow molecules, the ***achiral*** ferroelectric $smectic-AP_F$ phase can show very interesting phase transitions into ***tilted and chiral*** phases, as illustrated in Fig.1.3.4. and Fig.1.3.5. The spontaneous ***breaking of the mirror symmetry in phases formed by achiral molecules*** is illustrated in Fig.1.3.4. Let us suppose that we have a nontilted and polar smectic phase of achiral, bow-shaped molecules, as illustrated in Fig.1.3.4a. When the molecules spontaneously tilt in all layers in the same direction (i.e. synclinicaly), they can tilt either in the right or the left direction around the dipole moment of the molecule. This spontaneous tilt results in the breaking of the mirror symmetry: The structure in Fig.1.3.4b is a mirror image of the structure in Fig.1.3.4c.

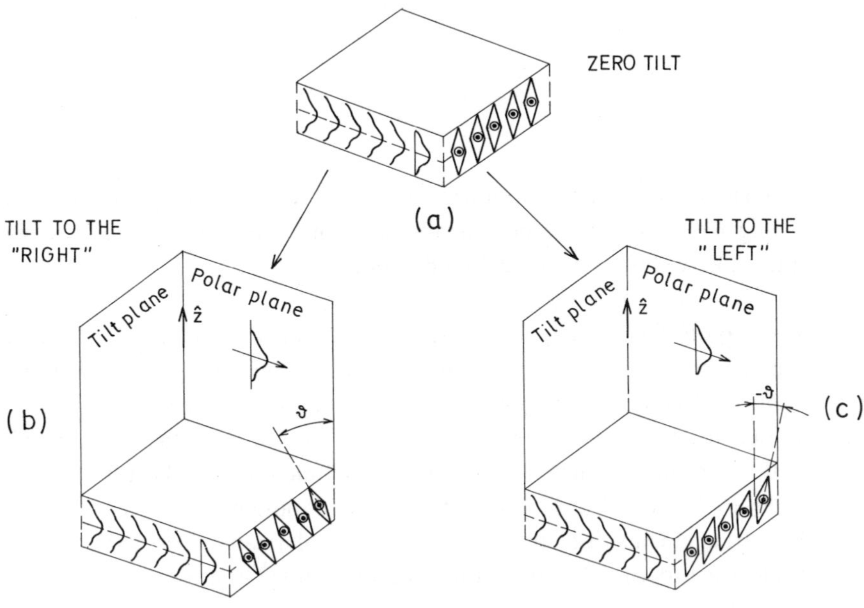

Fig.1.3.4. Spontaneous breaking of the mirror-symmetry of the phase formed by achiral, bow-shaped molecules. The originally nontilted smectic phase (a) is mirror-symmetric with the mirror plane in the plane of the molecules. When the molecules spontaneously tilt, there are two possibilities: molecules can tilt by rotating around the direction of the dipole to the "right" (b) or to the "left" (c). The mirror-symmetry of (a) is broken, because (b) and (c) are mirror images, which are clearly different. The handedness of the two conformations (b) and (c) are therefore opposite.

It is straightforward to show that by symmetry, the left and right conformations are both ferroelectric:

(i) A 180° rotation of the molecules around the direction of the arrow of the bow (i.e. y-direction) yields $P_x = 0$ and $P_z = 0$ because:

$$\vec{P} = \begin{pmatrix} P_x \\ P_y \\ P_z \end{pmatrix} \rightarrow \begin{pmatrix} -P_x \\ P_y \\ -P_z \end{pmatrix} \quad \Rightarrow \quad \vec{P} = \begin{pmatrix} 0 \\ P_y \\ 0 \end{pmatrix}$$

(ii) The tilted layer has no mirror plane and the macroscopic polarization along the arrow of the bow is allowed. The structure is chiral and will most likely show a helicoidal modulation.

There are also other possible phases formed of tilted and polar smectic layers of bow-shaped molecules, which are summarized in Fig.1.3.5. Starting from a single non-tilted polar smectic layer, the layers can be stacked on top of each other in four different ways: The molecules in layers can tilt either synclinically (i.e. in all layers in the same direction) or anticlinically (i.e. the tilt alternates from layer to layer). Within each of these two possibilities, there are two possible arrangements of the spontaneous polarization: ferroelectric or antiferroelectric ordering. We therefore obtain four tilted phases:

(i) The synclinic ferroelectric $smectic-C_S P_F$ phase. The phase is chiral.

(ii) The synclinic antiferroelectric $smectic-C_S P_A$ phase. The phase is achiral.

(iii) The anticlinic ferroelectric $smectic-C_A P_F$ phase. This phase is achiral.

(iv) The anticlinic antiferroelectric $smectic-C_A P_A$ phase. The phase is chiral and is equivalent to the $smectic-C_A^*$ phase.

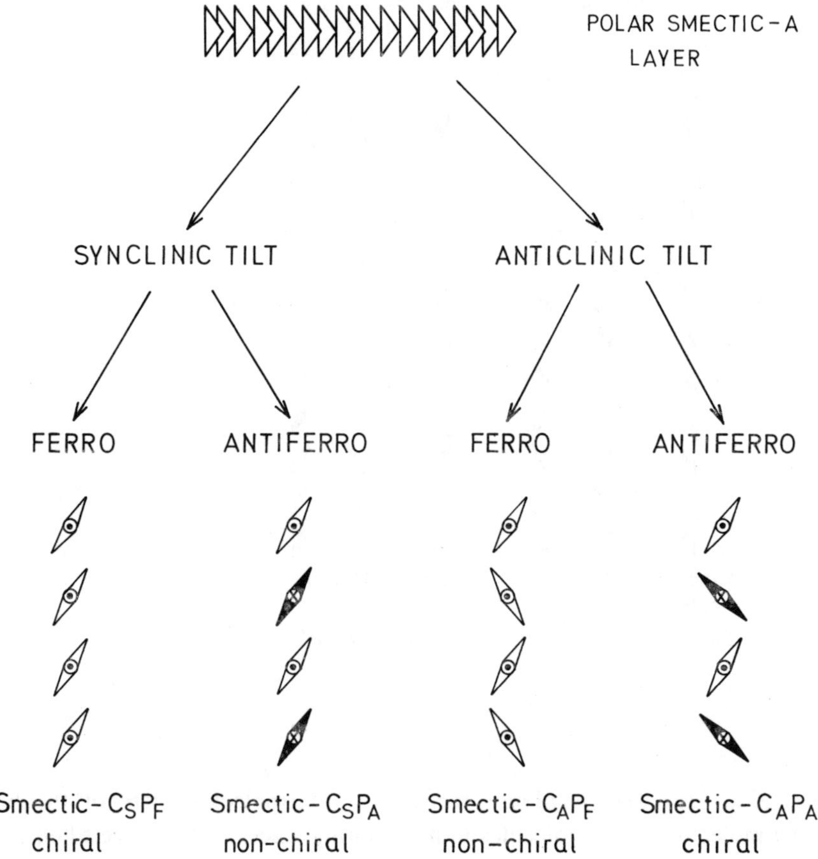

Fig.1.3.5. Four possible phases, formed by stacking polar tilted layers of bow-shaped molecules on top of each other.

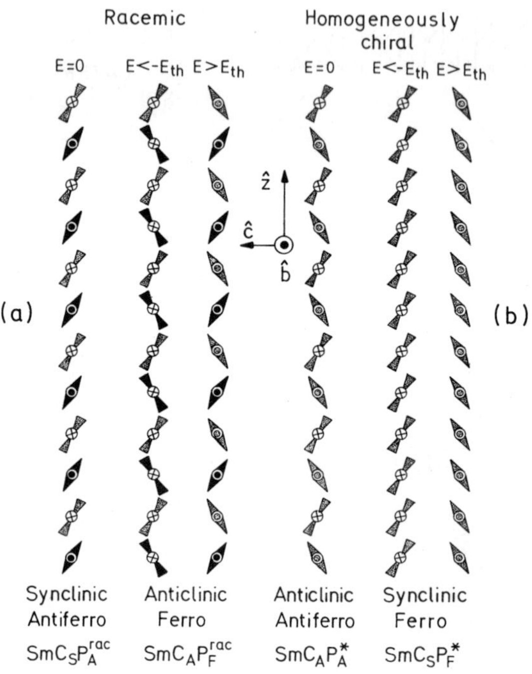

Fig.1.3.6. (a) The synclinic "racemic" antiferroelectric state of the bow-shaped molecules. The molecules in neighboring smectic planes are tilted in the same direction (synclinic tilt), whereas the directions of the spontaneous polarization in the neighboring layers are opposite (antiferroelectric ordering in $\vec{P}$). This is the $smectic-C_sP_A$ phase, or the "racemic" synclinic phase, because neighboring layers have opposite handedness. (b) The anticlinic "chiral" antiferroelectric phase. Both the tilt direction and the polarization direction are different for neighboring layers. This is the $smectic-CP_A$ phase, which is anticlinic in tilt (alternating tilt) and antiferroelectric in $\vec{P}$, and is therefore equivalent to the antiferroelectric $smectic-C_A^*$ phase of chiral smectics. After Link et al.(1997).

Using free standing films of these bow-shaped molecules, Link et al. (1997) have shown that the antiferroelectric smectic ordering is in fact the ground state of these materials. They have identified the antiferroelectric synclinic and antiferroelectric anticlinic ground states, which are shown in Fig.1.3.6. Similar structures are observed in crystals of achiral molecules, which do not have liquid crystalline phases (for a review, see Brown et al., 1996).

The combination of chirality and polarity in tilted smectic liquid crystals therefore results in many interesting phases, where these effects interplay. The thermodynamics of these phases has been discussed recently by Roy et al. (1999), and Čepič et al. (1999) who found several different stable phases, which can be formed by bow-like achiral molecules.

Note 1.3.1. In an optical experiment on free-standing films, Link et al. (1999) have found that surface layers of some antiferroelectric liquid crystals [4-(1-trifluoromethylhexyloxy-carbonyl, TFMHPOBC] as well as ferroelectric liquid crystals (DOBAMBC) have already in the *smectic – A* phase an anticlinic ordering that can be switched by an electric field into the synclinic orientation. The model for the tilt and polarization in such structures is shown on the figure below.

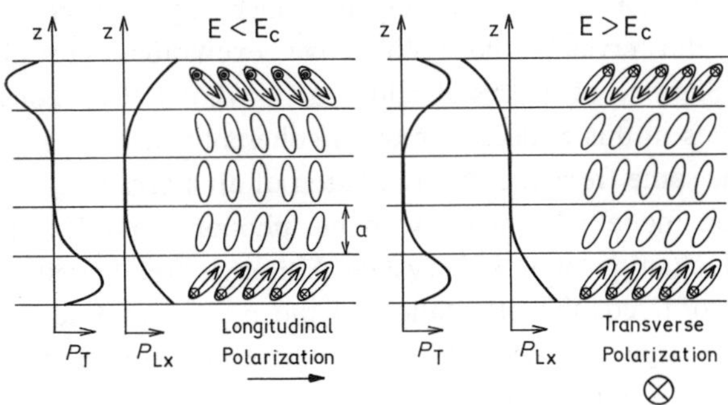

1.4. Broken Symmetry and Elementary Excitations

The concept of symmetry and broken symmetry is one of the most powerful and intriguing concepts of modern physics. Symmetry breaking can be met in very diverse branches of physics, such as elementary particle physics, astrophysics and cosmology, condensed matter physics, etc. (see, for example, Shubnikov and Koptsik, 1974; Michel, 1980; Boccara, 1981). Here, we shall discuss only briefly the concept of the broken symmetry in liquid crystals by comparing it to solid-state physics, where these concepts have been used very extensively.

The consequences of the orientational order in liquid crystals are most easily understood by comparing it to the positional order in solid crystals, as shown in Fig.1.4.1. In solids, a given crystal belongs to one of the 230 space groups. These groups all show distinct *positional and translational order* of constituent units, i.e. individual atoms or ions that form a solid crystal. On the other hand, liquid crystals are positionaly and transitionally disordered (nematics) or transitionally ordered (smectics), but are characterized by orientational order of their constituent units: elongated rod-like, or bow-shaped or disc-like molecules. We therefore define their state by their ***molecular orientation***, which in turn results in the concept of ***orientational order***. Having introduced this concept, we are free to compare the consequences of the positional order in solid crystals and orientational order in liquid crystals.

Another extremely useful concept in describing a given state of matter are the collective excitations. This term relates to motions in a matter, where a large number of constituents participate coherently and cooperatively. For example, in solid crystals these collective excitations are phonons. They represents oscillations of a crystal lattice as a whole. Each constituent of the lattice cooperates in this collective motion coherently with its neighbors, which implies also the same frequency of oscillation of all atoms in a crystal lattice for a given phonon mode. The analogy to the phonons in solid crystals are the orientational fluctuations in liquid crystals. Although liquid crystals are generally positionally disordered, the orientation of the molecules fluctuates coherently over extremely large distances. We have therefore time and space coherent motion of the direction of the average orientation of liquid crystal molecules. Whereas collective excitations are usually underdamped and oscillatory in solid crystals, the collective modes in liquid crystals are always overdamped. There is no experimental evidence for underdamped modes in liquid crystals. This

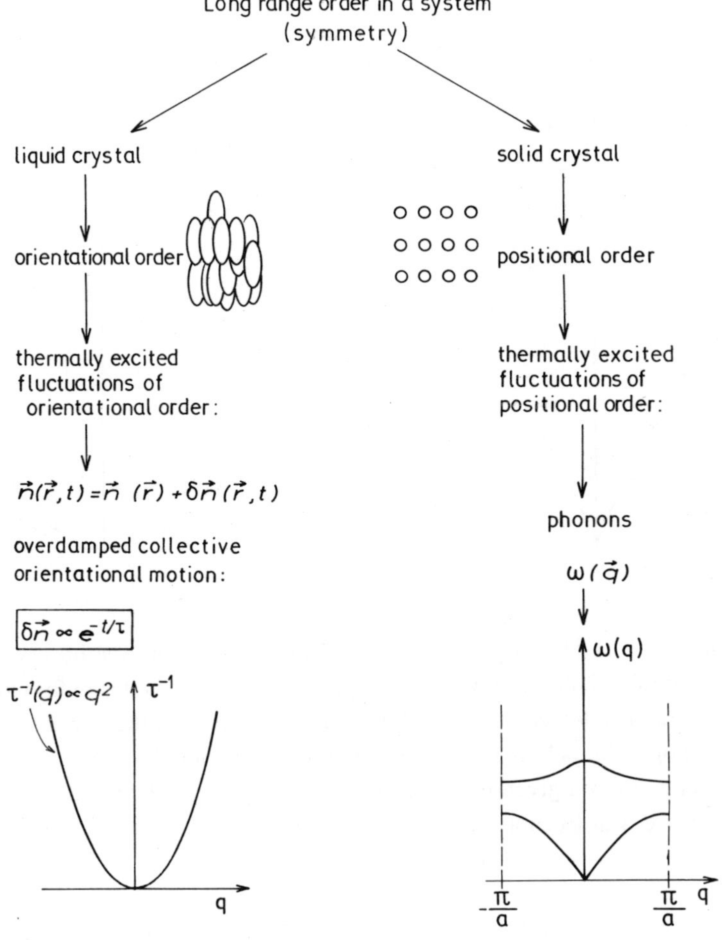

Fig.1.4.1. The comparison of positional and orientational order in solids and liquids.

indicates that in liquid crystals the inertial forces are much smaller than the viscous forces, that originate from the positional disorder of these phases.

As a consequence of their collective nature, we can introduce the concept of a dispersion relation for orientational fluctuations in liquid crystals. This is shown in Fig.1.4.2. in comparison to the phonon dispersion in solid crystals.

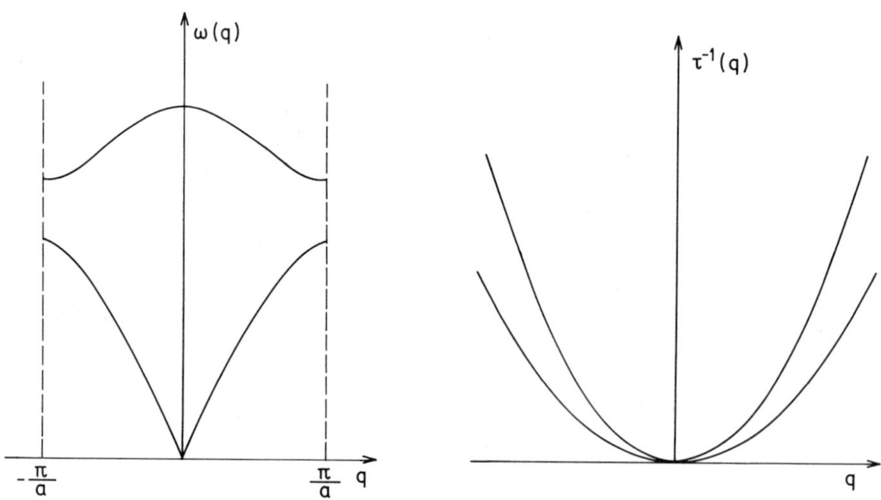

Fig.1.4.2. (a) The dispersion of the phonon modes in a solid crystal with a diatomic unit cell. At each wave-vector $\vec{q}$ we have two modes with two different frequencies. The lower branch is the acoustic branch and the upper branch is the so-called optic branch. (b) The dispersion of orientational fluctuations in nematic liquid crystal. For a given wave-vector $\vec{q}$ we have two relaxation modes because of two degrees of freedom for director reorientation. One is bend-twist mode and the other is splay-twist mode.

This was introduced by de Gennes (1968, 1969) and experimentally first confirmed by the light scattering experiments of Litster and Stinson (1970). For a given wave-vector $\vec{q}$, we may observe in solid crystals several distinct eigenmodes of a crystal lattice (phonons). Similarly, we can observe in nematic liquid crystals two director eigenmodes for a given wave-vector. These represent the bend-twist and splay-twist modes that both merge into the twist mode for fluctuations with a wave-vector perpendicular to the director.

It is interesting to note the similarities and the differences between the phonon spectrum of solid crystals and the spectrum of orientational fluctuations of a nematic liquid crystal. They both show the presence of zero-frequency modes. In solid crystals, the long-wavelength limit of acoustic modes always results in a zero-frequency mode. This is similar to the spectrum of nematics, where the long-wavelength limit of orientational fluctuations also results in a zero-frequency mode. In contrast to linear dispersion of acoustic modes in solids $\omega \propto q$, the dispersion of normal modes in nematic crystals is parabolic, $\tau^{-1} \propto q^2$, which is characteristic of overdamped systems.

The existence of zero frequency modes is closely related to the spontaneous breaking of a continuous symmetry of a system. This relation was first noted by Goldstone, Salam and Weinberg (1962), who formulated the famous ***Goldstone theorem***: Whenever a continuous symmetry group of a system in a ground state is spontaneously broken, a massless boson exists in the state of broken symmetry. The significance of this massless Goldstone boson for the theory of nonrelativistic many-body systems was soon recognized by Lange (1965, 1966), and later by Schneider and Meier (1973) who formulated the non-relativistic version of the Goldstone theorem: If the ground state of a nonrelativistic many-body system is a system with broken continuous symmetry, a gapless branch of collective excitations of the system exists, which try to restore the lost symmetry. This means that in the limit of zero wave-vector, the frequency of collective excitations vanishes:

$$\vec{q} \to 0 \Rightarrow \omega \to 0 \tag{1.4.1}$$

The existence of this zero-frequency Goldstone mode is responsible for some intriguing physical properties of condensed matter. For example, the solid phase of a crystal is a state of a spontaneously broken continuous translational symmetry of the isotropic liquid. As pointed out by P.W.Anderson (1981), see also Forster (1975), one of the most remarkable results of this spontaneous symmetry breaking is the rigidity of solids: it is this zero frequency, infinite wavelength mode which is responsible that all the atoms in a crystal move coherently by the same distance when we push the crystal in a certain direction.

This is in sharp contrast to the isotropic and disordered liquid which just flows away when we try to push it in a certain direction (see Fig. 1.4.3).

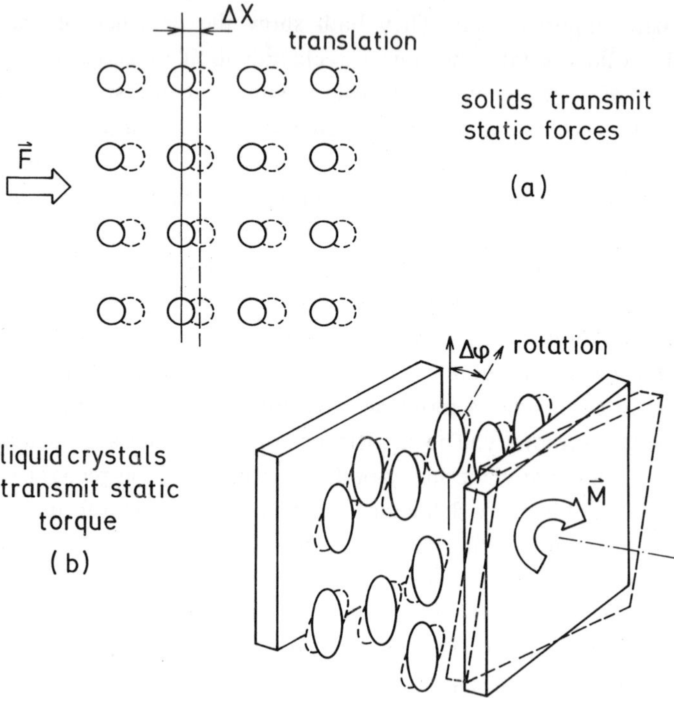

Fig.1.4.3. The difference between a solid crystal and a liquid crystal. (a) The solid crystal is a state of spontaneously broken continuous translation symmetry of the isotropic phase. It transmits static external forces, which tend to move the atoms out of their equilibrium places. (b) The nematic liquid crystal is a state of spontaneously broken continuous orientational symmetry of the isotropic phase of elongated molecules. It transmits static external torque, which tend to rotate the director field out of equilibrium direction.

The concept of a broken symmetry and a Goldstone mode in a quasi-disordered many body system was given independently in several alternate formulations (Orsay Liquid Crystal Group (1969); M.J.Stephen (1970);

D.Forster et al.(1971); H.W.Huang (1971); F.Jähnig and H.Schmidt (1971); P.C.Martin et al.(1972). They proved the existence of the so-called hydrodynamic Goldstone modes in liquid crystals which originate from the spontaneously broken continuous symmetry of an isotropic liquid. These hydrodynamic Goldstone modes are coherent and long-lived excitations even though the microscopic collision times in the system are short. For an extensive discussion of this problem the reader should refer to the book of D.Forster (1975).

We have already briefly mentioned that the nature of a zero-frequency Goldstone mode in liquid crystals can be most easily illustrated in the case of collective excitations in nematic liquid crystals. In the isotropic phase of liquid crystals both the centers of gravity and the long-axes liquid crystalline molecules are scattered in space without any long-range positional and orientational order as shown in Fig.1.4.4.

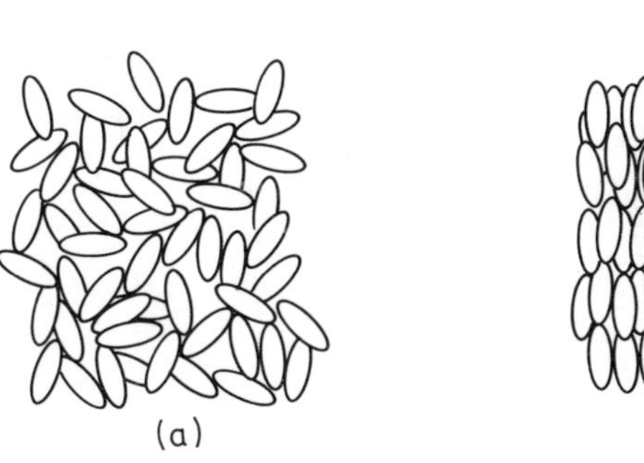

Fig.1.4.4. (a) The isotropic phase. (b) The nematic phase of a liquid crystal.

The isotropic liquid is thus invariant to a continuous group of rotations and a continuous group of translations in three dimensions. When the isotropic nematic is cooled below the clearing point, the long molecular axes align spontaneously in a certain direction and this spontaneous alignment is described by a director field $\vec{n}(\vec{r},t)$, as shown in Fig.1.4.4(b). The presence of this director field breaks the continuous rotational symmetry of the isotropic liquid, just as the appearance of a nonzero magnetization in ferromagnetic materials. Whereas the rotational symmetry of the nematic phase is reduced to $D_{\infty h}$, the free energy of the nematic phase is still invariant to the group of continuous rotations in three dimensions. This means that the nematic phase is a phase of a broken continuous rotational symmetry. As a consequence, a zero frequency mode should exist, which would represent small rotations of the system, which try to restore the broken rotational symmetry.

The nature of the Goldstone mode in a nematic liquid crystal can be illustrated by considering the orientational fluctuations in the limit of small wave-vectors. Fig.1.4.5. shows the so-called bend-wave mode, which represents small periodic bend-like elastic distortions with a wave-vector $\vec{q}$. In the limit of infinite wave-length, $\vec{q} \to 0$, this distortion reduces to a uniform rotation of the sample, as shown in Fig.1.4.5(b).

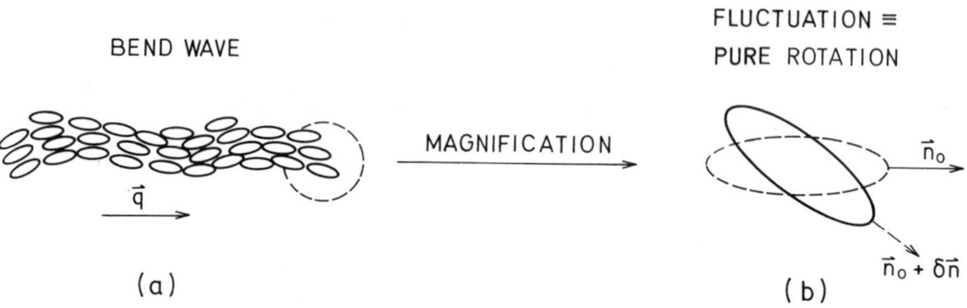

Fig.1.4.5. (a) Bend-like distortion in a nematic liquid crystal. In the limit of large wave-lengths, this distortion reduces to a uniform rotation, shown in (b). This is the Goldstone mode, restoring the broken rotational symmetry of the isotropic phase.

Here the axis of rotation is perpendicular to the plane of original distortion. Because a uniform rotation of the sample as a whole does not cost any energy, the relaxation time of this mode is infinitely long, which is characteristic of a Goldstone mode.

Finally, we should here mention that the zero frequency Goldstone mode in nematic liquid crystals is responsible for the transmission of ***static torque*** (see Fig.1.4.3). This is similar to solids, where the Goldstone modes are responsible for the transmission of ***static forces***. The Goldstone mode in liquid crystals was first observed in a light scattering experiments of Durand et al., see Orsay Liquid Crystal Group (1969), performed in nematic liquid crystals. The dispersion relation for these hydrodynamic modes is shown in Fig.1.4.6. It is parabolic and gapless.

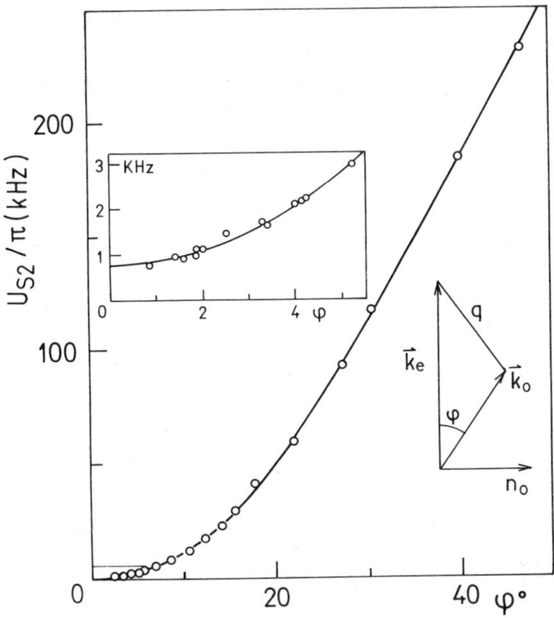

Fig.1.4.6. The parabolic and gapless dispersion relation for hydrodynamic modes in nematic liquid crystals (after Orsay Liquid Crystal Group, 1969). U_{S2} is the relaxation rate. The inset shows the scattering wave-vector.

The spontaneous breaking of a symmetry of the disordered phase is accompanied by another very interesting phenomenon, which is the critical slowing down of a soft mode in the high temperature, disordered phase. The soft mode is a collective mode that becomes unstable at the transition to the low temperature, ordered phase. The relaxation rate (or frequency) of this mode approaches zero for a specific wave-vector $\vec{q}_c$ in the reciprocal space, which is called the critical wave-vector

$$T \to T_c: \quad \tau^{-1}(\vec{q}_c) \to 0 \qquad (1.4.2)$$

The critical wave vector $\vec{q}_c$ equals zero in non-chiral systems and is usually finite in chiral systems.

In liquid crystals, the soft mode was first observed in the light scattering experiments performed by Litster and Stinson (1970) near the isotropic-nematic transition. Their results are shown in Fig.1.4.7. One can clearly see the softening of the soft mode in the isotropic phase of a nematic crystal. The softening is however incomplete because this phase transition is of first order and the ordered phase is stable before the complete softening takes place.

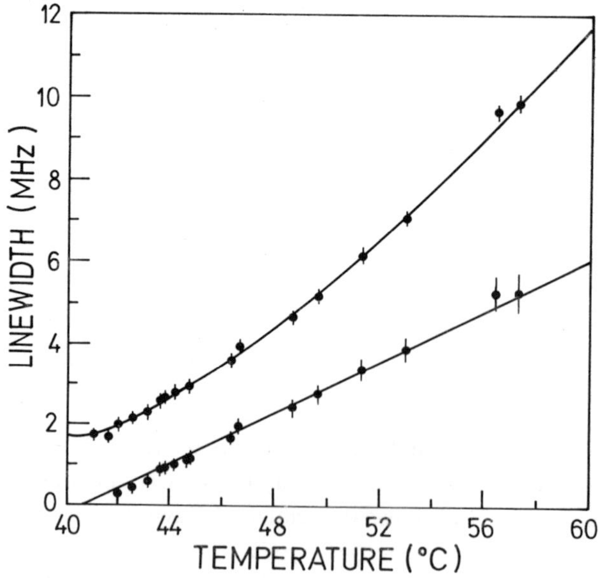

Fig.1.4.7. The soft mode, observed in the isotropic phase of a nematic liquid crystal (after Litster and Stinson, 1970). The soft mode here represents fluctuations of the orientational order.

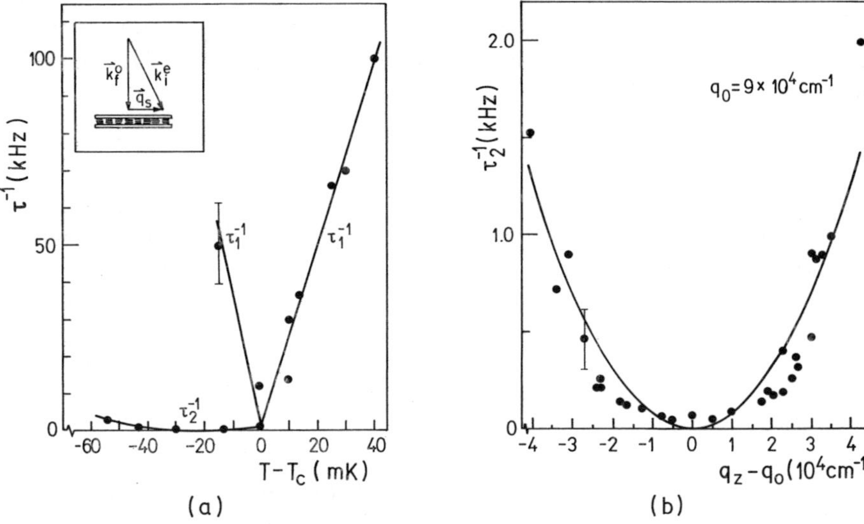

Fig.1.4.8. (a) The splitting of the soft mode into the phase and amplitude modes near the $smectic-A \rightarrow smectic-C^*$ transition in ferroelectric liquid crystals (Muševič et al., 1988). (b) The dispersion relation for the hydrodynamic director (phason) modes in the ferroelectric $smectic-C^*$ phase. Note that the nature of the soft mode is here different as in the nematic liquid crystals. The soft mode in ferroelectric liquid crystals represents collective fluctuations of the tilt angle whereas the degree of the orientational order (i.e. nematic order) remains nearly the same in $smectic-A$ and $smectic-C^*$ phases.

There are therefore two characteristic modes which are accompanying the second order phase transition, where the continuous symmetry group is spontaneously broken. Whereas the soft mode is a symmetry breaking mode, which critically slows down on approaching the phase transition from above, the Goldstone mode is a zero frequency mode that tries to restore the broken symmetry. The concept of the soft mode was first introduced to the *smectic-A* - *smectic-C** phase transition by Blinc in 1975 and 1976 and generalized by Blinc and Žekš in 1978. In ferroelectric and antiferroelectric liquid crystals, the critical behavior and the dispersion of these modes were first observed in the light scattering experiments, performed by Muševič et al. (1988, 1996b). The critical slowing down and the dispersion relation of hydrodynamic (i.e. phason) modes in ferroelectric liquid crystals are shown in Fig.1.4.8. For non-chiral smectics, the critical fluctuations were studied by Delaye and Keller (1976), Delaye (1979) and Galerne et al. (1972).

Note 1.4.1. Chemical structure and phase transitions in antiferroelectric liquid crystals: The chemical structure and the phase transition temperatures of the first antiferroelectric liquid crystal 4-(1-methyl-heptyloxycarbonyl-phenyl) 4'-octylbiphenyl-4-carboxylate (MHPOBC) are shown below.

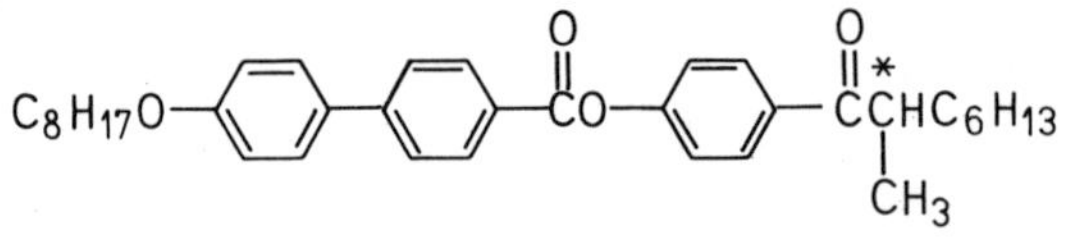

$$C_8H_{17}O-C_6H_4-C_6H_4-\overset{O}{\overset{\|}{C}}O-C_6H_4-\overset{O}{\overset{\|}{C}}-\overset{*}{C}H(CH_3)C_6H_{13}$$

$$\text{MHPOBC}\ \ C \overset{32°C}{\longleftrightarrow} SmI_A^* \overset{65°C}{\longleftrightarrow} SmC_A^* \overset{116.7°C}{\longleftrightarrow} SmC_\gamma^* \overset{118.4°C}{\longleftrightarrow} SmC^* \overset{119.3°C}{\longleftrightarrow} SmC_\alpha^* \overset{120.7°C}{\longleftrightarrow} SmA \overset{145°C}{\longleftrightarrow} I$$

The spontaneous polarization of MHPOBC is rather high, i.e. 112nC/cm^2 and is due to the two C = O dipole moments and the chiral group coupled to the rigid core of the molecule. This high spontaneous polarization is characteristic of antiferroelectric liquid crystals.

Chapter 2

Phase Transitions and Spontaneously Broken Symmetries in Ferroelectric Liquid Crystals

2.1 Introduction

The point symmetry of each layer of the chiral *smectic* $-$ *A* phase is D_∞. This group contains the following symmetry elements: Rotations $C(\varphi)$ through an arbitrary angle φ around the layer normal (i.e. C_∞ axis) and a rotation by π around a direction perpendicular to the layer normal. There is no mirror plane in the chiral *smectic* $-$ *A* phase, because this symmetry operation transforms right-handed objects into left-handed and vice versa.

At a second order phase transition, the symmetry of the high-temperature phase is spontaneously lowered, which means that some of the symmetry elements are lost (for illustration see Fig.2.1.1.). The thermodynamics of systems in the vicinity of the second order phase transitions is most elegantly described by the Landau theory (see Appendix 1) and was first introduced into the field of liquid crystals by P.G. de Gennes (1974). This theory is based exclusively on symmetry considerations and is therefore universal in the sense that it can be applied to very different physical systems.

According to the Landau theory, the symmetry of a given phase is spontaneously lowered at a second order phase transition by the condensation of the soft mode which transforms according to a given ***irreducible representation*** of the symmetry group of the higher temperature phase. The condensed normal coordinates of the soft mode represent the order parameter of the transition. For a given crystal structure, which is characterized by its ***symmetry group***, there are in general ***several irreducible representations*** and consequently there would be also ***several different possibilities of lowering the symmetry*** of the phase. Each of them represents a distinct second order phase transition.

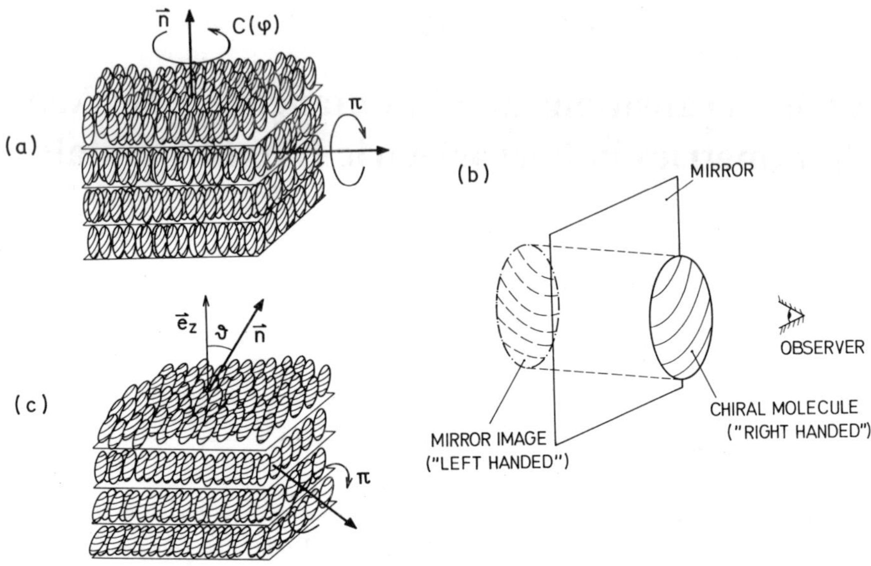

Fig.2.1.1. (a) The layered arrangement of chiral liquid crystalline molecules in the $smectic-A$ phase. When the structure is rotated around the layer normal $\vec{e}_z$ for an arbitrary angle φ, it transforms into itself. This will happen also if the structure is rotated by an angle π around an arbitrary axis, perpendicular to the layer normal. (b) A reflection in any mirror plane transforms "right handed" cigar-like molecules (as indicated by twisted lines, painted on the molecules) into "left-handed" molecules. Reflection is therefore *not a symmetry element* of liquid crystalline phases, composed of chiral molecules. (c) The tilted $smectic-C^*$ phase *cannot* be rotated by an arbitrary angle φ around the layer normal $\vec{e}_z$. Only a single rotation by π around an axis, perpendicular to the plane of the tilt is allowed. Obviously, some symmetry elements have been lost on cooling from the $smectic-A$ phase into the $smectic-C^*$ phase.

The symmetry aspects of the onset of ferroelectricity in chiral and tilted smectic phases were first discussed by Indenbom, Pikin and Loginov in 1976 (see also Pikin, 1991a). They analyzed all different possibilities of the spontaneous breaking of the symmetry of the $smectic-A$ phase and for the first time presented the irreducible representations of the D_∞ point group. These are listed in Table 2.1.1.

Representations of the group D_∞	Parameters of the transition	$C(\varphi)$	$u_2(\varphi)$	Symmetry of the phase formed
A_1	-	1	1	-
A_2	n_z	1	-1	C_∞
E_1	(P_x, P_y) or $(-n_z n_y, n_z n_x)$	$2\cos\varphi$	0	C_2
E_2	$n_x^2 - n_y^2, 2n_x n_y$	$2\cos 2\varphi$	0	D_2
E_n		$2\cos n\varphi$	0	D_n

Table 2.1.1. Irreducible representations of the D_∞ point group, order parameters of the transitions, characters of the irreducible representations and the resulting symmetry of the low-temperature phase. $u_2(\varphi)$ denotes a rotation by π around an axis, which is at an angle φ with respect to the x-axis which lies in the smectic plane. $\vec{P} = (P_x, P_y)$ is the spontaneous polarization in the smectic layers, $\vec{n} = (n_x, n_y, n_z)$ is the director and the z axis is along the normal to the smectic layers. After Indenbom et al. (1976).

Besides the trivial identity representation A_1, which leaves the symmetry of the phase unchanged, there are four different irreducible representations of the chiral $smectic-A$ phase :

(i) The one-dimensional representation A_2 induces a phase transition $D_\infty \to C_\infty$ and results in a symmetry change, as illustrated by the two chiral objects shown in Fig.2.1.2. The two-fold axes of the D_∞ point group are lost and the symmetry of the layers is reduced to C_∞. The order parameter of the transition is n_z, i.e. the z-component of the director. If the molecules are polar and have a dipole moment along the long molecular axis, this transition would correspond to a creation of a ferroelectric "longitudinal" order within a single molecular layer. The

appearance of ferroelectricity in the bulk sample would however still be questionable because of the strong interlayer electrostatic interaction. This kind of longitudinal ferroelectric phase is a proper ferroelectric phase and has never been observed in the bulk.

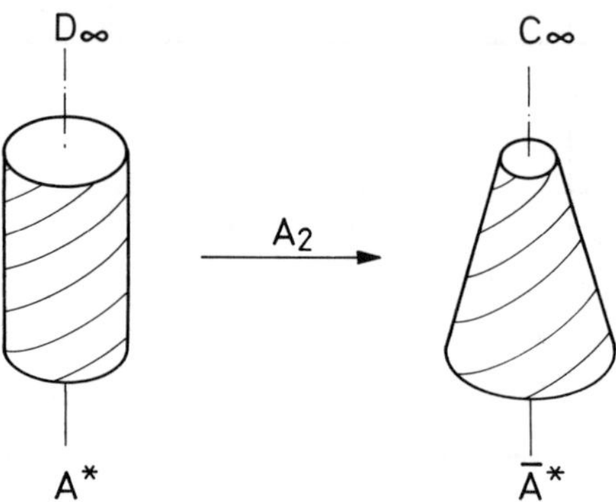

Fig.2.1.2. Hypothetical phase transition to the ferroelectric $smectic-A$ phase, as induced by the A_2 irreducible representation of the point group D_∞, leading to longitudinal ferroelectric phase.

(ii) The two-dimensional representation E_1 induces a symmetry change $D_\infty \to C_2$, as indicated in Fig.2.1.3. It represents tilting of chiral objects in the low-temperature phase and actually describes the $smectic-A$ - $smectic-C^*$ phase transition. In the *smectic-*C^* phase there is only one two-fold rotation axis along the direction perpendicular to the plane of tilt. Whereas the longitudinal dipole moments average-out, the transverse dipole moments can be finite, and in-plane ferroelectricity is therefore allowed. The order parameters of the transition are the two-component combinations

$$\vec{\xi} = (\xi_x, \xi_y) = (-n_z n_x, n_z n_y) \qquad \text{or} \qquad \vec{P} = (P_x, P_y). \qquad (2.1.1)$$

Because the dipole-dipole interactions are far too small to drive this phase transition, the primary order parameter of the transition is the tilt vector $\vec{\xi}$ and the secondary order parameter is the spontaneous polarization $\vec{P}$. The electric polarization appears in the *smectic-*C^* phase merely because of the coupling between the polarization and the tilt and is due to steric forces. The ferroelectric *smectic-C^** phase therefore formally belongs to the class of improper ferroelectrics.

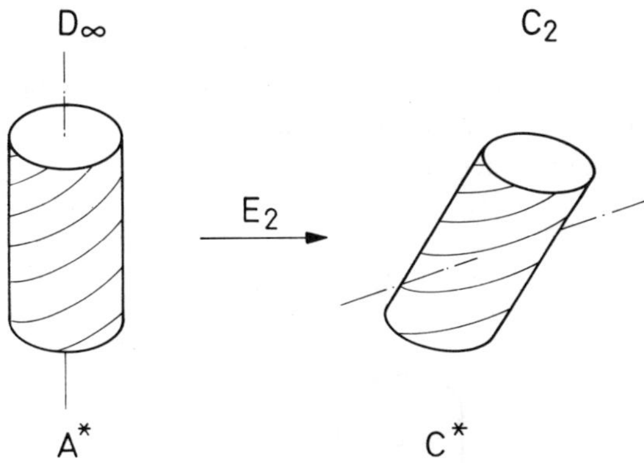

Fig.2.1.3. The *smectic-A* - *smectic-C^** phase transition, as induced by the irreducible representation E_1, leading to in-plane ferroelectricity.

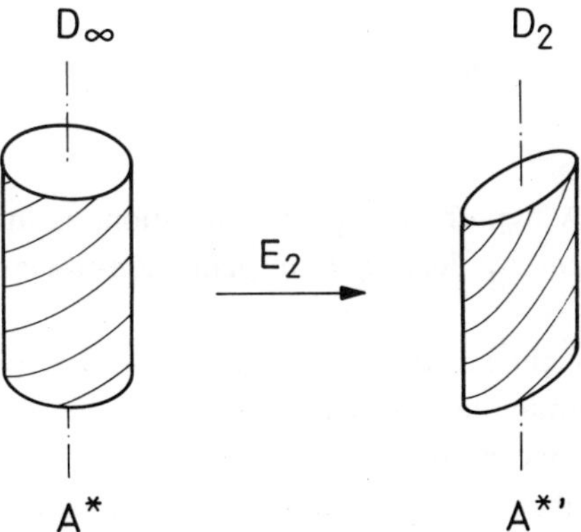

Fig.2.1.4. Phase transition from the chiral *smectic-A* phase, as induced by the irreducible representation E_2, leading to in-plane quadrupolarly ordered structure.

(iii) The two dimensional irreducible representation E_2 induces a symmetry change $D_\infty \to D_2$, as indicated in Fig.2.1.4. The resulting phase is non-polar and quadrupolarly-ordered with the order parameter $\left(n_x^2 - n_y^2, 2n_x n_y\right)$.

(iv) The irreducible representations of the type $E_n, n \geq 3$ induce symmetry changes $D_\infty \to C_n$. In the low-temperature phase there is a n-fold axis in direction of the layer normal, as indicated in Fig.2.1.5. The in-plane ordering is non-polar.

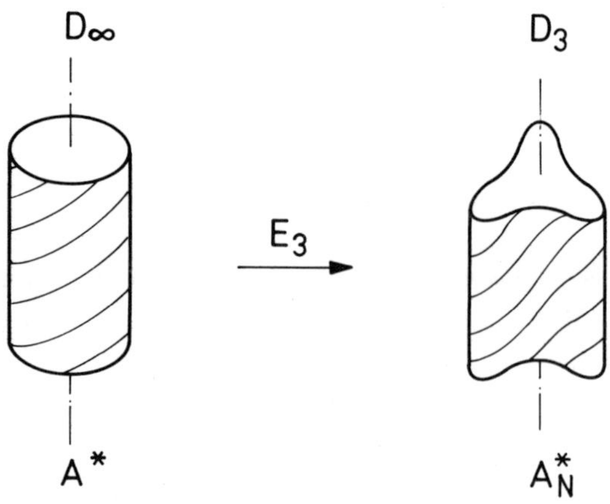

Fig.2.1.5. A hypothetical phase transition, as induced by the irreducible representation E_3, leading to in-plane non-polar ordering.

This leads to a conclusion that the *smectic-A* - *smectic-C** phase transition is induced by the irreducible representation E_1 of the point group D_∞. As far as group theory is concerned, the *smectic-A* - *smectic-C** phase transition is equally well described by either choosing the spontaneous polarization or the spontaneous tilt of the molecules as the order parameter. However, for physical reasons the tilt is chosen as the primary order parameter and the polarization is considered as a small perturbation to the *smectic-C** phase. It arises because of the coupling to the primary order parameter.

One of the most interesting aspects of the *smectic* $-A$ - *smectic-*C^* phase transition is the fact that a continuous rotational symmetry of the point group D_∞ is spontaneously broken into a discrete rotational symmetry of the C_2 point group. As a result, a new symmetry restoring branch of excitations appears in the ferroelectric phase. In view of the continuous broken symmetry, this new branch is gaplesss and a zero frequency, symmetry restoring Goldstone mode should exists at the critical wave-vector.

2.2. Landau Theory of the Smectic-A→Smectic-C* Phase Transition

Soon after the discovery of ferroelectricity in liquid crystals, the Landau theory of the *smectic-A* - *smectic-*C^* phase transition was first derived by Blinc (1975, 1976) for the non-chiral case and by Indenbom et al. (1976) for the chiral case. They have shown that the order parameters of the transition are the two-component combinations of the molecular tilt $(n_x n_z, n_y n_z)$ or electric polarization $\vec{P}=(P_x, P_y)$, shown in Fig.2.2.1.

Because the dipole-dipole interactions are far too small to drive this phase transition, the primary order parameter of the transition is the tilt vector and the secondary order parameter is the spontaneous polarization $\vec{P}$. The electric polarization appears in the *smectic-*C^* phase merely because of the coupling between the polarization and the tilt. The ferroelectric *smectic-*C^* phase therefore formally belongs to the class of improper ferroelectrics.

Indenbom et al. (1976) have shown that there are no third order invariants in the free-energy expansion in terms of the order parameters of the *smectic-A* - *smectic-*C^* phase transition. This transition may therefore be of second order, as indeed observed in most of single-compound ferroelectric liquid crystals. The lack of mirror symmetry in these systems, however allows for the existence of the so-called *Lifshitz term*

$$\xi_x \frac{\partial \xi_y}{\partial z} - \xi_y \frac{\partial \xi_x}{\partial z} \qquad (2.2.1)$$

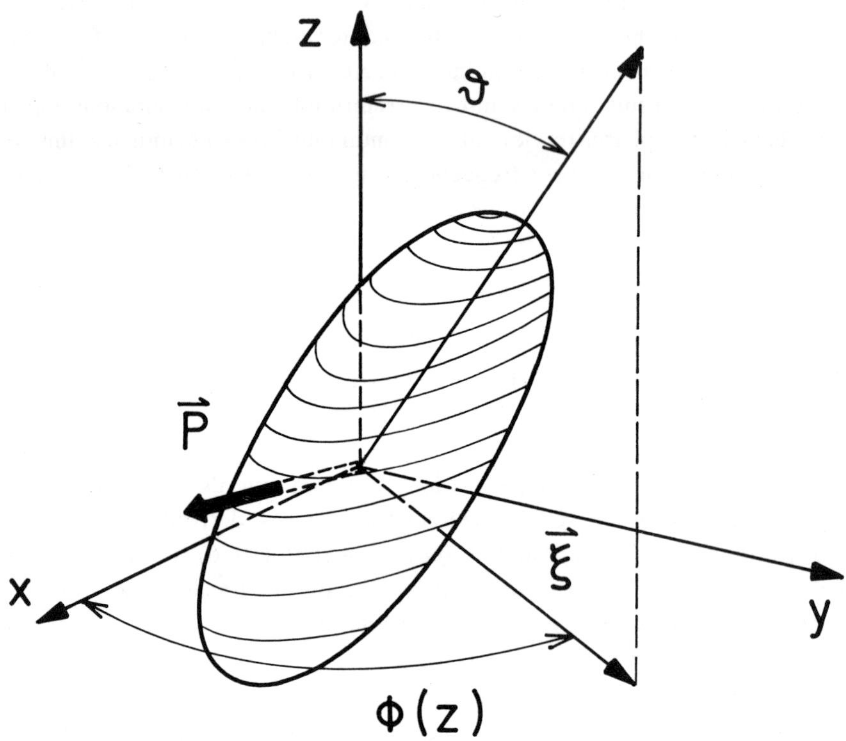

Fig.2.2.1. The order parameters of the *smectic-A* - *smectic-C** transition: the in-plane projection of the molecular tilt $\vec{\xi} = (\xi_x, \xi_y)$ and the electric polarization $\vec{P} = (P_x, P_y)$. Note that in the limit of small tilt angle, the tilt vector $\vec{\xi}$ equals $\vec{\xi} = (\xi_x, \xi_y) \approx (n_z n_x, n_z n_y)$. Also note that the local direction of the long molecular axis can be specified by the magnitude of the tilt angle θ and the phase of the tilt angle $\Phi(z)$.

A similar term can be formulated for the spontaneous polarization as

$$P_x \frac{\partial P_y}{\partial z} - P_y \frac{\partial P_x}{\partial z} \tag{2.2.2}$$

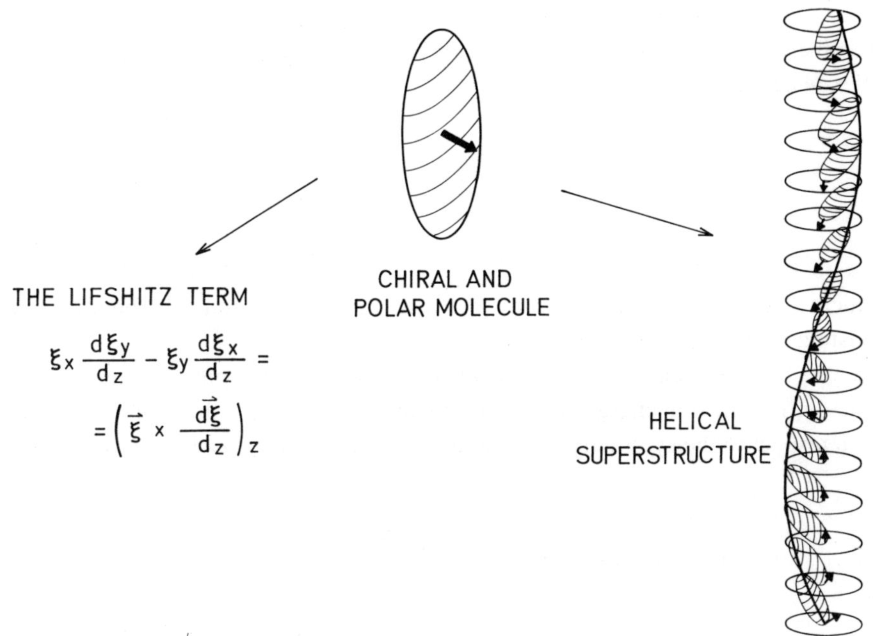

Fig.2.2.2. Molecular chirality gives rise to a helical modulation of a smectic structure. The Lifshitz term appears in the free-energy density expansion due to the absence of mirror symmetry.

The Lifshitz term is of chiral nature and induces a helicoidal precession of the molecular tilt, as we move along the layer normal. Although the length scale of this precession is in the mesoscopic range, it subtly reflects the fact that we have here chiral liquid crystalline molecules, which are the basic building blocks. It is interesting that the periodicity of the resulting helix will be, in general, incommensurate with the underlying one-dimensional translational (smectic) periodicity of the *smectic-A* phase.

The symmetry properties of the high-temperature phase allow for two types of coupling terms between the molecular tilt and polarization: a bilinear "piezoelectric-like" coupling

$$P_x\xi_y - P_y\xi_x \tag{2.2.3}$$

which results in a proportionality between the polarization and the tilt, and a "flexoelectric" term

$$P_x \frac{\partial \xi_x}{\partial z} + P_y \frac{\partial \xi_y}{\partial z} \tag{2.2.4}$$

which results in a proportionality between the polarization and bending and twisting of the director field in space.

In the vicinity of the *smectic-A* $\rightarrow$ *smectic-C** phase transition, the nonequilibrium free energy density is expanded in terms of the scalar invariants with respect to the symmetry operations of the chiral *smectic-A* phase as:

$$g = g_A + \tfrac{1}{2}a\left(\xi_x^2 + \xi_y^2\right) + \tfrac{1}{4}b\left(\xi_x^2 + \xi_y^2\right)^2 -$$

$$-\Lambda\left(\xi_x \frac{d\xi_y}{dz} - \xi_y \frac{d\xi_x}{dz}\right) + \tfrac{1}{2}K_3\left[\left(\frac{d\xi_x}{dz}\right)^2 + \left(\frac{d\xi_y}{dz}\right)^2\right] +$$

$$+ \tfrac{1}{2\varepsilon}\left(P_x^2 + P_y^2\right) - \mu\left(P_x \frac{d\xi_x}{dz} + P_y \frac{d\xi_y}{dz}\right) + C\left(P_x\xi_y - P_y\xi_x\right) \tag{2.2.5}$$

Here g_A is the equilibrium free-energy density of the *smectic-A* phase, $\xi_x = n_x n_z$, $\xi_y = n_y n_z$, and $\vec{n} = (n_x, n_y, n_z)$ is the director. The coefficient $a = \alpha(T - T_o)$ is temperature dependent, whereas all other coefficients are assumed to be constant. In particular, the coefficient of the P^2 term is constant, which reflects the assumption that the ***dipole-dipole interactions do not drive*** the *smectic-A* - *smectic-C** transition. The expansion does not include derivative terms in the polarization and the system is considered to be homogeneous in the x-y plane. The coefficient K_3 is the twist elastic constant of the *smectic-C** phase and should not be mixed with the nematic bend elastic constant K_3. Λ is the coefficient of the Lifshitz term which is responsible for the spontaneous twisting of the *smectic-C** phase. μ and C are the coefficients of the flexo-electric and piezo-electric bilinear coupling between the tilt and polarization and ε is the dielectric constant of the *smectic-A* phase. The coefficient b of the fourth order term in the primary order parameter is positive, as most transitions between the *smectic-A* and *smectic-C** phases are found to be of second order.

Let us now discuss the stability limits of the *smectic-A* phase. This can be most conveniently analyzed by introducing the inhomogeneous fluctuations in the form of "left" $(q<0)$ or "right" $(q>0)$ handed helical tilt and polarization waves, as shown in Fig.2.2.3.

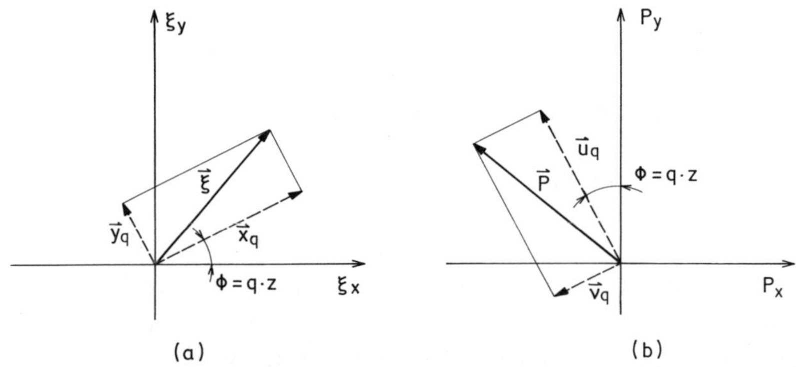

Fig.2.2.3. In the *smectic-A* phase, fluctuations of tilt $\xi(z,t)$, (a), and polarization $\vec{P}(z,t)$, (b), are taken in the form of helical, elliptically polarized waves with a wave-vector $\vec{q}=(0,0,q)$ along the layer normal. Note that for $x_q = y_q$ and $u_q = v_q$ these waves reduce to circularly polarized orientational waves.

$$\xi_x = \sum_q \left(x_q \cos qz - y_q \sin qz\right) \tag{2.2.6a}$$

$$\xi_y = \sum_q \left(x_q \sin qz + y_q \cos qz\right) \tag{2.2.6b}$$

$$P_x = \sum_q \left(-u_q \sin qz - v_q \cos qz\right) \tag{2.2.6c}$$

$$P_y = \sum_q \left(u_q \cos qz - v_q \sin qz\right) \tag{2.2.6d}$$

The helical form of the fluctuations has been chosen, as it reflects the mirror-less symmetry of the *smectic-*C^* phase. Let us now determine with the help of the stability analysis, the temperature and the wave-vector, where the *smectic-A* - *smectic-*C^* transition occurs.

The introduction of inhomogeneous fluctuations (Eq.2.2.6.a-d) changes the free-energy of the system. Above the phase transition temperature T_c , this change is expressed in the q-representation as:

$$\begin{aligned} G &= \tfrac{1}{L}\int_0^L \Delta g\, dz = \\ &= \sum_q \Big[\left(\tfrac{1}{2}a(T) - \Lambda q + \tfrac{1}{2}K_3 q^2\right)\left(x_q^2 + y_q^2\right) + \frac{1}{2\varepsilon}\left(u_q^2 + v_q^2\right) \\ &\quad - \left(\mu q + C\right)\left(x_q u_q + y_q v_q\right)\Big] \end{aligned} \tag{2.2.7}$$

We have considered only the harmonic part of the free-energy G and higher order terms have been neglected. Note that for the stability analysis it is important that the second derivative of the free energy with respect to the order parameter is positive, i.e. $\partial^2 G/\partial x^2|_{x=0} = 0$ Let us now determine the possible minima of this free energy. The equilibrium values of the polarization and the tilt are determined by

$$\left.\frac{\partial G}{\partial x_q}\right|_{x_q^{(0)}} = \left.\frac{\partial G}{\partial y_q}\right|_{y_q^{(0)}} = \left.\frac{\partial G}{\partial u_q}\right|_{u_q^{(0)}} = \left.\frac{\partial G}{\partial v_q}\right|_{v_q^{(0)}} = 0 \tag{2.2.8}$$

where for $T < T_c$ quartic terms have to be included. One solution of this system of equations corresponds to the *smectic-A* phase with zero tilt and polarization which exists at all temperatures:

$$x_q^{(0)} = y_q^{(0)} = u_q^{(0)} = v_q^{(0)} = 0 \tag{2.2.9}$$

In order to decide whether this solution is stable for a given temperature, we have to determine its stability against small fluctuations of the form (Eq.2.2.6). Physically this means that the solution should be stable against thermal fluctuations of the order parameter, which are always present at finite temperature. This is most conveniently done by rewriting the harmonic part of the nonequilibrium free-energy of the *smectic-A* phase in a matrix form, as first introduced by Čepič (1993).

$$G = \sum_q \vec{\Psi}_q^+ \cdot \underline{D}(q) \cdot \vec{\Psi}_q \tag{2.2.10}$$

Here, the amplitudes of the fluctuations (Eq.2.2.6.a-d) are conveniently arranged in a four-dimensional vector $\vec{\Psi}_q$

$$\vec{\Psi}_q = \begin{pmatrix} x_q \\ y_q \\ u_q \\ v_q \end{pmatrix} \tag{2.2.11}$$

and $\vec{\Psi}_q^+$ is its Hermitian conjugate. We have here introduced the so-called dynamical matrix $\underline{D}(q)$,

$$\underline{D}(q) = \begin{bmatrix} A_q & 0 & -D_q & 0 \\ 0 & A_q & 0 & -D_q \\ -D_q & 0 & C_q & 0 \\ 0 & -D_q & 0 & C_q \end{bmatrix} \tag{2.2.12}$$

with elements $A_q = \frac{1}{2}a(T) - \Lambda q + \frac{1}{2}K_3 q^2$, $D_q = \frac{1}{2}(\mu q_c + C)$ and $C_q = 1/(2\varepsilon)$. This matrix has a similar role as the well known dynamical matrix in the analysis of dynamical properties of solids (see, for example Ziman 1962) and is formed of second derivatives:

$$D_{ij} = \frac{\partial^2 G}{\partial \Psi_{iq} \partial \Psi^*_{jq}} \qquad (2.2.13)$$

In the *smectic-A* phase, the equilibrium value of the tilt angle is zero. This means that the harmonic part of the free-energy should always increase if we add an excitation with a wave-vector $\vec{q}$. Mathematically, this has the consequence that the eigenvalues λ_i of the dynamical matrix should be positive (i.e. the second derivatives of G with respect to Ψ). The stability limit of the phase is thus determined by the temperature T_c and the wave-vector $\vec{q}_c$, where the smallest eigenvalue of $\underline{D}(q)$ becomes zero. After a straightforward calculation (Blinc and Žekš, 1978) the *smectic-A* - *smectic-C** phase transition temperature is

$$T_c = T_o + \frac{1}{\alpha}\left(\varepsilon C^2 + \left(K_3 - \varepsilon\mu^2\right) q_c^2\right) \qquad (2.2.14)$$

whereas the critical-wave vector, where the phase transition occurs, is

$$q_c = \frac{\Lambda + \varepsilon\mu C}{K_3 - \varepsilon\mu^2} \qquad (2.2.15)$$

In the above expression (Eq.2.2.14), T_o is the *smectic-A* - *smectic-C** phase transition of the racemic (nonchiral) mixture of right- and left-handed enantiomers of a given substance. The phase transition temperature for the racemate is always lower than the corresponding phase transition temperature for a pure chiral substance. This is due to: *(i)* The chiral coupling between the tilt and polarization (εC^2 term in Eq.2.2.14) and *(ii)* The elastic energy of spontaneously twisted *smectic-C** structure (q_c^2 term in 2.2.14) which has to be gained to unwind the helix at the phase transition into the *smectic-A* phase. The total shift $T_c - T_o$ is of the order of 100mK. It is rather small, as expected in view of the smallness of the chiral terms in the free-energy expansion 2.2.5.

The period of the helix $p = 2\pi/q_c$ is given by the ratio between the chiral twisting terms and the elastic restoring terms, both of which contain flexoelastic contributions (Eq.2.2.15). Within the above simple model, the period of the helix of the *smectic-C** phase is temperature independent. Typically, it is of the order of the wave-length of visible light. The period can also be adjusted by mixing right and left-handed enantiomers. It is interesting that due to different signs of the Λ and $\varepsilon\mu C$ terms in Eq.2.2.15, non-helical structures with $q_c = 0$ and finite spontaneous polarization P_s are also possible. On the other hand, for nonchiral systems Λ and C are identically zero and the phase is always homogeneous with $q_c = 0$ and $P_s = 0$.

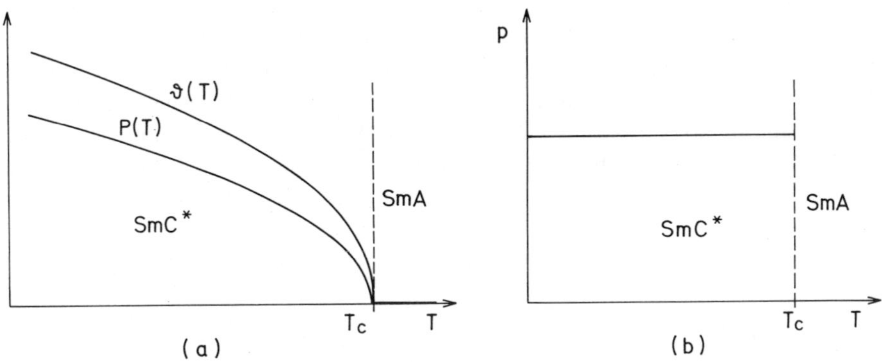

Fig.2.2.4. Temperature dependence of the tilt angle and spontaneous polarization (a) and the period of the helix in a ferroelectric smectic liquid crystal within a simple Landau theory (b).

Below T_c, another solution is thermodynamically stable, shown in Fig.2.2.4. It has a finite value of the tilt and the polarization and describes the helicoidally modulated *smectic-C** phase:

$$\begin{aligned} T < T_c \qquad & \xi_x = \theta_\circ \cos q_c z \quad , \quad \xi_y = \theta_\circ \sin q_c z \\ & P_x = -P_\circ \sin q_c z \quad , \quad P_y = P_\circ \cos q_c z \end{aligned} \tag{2.2.16}$$

The equilibrium tilt and the polarization are here given by

$$\theta_\circ = \sqrt{(\alpha/b)(T_c - T)}\ ,\ P_\circ = \varepsilon(\mu q_c + C)\cdot\theta_\circ \tag{2.2.17}$$

The temperature dependence of the tilt, polarization and the period of the helix are shown schematically in Fig.2.2.4.

Note 2.2.1. Tilt angle in ferroelectric liquid crystals:
The magnitude of the tilt angle in a typical ferroelectric liquid crystal DOBAMBC is shown on the figure below (after Huang and Dumrongrattana, 1986). The temperature dependence of the tilt angle does not follow the predicted square root dependence (Eq.2.2.18). Instead, the exponent of tilt angle is close to 0.33, which is the value expected for the XY universality class. Huang and Viner (1982) have shown that this is an effective exponent, which is due to the tricriticality of the system and has no relation to the universality class of the order parameter. For details see Section 2.5.

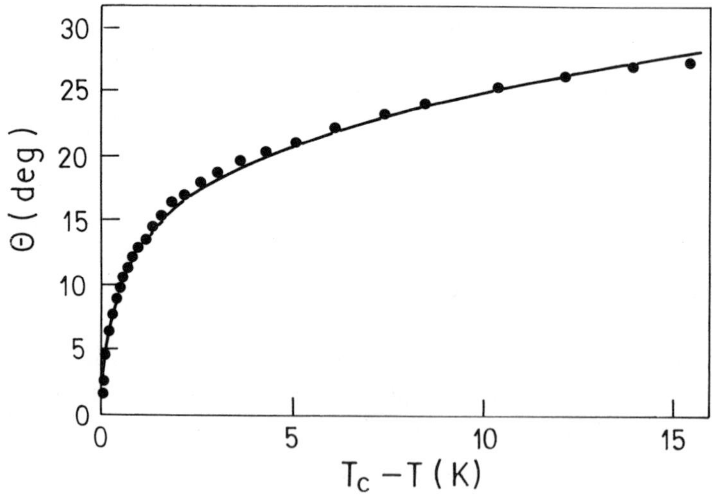

Note 2.2.2. Period of the helix in ferroelectric liquid crystals:
Most of the liquid crystals that exhibit a second order *smectic-A* - *smectic-C** phase transition show a characteristic temperature dependence of the helical period, which is

shown below for DOBAMBC (Muševič et al., 1984). Close to the *smectic-A* phase, the helix is usually short, i.e. one micron or less, and increases by decreasing temperature to a maximum value of several microns at approximately one degree below the phase transition. Then, the helix decreases gradually, as we go well below the phase transition temperature. This non-monotonous temperature dependence of the helix is in sharp contrast to the predictions of a simple Landau theory (Eq.2.2.15), which predicts temperature independent period of the helix. The helical period can be measured either using a microscope or using laser diffraction. In the first type of experiments, one observes the periodic, stripe-like pattern in thick, homogeneously aligned sample (see Fig.10.2.1).Here, stripes arise from the periodicity of the index of refraction due to helical modulation of the ferroelectric phase. More precise measurements of the helical period can be obtained from the light diffraction experiments. Here, the laser light of appropriate wavelength is diffracted from the periodic ferroelectric structure, which forms a phase grating. The condition for the Bragg diffracted intensity maximum in homogeneously aligned sample is

$$\frac{2\pi}{\lambda_\circ} \cdot \sin\vartheta_m = m \cdot \frac{2\pi}{p_\circ}, \qquad m=1,2,3$$

Here, $\lambda_\circ$ is the wavelength of laser light in vacuum, $p_\circ$ is the period of the helix and ϑ_m is the angle of diffraction of m-th maximum. The input laser beam is directed perpendicularly to the helical axis. Bragg reflection can also be observed in homeotropic geometry.

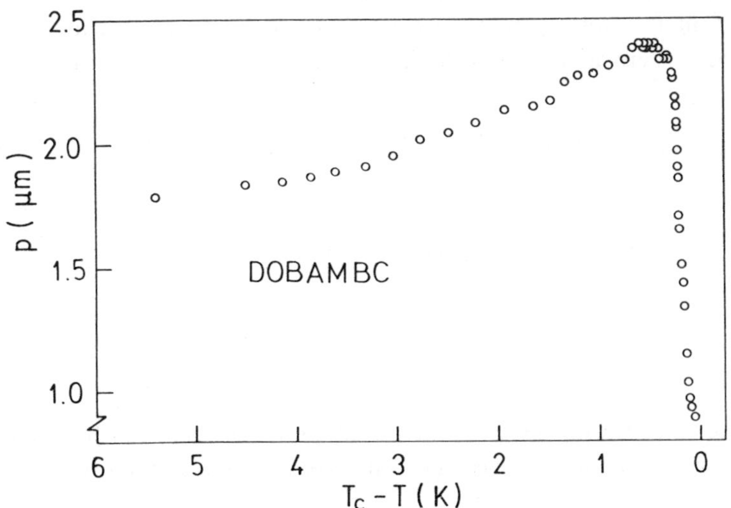

2.3. Order Parameter Dynamics, Soft Modes and Gapless Phasons

One of the most striking phenomena, which can be observed near the *smectic-A* - *smectic-C** phase transition is the critical dynamics of the order parameter fluctuations, which is associated with the spontaneous breaking of the point symmetry of the *smectic-A* phase at T_c. The symmetry breaking is here the result of the spontaneous collective tilting of molecular directors in the *smectic-C** phase. This reduces the original D_∞ point symmetry of the *smectic-A* phase to the C_2 symmetry of the chiral *smectic-C** layers. Thus the *continuous point symmetry* of the *smectic-A* phase is spontaneously broken. This has some very interesting consequences for the order parameter dynamics and is in particular reflected in the spectrum of the so-called elementary excitations of the system. According to the ***Goldstone theorem*** (Goldstone et al., 1962; Bludman and Klein, 1963; Lange, 1966; Forster, 1975; Blinc and Žekš, 1974a), a continuous symmetry breaking results in the appearance of a zero frequency, symmetry restoring ***Goldstone mode***. This mode exists in the *smectic-C** phase and tries to restore the broken continuous symmetry of the *smectic-A* phase. The excitation spectrum of the one branch of the order parameter fluctuations in the *smectic-C** phase is thus expected to be gapless.

The central idea of the conventional theory of the spontaneous symmetry breaking at the phase transition point as first proposed by van Hove (1954) and Landau and Khalatnikov (1954) is the critical slowing down of the order parameter excitations at T_c. This implies the existence of a mode $\tau_s^{-1}(\vec{q},T)$ in the order parameter excitation spectrum, the relaxation rate of which goes to zero as we approach the phase transition point from above

$$\tau_s^{-1}(\vec{q},T)=A\cdot(T-T_c)+B\cdot(\vec{q}-\vec{q}_c)^2 \quad , \quad T>T_c \tag{2.3.1}$$

Thus for $T=T_c$ and $\vec{q}=\vec{q}_c$ the relaxational rate of this mode equals zero:

$$\tau_q^{-1}(\vec{q}_c,T_c)=0 \tag{2.3.2}$$

This is the so-called ***soft mode*** of the transition. It is a symmetry breaking mode which becomes soft and freezes out (condenses) at the phase transition point T_c

(see Fig.2.3.1). The freezing of this mode therefore breaks the symmetry of the higher temperature phase.

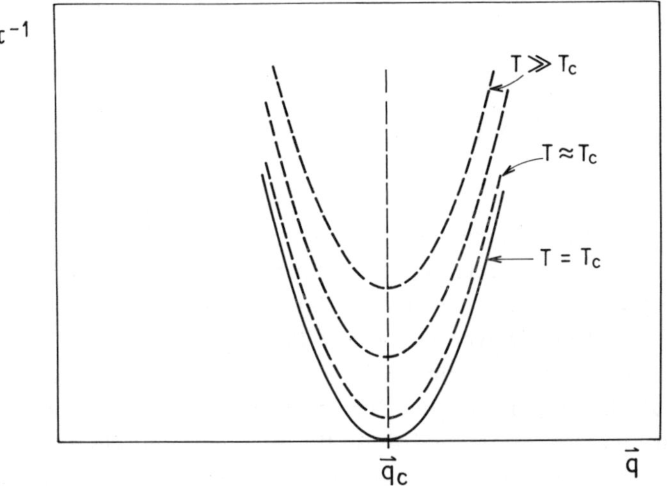

Fig.2.3.1. "Freezing" of the soft mode in the reciprocal space. The lowest mode on the dispersion curve $\tau^{-1}(\vec{q},T)$ condenses first and is therefore the soft mode of the transition. It freezes at a finite wave-vector $\vec{q}_c$ and gives rise to a modulated structure.

The concept of the soft mode was first introduced to the *smectic-A* - *smectic-C** phase transition by Blinc in 1975 and 1976 for the homogeneous case and generalized by Blinc and Žekš in 1978 for the case of a modulated structure. The theory is based on the free-energy expansion (Eq.2.2.5) and analyzes the spectrum of the order parameter fluctuations within the Landau-Khalatnikov equations of motion. Following this approach, the instantaneous value of the order parameter $\xi(z,t)$ is

$$\bar{\xi}(z,t)=\bar{\xi}_{\circ}(z)+\delta\bar{\xi}(z,t) \tag{2.3.3}$$

Here, $\xi_\circ(z)$ denotes the equilibrium value of the order parameter and $\delta\vec{\xi}(z,t)$ is the time-fluctuating part. One should note that in the *smectic-A* phase, the equilibrium value of the order parameter equals zero, $\xi_\circ = 0$. Similarly, the polarization field $P(z,t)$, which is the secondary order parameter of the transition, is

$$\vec{P}(z,t) = \vec{P}_\circ(z) + \delta\vec{P}(z,t) \qquad (2.3.4)$$

The dynamical properties of the helical structure can be most conveniently analyzed by introducing a reference frame that rotates in the same manner as the helix, as we move along the smectic layer normal (Martinot-Lagarde and Durand, 1980, 1981; Drevenšek et al., 1990). The helicoidally modulated structure then appears uniform to an observer in such a system and the mathematical description of the dynamical phenomena is simplified. We introduce two unit vectors $\vec{e}_{//}$ and $\vec{e}_\perp$ in the rotating frame. Here, the vector $\vec{e}_{//}$ is parallel to the direction of the local tilt in the *smectic-C** phase and $\vec{e}_\perp$ is perpendicular. The two unit vectors $\vec{e}_{//}$ and $\vec{e}_\perp$ are related to the unit vectors $\vec{e}_x$ and $\vec{e}_y$ in the laboratory system by

$$\vec{e}_{//} = \vec{e}_x \cdot \cos\Phi(z) + \vec{e}_y \cdot \sin\Phi(z) \qquad \text{2.3.5a.}$$

$$\vec{e}_\perp = -\vec{e}_x \cdot \sin\Phi(z) + \vec{e}_y \cdot \cos\Phi(z) \qquad \text{2.3.5b.}$$

As illustrated in Fig.2.3.2., the $\vec{e}_{//}$ unit vector points along the local direction of the tilt, the $\vec{e}_\perp$ vector is perpendicular to it and they form together with the layer normal $\vec{e}_z$ a positively oriented triad, $\vec{e}_{//} \times \vec{e}_\perp = \vec{e}_z$.

In the rotating reference frame, the time-fluctuating parts of both $\xi(z,t)$ and $P(z,t)$ are decomposed into components parallel to the equilibrium value of the order parameter $(\delta\xi_{//} \text{ and } \delta P_{//})$ and the perpendicular components $(\delta\xi_\perp \text{ and } \delta P_\perp)$, as illustrated in Fig.2.3.3. and 2.3.4. The parallel fluctuating components represent the so-called ***amplitude excitations***, as they change the magnitude of the order parameters, i.e. the tilt and polarization. The perpendicular components represent ***phase excitations***, as they change the phase profile and consequently the direction of the director and polarization fields.

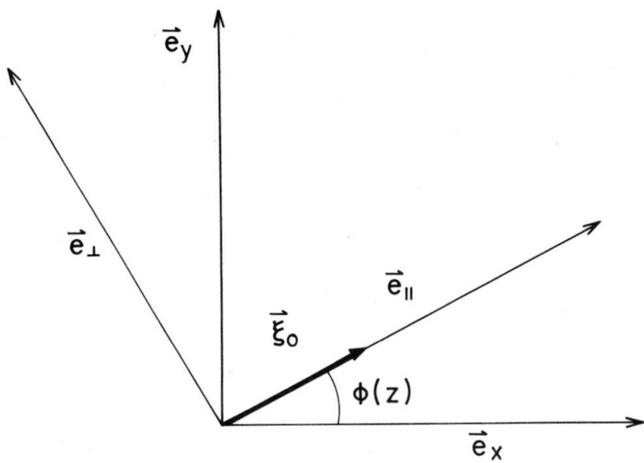

Fig.2.3.2. The definition of the helical (rotating) reference frame

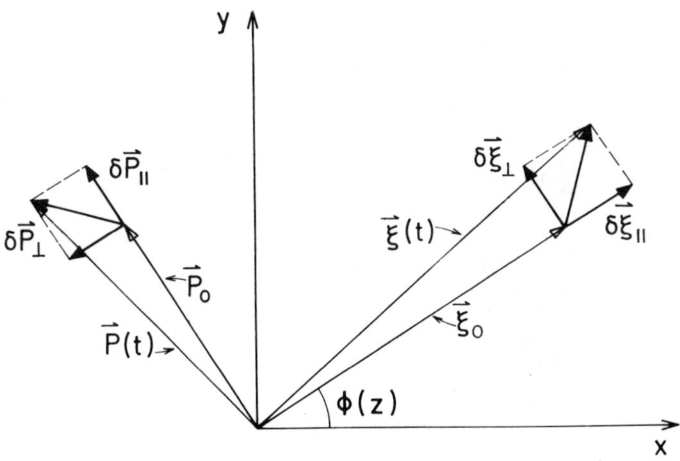

Fig.2.3.3. The instantaneous value of the tilt and polarization in the chiral *smectic-C** phase.

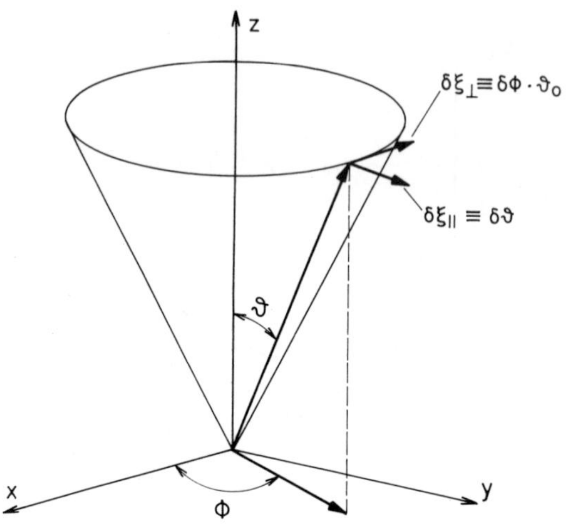

Fig.2.3.4. The fluctuations of the tilt can be decomposed into two components.

The time-fluctuating part of the order parameter is expanded in a Fourier series:

$$\delta\xi_{//}(z,t)=\sum_{q=-\infty}^{\infty}\delta\xi_q^{//}(t)\cdot e^{iqz} \tag{2.3.6a}$$

$$\delta\xi_{\perp}(z,t)=\sum_{q=-\infty}^{\infty}\delta\xi_q^{\perp}(t)\cdot e^{iqz} \tag{2.3.6b}$$

$$\delta P_{//}(z,t)=\sum_{q=-\infty}^{\infty}\delta P_q^{//}(t)\cdot e^{iqz} \tag{2.3.6c}$$

$$\delta P_{\perp}(z,t)=\sum_{q=-\infty}^{\infty}\delta P_q^{//}(t)\cdot e^{iqz} \tag{2.3.6d}$$

where $\delta\xi^i_{-q} = (\delta\xi^i_q)^*$ and $\delta P^i_{-q} = (\delta P^i_q)^*$ for $i = \| \text{ or } \perp$, in order to keep the physical variables real and the wave-vector q is given in the rotating frame.

This leads to the following expression for the nonequilibrium value of the free energy

$$\Delta G = \frac{1}{L}\int_0^L g(z)dz - G_\circ = \sum_q \left(\tfrac{1}{2}a(T) + \tfrac{3}{2}b\theta_\circ^2 - \Lambda q_c + \tfrac{1}{2}K_3(q_c^2 + q^2)\right)\cdot\left|\delta\xi_q^{//}\right|^2 +$$

$$+\sum_q \left(\tfrac{1}{2}a(T) + \tfrac{1}{2}b\theta_\circ^2 - \Lambda q_c + \tfrac{1}{2}K_3(q_c^2 + q^2)\right)\cdot\left|\delta\xi_q^{\perp}\right|^2 + \frac{1}{2\varepsilon}\sum_q\left(\left|\delta P_q^{//}\right|^2 + \left|\delta P_q^{\perp}\right|^2\right) +$$

$$+\tfrac{1}{2}(\mu q_c + C)\sum_q\left[\left(\delta P_q^{//}\cdot\delta\xi_q^{//*} + \delta P_q^{//*}\cdot\delta\xi_q^{//}\right) + \left(\delta P_q^{\perp}\cdot\delta\xi_q^{\perp *} + \delta P_q^{\perp *}\cdot\delta\xi_q^{\perp}\right)\right] -$$

$$- i(K_3 q_c - \Lambda)\sum_q q\left(\delta\xi_q^{//}\cdot\delta\xi_q^{\perp *} - \delta\xi_q^{//*}\cdot\delta\xi_q^{\perp}\right) +$$

$$+ i\frac{\mu}{2}\sum_q q\left[\left(\delta P_q^{//}\cdot\delta\xi_q^{\perp *} - \delta P_q^{//*}\cdot\delta\xi_q^{\perp}\right) - \left(\delta P_q^{\perp}\cdot\delta\xi_q^{//*} - \delta P_q^{\perp *}\cdot\delta\xi_q^{//}\right)\right] \quad (2.3.7)$$

$G_\circ$ is the equilibrium free energy, $\theta_\circ$ is the equilibrium value of the tilt angle (Eq.2.2.17), and q_c is the critical vector of the transition, which was defined before (Eq.2.2.15). The above equation thus expresses the increase of the free-energy of the system, when a plane-wave excitation with a wave-vector q has been created. As we can see, there is no mixing of waves with different wave-vectors q and q' and each of these waves contributes independently to the free-energy. This leads to a conclusion that plane-wave-like excitations of the form e^{iqz} are the normal modes or elementary excitations of the system. It is then natural to consider the eigendynamics of the system for each wave-vector q separately.

Within the Landau-Khalatnikov dynamical equations, the nonequilibrium part of the order parameter $\delta\xi(z,t) = \xi(z,t) - \xi_\circ$ is driven back to the equilibrium by the "thermodynamical restoring force" $\partial G(\xi,t)/\partial\xi$:

$$\frac{d\delta\xi}{dt} = -\Gamma\frac{\partial G(\xi,t)}{\partial\xi} \quad (2.3.8)$$

Here Γ is the kinetic coefficient, which is in liquid crystals just the inverse of the general viscosity coefficient. A similar dynamical equation can be written for the polarization field. There are no inertial terms in the dynamical equation 2.3.8. These terms are of the form $J\,\partial^2\xi/\partial t^2$, (where J is the moment of inertia) and are very small compared to viscous damping in liquid crystals.

After neglecting the inertial effects, the Landau-Khalatnikov equations of motion for the tilt and polarization can be written conveniently in a matrix form using the free-energy expansion (Eq.2.3.7),

$$\frac{d\vec{\Psi}_q}{dt} = -2\underline{\Gamma}\cdot\underline{D}(q)\cdot\vec{\Psi}_q \tag{2.3.9}$$

$\Psi_q = \left(\delta\xi_q^{II}, \delta\xi_q^{\perp}, \delta P_q^{II}, \delta P_q^{\perp}\right)$ is a 4-dimensional vector composed of the fluctuating part of the order parameter and $\underline{\Gamma}$ is a matrix composed of the reciprocal of the corresponding generalized viscosity coefficients

$$\underline{\Gamma} = \begin{bmatrix} \Gamma_1 & 0 & 0 & 0 \\ 0 & \Gamma_1 & 0 & 0 \\ 0 & 0 & \Gamma_2 & 0 \\ 0 & 0 & 0 & \Gamma_2 \end{bmatrix}, \quad \Gamma_1 = 1/\gamma_1 \text{ and } \Gamma_2 = 1/\gamma_2 \tag{2.3.10}$$

Here γ_1 and γ_2 are the viscosity coefficients for the motion of the director and the polarization, respectively (see, for example, Carlsson and Žekš, 1989; Skarp, 1988). For the sake of simplicity, the two viscosity coefficients for the phase and amplitude fluctuations of both the director and polarization excitations have been set equal and we have neglected all cross-coupling terms between the two motions, so that $\underline{\Gamma}$ is diagonal. The dynamical matrix $\underline{D}(q)$, which governs the dynamics of fluctuations is then

$$\underline{D}(q)=\begin{bmatrix} A_q & iE_q & -D_q & -iF_q \\ -iE_q & B_q & iF_q & -D_q \\ -D_q & -iF_q & C_q & 0 \\ iF_q & -D_q & 0 & C_q \end{bmatrix} \tag{2.3.11}$$

with elements

$$A_q = \tfrac{1}{2}a(T)+\tfrac{3}{2}b\theta_\circ^2 - \Lambda q_c + \tfrac{1}{2}K_3\left(q_c^2+q^2\right) \tag{2.3.12a}$$

$$B_q = \tfrac{1}{2}a(T)+\tfrac{1}{2}b\theta_\circ^2 - \Lambda q_c + \tfrac{1}{2}K_3\left(q_c^2+q^2\right) \tag{2.3.12b}$$

$$C_q = 1/(2\varepsilon) \tag{2.3.12c}$$

$$D_q = \tfrac{1}{2}(\mu q_c + C) \tag{2.3.12d}$$

$$E_q = q(K_3 q_c - \Lambda) \tag{2.3.12e}$$

$$F_q = \tfrac{1}{2}\mu q \tag{2.3.12f}$$

Note that $\underline{D}(q)$ for $T > T_c$ and $q = 0.$ equals to the expression 2.2.12 for $q = q_c$. The elementary excitations in liquid crystals are of the form

$$\vec{\Psi}_q(t) = \vec{\Psi}_q^\circ \cdot e^{-t/\tau} \tag{2.3.13}$$

Here τ_q^{-1} is the relaxation rate of the plane-wave excitation with the wave-vector q and the amplitude Ψ_q° . The Landau-Khalatnikov equations (Eq.2.3.9) are thus reduced to a problem of finding the eigenvalues and eigenvectors of the 4x4 matrix

$$\left(2\cdot\underline{\Gamma}\cdot\underline{D}(q)-\lambda\cdot\underline{I}\right)\vec{\Psi}_q^\circ = 0 \tag{2.3.14}$$

Here, the eigenvalue $\lambda = \tau_q^{-1}$ defines the relaxation rate for a given eigenvector Ψ_q°. It is straightforward to show that the dynamical matrix $\underline{D}(q)$ is Hermitian,

$$\underline{D}_q^+ = \underline{D}_q \tag{2.3.15}$$

which means that the eigenvalues λ_i are always real, whereas the corresponding eigenvectors are orthogonal. A nontrivial solution of the above set of equations (Eq.2.3.14) for the eigenvalues can be found if the determinant, associated with this system equals zero:

$$\left|2 \cdot \underline{\Gamma} \cdot \underline{D}(q) - \lambda \cdot \underline{I}\right| = 0 \tag{2.3.16}$$

This leads to a fourth-order polynomial for the unknown eigenvalue λ and we shall therefore obtain four possible eigenfrequencies $\lambda_i = \tau_i^{-1}(q)$. Physically, this means, that for a given wave-vector q we have in general four different eigenmodes with four different relaxation rates. In other words, for a given temperature, there will be in general four dispersion branches of elementary excitations $\tau_i^{-1}(q)$, $i = 1...4$ in the reciprocal space, because there are four degrees of freedom (two for director and two for polarization). These dispersion branches will be temperature dependent, because the dynamical matrix $\underline{D}(q)$ is temperature dependent, too.

Let us now calculate these eigenvalues. As the elements of the dynamical matrix (Eq.2.2.11) depend on the equilibrium value of the molecular tilt, it is obvious that the order parameter dynamics in the *smectic-A* will be quite different from the dynamics in the *smectic-C** phase. We shall therefore perform the calculation in two separate steps.

In the *smectic-A phase* the equilibrium tilt and the polarization equal zero and we obtain four relaxation rates for the plane-wave excitation $\Psi_q(t)$, which are grouped into two doubly degenerate modes

$$1/\tau_\pm(q) = \Gamma_1 A_q + \Gamma_2 C_q \pm \sqrt{(\Gamma_1 A - \Gamma_2 C_q)^2 + 4\Gamma_1\Gamma_2 D_q^2} \tag{2.3.17}$$

with

$$A_q(T) = \tfrac{1}{2}\alpha(T - T_c) + \tfrac{1}{2}K_3 q^2 + \tfrac{1}{2}(\mu q_c + C)^2 \tag{2.3.18a}$$

$$C_q = 1/(2\varepsilon) \tag{2.3.18b}$$

$$D_q = \tfrac{1}{2}(\mu q_c + C) \tag{2.3.18b}$$

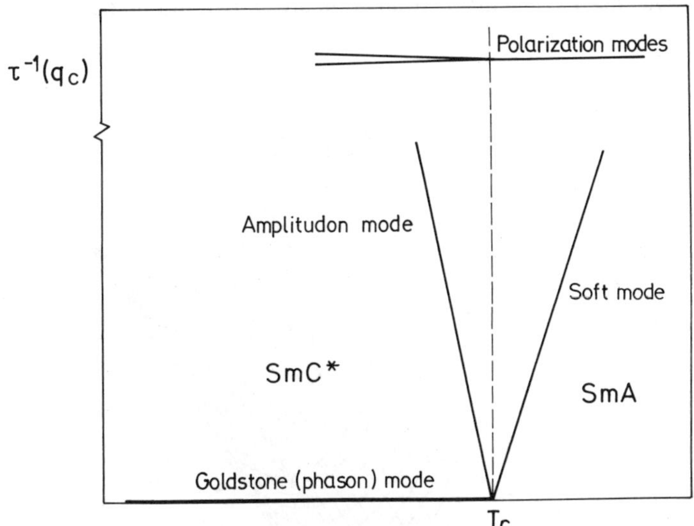

Fig.2.3.5. Critical behavior of the tilt-fluctuation mode in the vicinity of the *smectic-A - smectic-C** transition. Both the non-critical polarization mode and the critical tilt-fluctuation mode split in the *smectic-C** phase due to the broken D_∞ symmetry. The drawing is schematic and not to scale. Note that q is given in a laboratory frame.

The eigenvector of the doubly degenerate solution $1/\tau_-$ corresponds to an *in-phase collective excitation* of the tilt and polarization (Blinc and Žekš, 1978) and is conveniently denoted as a *director fluctuation*. Here, the tilt and polarization fluctuate in-phase, i.e. they both either increase or decrease coherently. On the other hand, the other doubly degenerate solution $1/\tau_+$ corresponds to *an out-of phase fluctuation* of these two order parameters and is usually denoted as the *polarization fluctuation*.

It is $1/\tau_-$ that critically slows-down and condenses at the *smectic-A* - *smectic-C** transition. This is therefore the soft mode of the transition, the relaxation rate of which vanishes at $q = 0$ and $T = T_c$ in the helical frame

$$1/\tau_- = 0 \tag{2.3.19}$$

The temperature dependence of the two modes at the critical wave-vector $\vec{q}_c$ near the *smectic-A* - *smectic-C** transition are shown in Fig.2.3.5.

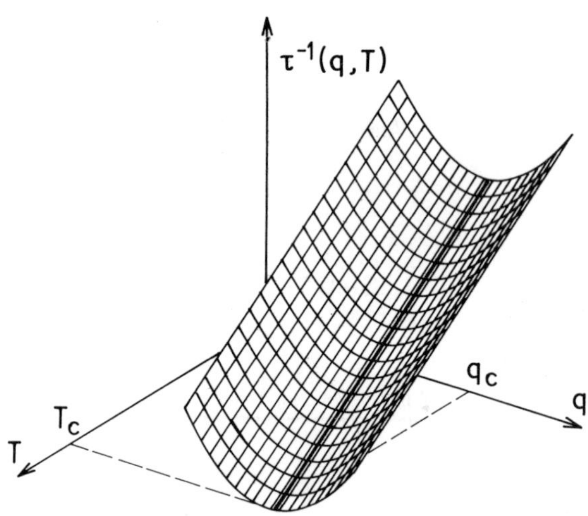

Fig.2.3.6. Condensation of the soft mode in the $(\vec{q}, T)$ space. The shaded surface represents the dispersion relation (Eq.2.3.20) for the order parameter excitations. Note that q is given in the laboratory frame.

In the vicinity of the phase transition we expand Eq.2.3.17 and obtain the dispersion relation for the soft mode:

$$\frac{1}{\tau_-} = \frac{\alpha}{\gamma_1}\left(T - T_c\right) + \frac{K_3}{\gamma_1}\cdot q^2 \tag{2.3.20}$$

As expected (Eq.2.3.1), the dispersion relation for the soft mode is strongly temperature dependent. It is parabolic in q and has a finite gap for all temperatures except at T_c, as shown in Fig.2.3.6.

Whereas the director fluctuations show critical behavior, the polarization fluctuations in the *smectic-A* phase are noncritical. This means that their relaxation rate is nearly independent of temperature and wave-vector:

$$\frac{1}{\tau_+} = \frac{1}{\gamma_2 \varepsilon} + \frac{\varepsilon\left(\mu q_c + C\right)^2}{\gamma_1} \tag{2.3.21}$$

It is interesting to look at the nature of the elementary excitations that show critical behavior for $T > T_c$. By considering the $q = 0$ excitation (Eqs.2.3.5. and 2.3.6.), it is easy to see that the critical excitation is a monochromatic plane wave of the form

$$\delta\bar{\xi}\left(z,t\right) \approx e^{-t/\tau(q=0)}\cdot\left(\cos q_c z\,,\sin q_c z\right) \tag{2.3.22}$$

It represents a helicoidal wave with the wave-vector q_c, as shown in Fig.2.3.7. Once it is created by a thermal excitation of the system, this wave relaxes back to zero exponentially. The relaxation time $\tau_-(T)$ of this excitation is strongly temperature dependent. For temperatures close to T_c the relaxation time increases and finally diverges at T_c. This means that at the phase transition temperature, the critical elementary excitation would never relax back to zero, once it is created. It will therefore freeze at the phase transition and result in a static, time-independent helicoidal structure, which is the chiral *smectic-C** phase.

There is an interesting and a very important issue, which is connected to the wave-vector q of the elementary excitation and the nature of these excitations in chiral systems. As we have seen, an excitation with $q = 0$ in the rotating frame appears as a helical wave in the laboratory system. At first sight this is a

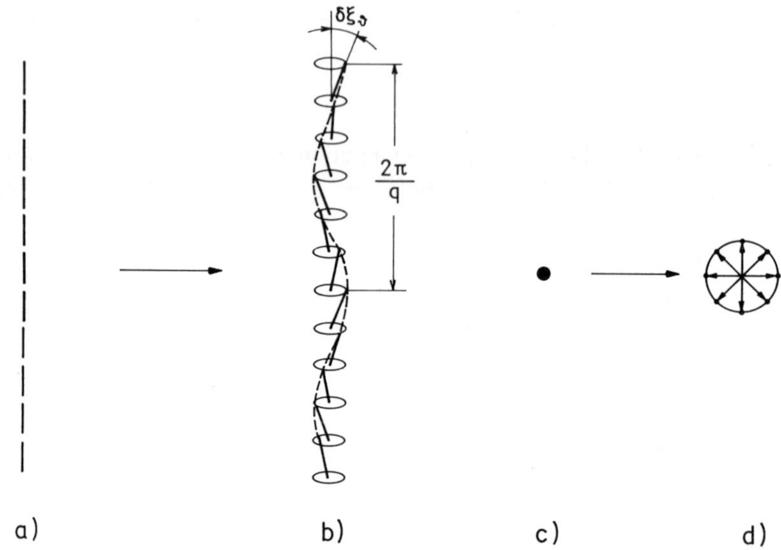

Fig.2.3.7. The elementary excitations in the *smectic-A* phase of a chiral liquid crystal. (a) The equilibrium *smectic-A* phase and (b) an elementary excitation with a wave-vector q, superposed to the equilibrium *smectic-A* phase. (c) and (d) show the situations (a) and (b), as viewed along the smectic normal.

confusing result, as the helical wave has a finite wave-vector $q_c \neq 0$. The reason for this apparent inconsistency lies in the definition of the helical reference frame, which was introduced when discussing the order parameter fluctuations (Eqs.2.3.5 and 2.3.6). In this helical frame an excitation with the wave-vector q appears as an excitation with the wave-vector $Q = (q_c \pm q)$ in the laboratory system. As a result, the condensation of the soft mode is observed at the wave-vector $Q = q_c$ in the laboratory system, as shown in Fig.2.3.7. It should be stressed that the condensation of the soft-mode occurs at a general point of the *q*-space, which is similar to the case of incommensurate crystals. It turns out that much of the analytical framework developed for the physics of incommensurate systems can be used in chiral liquid crystals (see Chapter 3, Symmetry breaking by external fields).

Whereas in the *smectic-A* phase there are two degenerate branches of excitations, this degeneracy is removed below T_c due to the breaking of the axial D_∞ symmetry of the *smectic-A* . phase. It turns out that for $T < T_c$ and for an arbitrary wavevector q of the excitation, the eigensolutions of the dynamical matrix cannot be expressed in a closed form. We can find analytical solutions only for the excitations with $q=0$ in the rotating frame. These are:

$$\frac{1}{\tau_{1,2}}(q=0)=\Gamma_1 A_\circ+\Gamma_2 C_\circ \pm\sqrt{(\Gamma_1 A_\circ+\Gamma_2 C_\circ)^2-4\Gamma_1\Gamma_2\left(A_\circ C_\circ-D_\circ^2\right)} \tag{2.3.23a}$$

$$\frac{1}{\tau_{3,4}}(q=0)=\Gamma_1 B_\circ+\Gamma_2 C_\circ \pm\sqrt{(\Gamma_1 B_\circ+\Gamma_2 C_\circ)^2-4\Gamma_1\Gamma_2\left(B_\circ C_\circ-D_\circ^2\right)} \tag{2.3.23b}$$

where

$$A_\circ=\alpha(T_c-T)+\tfrac{1}{2}\varepsilon(\mu q_c+C)^2 \tag{2.3.24a}$$

$$B_\circ=\tfrac{1}{2}\varepsilon(\mu q_c+C)^2 \tag{2.3.24b}$$

$$C_\circ=1/(2\varepsilon) \tag{2.3.24c}$$

$$D_\circ=\tfrac{1}{2}(\mu q_c+C) \tag{2.3.24c}$$

Similar to the *smectic-A* phase, the eigensolutions can be for $q=0$ further grouped in two pairs of excitations: a pair of *polarization excitations* $1/\tau_1$, $1/\tau_3$, which are nearly temperature and wave-vector independent and a pair of *director excitations* $1/\tau_2$, $1/\tau_4$, which show both a temperature and a wave-vector dependence.

The approximate eigensolutions of the dynamical matrix for $q\neq 0$ can be found by considering the q-dependent terms of the dynamical matrix as a small perturbation. Within the first order expansion in the wave-vector q in the vicinity of T_c we find that the ***director excitations*** $1/\tau_{2,4}$ (Eqs.2.3.23a and b) split into a branch of ***amplitudon*** and a branch of ***phason excitations***. The *amplitudon branch* of director excitations represents plane-wave fluctuations of the magnitude of the tilt angle, and is shown in Fig.2.3.8. The dispersion relation of the amplitudon excitations is both temperature and wave-vector dependent and is a kind of a mirror-image of the soft mode in the *smectic-C** phase:

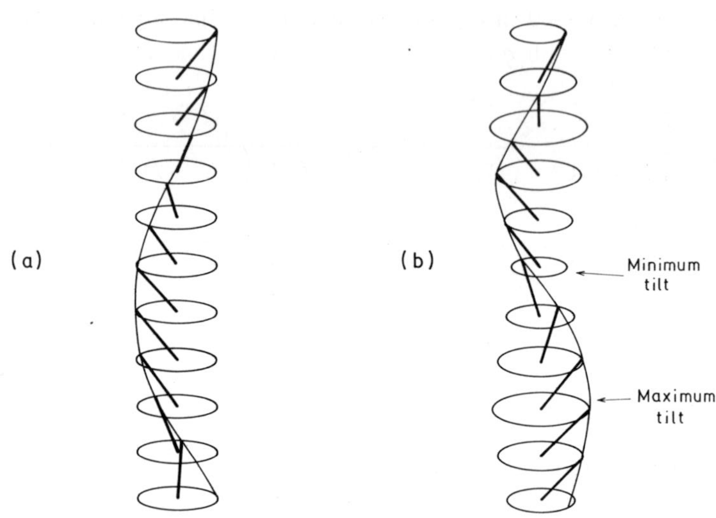

Fig.2.3.8. The amplitude excitation in the *smectic-C** phase of a ferroelectric liquid crystal. The equilibrium structure is shown in (a), whereas in (b) a plane-wave-like excitation of the magnitude of the tilt angle is added. The amplitudon is therefore an "umbrella"-like fluctuation of the helical structure.

$$\frac{1}{\tau_A(q)} = \frac{2\alpha}{\gamma_1} \cdot (T_c - T) + \frac{K_3}{\gamma_1} q^2 \qquad (2.3.25)$$

The dispersion of the *phason* excitations is

$$\frac{1}{\tau_{PH}(q)} = \frac{K_3}{\gamma_1} q^2 \qquad (2.3.26)$$

As one can see from the above expressions, the amplitudon excitations have finite relaxation rates for all temperatures except at T_c, where they merge into the soft mode branch of excitations. On the contrary, the phason branch of excitations is gapless for all temperatures and merges into the soft and amplitude branch at T_c, as shown in Fig.2.3.5.

The phase mode with $q = 0$ in a rotating frame is a zero frequency, symmetry restoring Goldsone mode, shown in Fig.2.3.9.

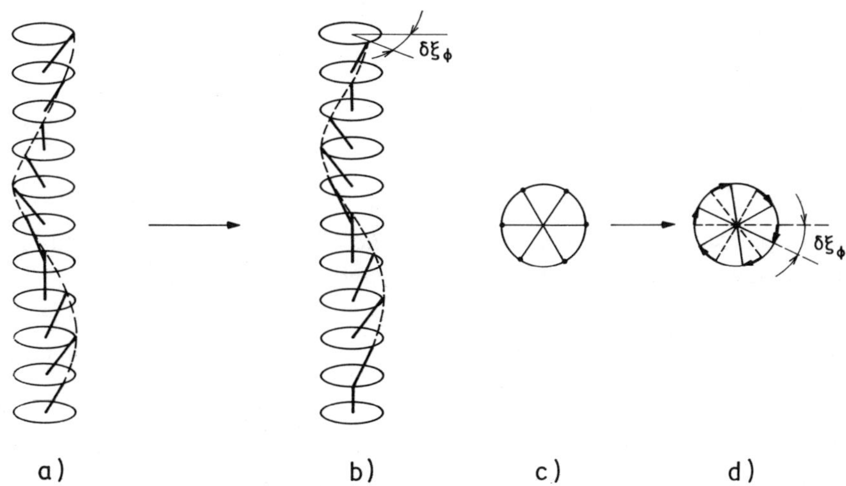

Fig.2.3.9. The Goldstone excitation in the *smectic-C** phase. The equilibrium *smectic-C** structure is shown in (a). The effect of the Goldstone excitation on the equilibrium structure is shown in (b). (c) and (d) show the same excitation, as observed along the helical axis.

It appears in the *smectic-C** phase as a result of the spontaneous breaking of the continuous D_∞ symmetry of the *smectic-A* phase. It should be stressed that this mode is observable at the wave-vector $q = q_c$ in the laboratory system. The nature of this mode can be most clearly understood by considering the effect of this excitation on the properties of the structure. By taking the $q = 0$ component in the expansion (Eq.2.3.6) the Goldstone excitation corresponds to a rotation through a small phase angle $\delta\Phi$

$$\delta\xi_\perp = \delta\theta^\perp_{q=0} = \theta_\circ \cdot \delta\Phi \tag{2.3.27}$$

which is constant throughout the sample. It is then clear that the Goldstone excitation represents an infinitesimal rotation by an angle $\delta\Phi$ of the sample as a whole, as shown in Fig.2.3.9. As this costs zero energy, there is no restoring force for this kind of fluctuation and the relaxation time of this excitation is infinite.

The two ***polarization branches*** are ***nearly degenerate*** in the *smectic-C** phase. The first order expansion in $(T_c - T)$ gives *T*-independent and degenerate eigenfrequencies:

$$\frac{1}{\tau_{P\pm}} \approx \frac{1}{\gamma_2 \varepsilon} + \frac{\varepsilon(\mu q_c + C)^2}{\gamma_1} \tag{2.3.28}$$

This degeneracy is removed in the *smectic-C** phase by the second-order expansion, which gives us a splitting $\Delta(T)$ between the two modes

$$\Delta = \frac{1}{\tau_{P+}} - \frac{1}{\tau_{P-}} \approx \frac{2\alpha(T_c - T)}{\gamma_1} \tag{2.3.29}$$

Note that this splitting is of the same order as the splitting between the phason and the amplitudon branches and is independent of the wave-vector.

To summarize, the collective excitations of the coupled director-polarization field in the vicinity of the *smectic-A* - *smectic-C** phase transition are characterized by the following:

(i) The excitations are ***plane, monochromatic waves*** both in the *smectic-A* and *smectic-C** phases

(ii) In the *smectic-A* phase there are ***two branches of doubly degenerate*** excitations: the ***polarization*** and the ***director branch*** of excitations. The ***polarization branch*** here represents the out-of phase collective fluctuations of the polarization and the tilt. This branch is temperature and wave-vector nearly independent and the relaxation rates are in the 100MHz region. The ***director branch*** represents the in-phase fluctuations of the tilt and polarization. It shows ***critical slowing down*** at T_c and a parabolic dispersion, centered at q_c in the laboratory system:

$$\frac{1}{\tau_S}(q,T)=\frac{\alpha}{\gamma_1}(T-T_c)+\frac{K_3}{\gamma_1}(q-q_c)^2 \qquad (2.3.30)$$

The relaxation rates of the modes, belonging to this branch are in the 10Hz-1MHz range, and strongly depend on the wave-vector and temperature. The most prominent mode, that belongs to the director branch of excitations is the ***soft-mode***. This is the mode the frequency of which equals zero at the phase transition point.

(iii) For $T<T_c$ the degeneracy of the polarization and director branches is removed and four branches of excitations emerge in the *smectic-*C^* phase. The soft mode branch splits into the ***amplitudon***

$$\frac{1}{\tau_A}(q,T)=\frac{2\alpha}{\gamma_1}(T_c-T)+\frac{K_3}{\gamma_1}(q-q_c)^2 \qquad (2.3.31)$$

and ***phason excitations***

$$\frac{1}{\tau_{PH}}(q)=\frac{K_3}{\gamma_1}(q-q_c)^2 \qquad (2.3.32)$$

Similarly, the polarization branch splits into two separate branches. The splitting is in both cases of the same order of magnitude. However, due to the large relaxation rates of the polarization branches (100MHz), the relative splitting is here very small (estimate from the Eq.2.3.28 gives $\Delta f/f_{\circ}\approx 10^{-2}$) and is experimentally difficult to observe. On the other hand, the director excitations have low relaxation rates in the kHz to MHz region and the splitting between the amplitudons and phasons should be more easily observed. However, the amplitudes of the amplitude fluctuations are much smaller than the amplitudes of phason excitation. This imposes severe experimental difficulties in observing both director excitation branches in the *smectic-*C^* phase.

There has been considerable discussion about the number of the excitation branches in the *smectic-C** phase, in view of the fact that the tilt is the primary order parameter and the polarization is the secondary order parameter (Brand and Pleiner, 1991; Žekš and Blinc, 1992). It should be stressed that the number and the nature of elementary excitations in a system is governed by the symmetry and the number of the components of the order parameter.

The first evidence for a critical behavior near the *smectic-A* - *smectic-C** phase transition was given in an elegant experiment, performed by Garoff and Meyer in 1977 (see Garoff, 1977a; Garoff and Meyer, 1977b). They measured the linear electrooptic response of the *smectic-A* phase of a chiral liquid crystal, i.e. a change in the birefringence due to a probing, time-depended external electrical field. The electric field couples to the $q = 0$ eigenmodes of the system and induces a finite, collective tilt of the molecules in the *smectic-A* phase. This is called the "electroclinic" effect. By approaching the phase transition from above, Garoff and Meyer observed a clear increase of the amplitude of the linear response, as presented in Fig.2.3.10.

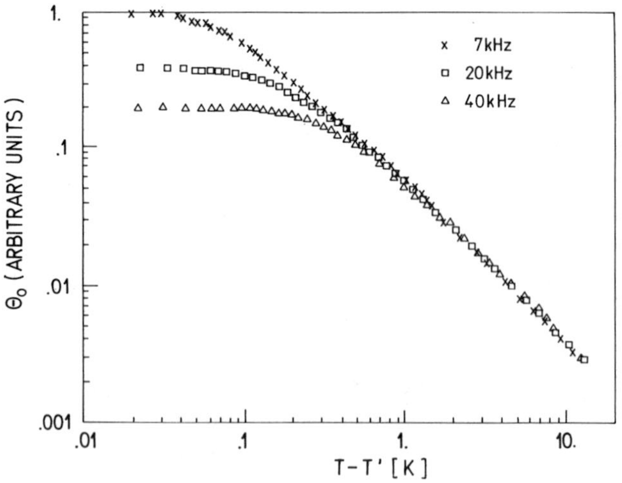

Fig.2.3.10. The nearly "critical" increase of the electroclinic response in the *smectic-A* phase of DOBAMBC (after Garoff and Meyer, 1979). T' equals to T_c in our notation.

The increase of the magnitude of the linear electrooptic response is larger for lower frequencies, which is consistent with expectations. For very low frequencies, the crystal can easily follow the oscillations of the external field. However, by increasing frequency of a driving field, the crystal cannot follow the fast changes of the field any more, which results in a decrease of the linear response. The theory of this linear electrooptic response is presented in Chapter 5, Optical properties.

From the measurements of Garoff and Meyer one can clearly observe, that the "divergence" (i.e. strong increase) of the susceptibility at $q = 0$ in the vicinity of the phase transition is incomplete. Instead of a divergent, Curie-Weiss behavior of the linear susceptibility, we have here a suppressed "divergence" in a temperature interval of approximately 100mK. This is in agreement with the fact that the experiment probes the susceptibility of the system at $q = 0$, which is a non-critical wavevector for a chiral system. The divergence and the Curie-Weiss law should therefore be observable at $q = q_c$ in a reciprocal space, which is accessible with a quasielastic light scattering experiment.

The first observation of the critical dynamics for the critical wavevector in the vicinity of the *smectic-A* - *smectic-C** phase transition was reported by Muševič et al. (1988) using the quasielastic light scattering technique. The light-scattering results of Muševič et al. are presented in Fig.2.3.11. We can see a critical behavior of the soft mode in the *smectic-A* phase and a splitting of the soft mode into an amplitude and phase mode in the *smectic-C** phase. Furthermore, this experiment demonstrated for the first time the existence of the phase excitations in the *smectic-C** phase. The dispersion relation of the phase modes is shown in Fig.2.3.12. It clearly indicates the existence of a nearly gapless phase (Goldstone) mode. Later these experiments were extended to chemically more stable compounds (Drevenšek et al., 1990) that clearly revealed the condensation of the soft mode at finite wave-vector in the *smectic-A* phase, as shown in Fig.2.3.13.

There is a number of other, predominantly dielectric experiments on the order parameter dynamics near the *smectic-A* - *smectic-C** phase transition, which qualitatively confirm the predictions of the Landau theory (see, for instance Vallerien et al., 1989; Levstik et al., 1987,1988 and 1990; Kremer et al., 1990). Whereas the dynamics of the director branch of excitations is consistent with the predictions of the simple Landau theory, the dynamics of the high-frequency polarization modes is still a matter of some experimental controversy. These high frequency excitations have been first observed in the 100MHz region

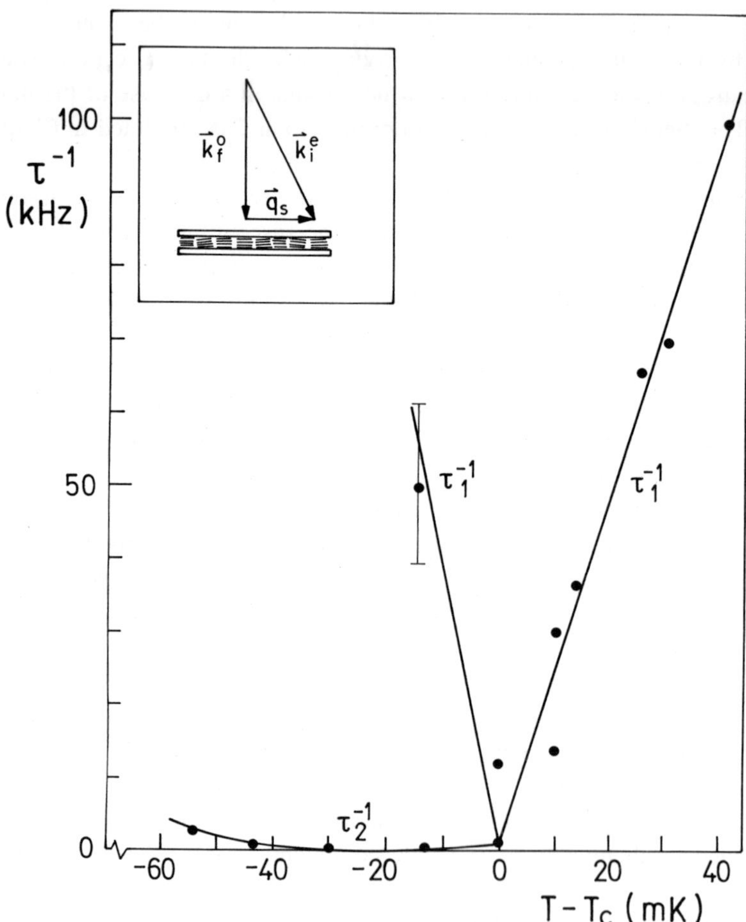

Fig.2.3.11. The first observation of the critical slowing down of the soft mode (τ_1^{-1}) in the *smectic-A* phase and its splitting into the amplitude (τ_1^{-1}) and phason (τ_2^{-1}) mode in the *smectic-C** phase of DOBAMBC (after Muševič et al., 1988).

of the dielectric response measurements by Benguigui (1982). Latter on, extensive dielectric measurements in the GHz region have clearly shown (Kremer et al., 1990, Gouda, 1992a) the noncritical behavior of these modes. An exception is here the four-wave mixing experiment of Lallane et al. (1989) who reported critical behavior of the polarization modes. The experiment is at variance with the results of the optical Kerr-effect experiments, performed by O'Brien et al. (1993), and Miyachi et al (1997), who observed a non-critical behavior of the polarization modes near the *smectic-A* - *smectic-C** transition.

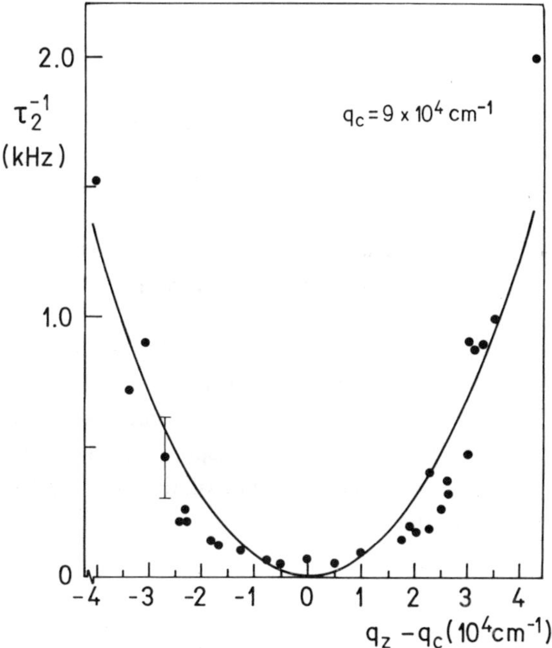

Fig.2.3.12. The first observation of a nearly gapless phason dispersion in the *smectic-C** phase of DOBAMBC (after Muševič et al., 1988). The dispersion was measured at $T_c - T = 0.1K$, the scattering wave-vector is along the helical axis and the zero of the x-axis is at a Bragg angle. The small residual gap of approximately 50Hz is due to a mismatch between the directions of the critical and scattering wave-vector.

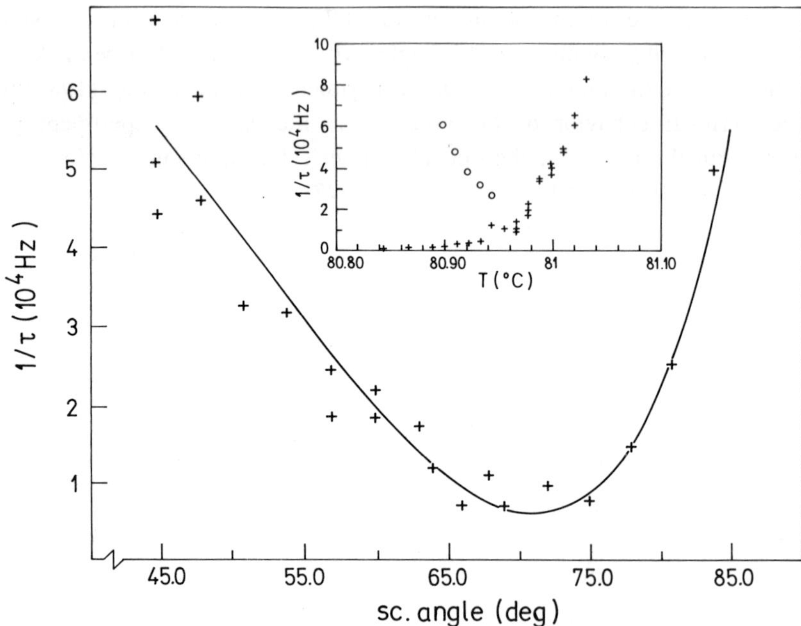

Fig.2.3.13. The dispersion of the soft mode branch of excitations in the *smectic-A* phase of CE-8 (Drevenšek et al., 1990). The measurement was performed very close to the phase transition. The minimum of the dispersion is at a finite angle of the detector, which corresponds to the critical wave-vector q_c. The inset shows the observed splitting of the soft mode into phase and amplitudon modes, respectively.

2.4. Adiabatic Approximation for the Coupled Director-Polarization Modes

For most practical reasons, the polarization terms in the free-energy density expansion 2.2.5.. can be eliminated by partially minimizing the free energy density with respect to P_x and P_y. This is the so-called adiabatic approximation.

Let us start with the expansion 2.2.5. of the nonequilibrium free energy density in the vicinity of the *smectic-A - smectic-C** phase transition

$$g = g_A + \tfrac{1}{2}a\left(\xi_x^2 + \xi_y^2\right) + \tfrac{1}{4}b\left(\xi_x^2 + \xi_y^2\right)^2 -$$

$$-\Lambda\left(\xi_x \frac{d\xi_y}{dz} - \xi_y \frac{d\xi_x}{dz}\right) + \tfrac{1}{2}K_3\left[\left(\frac{d\xi_x}{dz}\right)^2 + \left(\frac{d\xi_y}{dz}\right)^2\right] +$$

$$+ \tfrac{1}{2\varepsilon}\left(P_x^2 + P_y^2\right) - \mu\left(P_x \frac{d\xi_x}{dz} + P_y \frac{d\xi_y}{dz}\right) + C\left(P_x\xi_y - P_y\xi_x\right) \qquad (2.4.1)$$

We have already mentioned that the collective tilt of the molecules is the primary order parameter of the *smectic-A - smectic-C** phase transition and the polarization is only a secondary order parameter that is linearly coupled to the molecular tilt. Ferroelectric liquid crystals are therefore "improper ferroelectrics" as the onset of spontaneous polarization is driven by intermolecular rather than dipolar forces.

The fact that the spontaneous polarization is a secondary order parameter is responsible for the noncritical behavior of polarization modes, as we have seen in the previous chapter. The relaxation rates of the polarization modes are much higher than the relaxation rates of the director modes. This gives us the possibility of using the so-called adiabatic approximation, which is often used in physics. Within the adiabatic approximation we assume that the spontaneous polarization always follows the slow collective motion of the director modes. This means that the polarization is always in equilibrium with the director field, at least on the time scales of the director motion. Mathematically, this is described by the minimization of the free-energy density (Eq.2.4.1) with respect to both components of polarization

$$\left.\frac{\partial g}{\partial P_x}\right|_{P_0} = 0 \qquad (2.4.2a)$$

$$\left.\frac{\partial g}{\partial P_y}\right|_{P_\circ} = 0 \tag{2.4.2b}$$

This leads to a set of equations, describing the relation between the polarization and the tilt in equilibrium

$$P_x = \varepsilon\mu \frac{d\xi_x}{dz} - \varepsilon C \xi_y \tag{2.4.3a}$$

$$P_y = \varepsilon\mu \frac{d\xi_y}{dz} + \varepsilon C \xi_x \tag{2.4.3b}$$

After inserting the equilibrium values of the spontaneous polarization (2.4.3.a-b) into Eq.2.4.1, the free-energy density becomes a function of the tilt vector $\xi = (\xi_x, \xi_y)$

$$\begin{aligned} g = g_A + \tfrac{1}{2}\tilde{a}\left(\xi_x^2 + \xi_y^2\right) + \tfrac{1}{4} b\left(\xi_x^2 + \xi_y^2\right)^2 - \\ - \tilde{\Lambda}\left(\xi_x \frac{d\xi_y}{dz} - \xi_y \frac{d\xi_x}{dz}\right) + \tfrac{1}{2}\tilde{K}_3\left(\left(\frac{d\xi_x}{dz}\right)^2 + \left(\frac{d\xi_y}{dz}\right)^2\right) \end{aligned} \tag{2.4.4}$$

Here the expansion coefficients have been renormalized:

$$\tilde{a}(T) = a(T) - \varepsilon C^2, \; T_\circ \equiv T_c \tag{2.4.5a}$$

$$\tilde{\Lambda} = \Lambda + \varepsilon\mu C \tag{2.4.5b}$$

$$\tilde{K}_3 = K_3 - \varepsilon\mu^2 \tag{2.4.5c}$$

In the following, we shall use the same notation for the expansion coefficients, as before renormalization. Let us now transform the free-energy density into a helical reference frame, or what is the same, let us express the tilt vector in terms of the component $\xi_{//}$ that is parallel to the direction of the tilt and a perpendicular component $\xi_{\perp}$, as shown in Fig.2.3.3. of the previous Section 2.3.

$$\xi_x = \xi_{\|} \cos\Phi - \xi_{\perp} \sin\Phi \tag{2.4.6a}$$

$$\xi_y = \xi_{\|} \sin\Phi + \xi_{\perp} \cos\Phi \tag{2.4.6b}$$

Here, $\Phi(z) = q_c z$ and q_c is the wave-vector of the helix. This means that the helical reference frame "follows" the twisting of the chiral *smectic-C** phase, as we move along the helical axis. One can imagine, that in such a reference frame, the helical *smectic-C** appears homogeneous. This means that the equations of motion are simpler in such a helical reference frame. Indeed, after inserting Eq.2.3.6a-b into Eq.2.3.4., we obtain the following free-energy density in the helical reference frame

$$g = g_A + \left(\tfrac{1}{2}a(T) - \tfrac{1}{2}K_3 q_c^2\right)\left(\xi_{//}^2 + \xi_{\perp}^2\right) + \tfrac{1}{4}b\left(\xi_{//}^2 + \xi_{\perp}^2\right)^2 + \tfrac{1}{2}K_3\left[\left(\frac{d\xi_{//}}{dz}\right)^2 + \left(\frac{d\xi_{\perp}}{dz}\right)^2\right] \tag{2.4.7}$$

Let us now discuss the dynamics of chiral *smectic-C** phase in the adiabatic approximation. We introduce the two components of the fluctuating part of the molecular tilt:

$$\xi_{//} = \xi_{\circ} + \delta\xi_{//} \tag{2.4.8a}$$

$$\xi_{\perp} = \delta\xi_{\perp} \tag{2.4.8b}$$

and the free energy density Eq.2.4.7. is

$$T > T_c^*:\ g = g_A + \tfrac{1}{2}a(T)\left(\delta\xi_{//}^2 + \delta\xi_{\perp}^2\right) + \tfrac{1}{2}K_3\left[\left(\frac{d\xi_{//}}{dz}\right)^2 + \left(\frac{d\xi_{\perp}}{dz}\right)^2\right] \tag{2.4.9a}$$

$$T < T_c^*: \; g = g_A - a(T)\delta\xi_{//}^2 + \tfrac{1}{2}K_3\left[\left(\frac{d\xi_{//}}{dz}\right)^2 + \left(\frac{d\xi_{\perp}}{dz}\right)^2\right] \tag{2.4.9b}$$

We se that the symmetry between the two fluctuating components is removed below T_c, where we have amplitude and phase fluctuations instead of doubly degenerate soft mode fluctuations above T_c. The dispersion relation for the excitations is calculated by introducing the plane-wave expansion for the two fluctuating components of the tilt:

$$\delta\xi_{//} = \sum_q \delta\xi_q^{//} \cdot e^{iqz} \tag{2.4.10a}$$

$$\delta\xi_{\perp} = \sum_q \delta\xi_q^{\perp} \cdot e^{iqz} \tag{2.4.10b}$$

After integrating, the total free-energy $G = \frac{1}{L}\int_0^L g(z)dz$ is

$$T > T_c^* \quad G = G_A + \tfrac{1}{2}\sum_q \left(a(T) + K_3 q^2\right)\left(\delta\xi_q^{//}\delta\xi_q^{//*} + \delta\xi_q^{\perp}\delta\xi_q^{\perp*}\right) \tag{2.4.11a}$$

$$T < T_c^* \quad G = G_A + \tfrac{1}{2}\sum_q \left[\left(-2a(T) + K_3 q^2\right)\delta\xi_q^{//}\delta\xi_q^{//*} + K_3 q^2 \xi_q^{\perp}\delta\xi_q^{\perp*}\right] \tag{2.4.11b}$$

The total free energy due to fluctuations can be written using the dynamical matrix $\underline{D}(q)$ as

$$G = G_A + \sum_q \Psi_q^+ \underline{D}(q)\Psi_q \tag{2.4.12}$$

Here, $\Psi_q = \begin{bmatrix} \delta\xi_q^{//} \\ \delta\xi_q^{\perp} \end{bmatrix}$ and $\Psi_q^+ = \left[\delta\xi_q^{//} \quad \delta\xi_q^{\perp}\right]$. The dynamical matrix is

$$T > T_c^*: \ \underline{D}(q) = \begin{bmatrix} A_q & 0 \\ 0 & A_q \end{bmatrix} \qquad A_q = \tfrac{1}{2}\left(a(T) + K_3 q^2\right) \tag{2.4.13a}$$

$$T < T_c^*: \ \underline{D}(q) = \begin{bmatrix} B_q & 0 \\ 0 & C_q \end{bmatrix} \qquad B_q = -a(T) + \tfrac{1}{2} K_3 q^2 \tag{2.4.13b}$$

$$C_q = \tfrac{1}{2} K_3 q^2$$

In the *smectic* − *A* phase we have two degenerate branches of plane-wave excitations. This corresponds to soft mode excitations that have parabolic dispersion relation. In the ferroelectric *smectic-*C^* phase this degeneracy is removed, which can be seen from the difference of the diagonal elements of the dynamical matrix (Eq.2.4.13b). The soft mode therefore splits into an amplitude and phase mode. Both show parabolic dispersion relation. The difference is that the amplitude excitations have a frequency gap that increases with decreasing temperature, whereas the phase excitations are gapless. The dynamics can be calculated in a straightforward way following the formalism of Landau and Khalatnikov and the dynamical matrix given by the Eq.2.4.13a-b.

2.5. Beyond the Classical Landau Theory

There is a systematic and a definite disagreement between the predictions of the Landau-theory, presented in the previous chapters and the experiments on simple (i.e. single component) ferroelectric liquid crystals. As an example, we shall here analyze the discrepancies reported for the typical ferroelectric liquid crystal (n-decyloxybenzylidene)p-amino(2-methylbutyl)cinnamate (DOBAMBC). The inconsistencies are listed in Table 2.5.1. and are as follows:

(i) The predicted critical exponent for the tilt angle is $\beta = 0.5$, whereas measurements, performed by different groups with different experimental methods report on a crossover of this critical exponent from the value of $\beta \approx 0.25$ in a region far from T_c to a value of $\beta \geq 0.5$ in a region close

to T_c. The crossover temperature is around $T_c - T = 1K$ (Dumrongrattana et al., 1986; Seppen et al., 1988).

(ii) The ratio of the polarization versus tilt, $P_\circ/\theta_\circ$, is expected to be temperature independent within the simple Indenbom-Pikin model. This ratio, however, has been found to be strongly temperature dependent in a temperature interval of 1K below T_c (Dumrongrattana and Huang, 1986a; Dumrongrattana et al., 1986b).

(iii) The period of the helix is predicted to be temperature independent in the $smectic - C^*$ phase. Light scattering experiments and polarizing microscope measurements in DOBAMBC, CE-8 and other similar substances revealed that in fact the period of the helix shows a strong temperature dependence in a temperature interval 1K below T_c (Muševič et al., 1984; Ostrovski et al., 1978).

(iv) The dielectric susceptibility is expected to be constant below T_c, except for a small, cusp-like anomaly from the $q_c \neq 0$ soft mode at T_c. The experiment, on the other hand, reveals a strong temperature dependence of the dielectric response in the $smectic - C^*$ phase (Levstik et al., 1987).

(v) The *(H,T)* phase diagram of the $smectic - C^*$ ferroelectric liquid crystal in a strong, transverse magnetic field, applied perpendicular to the helix, is predicted to exhibit a temperature independent line of the critical magnetic field H_c for the unwinding of the helix, i.e. the $smectic - C^*$ - $smectic - \overline{C}$ transition. The experiment (Muševič et al., 1982) on the other hand reports a strong temperature dependence of this critical field, accompanied by the appearance of a small region of a reentrant, modulated $smectic - C^*$. The width of this reentrant region is approximately $0.1K$ from the λ-line. The Lifshitz point, where the $smectic - A$, modulated $smectic - C^*$ and unwound $smectic - \overline{C}$ meet, seems to run-away to very large magnetic fields.

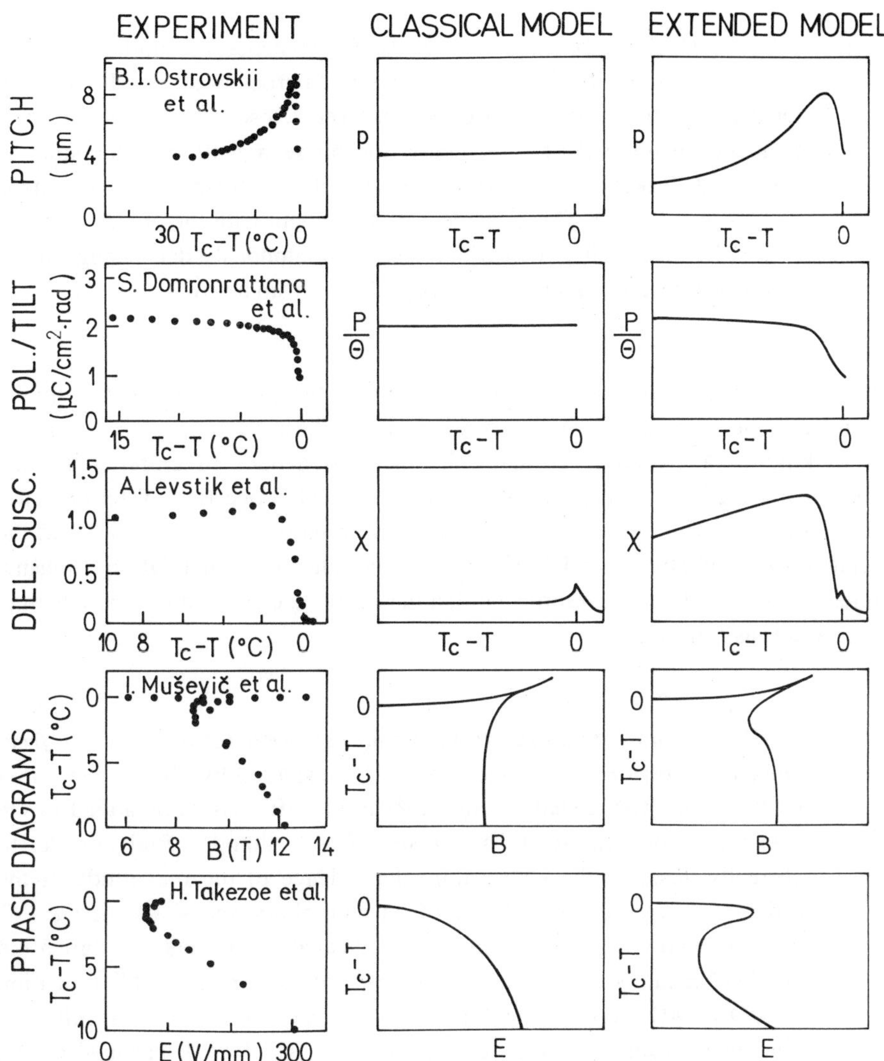

Table 2.5.1. Experimentally determined temperature dependences of the helical period (Ostrovski et al., 1978), the polarization versus tilt ratio (Dumrongrattana and Huang, 1986), the dielectric susceptibility (Levstik et al., 1987) and the phase diagrams in external magnetic (Muševič et al., 1982) and electric fields (Takezoe et al., 1984). The predictions of a simple Landau model and the so-called extended model are shown in parallel columns (after Žekš et al., 1988).

(vi) The *(E,T)* phase diagram reveals the existence of similar reentrant phenomena when an electric field is applied perpendicular to the helix. A reentrant $smectic-C^*$ phase is observed close to T_c, again in a temperature interval of 0.1K below the phase transition temperature in zero field (Takezoe et al., 1984; Levstik et al., 1991a; Dumrongrattana and Huang, 1986c). This is in clear disagreement with the predictions of a simple Landau theory, which predicts a monotonously decreasing line of the critical electric field, as we approach T_c from below.

It is very symptomatic that the deviations from the predictions of the simple Landau theory appear in a 1K interval below the phase transition. This seems to be a kind of a crossover point in the thermodynamics of chiral and tilted smectics, which has consequences for the physical properties of ferroelectric liquid crystals. The extensive theoretical work initiated by Žekš (1984) and the related work of Huang and Viner (1982) has clarified much of the unusual phenomena, listed above. It turns out that these can be classified into two distinct categories, as follows:

(i) ***Phenomena originating from the higher-order tilt terms in the free-energy expansion***. This effect was first considered by Huang and Viner (1982). They found that a sixth order term $c \cdot \theta^6$ has to be added to the free-energy of a non-chiral tilted $smectic-C$ liquid crystal in order to correctly describe the temperature dependence of the magnitude of the order parameter in the vicinity of an achiral $smectic-A$ - $smectic-C$ phase transition. This has been used in the analysis of the tilt, polarization and heat capacity data of the chiral $smectic-A$ - $smectic-C^*$ transition in DOBAMBC (Huang and Dumrongrattana, 1986). As a result, one obtains a certain crossover temperature T_{cr}, which is determined by the magnitude of sixth order terms and has the following role: Far below this crossover temperature, the tilt exponent β tends to its tricritical value $\beta = 0.25$, whereas close to T_c the exponent tends to mean-field value $\beta = 0.5$, predicted by the Landau theory. This is shown as an example in Fig.2.5.1. and explains the experimental observations *(i)*

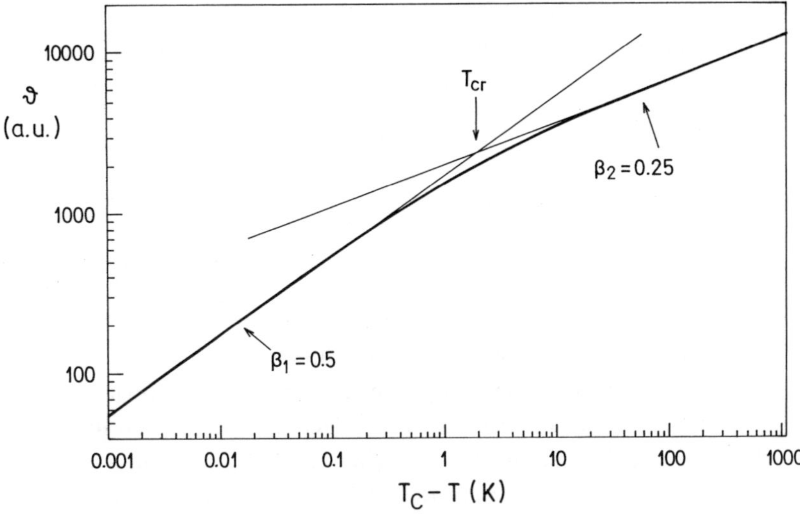

Fig.2.5.1. Characteristic temperature dependence of the magnitude of the tilt angle for a free-energy density where the sixth-order terms in the tilt angle have been included. One can clearly see that the crossover temperature T_{cr} is around $T_c - T \approx 1K$. Close to T_c the exponent for the tilt angle approaches the classical Landau value $\beta = 0.5$, whereas far from the transition, the exponent tends to the tricritical value of $\beta = 0.25$ (Huang and Viner, 1982).

(ii) ***Phenomena related to the polar-chiral and the quadrupolar-achiral nature*** of the $smectic - A$ - $smectic - C^*$ transition. The polar chiral nature of this phase transition is most vividly demonstrated in the temperature dependence of the period of the helix in a typical ferroelectric $smectic - C^*$ crystal, as shown in *Note 2.2.2*. As the helical structure in a chiral $smectic - C^*$ phase is the result of a balance between the chiral (Lifshitz) and elastic forces, a consistent theory should explain at least qualitatively the temperature dependence of the period of a helix. Since the chirality of the transition is intimately related to the spontaneous polar ordering below T_c, it is natural to look in details at the nature of transverse polar ordering in the chiral $smectic - C^*$ phase. This was first done by Žekš in 1984, who realized that in ferroelectric liquid crystals there is a competition between the ***quadrupolar*** (i.e. non-polar)

and ***polar ordering*** of transverse molecular dipole moments. The mechanism can be most easily understood by considering a general form of the potential $V(\Psi)$, which governs the rotation of a single liquid crystalline molecule around its long axis in the $smectic-C^*$ phase:

$$V(\Psi) = -a_1\theta \cos\Psi - a_2\theta^2 \cos 2\Psi \qquad (2.5.1)$$

Here, Ψ is the angle that measures the deviation of the orientation of the molecular dipole moment from the direction of the C_2 axis in the $smectic-C^*$ phase. The first part of the potential is shown in Fig.2.5.2a and favors a polar ordering of the molecular dipoles. The magnitude of this potential is linear in θ, i.e. it changes its sign by the transformation $\theta \to -\theta$. The second term in Eq.2.5.1. is shown in Fig.2.5.2b and favors quadrupolar ordering of the dipoles. Here, $+\theta$ and $-\theta$ directions of the tilt are energetically equivalent. In lowest order, the magnitude of this potential is therefore proportional to the square of the tilt angle. The resulting potential $V(\Psi)$ will be a mixture of the polar-favoring and a quadrupolar-favoring potentials, as shown in Fig.2.3.2.c. This will result in a competition between the polar and quadrupolar ordering of the local dipole moments. Because both of these potentials are temperature dependent, the result of their competition will be temperature dependent, too. Close to T_c the tilt angle θ is small and the polar part of the potential, which is linear in θ, will be much larger than the quadrupolar part, which is quadratic in θ. In this temperature region we shall therefore have predominantly a polar ordering of the transverse dipoles. On the other hand, by reducing the temperature, the quadrupolar part of the potential will increase strongly and finally induce a predominantly quadrupolar ordering of the dipoles. There will be therefore a crossover from the polar-like to the quadrupolar-like ordering of the transverse dipole moments that was indeed observed in NMR and NQR experiments in chiral smectic phases, which are presented in Chapter 7 (Luzar et al., 1984).

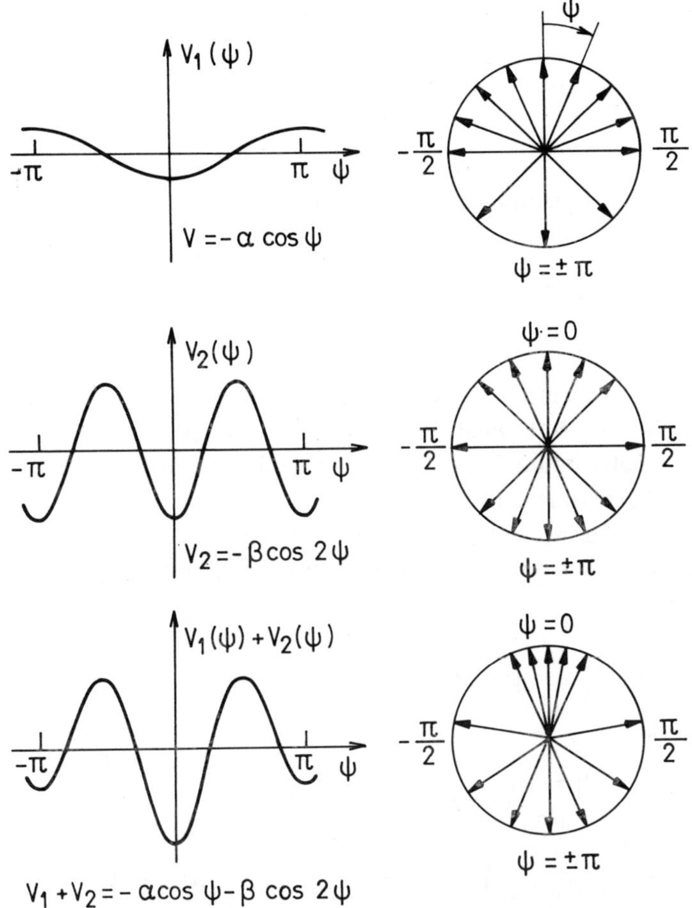

Fig.2.5.2. The potential for the rotation of a single molecule around its long axis and the corresponding orientational probability distributions of the transverse molecular dipole. (a) The polar part of the potential favors "biased" rotation, resulting in a polar distribution. (b) The quadrupolar part results in a non-polar distribution of the local polarization. (c) The total potential is a mixture of both (a) and (b) (after Žekš et al., 1988).

This physical mechanism of the competition between polar and quadrupolar ordering of permanent electric dipoles is mathematically described by including in the Landau free-energy density in addition to the bilinear (polar) term of chiral origin

$$g_{polar} = C\left(P_x\xi_y - P_y\xi_x\right) \tag{2.5.2}$$

also a biquadratic (quadrupolar, non-chiral) coupling term

$$g_{quad} = \tfrac{1}{2}\Omega\left(P_x\xi_y - P_y\xi_x\right)^2 \tag{2.5.3}$$

One of the consequences of the presence of the biquadratic term in the free-energy expansion is a qualitatively different temperature behavior of the spontaneous polarization and the period of the helix, which is very close to the one observed experimentally.

Let us now summarize: The free-energy density of the so-called generalized Landau model for the ferroelectric liquid crystals is

$$\begin{aligned} g(z) = {} & g_A + \tfrac{1}{2}a\left(\xi_x^2+\xi_y^2\right) + \tfrac{1}{4}b\left(\xi_x^2+\xi_y^2\right)^2 + \tfrac{1}{6}c\left(\xi_x^2+\xi_y^2\right)^3 - \\ & -\Lambda\left(\xi_x\frac{d\xi_y}{dz} - \xi_y\frac{d\xi_x}{dz}\right) + \tfrac{1}{2}K_3\left[\left(\frac{d\xi_x}{dz}\right)^2 + \left(\frac{d\xi_y}{dz}\right)^2\right] - \\ & -d\left(\xi_x^2+\xi_y^2\right)\left(\xi_x\frac{d\xi_y}{dz} - \xi_y\frac{d\xi_x}{dz}\right) + \frac{1}{2\varepsilon}\left(P_x^2+P_y^2\right) + \\ & +\tfrac{1}{4}\eta\left(P_x^2+P_y^2\right)^2 - \mu\left(P_x\frac{d\xi_x}{dz} + P_y\frac{d\xi_y}{dz}\right) + \\ & +C\left(P_x\xi_y - P_y\xi_x\right) - \tfrac{1}{2}\Omega\left(P_x\xi_y - P_y\xi_x\right)^2 \end{aligned} \tag{2.5.4}$$

Similarly as before, only the quadratic coefficient $a = \alpha\left(T - T_{\circ}\right)$ is temperature dependent and $T_{\circ}$ is the $smectic - A$ - $smectic - C$ transition temperature for the racemic (achiral) substance. The sixth order term $\frac{1}{6}c\left(\xi_x^2+\xi_y^2\right)^3$ has been added to correctly describe the temperature dependences of the tilt and heat capacity

(Huang and Viner, 1982). The d-term describes the coupling between the Lifshitz term and the magnitude of the order parameter. It therefore renormalizes the magnitude of the Lifshitz coefficient and induces a monotonous temperature dependence of the helical period far from T_c. The fourth order term $\frac{1}{4}\eta\left(P_x^2+P_y^2\right)^2$ has to be added for stability reasons and the bilinear C-term and the biquadratic Ω term are most relevant for the nature of the transition, as described before.

Let us briefly outline the general features of the generalized model. By introducing the ansatz for the primary and the secondary order parameters

$$\xi_x=\theta_\circ\cdot\cos qz \qquad \xi_y=\theta_\circ\cdot\sin qz \tag{2.5.5a}$$

$$P_x=-P_\circ\cdot\sin qz \qquad P_y=P_\circ\cdot\cos qz \tag{2.5.5b}$$

the free-energy density is rewritten as

$$g(z)=g_A+\tfrac{1}{2}a\theta_\circ^2+\tfrac{1}{4}b\theta_\circ^4+\tfrac{1}{6}c\theta_\circ^6-\Lambda q\theta_\circ^2+\tfrac{1}{2}K_3q^2\theta_\circ^2+\frac{1}{2\varepsilon}P_\circ^2-$$
$$-\mu qP_\circ\theta_\circ-CP_\circ\theta_\circ-\tfrac{1}{2}\Omega P_\circ^2\theta_\circ^2+\tfrac{1}{4}\eta P_\circ^4-dq\theta_\circ^4 \tag{2.5.6}$$

It is a function of the magnitude of the spontaneous polarization, the magnitude of the tilt angle and the wave-vector q of the helix. After minimizing Eq.2.5.6 with respect to q, we obtain the equation governing the temperature behavior of the wave vector of the helix (Žekš, 1984)

$$q=\frac{1}{K_3}\left(\Lambda+\mu\frac{P_\circ}{\theta_\circ}+d\theta_\circ^2\right) \tag{2.5.7}$$

This equation can be interpreted as a renormalization of the "bare" Lifshitz coefficient Λ by two independent mechanisms:

(i) The ratio $P_\circ/\theta_\circ$ renormalizes the Lifshitz coefficient via the flexoelectric coupling term $\mu \cdot \vec{P} \cdot d\vec{\xi}/dz$. In the simple Landau-free energy expansion (2.2.5) this term is temperature independent. Within the generalized model, however, this ratio is temperature dependent. For a given equilibrium angle $\theta_\circ$, the fine details of the balance between the polar and quadrupolar ordering in general change the magnitude of $P_\circ$, which is now ***in general not proportional*** to $\theta_\circ$. We realize that the competition between the polar and quadrupolar ordering finally influences the period of the helix via the flexoelectric coupling.

(ii) The Lifshitz coefficient Λ is also renormalized by the magnitude of the tilt angle. Depending on the sign of the coefficient *d,* this will introduce either a monotonously increasing or decreasing period of the helix with decreasing temperature. Both cases can be observed in the $smectic-C^*$ phases of different substances.

It is important to note that the signs of the coefficients $\Lambda, \mu, C, \text{and } d$ are arbitrary and will depend on whether the helical structure is left- or right-handed and whether the crystal is a (+) or a (-) substance according to the nomenclature of Clark and Lagerwall (see Fig.1.1.6). We can therefore have a number of different temperature behaviors of the helical period in different ferroelectric liquid crystals. As an example, we present in Fig.2.5.3. the temperature dependence of the period of the helix in DOBAMBC as determined from the light-diffraction experiments (circles) of Muševič et al. (1984) and an independent set of measurements of the tilt angle and $P_\circ/\theta_\circ$ (crosses, Huang and Dumrongrattana, 1986). One can see that the agreement is remarkable and can very well explain the observations of the strong temperature dependencies of the helical period in most $smectic-C^*$ materials. As a consequence, this can explain the existence of reentrant phases, observed in high magnetic or electric fields (see Chapter 3, Symmetry breaking by external fields).

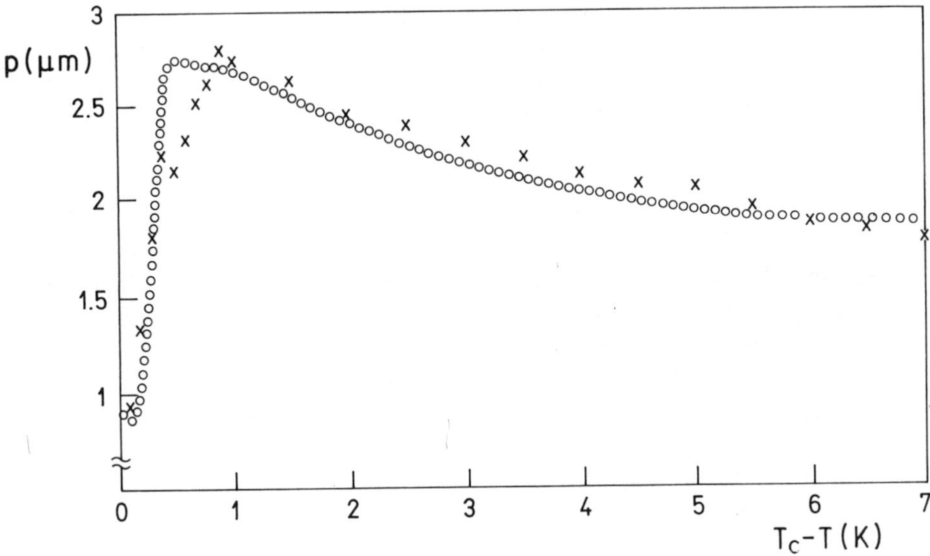

Fig.2.5.3. The period of the helix in DOBAMBC as determined from the light diffraction measurements (circles) and the measurements of the tilt and the ratio $P_\circ/\theta_\circ$ (crosses, Huang and Dumrongrattana, 1986). From the measurements of $\theta_\circ$ and $P_\circ/\theta_\circ$, the period of the helix was determined using Eq.2.5.7. and fitted to the best choice of the three unknown coefficients.

After minimization of the free-energy density Eq.2.5.6 with respect to $P_\circ$ and $\theta_\circ$ and by inserting the expression for the wave-vector q (Eq.2.5.7), we obtain a system of two coupled equations for the unknown $P_\circ$ and $\theta_\circ$. The "tilt" equation is

$$\left(a-\frac{\Lambda^2}{K_3}\right)\theta_\circ+\left(b-\frac{4\Lambda d}{K_3}\right)\theta_\circ^3+\left(c-\frac{3d^2}{K_3}\right)\theta_\circ^5-$$

$$-\Omega\theta_\circ P_\circ^2-\left(C+\frac{\Lambda\mu}{K_3}\right)P_\circ-\frac{3d\mu}{K_3}\theta_\circ^2 P_\circ=0 \tag{2.5.8}$$

whereas the "polarization" equation is

$$\left(\frac{1}{\varepsilon}-\frac{\mu^2}{K_3}\right)P_\circ-\left(C+\frac{\Lambda\mu}{K_3}\right)\theta_\circ-\Omega P_\circ\theta_\circ^2+\eta P_\circ^3-\frac{\mu d}{K_3}\theta_\circ^3=0 \qquad (2.5.9)$$

In general, these two equations have to be solved numerically. The polarization is here not simply proportional to the tilt as in the case of a simple Landau expansion 2.2.5. Instead, the ratio $P_\circ/\theta_\circ$ is generally temperature dependent. This can be most easily seen for the homogeneous case, i.e. $q=0$ and $\mu=0$. The polarization equation is then

$$\left(1/\varepsilon-\Omega\theta_\circ^2\right)P_\circ+\eta P_\circ^3-C\theta_\circ=0 \qquad (2.5.10)$$

Close to T_c, both $P_\circ$ and $\theta_\circ$ are small and we have the limiting value

$$P_\circ/\theta_\circ \to \varepsilon/C \qquad (2.5.11)$$

whereas far from T_c, $\theta_\circ$ is large and we have another limiting value

$$P_\circ/\theta_\circ \to \sqrt{\Omega/\eta} \qquad (2.5.12)$$

We have therefore a crossover from one value at the phase transition to another far from T_c, as indeed observed in the experiments (see Dumrongrattana et al., 1986).

In conclusion, the generalized Landau model of the *smectic* $-A$ - *smectic* $-C^*$ phase transition can explain most of the observed "anomalous" behavior, which was for many years found to be inconsistent with the simple Landau theory as introduced by Indenbom and Pikin. This led to a first set (Carlsson et al.,1988) of the coefficients of the generalized Landau free-energy expansion, which is based on the analysis of the experimental data on

DOBAMBC, as obtained by Huang and Dumrongrattana (1986). and Carlsson et al. (1988). The two updated sets of the coefficients are listed in Table 2.5.2.

Parameter	$\alpha[N/m^2K]$	$b[N/m^2]$	$c[N/m^2]$	$\Lambda[N/m]$	$K_3[N]$
Carlsson et al.	3.5×10^4	8.0×10^5	9.0×10^6	-1.4×10^{-5}	3.4×10^{-12}
Huang et al.	4.6×10^4	5.2×10^5	9.0×10^6	2.3×10^{-5}	2.5×10^{-12}

Parameter	$d[N/m]$	$\varepsilon[As/Vm]$	η $[Nm^5/A^4s^4]$	$\mu[V]$	$C[V/m]$	Ω $[Nm^2/A^2s^2]$
Carlsson et al.	-7.0×10^{-5}	2.7×10^{-13}	2.2×10^{22}	-0.2	-4.1×10^7	9.4×10^{13}
Huang et al.	2.5×10^{-5}	2.6×10^{-11}	3.8×10^{19}	-0.2	2.8×10^6	5.7×10^{11}

Table 2.5.2. Values of the material parameters of DOBAMBC, as calculated by Carlsson et al. (1988) and Huang and Dumrongrattana (1986) (after Carlsson et al., 1988).

Note 2.5.1. Ratio of polarization and tilt angle in ferroelectric liquid crystals:
This ratio has been first measured by Dumrongrattana and Huang in DOBAMBC (1986) and later by Huang et al.(1987a) on DOBA-1-MPC. In both cases, high resolution tilt and polarization measurements have definitely shown that this ratio is not temperature independent, as follows from a simple Landau theory. An example is shown on the figure

below for DOBAMBC. The ratio is small close to the transition into the *smectic* $-A$ phase and sharply increases within $1K$. Then, the ratio saturates at a nearly constant value. This behavior is typical for ferroelectric liquid crystals showing a second order phase transition into the *smectic* $-A$ phase and can be well described within the generalized Landau model.

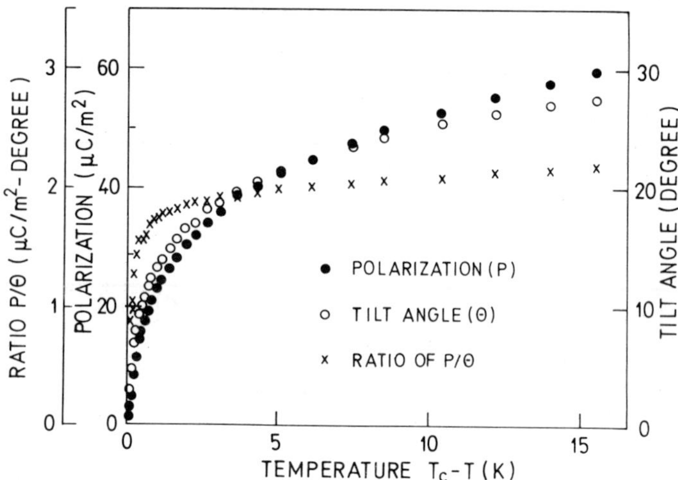

Chapter 3

Symmetry Breaking by External Fields

3.1. Introduction

When an external magnetic field is applied perpendicularly to the helical axis of a ferroelectric liquid crystal with a positive diamagnetic anisotropy, the molecules tend to align into the field direction. This will distort the originally "smooth" helical arrangement of the $smectic-C^*$ phase and induce the so-called soliton-like structure for fields lower than the critical field, as illustrated in Figure 3.1.1:

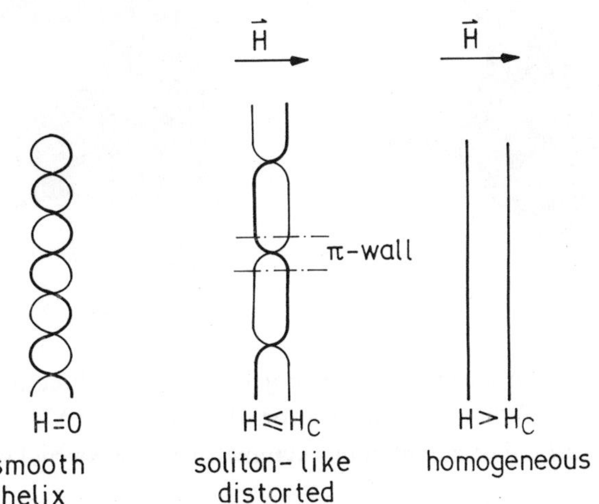

Fig.3.1.1. Magnetic field induced soliton-like helical structure and the unwinding of the $smectic-C^*$ phase by an external magnetic field.

One can see from Fig.3.1.1. that large and almost uniformly aligned domains are separated by the π-domain walls, where the phase of the order parameter changes by π on a very short distance. When the magnitude of the external field is increased, the domain walls become narrow, whereas the separation between the domain walls increases and diverges at some critical magnetic field H_c. Above the critical magnetic field, the structure is unwound and spatially homogeneous. This chiral but unwound phase is denoted $smectic-\overline{C}^*$ phase, to distinguish it from its modulated counterpart.

The soliton-like distortion of the $smectic-C^*$ phase, as induced by an external magnetic field, ***breaks the continuous symmetry*** of the unperturbed $smectic-C^*$ phase. This can be noted by considering the symmetry properties of the structures of the unperturbed and distorted $smectic-C^*$ phases, as shown in Figs.3.1.2a and 3.1.2b, respectively.

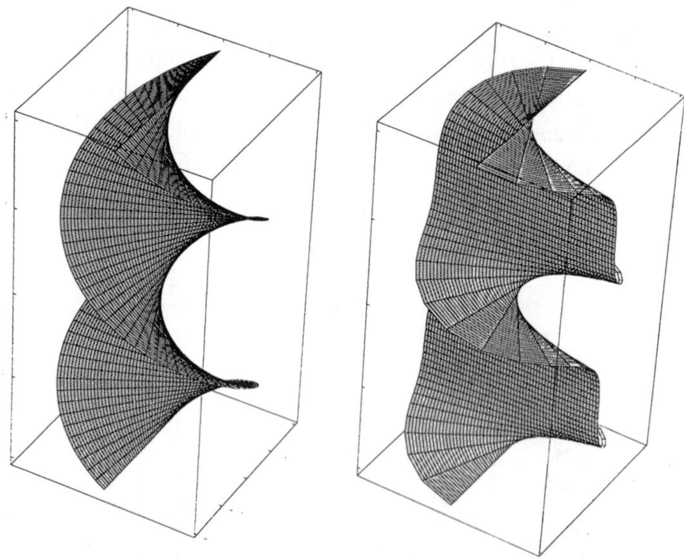

Fig.3.1.2. The surface of the projection of the director $\xi(z)$ of a left-handed $smectic-C^*$ helical structure for a) $H=0$ and b) $H \approx H_c$. This is an example of a breaking of a continuous symmetry by an external magnetic field.

In the unperturbed $smectic-C^*$ phase (Fig.3.1.2a, left), any infinitesimal translation of the structure along the z-axis can be combined with a suitable infinitesimal rotation around the z-axis to transform the system into itself. The unperturbed *smectic* $-C^*$ phase has thus a characteristic ***continuous helical symmetry***.

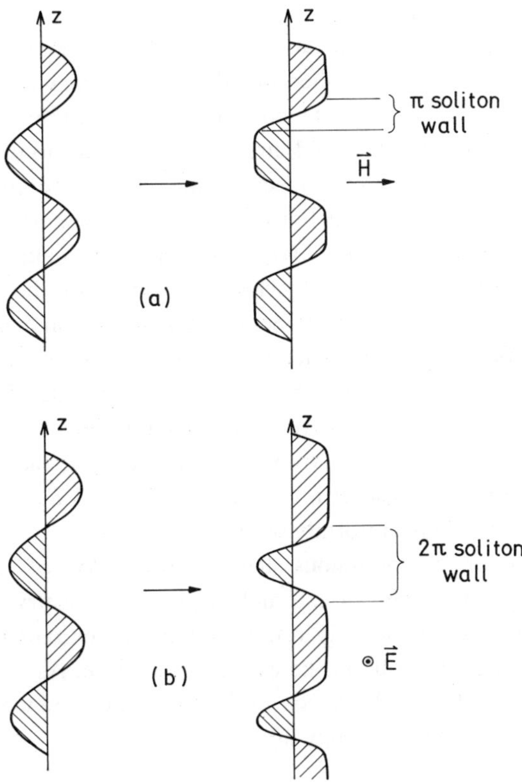

Figure 3.1.3. Comparison between π-soliton lattice and 2π-soliton lattice. (a) π-soliton lattice is induced by a transverse magnetic field. (b) 2π-soliton lattice is induced by a transverse electric field. In (a), the magnetic field is applied in the x-direction and the corresponding projection of the director field $\xi(z)$ is shown. In (b), the electric field is applied in a direction out of the paper and the projection of the director field $\xi(z)$ is shown. Note the difference in the translational properties of a magnetic and an electric field induced soliton lattice.

On the other hand, as one can see from Fig.3.1.2b-right, the distorted $smectic-C^*$ phase is invariant under the group of ***discrete transformations***, i.e. it is transformed into itself by a translation of $p_\circ/2$ along the z-axis, followed by a 180^0 rotation around the helical axis. The application of an external magnetic field therefore breaks the continuous symmetry of the $smectic-C^*$ phase and results in a discrete symmetry.

Very similar phenomena can be expected when an external DC or very low-frequency electric filed is applied in a direction, perpendicular to the helical axis. The field couples to the spontaneous polarization of the sample, and tends to align the polarization into the field direction. In contrast to the formation of a π-soliton structure in the case of a quadratic diamagnetic coupling, a 2π soliton lattice is formed in the case of a linear ferroelectric coupling, as one can see from Figures 3.1.3a and b. Here, large and uniformly aligned domains are separated by narrow domain walls, where the phase of the order parameter changes by 360^0. Within the large domains, the polarization is almost uniformly aligned into the field direction, whereas within the narrow domain walls, the polarization precesses by 360^0. By increasing the magnitude of the electric field homogeneous regions with the polarization preferentially aligned into the field direction grow at the expense of narrow domain walls. The distance between the domain walls becomes infinite at a critical electric field E_c where the helix is unwound and the sample is uniformly polarized.

Breaking of a continuous helical symmetry of the unperturbed $smectic-C^*$ phase by either an external magnetic or electric fields has some fundamental consequences and is reflected both in extraordinary static and dynamic properties of the soliton-like distorted structure. As we shall see, most of the properties of a distorted $smectic-C^*$ structure can be well understood by considering simple symmetry arguments.

Note 3.1.1. Diamagnetic properties of liquid crystals:
In paramagnetic and ferromagnetic materials, the permanent magnetic dipoles, that are present in a material, are aligned either spontaneously (ferromagnets) or by an external magnetic field (paramagnets). In diamagnetic materials, magnetic dipole moments are induced on the atomic or molecular scale by external magnetic field. They are aligned in such a way, that the total magnetic field is decreased. For example, when a benzene ring

is put in a magnetic field, electric currents will be induced in a ring due to changing magnetic field. According to the Lentz rule, these currents will tend to decrease the magnetic flux through the ring and the field is to a certain extent expelled to the exterior of the ring. This increases the magnetic energy of a system. The magnetic energy can be lowered if the ring changes the orientation, so that the lines of the magnetic field are in the plane of the ring. The magnetic flux is zero in this case. This explains why most liquid crystals, which are formed from several benzene rings, tend to align their long molecular axes into the direction of the field. The diamagnetic anisotropy is in this case positive, $\Delta\chi = \chi_{\parallel} - \chi_{\perp} > 0$. Some liquid crystals show negative diamagnetic anisotropy, as for example p-cyano-p'-alkylcyclohexyl-cyclohexanes (CCH) compounds (Seppen, 1987). Typical magnitude of diamagnetic anisotropy of liquid crystals is $\Delta\chi \approx 10^{-6}$ in SI units. Note, $\Delta\chi(SI) = 4\pi \cdot \Delta\chi(CGS)$. More experimental data were published by Stanarius (1998).

3.2. The Lifshitz Point in the *(H,T)* Phase Diagram

The Lifshitz point was first discussed by Hornreich, Luban, and Shtrikman in 1975 and introduced to the field of liquid crystals by A.Michelson (1977a,b). It is a triple point, where the disordered $smectic-A$, the homogeneously ordered $smectic-\overline{C}^*$ and the inhomogeneously ordered $smectic-C^*$ phases meet, as indicated in Fig.3.2.1.

Michelson (1977a,b) has proposed the possible physical realization of the Lifshitz point in the (H,T) phase diagram of a ferroelectric liquid crystal, where an external magnetic field is applied perpendicularly to the helical axis of the $smectic-C^*$ phase. If the diamagnetic anisotropy of the liquid crystalline molecules is positive, molecules tend to align along the field direction. For negative anisotropy, they tend to align in a plane perpendicular to the field. As we have seen, this results in the so-called soliton-like distortion of the helical structure, where long, almost uniformly aligned domains are separated by thin domain walls. The thickness of these π domain walls decreases with increasing field, whereas the distance between the domain walls increases. At a certain critical field H_c the distorted helical structure unwinds into the spatially homogeneous $smectic-\overline{C}^*$ structure, as shown in the (H,T) phase diagram in Figs.3.2.1a and b.

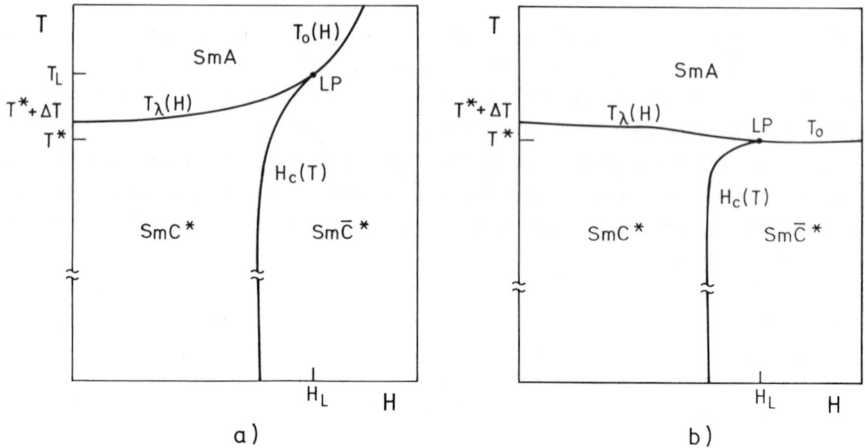

Fig.3.2.1. Lifshitz point (H_L, T_L) in the (H,T) phase diagram of a ferroelectric liquid crystal in an external magnetic field. (a) Positive diamagnetic anisotropy. (b) Negative diamagnetic anisotropy.

The unwound phase can be alternatively reached by decreasing the temperature T from the *smectic – A* phase of a ferroelectric liquid crystal. For a low external magnetic field, we cross at a certain temperature the phase transition line $T_\lambda(H)$ and enter the modulated *smectic* $-\,C^*$ phase. On the other hand, at high enough magnetic fields we cross the phase transition line $T_\circ(H)$ and enter the spatially homogeneous (unwound) *smectic* $-\,\overline{C}^*$ phase, as shown in Fig.3.2.1. The phase transition lines $T_\lambda(H)$, $T_\circ(H)$ and $H_c(T)$ thus merge into a ***Lifshitz point*** (H_L, T_L) where the disordered (*smectic – A*), the homogeneously ordered (*smectic* $-\,\overline{C}^*$) and the helicoidally ordered (*smectic* $-\,C^*$) phases meet.

The basic features of the (H,T) phase diagram of a ferroelectric liquid crystal in an external magnetic field can be derived from the simple Landau free-energy expansion

$$g(z) = g_A + \tfrac{1}{2}a(T)\left(\xi_x^2 + \xi_y^2\right) + \tfrac{1}{4}b\left(\xi_x^2 + \xi_y^2\right)^2 - \Lambda\left(\xi_x \frac{\partial \xi_y}{\partial z} - \xi_y \frac{\partial \xi_x}{\partial z}\right) + \\ + \tfrac{1}{2}K_3\left(\left(\frac{\partial \xi_x^2}{\partial z}\right)^2 + \left(\frac{\partial \xi_y^2}{\partial z}\right)^2\right) - \tfrac{1}{2}\Delta\chi H^2 \xi_y^2 \qquad (3.2.1)$$

Here g_A is the equilibrium free-energy of the $smectic-A$ phase, $\Delta\chi = \chi_{\parallel} - \chi_{\perp}$ is the diamagnetic anisotropy and the magnetic field is applied in the y-direction, $H = (0, H, 0)$. We assume that $a(T) = \alpha(T - T_R)$, α and b are positive constants, and T_R is the phase transition temperature for the racemic mixture. In Eq.3.2.1, Λ is the coefficient of the Lifshitz term, and K_3 is the torsional elastic constant. We assume that the system is homogeneous in the x-y plane, which is parallel to the smectic planes.

The phase transition boundaries between the $smectic-A$ phase and the modulated $smectic-C^*$ or homogeneous $smectic-\overline{C}^*$ phases can be obtained by the linear stability analysis (Michelson, 1977a), similarly to the one performed in Section 2.2. As a result, one obtains the so-called Lifshitz field H_L and the Lifshitz temperature T_L

$$H_L = \frac{2\Lambda}{\sqrt{K_3|\Delta\chi|}} \qquad T_L = T_R + 4\frac{\Lambda^2}{\alpha K_3} \tag{3.2.2}$$

which characterize the phase diagram and determine the position of the Lifshitz point in the (H, T) phase diagram. Below the Lifshitz field H_L, the $smectic-A$ and the modulated $smectic-C^*$ phases are separated by the λ -line, $T_\lambda(H)$

$$H < H_L \qquad T_\lambda(H) = T_R + \tfrac{1}{\alpha}\varepsilon C^2 + \Delta T\left(1 + \frac{\Delta\chi}{|\Delta\chi|}\frac{H^2}{H_L^2}\right)^2 \tag{3.2.3}$$

Here $\Delta T = \Lambda^2/\alpha K_3$ is a coefficient, which is a measure of the magnetic-field dependence of the λ-line and T_R is the phase transition temperature for the racemic mixture. As one can see from the above equation, this line increases with the field for positive $\Delta\chi$, whereas it decreases with the field for negative $\Delta\chi$. Along the λ-line the phase transition is of the second order. Above the Lifshitz field the $smectic-A$ phase transforms into the homogeneous $smectic-\overline{C}^*$ phase at the second order phase boundary $T_\circ(H)$

$$H > H_L \qquad T_o(H) = T_R + \tfrac{1}{\alpha}\varepsilon C^2 + 2\Delta T\left(1 + \frac{\Delta\chi}{|\Delta\chi|}\right)\frac{H^2}{H_L^2} \tag{3.2.4}$$

For positive $\Delta\chi \geq 0$ and beyond the Lifshitz point, the phase transition line increases quadratically with the field whereas for a negative diamagnetic anisotropy it is field independent .

The (H,T) phase diagrams for the cases of positive and negative diamagnetic anisotropies as derived from a simple Landau free-energy expansion (Eq.(3.2.1)) are shown schematically in Figs.3.2.1a and b. Here the most interesting feature is the position of the Lifshitz point (H_L, T_L) in the phase diagram, which can be located by the Eqs.3.2.2 and 3.2.3. It can be shown (Michelson, 1977a) that at the Lifshitz point the two phase boundaries $T_\lambda(H)$, and $T_\circ(H)$ merge tangentially into each other and to the line of the critical magnetic field $H_c(T)$, which separates the modulated and the unwound phases. It can also be shown that near the Lifshitz point, the line of the critical magnetic field $H_c(T)$ is of the first order (Michelson, 1977a).

Along the λ-line, the wave vector $q_c(H)$ of the modulated $smectic-C^*$ phase decreases smoothly with the field

$$H < H_L \qquad q_c(H) = q_c(0)\sqrt{1-(H/H_L)^4} \tag{3.2.5}$$

from the value $q_c(0) = \Lambda/K_{33}$ at $H = 0$ to a zero value at and above the Lifshitz field:

$$H > H_L \qquad q_c(H) = 0 \tag{3.2.6}$$

The field dependence of the wave-vector along the λ-line is shown in Fig.3.2.2. The line of the critical field $H_c(T)$, which originates in the Lifshitz point and separates the modulated $smectic-C^*$ and the homogeneous $smectic-\overline{C}^*$ phases, was first calculated numerically by Benguigui and Jacobs (1988) and Jacobs and Benguigui (1989). Analytically, it can be derived only in the so-called constant amplitude approximation (CAA). Within the CAA one assumes that the magnitude of the tilt angle does not change significantly with the external field, whereas the phase profile $\Phi(z)$ is allowed to vary with the field. It can be shown, that this is a reasonable approximation except within 10mK around the Lifshitz point (Kutnjak-Urbanc 1993a; Kutnjak-Urbanc and Žekš, 1993b).

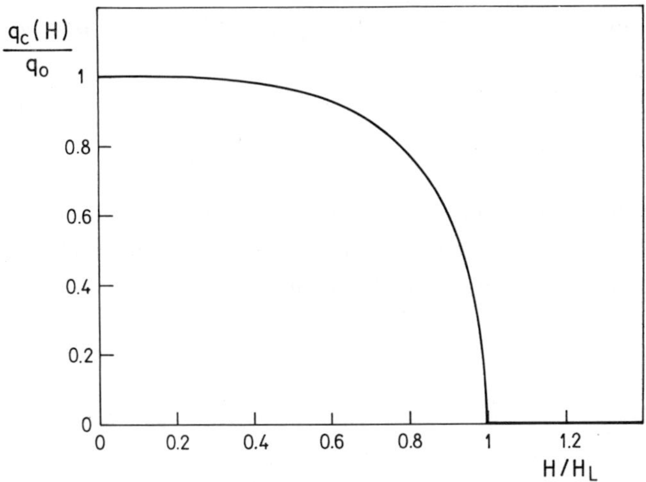

Fig.3.2.2. The magnetic field dependence of the modulation wave-vector $q_c(H)$ along the λ-line. The dependence is smooth and the wave-vector approaches the zero value in a continuous fashion.

Let us now calculate the line of the critical field H_c in the CAA. Because the magnitude of the tilt does not change, it is enough to consider the phase dependent part of the free-energy density, which is

$$g(z) = -\Lambda\theta^2 \frac{\partial\Phi(z)}{\partial z} + \tfrac{1}{2} K_3\theta^2 \left(\frac{\partial\Phi(z)}{\partial z}\right)^2 - \tfrac{1}{2}\Delta\chi H^2\theta^2 \sin^2\Phi(z) \qquad (3.2.7)$$

After the Euler-Lagrange minimization of the free-energy $G = \int g(z)dz$ with respect to $\Phi(z)$ and $\partial\Phi/\partial z$ we obtain the well-known sine-Gordon equation for $\Phi(z)$:

$$\frac{\partial^2\Phi(z)}{\partial z^2} + \left(\frac{\Delta\chi H^2}{2K_3}\right)\sin 2\Phi(z) = 0 \qquad (3.2.8)$$

The phase profiles $\Phi(z)$ which minimize the free-energy have the form of the so-called ***π-soliton lattice*** and are described by ***Jacobi's elliptic functions***

$$\sin\Phi_o = sn(u,k) \tag{3.2.9}$$

Here *sn(u,k)* is the Jacobi elliptic sine of the reduced coordinate $u = z/(\xi k)$, where $\xi = \sqrt{K_3/(\Delta\chi H^2)}$ is the magnetic coherence length. The modulus k of the Jacobi's elliptic function is defined by the equation

$$k = \frac{H}{H_c} E(k) \tag{3.2.10}$$

where $E(k) = \int_0^{\pi/2} \sqrt{1 - k^2 \sin^2 x}\ dx$ is the complete elliptic integral of the second kind (Byrd and Friedman, 1954) and H_c is the critical magnetic field for the unwinding of the helical structure:

$$H_c = \frac{\pi^2}{p_o} \cdot \sqrt{\frac{K_3}{|\Delta\chi|}} \tag{3.2.11}$$

Within the CAA approximation, the critical field is inversely proportional to the period p_o of the unperturbed helix, which is similar to the behavior of chiral nematics in an external magnetic field (de Gennes, 1968a; Meyer 1968; Durand et al., 1969). The critical magnetic field is here temperature independent because the period of the unperturbed helix is temperature independent within the simple free-energy expansion (Eq.3.2.1). From Eqs.3.2.2 and 3.2.11 and remembering that $p_o = 2\pi\ K_3/\Lambda$, we also obtain the relation $H_c = \frac{\pi}{4} H_L$. The Lifshitz point is therefore located at fields slightly higher than the low-temperature critical field and the phase diagram is "horn"-like, as shown in Fig.3.2.1.(a).

The phase profile $\Phi_o(z)$ which satisfies the sine-Gordon equation is shown schematically in Fig.3.2.3.(a) for different magnetic fields. For *H=0* the helical modulation is plane-wave-like with a linear dependence of the equilibrium phase, $\Phi_o = (2\pi / p_o)z$.

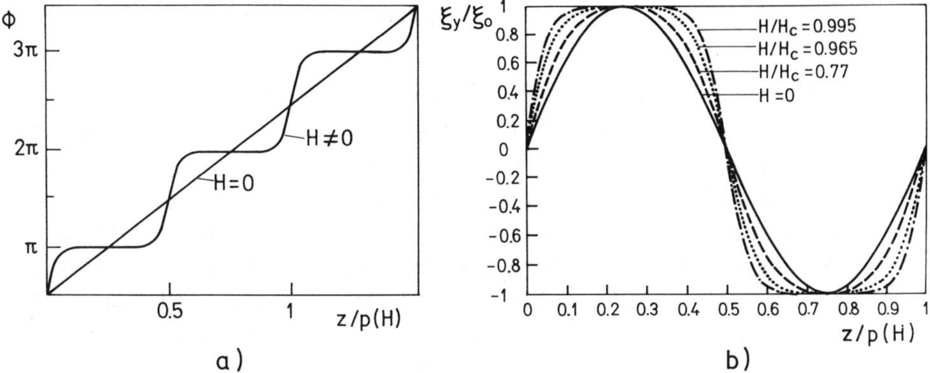

Fig.3.2.3. Solitons in the CAA approximation. (a) The phase profiles $\Phi_0(z)$ which are the solutions of the sine-Gordon equation for $H=0$ and $H \neq 0$, $\Delta\chi > 0$. The figure is schematic and illustrates the appearance of the π-soliton walls, where the phase changes by 180° on a very short distance. (b) The corresponding projections of the director $\xi(z)$ on the y-axis for different values of the magnetic field, which is applied in the y-direction. The profiles were calculated from Eqs.3.2.9 and 3.2.10..

For $H \neq 0$ the helical modulation becomes soliton-like, which is most pronounced in the narrow-soliton regime near H_c, as shown in Fig.3.2.3.(a). Large regions of nearly constant phase, where the molecules are almost homogeneously aligned into the field direction, are separated by narrow π-soliton walls. Within the soliton wall the phase changes rapidly by 180° on a very short distance. The corresponding projection of the helical, distorted $smectic-C^*$ phase is shown in Fig.3.2.3.(b). By increasing the magnetic field the width of the soliton walls decreases slightly, whereas the distance between the soliton walls increases and diverges at H_c. The phase transition is here of second order with the divergence of the helical period:

$$p(H) = p_o \left(\tfrac{2}{\pi}\right)^2 K(k) E(k) \tag{3.2.12}$$

Here $K(k)=\int_0^{\pi/2} dx/\sqrt{1-k^2\sin^2 x}$ is the elliptic integral of the first kind which diverges logarithmically at H_c (Byrd and Friedman, 1954). Modulus k is determined by Eq.3.2.10.

The constant amplitude approximation breaks down in the immediate vicinity of the Lifshitz point. In this region one has to consider both the amplitude and phase changes when an external field is applied. The corresponding phase profiles $\Phi_o(z)$ and the amplitude profiles $\theta_o(z)$ can be obtained only numerically (Yamashita, 1985, 1988 and 1992; Kutnjak-Urbanc, 1993). An example of the structure of a single, π-soliton wall close to the critical field is shown in Fig.3.2.4. (Kutnjak-Urbanc,1993a). In contrast to the "smooth" z-dependence of the phase profile $\Phi_o(z)$, obtained in the CAA approximation (see Fig.3.2.3.), the equilibrium amplitude profile $\theta_o(z)$ (solid curve) shows a characteristic depression in the center of the soliton (z=0) and a characteristic overshoot near its edges.

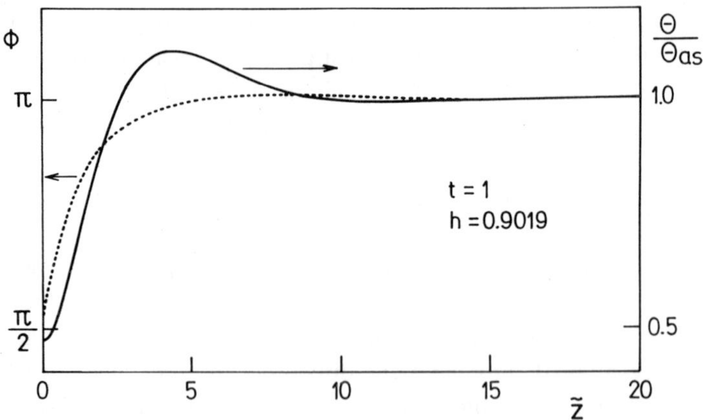

Fig.3.2.4. Solitons beyond the CAA approximation. A single π-soliton structure in a ferroelectric liquid crystal close to the Lifshitz point (Kutnjak-Urbanc, 1993a). The solid line represents the equilibrium magnitude of the tilt angle $\theta_o(z)$ and the broken line is the equilibrium phase profile $\Phi_o(z)$. The center of the soliton wall is at $\tilde{z}=0$. Note that $\theta_o(z)$ shows a slight overshoot above the asymptotic value θ_{as} and a quasi oscillatory dependence at the soliton edges. $t=(T_L-T)/T_c$ is the reduced temperature, $h=H/H_L$ is the reduced magnetic field and $\tilde{z}=q_c z$ is normalized coordinate.

The depression of the magnitude of the tilt angle is here coincident with a local fast variation of the phase. This results in excess elastic free energy that is minimized by lowering the local magnitude of the tilt angle.

As we have already mentioned, in the immediate vicinity of the Lifshitz point, the phase transition boundary between the modulated and homogeneous phases is of first order, whereas far below it is of second order. This means that by approaching the Lifshitz point along the line of the critical field H_c from below, we should reach the so-called ***tricritical point***, where the first order phase transition boundary changes into the second order one. This tricritical point was first introduced by Jacobs and Walker (1980) in their studies of the charge-density-waves in solid crystals.

The *(H,T)* phase diagram of a ferroelectric liquid crystal in an external field was first reported in high-magnetic field experiments on p-decyloxybenzilidene-p'-amino-2-methylbutyl cinnamate (DOBAMBC) (Muševič et al., 1982) and is shown in Fig.3.2.5. The phase boundaries were determined by static light scattering and dielectric spectroscopy and are in qualitative agreement with theoretical predictions. The $smectic-A$ - $smectic-C^*$ phase transition line was found to be practically field-independent within the resolution of 20mK up to 10T, which is in agreement with the expected increase of the λ-line. Near the λ-line, the $smectic-C^*$ - $smectic-\overline{C}^*$ phase transition is discontinuous, as can be seen from the magnetic field dependence of the wave vector of the helix, shown in Fig.3.2.6. for different temperatures (Muševič et al., 1982). Later on the experimental work on the *(H,T)* phase diagram was continued by A.Seppen (1987), who succeeded to observe the Lifshitz point in a mixture of chiral and racemic DOBAMBC. Later, related experiments were performed in a high frequency electric field (Wang et al., 1991) which couples quadratically to the dielectric anisotropy. Their results were similar to those reported by Muševič et al.(1982).

The magnetic field experiment in chiral DOBAMBC however revealed the existence of a reentrant phase very near the λ-line and indicated a "runaway" of the Lifshitz point in chiral DOBAMBC. A similar behavior was observed in electric-field experiments (Wang et al., 1991). Furthermore, the experiment clearly showed a very strong temperature dependence of the critical field H_c, in apparent disagreement with the predictions of a simple Landau model. The observation of a strong temperature dependence of the critical magnetic field thus addressed the question of the relevance of the simple Landau model and the reasons for its failure to describe the magnetic-field effects correctly.

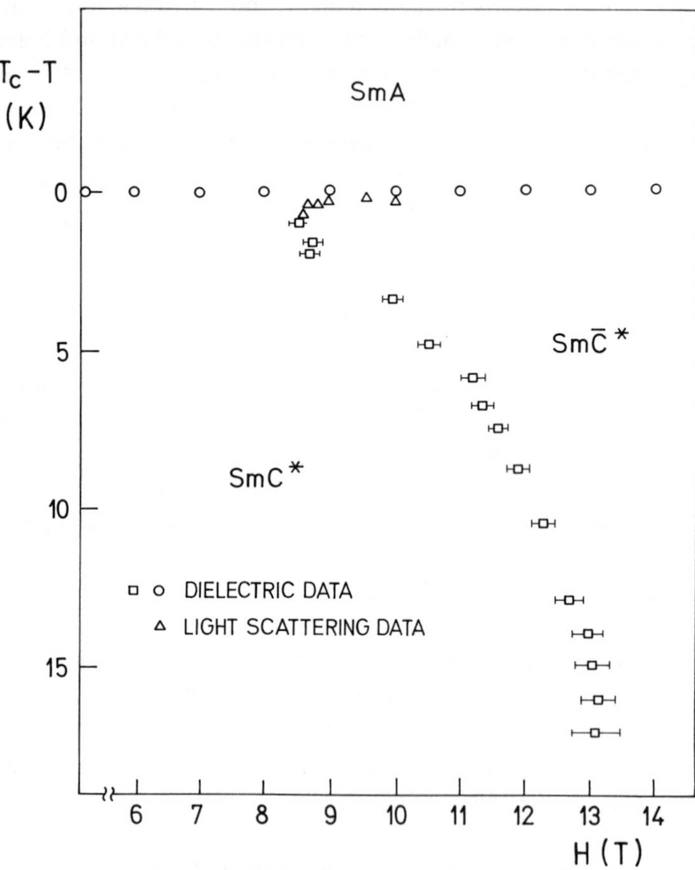

Fig.3.2.5. The *(H,T)* phase diagram of a ferroelectric liquid crystal p-decyloxybenzilidene-p'-amino-2-methylbutyl cinnamate (DOBAMBC) in an external magnetic field (Muševič et al., 1982). Note a very narrow region of a reentrant *smectic* $-C^*$ close to the λ-line.

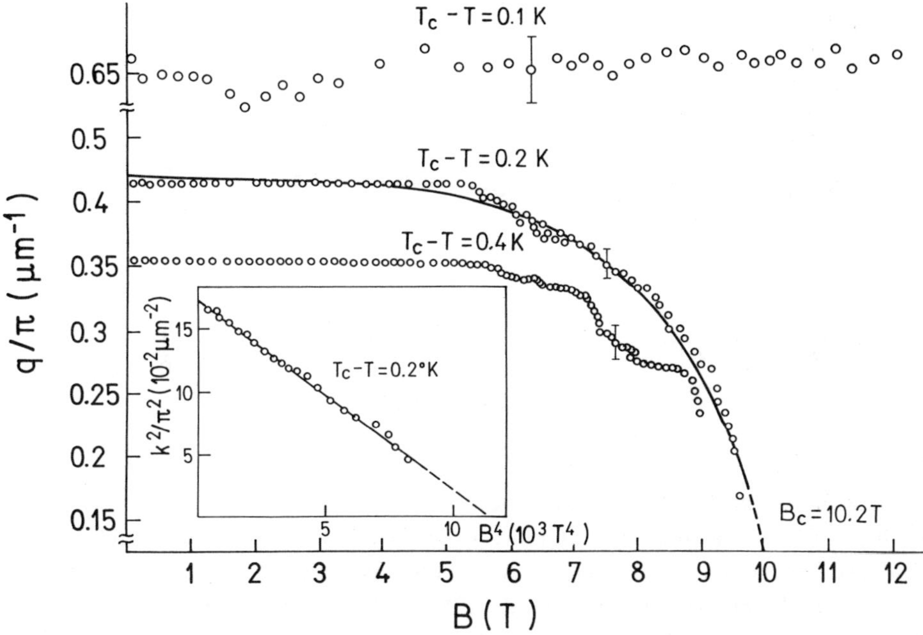

Fig.3.2.6. Magnetic field dependence of the wave vector of the helix in ferroelectric liquid crystal DOBAMBC for several temperatures near the λ-line. The inset shows the best fit to the Eq.3.2.5., indicating the close vicinity of the Lifshitz point. Note a "runaway" of the critical field for the measurement 0.1K below the λ-line. (Muševič et al., 1982).

The key question that any correct phenomenological model should answer is the temperature dependence of the period of the helix in ferroelectric liquid crystals, which shows in many cases a very characteristic and anomalous behavior in the vicinity of the $smectic - A$ - $smectic - C^*$ transition (Muševič et al., 1984). Because of the relation (Eq.3.2.11.) between the helical period and the critical magnetic field, this is reflected in the non-monotonous temperature dependence of the critical field and is therefore the reason for the observed re-entrance of the $smectic - C^*$ phase in the (H,T) phase diagram.

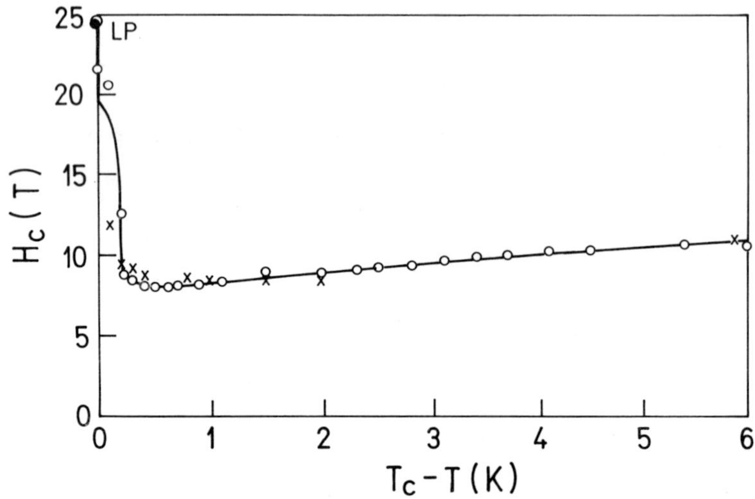

Fig.3.2.7. Temperature dependence of the critical magnetic field in DOBAMBC as calculated within the generalized Landau model (solid line, Kutnjak-Urbanc and Žekš, 1993). The experimental values of the critical field are marked by crosses (Muševič et al., 1982), whereas circles represent critical field, calculated from the helical period (Eq.3.2.11) (Muševič et al., 1984).

There are several models which can explain this reentrance (Benguigui and Jacobs, 1988; Jacobs and Benguigui 1989; Kutnjak-Urbanc and Žekš, 1993b). Among them, the so-called generalized Landau model, as first introduced by Žekš (1984) seems to give the most complete description of the thermodynamic properties of ferroelectric liquid crystals (Carlsson et al., 1988). Numerical studies of the observed reentrant behavior in magnetic and high-frequency electric fields have shown a good agreement with the generalized Landau model (Kutnjak-Urbanc and Žekš, 1993), as shown in Fig.3.2.7. As can be seen from this figure, the reentrant behavior can indeed be well described by the nonmonotonous temperature dependence of the helical period. This means that the CAA approximation is valid for all experimentally accessible temperatures and breaks down only in a very small temperature interval (10mK) near the Lifshitz point. Here, the generalized model predicts a strong increase of the critical field, as is indicated by the location of the Lifshitz point *(LP)* in Fig.3.2.7.

The *(H,T)* phase diagrams as predicted by the simple Indenbom-Pikin model and the generalized model are shown Fig.3.2.8.(a) and (b). The comparison clearly shows the validity of the generalized model, which can explain rather well the peculiar form of the experimentally observed phase diagram.

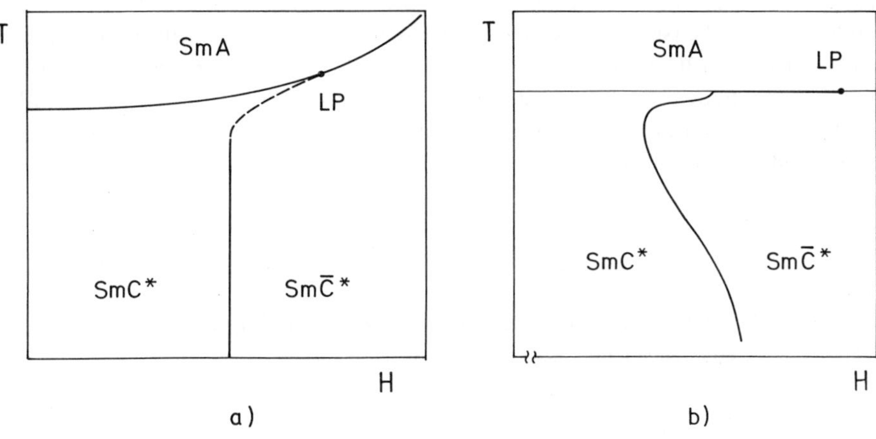

Fig.3.2.8. (a) The phase diagram of a ferroelectric liquid crystal DOBAMBC, as predicted by the simple Landau free-energy expansion first introduced by Indenbom and Pikin. (b) The *(H,T)* phase diagram, as predicted by the generalized form of the free-energy density. After Kutnjak-Urbanc and Žekš, 1992.

Note 3.2.1. Diamagnetic anisotropy of DOBAMBC:
In SI units, the diamagnetic coupling term in Eq.3.2.1. is (see also note 3.1.1)

$$g_m = -\frac{\Delta\chi}{\mu_\circ} \cdot (\vec{B} \cdot \vec{n})^2$$

Using the Eq.3.2.11. for the critical magnetic field, the diamagnetic anisotropy is

$$\Delta\chi = \mu_\circ \pi^4 \frac{K_{33}}{(p_\circ B_c)^2}$$

The elastic constant K_{33} has been measured in a dielectric experiment by Z.Kutnjak (1991) and he found $K_{33} = 1.5\times10^{-11}N$ for DOBAMBC and $K_{33} = 2\times10^{-11}N$ for DOBA-1-MPC. In a magnetic-field experiment of Muševič et al.(1982), it has been proven that the critical magnetic field for the unwinding of the helix is inversely proportional to the period of the helix. The product $p_\circ B_c$ was temperature independent and for DOBAMBC equals to $p_\circ \cdot B_c = 25T\cdot\mu m = 2.5\times10^{-7}Tm$. From this, we obtain the anisotropy of diamagnetic susceptibility of DOBAMBC $\Delta\chi(DOBAMBC) = 2.8\times10^{-6}$ which is of the same order as the diamagnetic susceptibility of nematic liquid crystals.

Note 3.2.2. Magnetic coherence lengths in ferroelectric liquid crystals:
The magnetic coherence length is in SI units expressed as

$$\xi(H) = \sqrt{\frac{\mu_\circ K_{33}}{\Delta\chi}}\cdot\frac{1}{B} = \frac{p_\circ}{\pi^2}\cdot\frac{B_c}{B}$$

It is equivalent to the definition of the magnetic coherence length in nematics (see de Gennes, 1974) and measures the distance, where the external magnetic field has significant effect on the alignment of the director. The critical magnetic fields for the unwinding of the helix of ferroelectric liquid crystals are large. For example in DOBAMBC the critical magnetic field is $B_c = 12.5T$ for the period of the helix of $p_\circ = 2\mu m$. The correlation length is maximum at the critical field. In this case, $\xi(H_c) = 0.2\mu m$ at $B = 12.5T$.

Note 3.2.3. Liquid crystal droplets can be levitated in a high magnetic field:
Due to the magnetic force on a diamagnetic fluid in an inhomogeneous magnetic field, liquid droplets can be levitated against gravity. The experiment is described in a paper by Mahajan et al.(1997).

3.3. Behavior in External DC Electric Fields

The (E,T) phase diagram of a ferroelectric liquid crystal in an static electric field, which is applied perpendicularly to the helix, was first discussed by Michelson and Cabib (1977e). They have shown that it is fundamentally different from its magnetic counterpart. This difference can be easily understood by noting that even a very small external DC electric field, which is applied along the layers of the $smectic-A$ phase breaks the symmetry of this phase and induces a spatially uniform polarization and tilt of the molecules. This means that the $smectic-A$ phase is stable only for $E=0$ in the (E,T) phase diagram, whereas for any finite field, a homogeneously and uniformly polarized $smectic-\overline{C}^*$ phase is stable. As a consequence, there will be only one phase boundary $T_c(E)$, which separates a spatially uniform $smectic-\overline{C}^*$ phase from the low temperature modulated $smectic-C^*$ phase, as shown in Fig.3.3.1. A triple point, where the $smectic-A$, $smectic-\overline{C}^*$ and $smectic-C^*$ phases meet is thus located at $T_c(E=0)$.

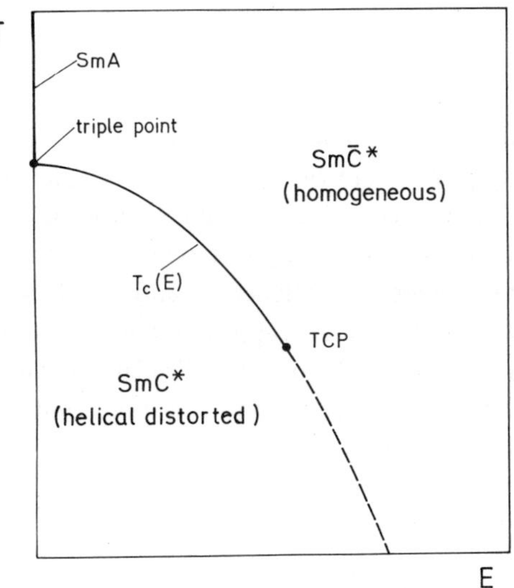

Fig.3.3.1. The (E,T) phase diagram of a ferroelectric liquid crystal in a DC electric field, which is applied parallel to the smectic layers. TCP indicates the tricritical point, where the second order phase transition boundary (solid line) changes into the first order phase transition boundary (dashed line).

The phase boundary $T_c(E)$ was first calculated by Michelson and Cabib (1977e) by considering the stability limits of the uniform $smectic-\overline{C}^*$ phase in an external electric field. Starting from the free energy density

$$g(z) = g_A + \tfrac{1}{2}a(T)\left(\xi_x^2 + \xi_y^2\right) + \tfrac{1}{4}b\left(\xi_x^2 + \xi_y^2\right)^2 - \Lambda\left(\xi_x \frac{\partial\xi_y}{\partial z} - \xi_y \frac{\partial\xi_x}{\partial z}\right) + \\ + \tfrac{1}{2}K_3\left(\left(\frac{\partial\xi_x^2}{\partial z}\right)^2 + \left(\frac{\partial\xi_y^2}{\partial z}\right)^2\right) - \varepsilon C E_y \xi_x \tag{3.3.1}$$

and neglecting the quadratic coupling of the electric field with the dielectric anisotropy, they found that for small electric fields $\vec{E} = (0, E, 0)$, the phase transition boundary decreases quadratically with the field

$$T_c - T_c(E) \propto E^2 \tag{3.3.2}$$

Michelson and Cabib also concluded that by following the phase transition boundary $T_c(E)$, the phase transition which is of second order for small fields becomes of first order at a certain value of the electric field. This is the so-called tricritical point in the (E, T) phase diagram, which is indicated in Fig.3.3.1.

Whereas the analysis of Michelson and Cabib is valid only for small values of the external field, the CAA approximation is expected to be valid in a large temperature-field interval below T_c. Similarly to the case of a magnetic field, we therefore neglect the amplitude variations and retain only the phase dependent part of the free energy

$$g(z) = -\Lambda\theta^2 \frac{\partial\Phi(z)}{\partial z} + \tfrac{1}{2}K_3\theta^2\left(\frac{\partial\Phi(z)}{\partial z}\right)^2 - EP\cos\Phi(z) \tag{3.3.3}$$

After minimization, we obtain again the sine-Gordon equation for $\Phi(z)$:

$$\frac{d^2\Phi(z)}{dz^2} - \frac{EP}{K_3\theta^2}\sin\Phi(z) = 0 \tag{3.3.4}$$

Note the difference between the sine-Gordon for the magnetic-field coupling, which includes the $\sin 2\Phi(z)$ term in contrast to $\sin\Phi(z)$ term in the above expression. As a result, the solution of 3.3.4 that minimizes 3.3.1. (Kutnjak-Urbanc, 1993a) is a one-dimensional ***2π-multisoliton lattice***

$$\cos\Phi_o = 2\frac{cn^2(u,k)}{dn^2(u,k)} - 1 \tag{3.3.5}$$

This is in contrast to the π-soliton lattice (Eq.3.2.9) we have met in the case of the quadratic diamagnetic coupling. An example of this electric field induced 2π -multisoliton lattice is shown in Fig.3.3.2.

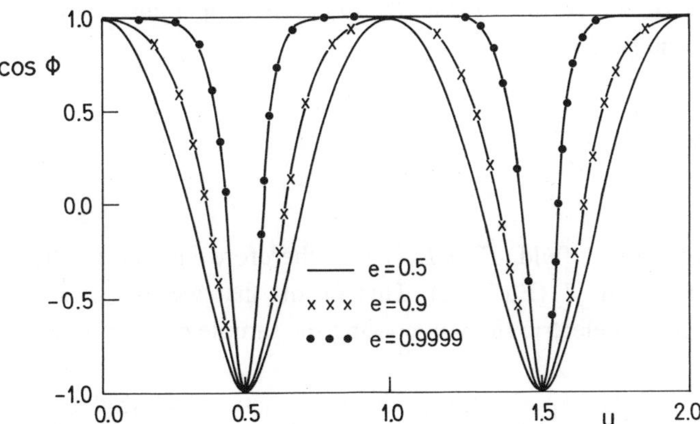

Fig.3.3.2. The projection of the tilt vector $\xi(z)$ on the x-axis for various values of a DC electric field, which is applied in the y-direction. Here, $e = E/E_c$ and the critical field is defined in the Eq.3.3.7. (after Kutnjak-Urbanc, 1990).

In the expression 3.3.5., $cn(u,k)$ is the Jacobi elliptic cosine and $dn(u,k)$ is the Jacobi delta amplitude of the reduced coordinate $u = z/(\xi_E k)$ (Byrd and

Friedman, 1954). $\xi_E = \sqrt{K_3\theta^2/EP}$ is the electric coherence length and the modulus k is determined by

$$k = \sqrt{\frac{E}{E_c}}E(k) \tag{3.3.6}$$

Here, E_c is the critical electric field for the unwinding of the *smectic* $-C^*$ phase, given by

$$E_c = \frac{\pi^4}{4}\cdot\frac{K_3\theta^2}{P}\cdot\frac{1}{p_o^2} \tag{3.3.7}$$

Because the period of the helix p_o is temperature independent within this simple free-energy expansion, and both θ and P have the same temperature dependence of the form $\theta \propto P \propto \sqrt{T_c - T}$, the temperature dependence of the critical electric field is

$$E_c \propto \sqrt{T_c - T} \tag{3.3.8}$$

The line of critical field $E_c = E_c(T)$ is therefore similar to the behavior of $T_c = T_c(E)$ close to T_c (Eq.3.3.2). This means that we expect a monotonously increasing critical electric field on cooling the sample deep into the *smectic* $-C^*$ phase.

Again, the experiments are in contradiction with this theory. A number of experiments, performed by different groups on different substances (Takezoe et al., 1978; Kondo et al., 1983; Takezoe et al., 1984; Rozanski and Kuczynski, 1984; Dumrongrattana and Huang, 1986c; Ezcurra et al., 1989; Kutnjak, 1991; Levstik et al., 1991a) clearly show ***reentrant phenomena***, as presented in Fig.3.3.3. This reentrance is characteristic for the ferroelectric materials with a non-monotonous temperature dependence of the helical pitch. According to Eq.3.3.7, one expects that it should be reflected in the temperature dependence of the critical electric field.

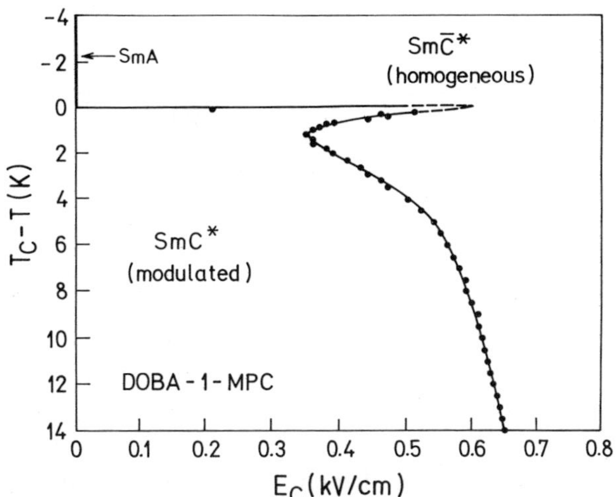

Fig.3.3.3. (E,T) phase diagram of DOBA-1-MPC in a DC electric field applied parallel to the smectic layers (Kutnjak, 1991).

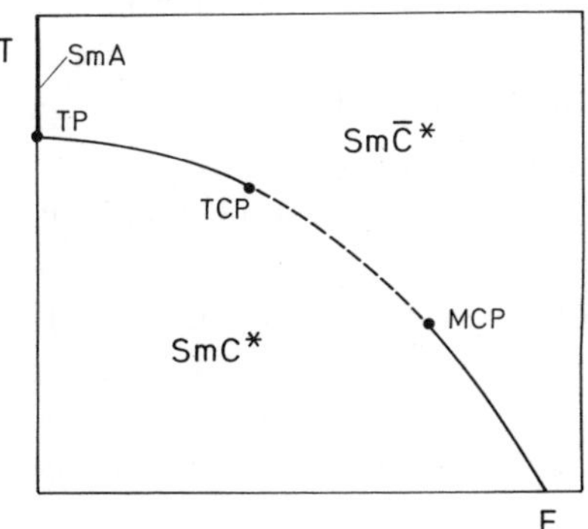

Fig.3.3.4. The (E,T) phase diagram of a ferroelectric liquid crystal in a transverse DC electric field, as predicted by a simple Indenbom-Pikin model. TCP denotes the tricritical point, whereas MCP denotes the multicritical point. At MCP, the phase transition changes from the first order to the continuous, soliton-like. Solid lines represent the second order and dashed lines represent the first order phase transition boundary. Compare to Fig.3.3.1.

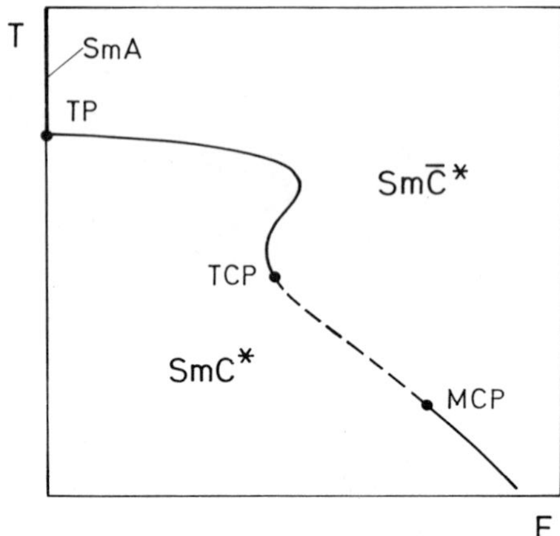

Fig.3.3.5. The (E,T) phase diagram as reproduced by the generalized Landau model (for details, see Kutnjak-Urbanc, 1993a). The drawing is purely schematic.

Similarly to the magnetic field diagram, the (E,T) phase diagram was extensively studied by Benguigui and Jacobs (1994) and by Kutnjak-Urbanc (1993). The theoretical considerations of both groups led to a semi-quantitative explanation of the experimental data. Generic forms of the (E,T) phase diagrams, as predicted by the simple Indenbom-Pikin and the generalized model (Kutnjak-Urbanc, 1993), are shown in Fig.3.3.4. and 3.3.5. A good qualitative agreement is observed between the experiment and the generalized model, first introduced by Žekš (1984).

There are other salient features of the (E,T) phase diagram, like for example the multicritical point on the $E_c(T)$ line, first introduced by Schaub and Mukamel (1983) or amplitude solitons beyond the CAA approximations, that have been extensively studied by M.Yamashita (1992). Similarly to the case of a magnetic field, these phenomena are expected to be observable in a temperature interval of the order of 10 mK close to T_c and therefore seem to be beyond the limit of the available experimental resolution for the known materials.

Note 3.3.1. Linear and quadratic coupling to the external electric field:
The soliton deformation of the ferroelectric helix in the presence of both linear and dielectric coupling to the external electric field has been discussed by Dmitrienko and

Belyakov (1980), Hudak (1983) and Kutnjak-Urbanc (1993a). The coupling term is

$$g_E = -\vec{P}\vec{E} - \tfrac{1}{2}\Delta\varepsilon\left(\vec{n}\vec{E}\right)^2$$

In the CAA approximation the phase profile can be obtained from the double sine-Gordon equation(Kutnjak-Urbanc, 1993a)

$$\tilde{\theta}^2 \frac{d^2\Phi}{dz^2} - \frac{E}{E^*}\tilde{\theta}\sin\Phi + \tfrac{1}{2}\Delta\tilde{\theta}^2\left(\frac{E}{E^*}\right)^2 \cdot \sin 2\Phi = 0$$

Here, $\tilde{\theta}$, E^* and Δ are dimensionless quantities (Kutnjak-Urbanc, 1993a). The shape of the phase profile $\Phi(z)$ depends on the relative strengths of both interactions. When the linear term is dominating, we obtain 2π soliton lattice. By increasing the strength of the quadratic coupling, each 2π domain splits into two π domains. These different regimes can be obtained, for example by varying the frequency of an external electric field. At low frequencies, the linear term is dominating, whereas at large frequencies the spontaneous polarization cannot follow the field and the quadratic coupling term is important, leading to π -soliton lattice.

3.4. Phason Dynamics in the π-Multisoliton Lattice

When an external magnetic field is applied in a direction perpendicular to the helical axis of a ferroelectric liquid crystal, it distorts the helical arrangement of liquid crystalline molecules and induces for fields lower than the critical field H_c a so-called soliton-like structure. This distortion of the helical structure breaks the continuous helical symmetry of the unperturbed $smectic - C^*$ phase, as we have shown in the Introduction to this Chapter.

This symmetry breaking by external fields has important consequences for the spectrum of the order parameter excitations in ferroelectric liquid crystals. Here we shall discuss the magnetic-field effects on the phason excitation spectrum only, although similar phenomena can be expected for the behavior of the soft and amplitudon excitations as well. Many consequences of the symmetry breaking by external fields can be understood by using symmetry arguments and similarities to other systems. For example, in the unperturbed *smectic* $-C^*$ phase, the phason excitations, which propagate along the helical axis exhibit a plane-wave behavior with a gapless and parabolic dispersion relation. Phason excitations are here just twist-bend-like elastic distortion waves, superposed onto the helical structure. When such an excitation propagates along the helix, it "sees" a smooth and uniformly twisted structure. All points of a smooth and undistorted helix are thus equivalent with respect to such an excitation, which results in the plane-wave nature of phason propagation in the unperturbed helix. However, when such a phason propagates in a magnetic-field distorted structure, it "sees" a regular and periodic array of π-soliton walls, resulting from the magnetic field induced breaking of the helical symmetry. This soliton-wall lattice thus results in a periodic perturbation of the potential, experienced by the phason, as it propagates along the distorted helix. The dynamics of a phason excitation in a soliton-like distorted *smectic* $-C^*$ phase is thus analogous to the motion of a particle in an external periodic potential, which is a well understood problem. As a result of the presence of the periodic potential, the concept of the Brillouin zone (BZ) has to be introduced, and the eigenfunctions of the phason excitations obtain the Bloch form. Perhaps the most striking consequence of the symmetry breaking by the external fields is the appearance of band gaps $G(q)$ in the excitation spectrum $\tau^{-1}(q)$. As shown schematically in Fig.3.4.1. the excitation spectrum becomes a periodic function in the reciprocal space and is similar to the energy spectrum of a particle in a periodic potential, Fig.3.4.2.

The influence of an external magnetic field on the phason dispersion relation can be analytically calculated in the constant amplitude approximation, thus neglecting the amplitudon contribution to the nonequilibrium free energy. The phase dependent part of the free-energy is

$$g(z) = -\Lambda\theta^2 \frac{\partial\Phi(z)}{\partial z} + \tfrac{1}{2} K_3\theta^2 \left(\frac{\partial\Phi(z)}{\partial z}\right)^2 - \tfrac{1}{2}\Delta\chi H^2\theta^2 \sin^2\Phi(z) \qquad (3.4.1)$$

Here the magnetic field $\vec{H} = (0, H, 0)$ is applied perpendicularly to the helical axis, which is in the z-direction. We have already seen in Section 3.2. that the minimization of the free-energy (Eq.3.4.1) leads to the sine-Gordon equation for the equilibrium phase profiles of the order parameter $\vec{\xi}$, which have been shown to have Jacobi elliptic forms.

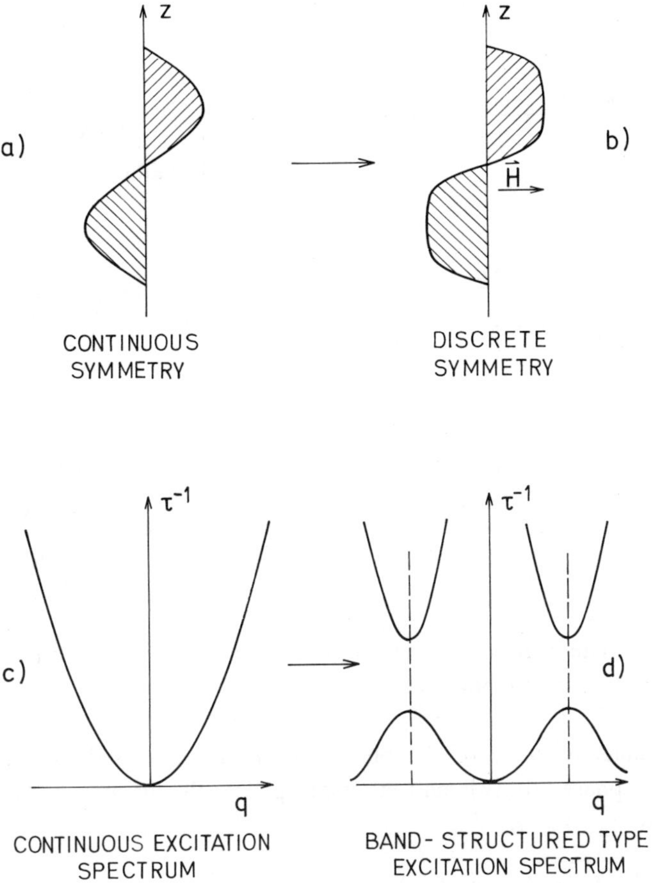

Fig.3.4.1. The interplay between the structures, their symmetries and the excitation spectrum. (a) The unperturbed, helicoidally modulated $smectic-C^*$ structure. By the application of a transverse magnetic field, this "smooth" structure is transformed into the π-soliton lattice, shown in (b). The consequences for the phason excitation spectrum are shown in (c) and (d). (c) The relaxation rate τ^{-1} in the unperturbed $smectic-C^*$ phase parabolically increases with increasing wave-vector. (d) In the soliton-like distorted structure we have Brillouin zones and a band gap appears between the acoustic-like and optic-like branches.

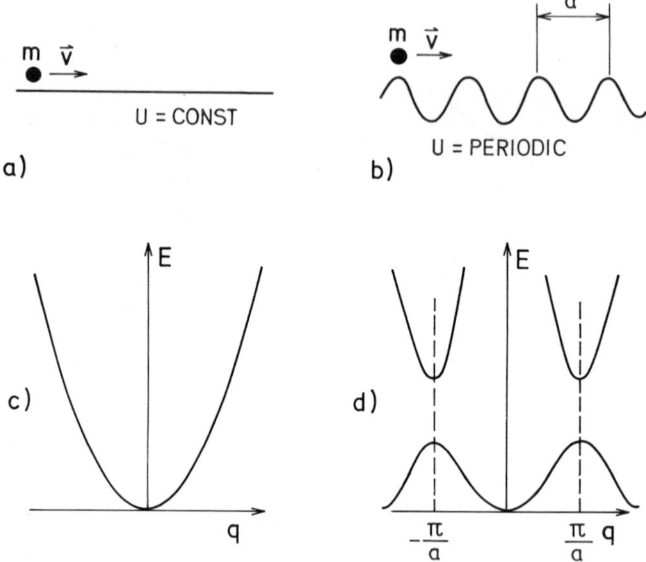

Fig.3.4.2. The energy spectrum of a particle, propagating in: (a, c) constant potential; (b, d) periodic potential.

The instantaneous value of the phase $\Phi(z,t)$ is expressed as a sum of the equilibrium phase $\Phi_o(z)$ and a small phase fluctuation $\delta\Phi(z,t)$ out of equilibrium:

$$\Phi(z,t) = \Phi_o(z) + \delta\Phi(z,t) \tag{3.4.2}$$

The dynamics of fluctuations is analyzed within the Landau-Khalatnikov dynamical equations. This leads to an overdamped form of the phason excitations

$$\delta\Phi(z,t) = \Psi(z) \cdot e^{-t/\tau(H)} \tag{3.4.3}$$

where $\tau(H)$ is the relaxation time of the phason. The form of the eigenfunction $\Psi(z)$, which satisfies the Landau-Khalatnikov dynamical equations is governed by the Lame's equation of order one:

$$\frac{d^2\Psi}{du^2}+\left[h_H-2k^2sn^2(u,k)\right]\Psi=0 \qquad (3.4.4)$$

Here $h_H=k^2\{\xi_H^2\gamma/[\tau(H)K_3]+1\}$ is the eigenvalue of the Lame's equation, γ is the rotational viscosity, $\xi_H=\sqrt{K_3/(|\Delta\chi|H^2)}$ is the magnetic coherence length and $sn(u,k)$ is the Jacobian sine amplitude of the argument $u=z/(\xi_H k)$ and the modulus k has been defined before (Eq.3.2.10).

The Lame's equation of order one is exactly equivalent to the Schrodinger equation describing the propagation of a particle in a periodic soliton-like potential

$$V(u)=-2k^2sn^2(u,k) \qquad (3.4.5)$$

The magnitude of this periodic potential is equal to $-2k^2$ and is in the limit of small fields proportional to H^2. Thus for $H\to 0$ the periodic potential vanishes, $V\to 0$, and we are in the plane-wave regime. For $H\neq 0$ the situation is substantially different. We see that in this case a small periodic perturbation appears in the equation of motion (Eq.3.4.4) which is similar to the well-known motion of an electron in a periodic potential. The emergence of this periodic potential for phason propagation in a distorted helix can be understood by considering that a phason represents a twist-bend like excitation. For zero field and continuous helical symmetry, all points on the helix are equivalent to each other and the phason "sees" a uniformly twisted structure, which is equivalent to the uniform structure. This results in a plane-wave, monochromatic nature of phason propagation and gapless dispersion relation. On the other hand, a phason propagating in a soliton lattice "sees" a periodic perturbation, which is a direct result of the presence of domain walls.

Following the analogy between the motion of the electron in a periodic potential and the phason in the soliton lattice, we can expect that the presence of an external magnetic field would have the following consequences for the phason dynamics:

i) One has to introduce the concept of a Brillouin zone. The width of the Brillouin zone (BZ) depends on the periodicity of the periodic potential, i.e. the period of the structure. This is $p_\circ/2$ and the first BZ is in the interval $(-q_c, q_c)$.

ii) Phason eigenfunctions will change from a monochromatic plane wave, characterized by a single wave-vector into to a Bloch wave, which is characterized by a set of wave-vectors.

iii) The phason excitation spectrum, which is continuous and gapless in zero field, will show for $H \neq 0$ a band-like structure with forbidden frequency gaps.

These phenomena are indeed reflected in the form of eigensolutions and the excitation spectrum of the Lame's equation of order one, which have been known since the work of Ch.Hermite in 1872 (Whittaker and Watson, 1935; Sutherland, 1973; Yamashita et al., 1981; Fan et al., 1970; Parsons and Hayes, 1974). The eigensolutions are

$$\Psi(u) \propto \frac{H(u-u_o)}{\Theta(u)} e^{-uZ(u_o)} \qquad (3.4.6)$$

where $H(u)$, $\Theta(u)$ and $Z(u)$ denote Jacobi's eta, theta and zeta functions, respectively (see, for example Byrd and Friedman, 1954). The relaxation rate of the phason $\tau^{-1}(H)$ is determined by the eigenvalue

$$\tau^{-1}(H, u_o) = \frac{\Delta\chi}{\gamma} \frac{H^2}{k^2} dn^2(u_o) \qquad (3.4.7)$$

Here the parameter u_o determines the wave-vector $\vec{q} = (0,0,q)$ of the phase excitation and shall be defined later in the text. If we look at the expressions 3.4.6. and 3.4.7., we realize that the expression for the eigenvalue represents a dispersion relation, or, equivalently the phason excitation spectrum. Here, the wave-vector, which is ascribed to a particular eigensolution, is determined by the transformation properties of the eigensolutions (Eq.3.4.6.). These have indeed a Bloch form and should therefore transform as

$$\Psi_q(\vec{r} + \vec{R}) = \Psi_q(\vec{r}) e^{i\vec{q}\vec{R}} \qquad (3.4.8)$$

when the system is translated over a lattice vector $\vec{R}$. This lattice vector in our case equals to a half-period of the helix, $|\vec{R}| = p_{\circ}/2$.

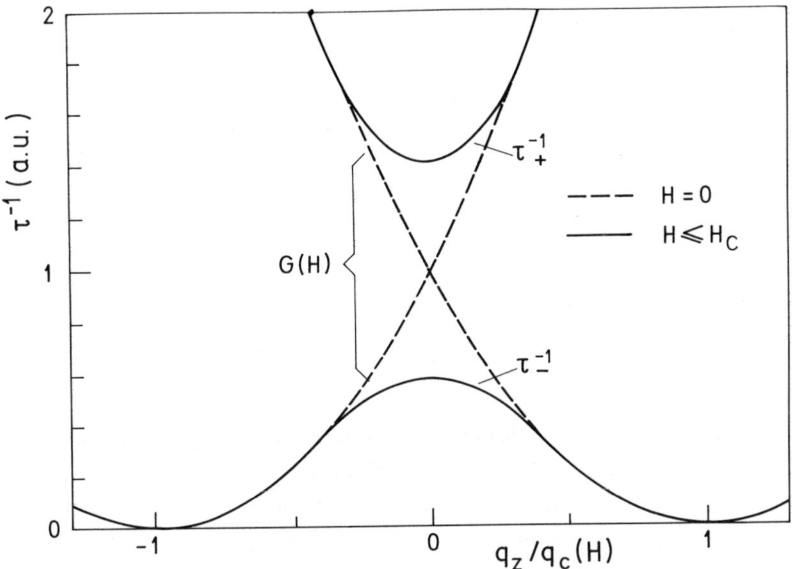

Fig.3.4.3. Splitting of the spectrum of the phason excitations in the presence of an external magnetic field. The dispersion relation splits into an optic-like band $\tau_+^{-1}(q)$ and an acoustic like band $\tau_-^{-1}(q)$ of excitations.

The analytic calculation of the excitation spectrum shows a surprising result, because ***only one forbidden gap*** *G(H)* appears in the phason spectrum, separating the so-called ***optic-like branch*** from the ***acoustic-like*** excitation branch. The structure of the spectrum is shown schematically in Fig.3.4.3. and it corresponds to the wave-vectors in the laboratory system. This single-gap in the excitation spectrum is in sharp contrast to the usually large number of forbidden gaps, which appear for example in the band theory of solids. The reason for this is the form of the periodic potential term $V(u) \propto sn^2(u)$, which was discussed in the early work of Landau and Lifshitz (1958) and later by Sutherland (1973) and by Cowley and Bruce (1978).

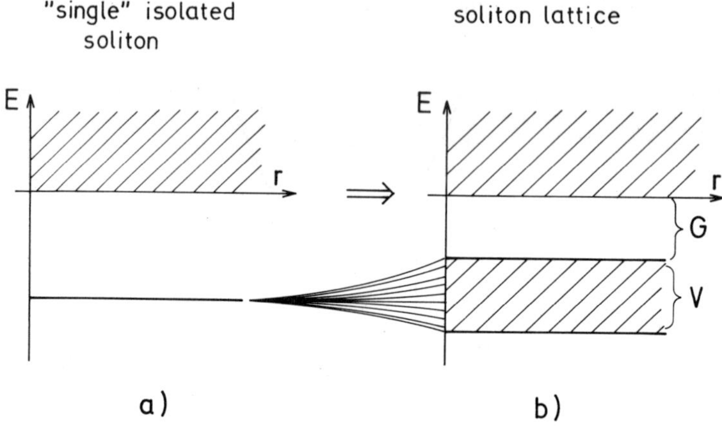

Fig.3.4.4. Phason spectrum of a soliton lattice, as explained in the "tight" binding scheme. (a) Energy level of a π-soliton lattice, where the soliton walls are far away and nearly isolated from each other. The soliton-like potential of a single soliton wall (Eq.3.4.5.) has only one bound state and a continuum of unbound levels. (b) When the soliton walls are brought together, the overlapping of bound states creates a "valence-like" band of states and a "conductive-like" band. There is only one forbidden gap in between, because there is only one bound state.

This potential is peculiar in the sense that it exhibits a single bound state and a semi-continuum of free states, separated by a single gap. If we consider the evolution of the band-structure in the tight-binding approximation, we see that the bound state evolves into the "valence" band, whereas the semi-continuum of the unbound states forms the "conduction" band, as shown in Fig.3.4.4. Nevertheless, there is always only one forbidden gap in between.

The width of the Brillouin zone is determined by the period of the potential term $V(u)$ which is half of the helical period, $p(H)/2$ and is field-dependent in view of the Equation (3.2.12). This means that the reciprocal lattice vector equals to

$$|\vec{K}| = \frac{2\pi}{(p(H)/2)} = \frac{4\pi}{p(H)} \tag{3.4.9}$$

and the corresponding Brillouin zone is $(-|K|/2,+|K|/2)\equiv(-q_c,+q_c)$. It should be stressed that the phason excitation, as introduced by Eq.(3.4.2) implies the use of a helical reference frame, because of the equilibrium phase Φ_o. As a consequence, the excitation with a wave-vector $\vec{q}$ in a helical reference frame is observable as an excitation with a wave-vector $\vec{q}\pm\vec{q}_c$ in the laboratory frame. This has the consequence that the Brillouin zone is shifted by $\pm\vec{q}_c$ and the zone boundaries and the band gap $G(H)$ are situated at $\vec{q}=0$ in the laboratory system, as shown in Figure 3.4.3.

Let us now analyze the eigenvalues and the corresponding wave-vectors in detail. The Bloch form of the eigensolutions (Eq.3.4.6.) implies that the value of the Jacobi zeta function $Z(u_\circ)$ is purely imaginary. This happens when the corresponding unknown parameter $u_\circ$ equals either

$$u_\circ=\begin{cases}0 & \text{..... optic - like branch}\\ K(k) & \text{..... acoustic - like branch}\end{cases} \tag{3.4.10}$$

The analysis of the eigenvalues gives for the ***optic-like branch*** of excitations the relaxation rates

$$\tau_+^{-1}(H)=\frac{\Delta\chi}{\gamma}\cdot\frac{H^2}{k^2}\cdot\frac{dn^2(y_o,k)}{cn^2(y_o,k)} \tag{3.4.11}$$

whereas the corresponding wave-vector is

$$q=\pm q_c\left\{1\pm\frac{2K(k)}{\pi}\left[Z(y_o,k')-\frac{sn(y_o,k')dn(y_o,k')}{cn(y_o,k')}+\frac{y_o\pi}{2K(k)K'(k)}\right]\right\} \tag{3.4.12}$$

Here $k'=\sqrt{1-k^2}$ is the complementary modulus, $K'(k)$ is the associated complete elliptic integral of the first kind and $y_o\in(-K',K')$. Similarly, we obtain for the ***acoustic-like branch*** of excitations the following relaxation rates:

$$\tau_{-}^{-1}(H)=\frac{\Delta\chi}{\gamma}\frac{1-k^2}{k^2}H^2\frac{cn^2(y_o,k')}{dn^2(y_o,k')} \tag{3.4.13}$$

and the wave-vectors

$$q=\pm q_c\left\{1\pm\frac{2K(k)}{\pi}\left[Z(y_o,k)-k'^2\frac{sn(y_o,k')cn(y_o,k')}{dn(y_o,k')}+\frac{\pi}{2K(k)K'(k)}y_o\right]\right\} \tag{3.4.14}$$

with $y_o \in (-K', K')$.

The analysis of the eigenfunctions of the Lame's equation shows that there are three distinct eigenfunctions with three distinct eigenvalues. At the edges of the Brillouin zone there are two real solutions which have the form of the Jacobian sine and cosine amplitudes, respectively. Their relaxation rates show an opposite field dependence: Whereas the relaxation rate of the first eigensolution $\Psi_1(u)$ increases with increasing field, the relaxation rate of the second eigensolution $\Psi_2(u)$ decreases with increasing field, as shown in Fig.3.4.5:

$$\Psi_1(u)=sn(u,k) \qquad \tau_{+}^{-1}(H)=\frac{\Delta\chi}{\gamma}\frac{1}{k^2}H^2 \tag{3.4.15}$$

$$\Psi_2(u)=cn(u,k) \qquad \tau_{-}^{-1}(H)=\frac{\Delta\chi}{\gamma}\frac{1-k^2}{k^2}H^2 \tag{3.4.16}$$

The magnetic-field induced gap, which is just the difference of the two relaxation rates is thus proportional to the square of the applied field:

$$G(H)=\tau_{+}^{-1}(H)-\tau_{-}^{-1}(H)=\frac{\Delta\chi}{\gamma}H^2 \tag{3.4.17}$$

The third distinct eigensolution is obtained in the center of the Brillouin zone. It has the form of a Jacobian delta amplitude $dn(u,k)$ and has a zero relaxation rate for all fields below H_C :

$$\Psi_3(u) = dn(u,k) \qquad \tau^{-1}(H) = 0 \tag{3.4.18}$$

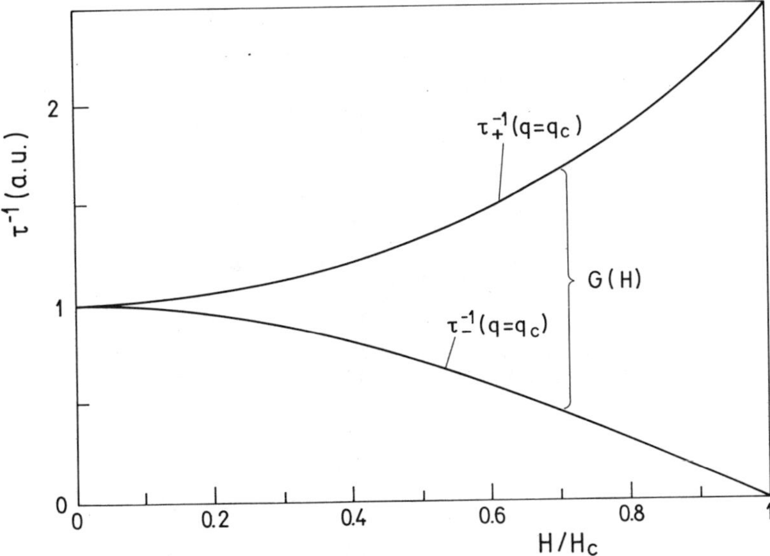

Fig.3.4.5. Magnetic field dependence of the relaxation rates of the optic-like τ_+^{-1} phason and the acoustic-like phason τ_-^{-1} at the edges of the Brillouin zone, as calculated from Eqs.(3.4.15) and (3.4.16).

It is interesting to discuss the physical interpretation of these eigenfunctions $\Psi_1(u)$, $\Psi_2(u)$ and Ψ_3, which are shown in Fig.3.4.6. The first one, $\Psi_1(u)$ (Eq.(3.4.13)) has the nodes in the center of the π-soliton walls and represents fluctuations of the shape of the domain walls, i.e. oscillations of the size of the commensurate region between the solitons. The second eigensolution $\Psi_2(u)$ has the nodes in the middle of the two neighboring domain walls and represents coherent, out-of-phase motion of the domain walls, i.e. an oscillation of the soliton lattice. The third eigensolution, $\Psi_3(u)$, is the zero-frequency, symmetry restoring Goldstone mode and represents the sliding of the helix as a whole along the z-direction.

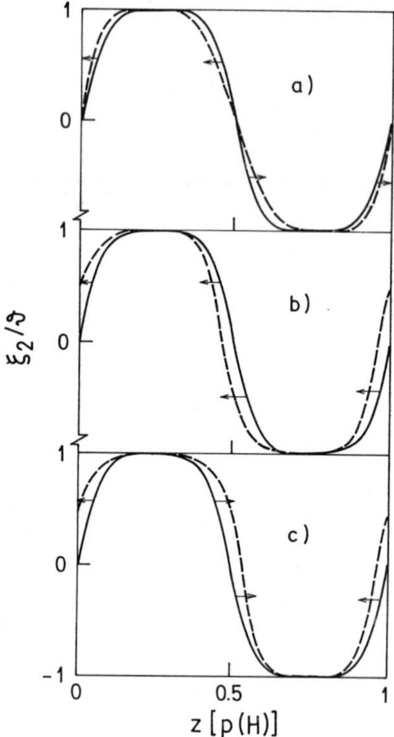

Fig.3.4.6. The effect of the phason excitations on the order parameter ξ_y as calculated near H_c. The solid line represents the equilibrium phase profile $\Phi_o(z)$ whereas the effects of fluctuations are shown by the dashed line. (a) Optic-like excitation at the edge of the BZ, representing the fluctuation of the shape of domain walls, i.e. the oscillations of the size of commensurate regions. (b) The Goldstone mode that represents sliding of the helix as a whole. (c) The acoustic-like fluctuation at the edge of the BZ, representing "breathing" of the neighboring domain walls (oscillations of the soliton lattice).

The first observation of the magnetic-field induced splitting of the phason excitation was reported in a quasielastic light scattering experiment in high magnetic fields (Muševič et al., 1992, 1994). The observed dispersions of the acoustic phason branch in a mixture of chiral and racemic 4-(2'-methylbutyl)phenyl 4'-n-octylbiphenyl-4-carboxylate (CE-8) at H=0, H=6.0T, H=6.65T and H=8.1T are shown in Fig.3.4.7(a), (b), (c) and (d), respectively.

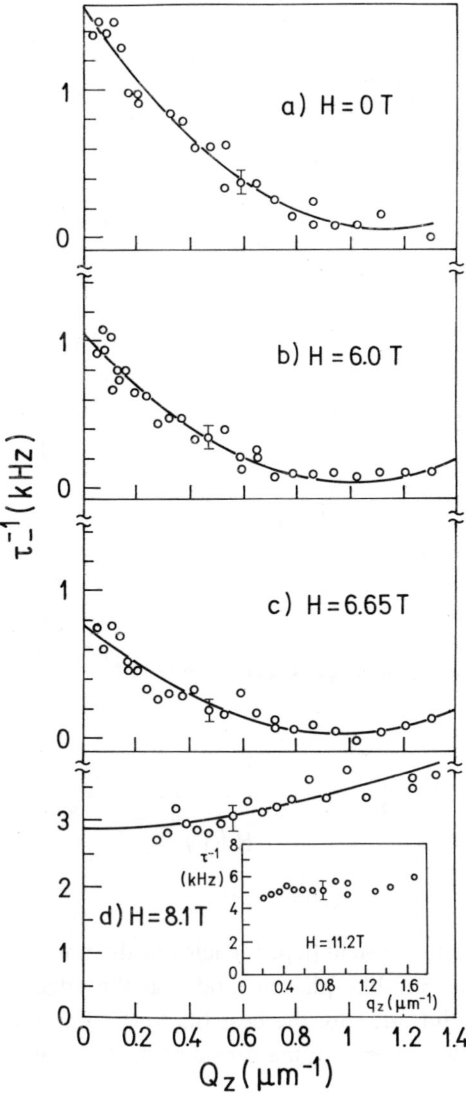

Fig.3.4.7. The dispersion of the acoustic-like phason branch in a mixture of chiral and racemic CE-8, as measured at different magnetic fields, 3K below T_c. (Muševič et al., 1994.). The critical magnetic field is 8T and the solid lines are the best parabolic fits.

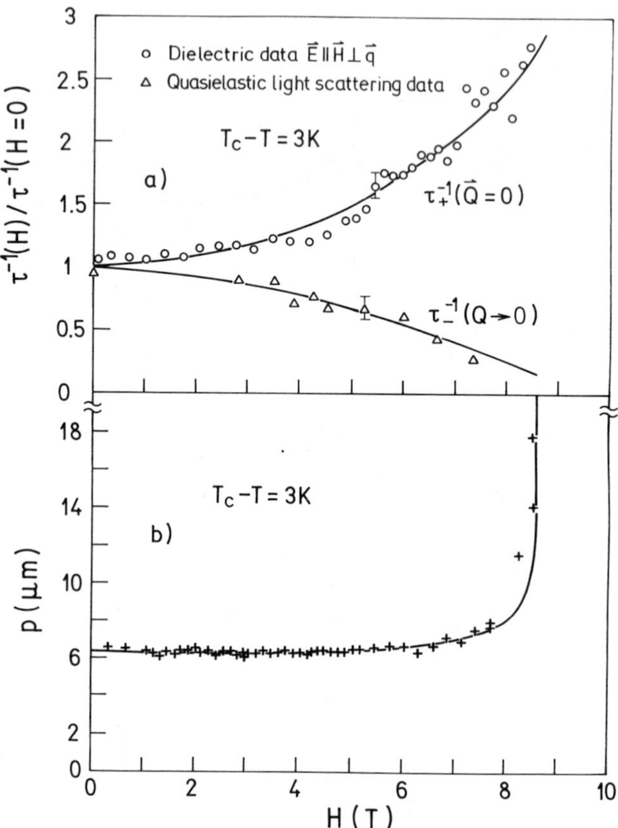

Fig.3.4.8. (a) Magnetic field dependencies of the normalized relaxation rates of the acoustic and optic-like phason modes at the edge of the BZ (Muševič 1992, 1994). The solid lines are the best fits to the Eqs.3.4.15. and 3.4.16. (b) The magnetic field dependence of the period of the helix in the same substance.

The critical magnetic field in this material was approximately 8T, and slightly depended on the sample, used in the experiment. One can observe the decrease of the phason relaxation rates at small wave-vectors, which is characteristic of the acoustic-like phason modes at the edge of the BZ. At the same time, the dispersion is slightly shifted towards smaller wave-vectors, indicating the shrinking of the BZ. Just above the critical field, see Fig.3.4.7(d), the dispersion is centered at $q = 0$ and shifted to higher frequencies. This indicates a phase transition from the modulated phase, where the zero frequency phason is at a finite wave-vector $q \neq 0$, to the homogeneous phase, where the dispersion is centered at $q = 0$. In the experiment, the magnetic field dependence of the optic-like phason modes at the edge of the BZ could not be determined by the quasielastic light scattering because of the small scattering intensity from these modes. Instead dielectric spectroscopy was used to test the optic-like phason modes at the edge of the zone.

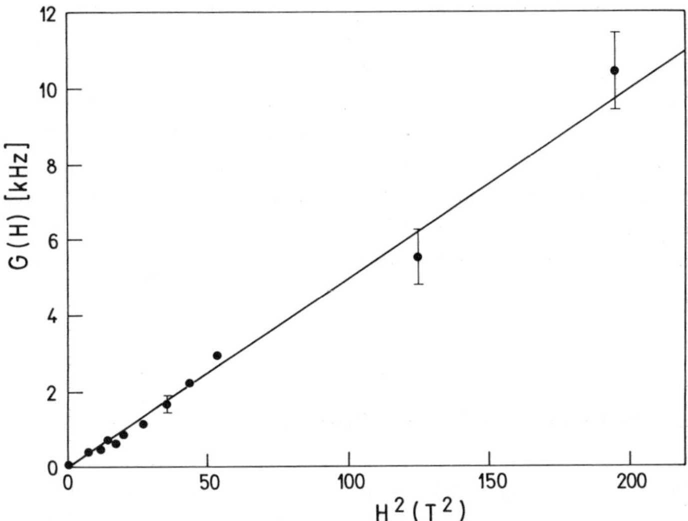

Fig.3.4.9. Magnetic field dependence of the phason band gap at the edge of the BZ (Muševič, 1994)

The resulting magnetic-field dependencies of the phason relaxation rates together with a fit to the theoretical predictions are shown in Fig.3.4.8. and clearly show the magnetic field induced splitting of the phason spectrum. The resulting gap *G(H)* is shown in Fig.3.4.9. up to the highest available field of 14T. It shows an excellent agreement with theoretical predictions. A particularly interesting feature is the disappearance of the zero-frequency Goldstone mode beyond the critical magnetic field H_c. This will be discussed in Section 3.6.

Note 3.4.1. Viscosity, K_{33} and diamagnetic anisotropy of CE-8:
The light scattering experiment in a magnetic field can give us values of important material parameters of liquid crystals. In the particular experiment (Muševič et al., 1992) a mixture of pure and racemic CE-8 has been used in order to decrease the critical magnetic field for the unwinding of the helix. In a 35%/65% of pure and racemic CE-8 the measured helical period was $p_\circ = 5.7\mu m$ at zero field and the critical magnetic field was $H_c = 8T$. From the measurements of phason dispersion, the authors have determined the phason diffusivity $K_{33}/\gamma = 1.1\times10^3 \mu m^2 s^{-1}$, whereas from the magnetic field dependence of the phason gap they have obtained $\Delta\chi/(\mu_\circ\gamma) = 48T^{-2}s^{-1}$. In an independent experiment on thin cells, Škarabot et al.(1999a) have determined the rotational viscosity of CE-8, $\gamma = 0.011 kgm^{-1}s^{-1}$. The twist elastic constant of CE-8 is therefore $K_{33} = 1.2\times10^{-11}N$, which is similar to the value determined by Kutnjak (1991) for DOBAMBC. Further, we determine from $\Delta\chi/(\mu_\circ\gamma) = 48T^{-2}s^{-1}$ the diamagnetic anisotropy of CE-8: $\Delta\chi(CE-8) = 0.63\times10^{-6}$. This can be compared to the value of diamagnetic anisotropy, obtained from the independent data on the critical magnetic field and the helical period. Following *Note 3.2.1.*, we obtain $\Delta\chi(CE-8) = 0.67\times10^{-6}$, which is in excellent agreement with light scattering data.

3.5. Phasons in a 2π- Multisoliton Lattice

Unlike the external magnetic field, a DC electric field applied in a direction perpendicular to the helical axis of the ferroelectric $smectic-C^*$ phase, creates a ***2π-soliton lattice***. Here, almost homogeneously aligned regions of the $smectic-C^*$ are separated by the 2π soliton walls, as shown in Fig.3.5.1. When we move along the layer normal, the phase of the order parameter changes by 2π in the domain wall and is then practically constant in the regions of nearly homogeneously aligned molecules.

Similar to magnetic-field effects, we have here again symmetry breaking by an external field. As a consequence of this symmetry breaking, we expect that the originally continuous dispersion relation for the order parameter excitations would become periodic in the reciprocal space and that band gaps would appear at the edge of the Brillouin zone.

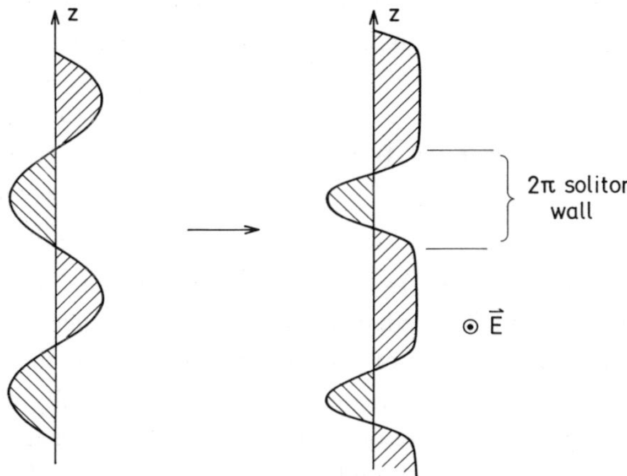

Fig.3.5.1. The application of a DC electric field distorts the originally "smoothly" modulated helical $smectic-C^*$ phase into a so-called 2π-soliton lattice.

Let us now calculate the phason dynamics in this 2π multisoliton lattice, which is induced by an external DC electric field in the $smectic-C^*$ phase. The nonequilibrium, phase-dependent part of the free-energy density, is in the CAA approximation

$$g(z) = -\Lambda\theta^2 \frac{\partial\Phi(z)}{\partial z} + \tfrac{1}{2} K_{33}\theta^2 \left(\frac{\partial\Phi(z)}{\partial z}\right)^2 - EP\cos\Phi(z) \qquad (3.5.1)$$

Here, a DC electric field $\vec{E} = (0, E, 0)$ is applied perpendicularly to the helical axis. We have already seen that the minimization of the free-energy (Eq.3.5.1) leads in the presence of an electric field to a 2π soliton lattice. The equilibrium phase profiles $\Phi_o(z)$ here satisfy the sine-Gordon equation and can be calculated analytically using Jacobi's elliptic functions (see Eq.3.3.5).

The dynamics of small phase fluctuations $\Phi(z,t) = \Phi_o(z) + \delta\Phi(z,t)$ around the equilibrium phase is governed by the Landau-Khalatnikov dynamical equations. It is easy to show from (Eq.3.5.1), using the Landau-Khalatnikov equations, that the phason excitations are again overdamped waves of the form

$\delta\Phi(z,t) = \Psi(z)e^{-t/\tau(E)}$. In analogy to the magnetic field case, these excitations satisfy a slightly modified equation of motion (Kutnjak-Urbanc, 1993)

$$\frac{d^2\Psi}{dv^2} + \left[h_E - 2k^2 sn^2(v,k)\right]\Psi = 0 \tag{3.5.2}$$

Here, $h_E = k^2\left[\xi_E^2\gamma/(\tau(E)K_{33}) + 1\right]$ is the eigenvalue, γ is the viscosity, $v = u - K(k)$ is modified reduced coordinate and other quantities have been defined before. The equation of motion is therefore again the Lame's equation of order one that we have met in the case of a magnetically distorted helix (Eq. 3.4.4.). The difference is here only in the reduced coordinate v, which is in the case of an electric field shifted for a quarter-period $K(k)$ of the elliptic functions, i.e. $v = u - K(k)$ and $u = z/(\xi_E k)$.

Following the formalism developed for the phason dynamics in a magnetically deformed helix, we find the general eigensolutions as

$$\Psi(v) \propto \frac{H(v - v_o)}{\Theta(v)} e^{-vZ(v_o)} \tag{3.5.3}$$

The eigenvalues, which determine the phason relaxation rates are

$$\tau^{-1}(E, v_o) = \frac{EP}{\gamma\theta^2} \cdot \frac{1}{k^2} dn^2(v_o) \tag{3.5.4}$$

The phason spectrum again consists of two dispersion branches, i.e. the acoustic-like and the optic-like branch, which are separated by a field dependent gap. The relaxation rates for the acoustic-like branch of phase excitations are obtained from the above equations (Eq.3.5.3. and 3.5.4.) as

$$\tau_-^{-1}(E) = \frac{EP}{\gamma\theta^2} \cdot \frac{1-k^2}{k^2} \cdot \frac{cn^2(y_o, k')}{dn^2(y_o, k')} \tag{3.5.5}$$

whereas the corresponding wave-vectors are

$$q = \pm q_c \left\{ \frac{1}{2} \pm i \frac{K(k)}{\pi} Z(K + iy_o) \right\} \tag{3.5.6}$$

Here $k' = \sqrt{1-k^2}$ is the complementary modulus, $K'(k)$ is the associated complete integral of the first kind and $y_o \in (-K', K')$. Similarly, the relaxation rates of the optic-like branch of excitations are

$$\tau_+^{-1}(E) = \frac{EP}{\gamma\theta^2} \cdot \frac{1}{k^2} \cdot \frac{dn^2(y_o, k)}{cn^2(y_o, k)} \tag{3.5.7}$$

whereas the corresponding wave-vectors are

$$q = \pm q_c \left\{ \frac{1}{2} \mp i \frac{K(k)}{\pi} Z(iy_o) \right\} \tag{3.5.8}$$

In spite of a great similarity between the phason dynamics in a magnetically-deformed and DC electric field-deformed $smectic-C^*$ phases, there is a fundamental difference, which is reflected in the width of a Brillouin zone (BZ). The width of the BZ of a $smectic-C^*$ phase in a transverse DC electric field is

$$\left[-\frac{q_c}{2}, +\frac{q_c}{2} \right] \tag{3.5.9}$$

and is therefore two times smaller than in the magnetically deformed $smectic-C^*$ phase. Formally this can be seen from the Eqs.(3.5.6) and (3.5.8) which define the wave-vectors in the first BZ. On the other hand, this can be understood by considering the periodicity of the periodic potential in the Lame's equation, as it is shown in Fig.3.5.2.

Figure 3.5.2(a) shows a π-multisoliton lattice, induced by an external magnetic field. The periodicity of the field-induced potential for the propagation of phason in the z-direction is determined by the separation between the

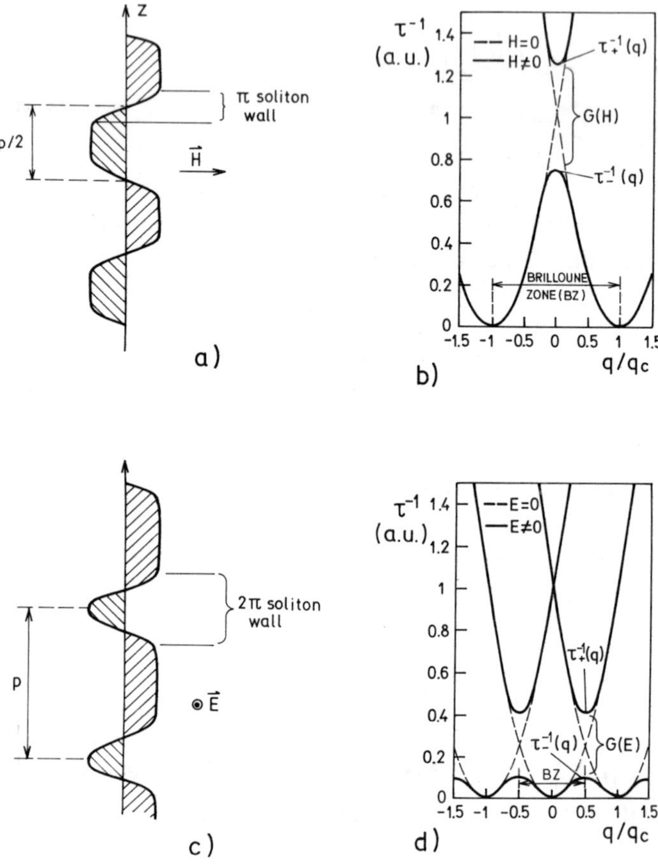

Fig.3.5.2. (a) The projection of a magnetically deformed helix on the (y,z) plane, with the magnetic field applied in the y-direction. The corresponding phason dispersion is shown in (b). (c) The projection of a DC electric field deformed helix, with an electric field applied in a direction out of the paper. The corresponding phason dispersion is shown in (d). Note the following important details: *(i)* The difference in translational periodicities of the magnetic and electric-filed distorted helices, (a) and (c). *(ii)* The resulting difference in the widths of the Brillouin zone, (b) and (d).

neighboring π-domain walls and equals $p_{\circ}/2$. On the other hand, the separation between the soliton walls in a 2π-multisoliton lattice is two times larger and equals to the helical period $p_{\circ}$, as shown in Fig.3.5.2(c). This means that the width of the corresponding Brillouin zone is in the 2π lattice two times smaller than in the π-soliton lattice. As a result, the phason dispersions in the reciprocal space are quite different, as shown in Figs 3.5.2(b) and (d), respectively. Whereas in the π-soliton lattice the band gaps appear at the center of the reciprocal space, these gaps are located at $\pm q_c/2$ in the case of a 2π-soliton lattice. This means that the phason gap in an electric field induced soliton lattice cannot be observed with dielectric spectroscopy, which couples to $\vec{q}=0$ modes. It can be observed only with the quasielastic light scattering spectroscopy.

Following the analogy to the phason dynamics in magnetic fields, we can expect that there would be again three distinct eigenfunctions at the edges and in the center of the BZ (Eqs.3.4.15,16,18). Similarly to the standing electron waves in a one dimensional periodic potential, which are observable for the wave-vectors at the edges of a Brillouin zone, these eigenfunctions should here represent standing phason waves. Mathematically, this is reflected in the fact that a phason eigenfunction (Eq.3.5.3), which is in general a complex Bloch function, obtains real values at these specific points. A detailed analysis indeed shows (Kutnjak-Urbanc, 1993) that there are standing phason waves at the edges and in the center of the BZ. Similarly to the magnetic case, the two modes at the edges of the BZ represent coherent fluctuations of the position and of the shape of domain walls, respectively. The third real eigensolution at the center of the BZ represents a pure translation of the 2π-soliton lattice and is thus the zero-frequency Goldstone mode. There is again a field dependent gap between the optic-like and the acoustic-like phason branches, which in this case increases linearly with the applied electric field.

$$G(E)=\tau_{+}^{-1}(E)-\tau_{-}^{-1}(E)=\frac{P}{\gamma\theta^{2}}E \tag{3.5.10}$$

Whereas the band gap (Eq.3.5.10) appears at a finite wave-vector $\pm q_c/2$, and is therefore observable only in a light scattering experiment, a non-monotonic behavior of the relaxation frequencies is obtained at $q=0$, as shown schematically in Fig.3.5.3. Let us remember that this point is accessible in a dielectric experiment.

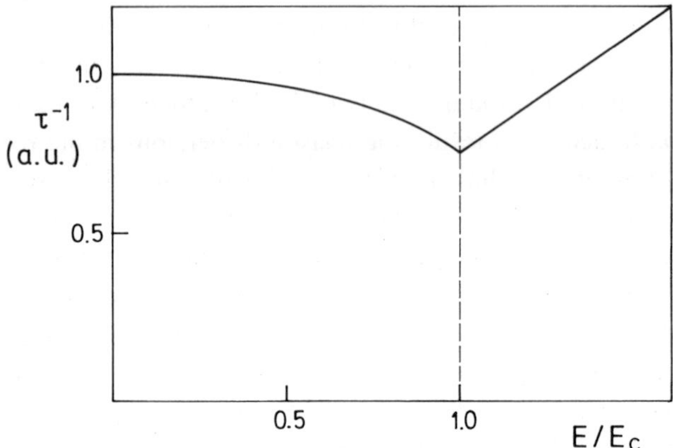

Fig.3.5.3. The electric-field-dependence of the phason relaxation rates at $q = 0$, as calculated from the Eq.3.5.4. The parameter $\nu_{\circ}$, which determines the wave-vector, was determined by considering the translational properties of the eigenfunctions (Eq.3.5.3), as discussed in Section 3.4.

There is a number of experiments reporting on the behavior of the phase modes in the $smectic - C^{*}$ phase in an electric field. Most of the dielectric experiments report monotonously increasing relaxation frequencies of the $\vec{q} = 0$ mode. This is expected only for fields larger than the critical electric field. An exception is here a linear-electrooptic response experiment by Pavel and Glogarova (1991a), who reported behavior reminiscent to the one predicted by theory.

The quasielastic light scattering experiment (Drevenšek et al., 1991) clearly shows the dispersion of the low-frequency, acoustic-like branch, which is practically not influenced by fields lower than the critical field, as shown in Fig.3.5.4(a). The experiment also shows a sudden jump of the observed phason dispersion to higher relaxation rates, when the electric field is larger than the critical. This jump is then followed by a gradual increase of the relaxation rates with increasing field, see Fig.3.5.4(b). These observations can be explained by the fact that below the critical field we observe a dominant, low-frequency dispersion branch. At the critical field this branch becomes unstable and disappears. Instead, the optical-like branch is observable at fields larger than the critical one. The relaxation rate of this branch should increase linearly with the magnitude of the external field, as indeed observed in the experiment. The experiment unfortunately failed to observe the predicted field-induced gap.

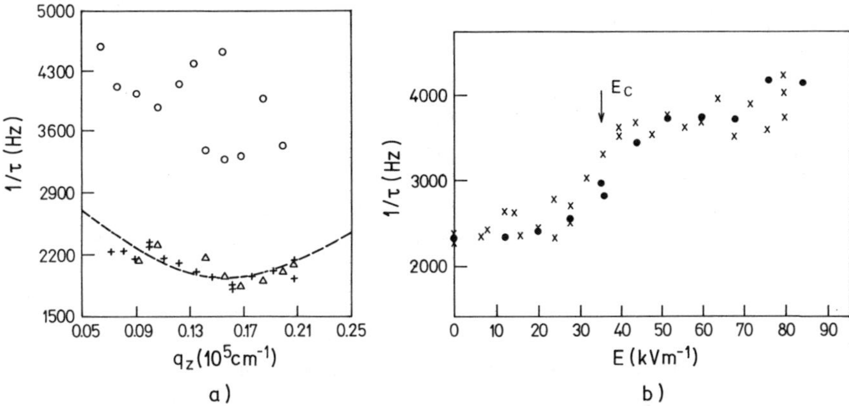

Fig.3.5.4. (a) Phason dispersion for fields, lower than the critical electric field, (symbols $+$ and Δ) and for a field far beyond the critical value, open circles. (b) Field-dependent phason relaxation rates at a finite wave-vector. The critical field is $35kVm^{-1}$. After Drevenšek et al., 1991.

3.6. Gapless Phason and Broken Symmetry

One of the most striking observations, which is related to the phason dynamics in a soliton lattice, is the existence of a zero frequency, symmetry restoring Goldstone mode. It has been shown analytically in Sections 3.4. and 3.5., that this mode exists in the unperturbed $smectic-C^*$, as well as in the soliton-like distorted $smectic-C^*$ phase. We have seen, however, that it does not exist in the unwound $smectic-\overline{C}^*$ phase, see Fig.3.6.1. The appearance of this mode is such a prominent physical phenomenon, that a question of particular interest naturally arises: Is there some fundamental reason, why a gapless, zero frequency phason mode should exist only for fields below the critical field?

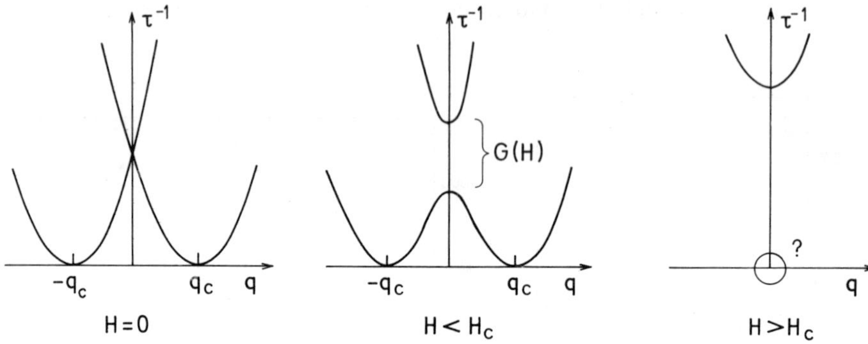

Fig.3.6.1. Phason dispersion in the (a) unperturbed $smectic - C^*$ phase, (b) field-induced, soliton-like deformed $smectic - C^*$ and (c) homogeneous and unwound $smectic - \overline{C}^*$ phase. Note that a zero frequency mode exists only for fields below the critical magnetic field.

The answer to this question has been given a long time ago by Goldstone, Salam and Weinberg (1962), who stated that a massless, zero frequency mode appears, whenever a ***continuous symmetry of a system is spontaneously broken*** (see also Forster, 1975; Bludman and Klein, 1963; Lange, 1966). In order to apply this theorem to our case, we have to consider the phase transition line, where a continuous symmetry of the high temperature phase is spontaneously broken. For a non-zero external field, this occurs at the phase transition from the $smectic - A$ phase into the tilted smectic phases (helical or homogeneous), see Fig.3.6.2.

Let us now discuss symmetry breaking in the *(H,T)* phase diagram of a ferroelectric liquid crystal. The symmetry of the $smectic - A$ phase in zero magnetic field is D_∞, i.e. we have a continuous rotational symmetry around the normal to the smectic layers. On the length scale of interest, this phase has as well a continuous translational symmetry in the z-direction. For zero external field, both the continuous rotational symmetry group D_∞ as well as the continuous translational symmetry of the $smectic - A$ phase are spontaneously broken at a $smectic - A$ - $smectic - C^*$ transition by the appearance of a helical structure. This is the basic reason for the existence of a zero-frequency symmetry restoring Goldstone mode in zero fields.

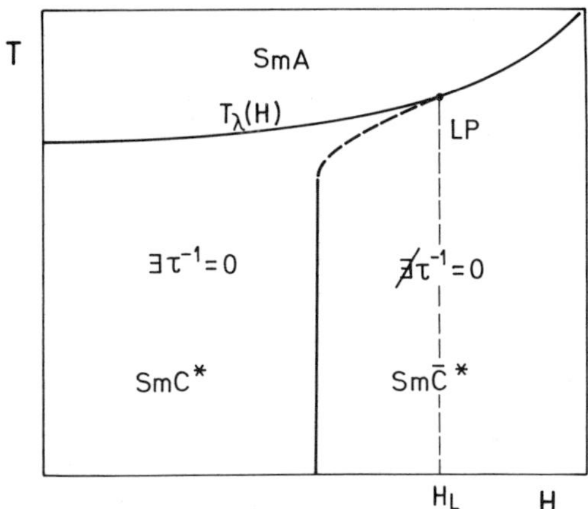

Fig.3.6.2. The regions in the *(H,T)* phase diagram of a ferroelectric liquid crystal in an external magnetic field, where a zero-frequency, symmetry restoring Goldstone mode exists.

In the presence of an external magnetic field applied perpendicularly to the helix, the symmetry of the $smectic-A$ phase is reduced to D_2. For fields below the Lifshitz field, the $smectic-A$ - $smectic-C^*$ phase transition is characterized by a spontaneous breaking of the discrete symmetry D_2 into C_2. However, at the same time the ***continuous translation symmetry*** of the $smectic-A$ phase is spontaneously broken into the ***discrete translational symmetry*** of the $smectic-C^*$, soliton-like lattice. This is the reason for the existence of the Goldstone mode in the soliton-like lattice in the presence of an external magnetic or electric field. Because the Goldstone mode is a symmetry restoring mode, it here emerges in a form of a ***sliding translational wave***, which tries to restore the broken translational symmetry of the $smectic-A$ phase. Finally, in a region beyond the Lifshitz field H_L, only the discrete point symmetry D_2 is spontaneously broken into C_2 at the $smectic-A$ - $smectic-\overline{C}^*$ phase transition whereas the translational symmetry remains unchanged. There is therefore no continuous symmetry breaking at this phase transition and a zero frequency Goldstone mode should not exist in the unwound $smectic-\overline{C}^*$. This is indeed consistent with experimental observations.

Chapter 4

Phase Transitions and Dynamics in Restricted Geometry

4.1. Introduction

Whereas the effects of symmetry breaking in ferroelectric liquid crystals by external electric or magnetic fields seem to be well understood, our understanding of the effects of confinement on the static and dynamic properties of liquid crystals is less complete. This is in particular true for the case of smectic phases of lower symmetry, confined to complex media like random porous matrices. Phase transitions in ferroelectric and antiferroelectric liquid crystals seem to be excellent candidates for these studies, because there has been a great progress in our understanding of these phases. It has been initiated not only by the discovery of the technological importance of these materials (Clark and Lagerwall, 1980, Clark and Lagerwall, 1984, Handschy and Clark 1984) but also by the variety of analogies to other systems.

The observations of liquid crystals in restricted geometries clearly show that their generally rather complex behavior is a consequence of the complex geometry. However, in most cases the degree of complexity of the constraining media (i.e. aerogels or other porous media) is so high that it does not enable one to identify the various sources of the complex response of the confined liquid crystal. It therefore seems reasonable to understand first a simple but extreme geometry like the case of two flat, solid surfaces separated by a very small distance. For nematic liquid crystals, phase transitions in this geometry have been extensively studied both from the experimental (Miyano, 1979; Tarczon and Miyano, 1980; van Sprang, 1983; Hsiung *et al.* 1986; Yokoyama, 1988; Sheng,1976; Allender *et al.* 1981; Sluckin and Poniewierski, 1985; Sluckin and Poniewierski 1990). In view of the applications of ferroelectric liquid crystals in flat panel displays, ferroelectric liquid crystals confined to planparallel plates separated by a micrometer distance have also been the subject of wide experimental and theoretical efforts (Clark and Lagerwall 1984; Pikin and Yoshino, 1981; Glogarova et al., 1983,1984a, b, c, 1993, 1995; Kondo *et al.*, 1982 and 1983; Kai *et al*, 1983; Yang *et al.*, 1991; Panarin *et al.*, 1994a, b). The aim of this Chapter is to discuss the effects of submicron confinement on the $smectic-A$ - $smectic-C^*$ phase transition and order parameter dynamics.

4.2. Phase Transitions in Thin Homogeneous Cells with Quadrupolar Anchoring

We shall first discuss the phase diagram of a ferroelectric liquid crystal in the most simple restricted geometry, where it is confined in a bookshelf geometry between two flat, plan-parallel plates, separated by a distance $d = 2L$, as shown in the inset to Fig.4.2.1 (see also Pikin and Yoshino, 1981). We consider a homogeneous and non-polar Rapini-Papoular surface coupling (Rapini and Papoular, 1969) and we neglect bending or chevron structure of smectic layers. The surface anchoring energy density is assumed to have a simple quadrupolar form

$$g_s = [\delta(x+L) + \delta(x-L)]\tfrac{1}{2} C_s \xi_x^2 \qquad (4.2.1)$$

C_s is constant throughout the interface and ξ_x is the out-of plane component of the tilt. For a positive value of the coupling constant, $C_s > 0$, parallel alignment of $\vec{\xi}$ with respect to the interface is favored. The homogeneous form of the surface anchoring energy in the *z-y* plane, which favors homogeneous ordering, is competing with the "bulk" elastic energy, which favors helicoidal ordering far from the surfaces. For a thin enough sample we may thus expect that the helical structure will be unwound by the surface. The role of the surface term is thus similar to the role of homogeneous magnetic or AC electric fields, which tend to unwind the helical $smectic-C^*$ structure. We can thus expect that the *(d,T)* phase diagram of a ferroelectric liquid crystal would resemble the *(H,T)* phase diagram because of similar couplings (Michelson, 1977a).

The phase boundaries in the *(d,T)* phase diagram of a ferroelectric liquid crystal can be determined by the stability analysis of the $smectic-A$ phase using the free energy density (Povše *et al.*, 1993)

$$g(x,z)=\tfrac{1}{2}a\left(\xi_x^2+\xi_y^2\right)-\Lambda\left(\xi_x\frac{d\xi_x}{dz}-\xi_y\frac{d\xi_x}{dz}\right)+\tfrac{1}{2}K\left[\left(\frac{d\xi_x}{dx}\right)^2+\left(\frac{d\xi_y}{dx}\right)^2\right]+$$

$$+\tfrac{1}{2}K\left[\left(\frac{d\xi_x}{dz}\right)^2+\left(\frac{d\xi_y}{dz}\right)^2\right]+\left[\delta(x+L)+\delta(x-L)\right]\cdot\tfrac{1}{2}C_s\xi_x^2 \qquad (4.2.2)$$

We have used here the one constant approximation for the elastic distortion in the *x* and *z* directions, respectively. The $smectic-A$ - $smectic-C^*$ transition for an unconfined bulk sample is at $T_c=T_\circ+\frac{1}{\alpha}Kq_c^2$, where $q_c=\Lambda/K$ is the wave-vector of an undeformed helix, $T_\circ$ is the phase transition temperature for a racemic mixture and $a=\alpha(T-T_\circ)$. The stability analysis leads to a system of Euler-Lagrange equations for the bulk

$$a\xi_x-2\Lambda\frac{d\xi_y}{dz}-K\frac{d^2\xi_x}{dx^2}-K\frac{d^2\xi_x}{dz^2}=0 \qquad (4.2.3a)$$

$$a\xi_y+2\Lambda\frac{d\xi_x}{dz}-K\frac{d^2\xi_y}{dx^2}-K\frac{d^2\xi_y}{dz^2}=0 \qquad (4.2.3b)$$

and the surface

$$K\left.\frac{d\xi_x}{dx}\right|_{x=L}+C_s\cdot\xi_x(L)=0 \qquad \left.\frac{d\xi_y}{dx}\right|_{x=L}=0 \qquad (4.2.4a)$$

$$-K\left.\frac{d\xi_x}{dx}\right|_{x=-L}+C_s\cdot\xi_x(-L)=0 \qquad \left.\frac{d\xi_y}{dx}\right|_{x=-L}=0 \qquad (4.2.4b)$$

These can be solved by the "ansatz"

$$\xi_x(x,z) = f(x)\cos(qz) \tag{4.2.5a}$$

$$\xi_y(x,z) = g(z)\sin(qz) \tag{4.2.5b}$$

As a result (Povše *et al.*, 1993), one obtains the thickness dependent stability limit of the $smectic-A$ phase $T_c(d)$, which is shown in Fig.4.2. and the thickness dependence of the critical wave-vector q_c along this phase transition boundary, which is shown in Fig.4.2.2.

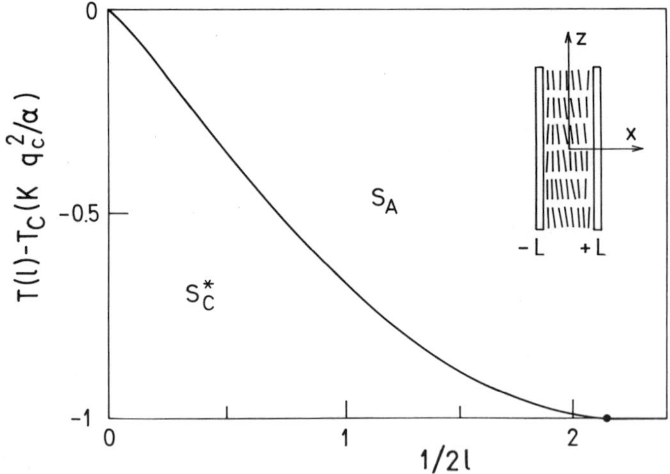

Fig.4.2.1. Shift of the $smectic-A$ - $smectic-C^*$ phase transition temperature as a function of the inverse cell thickness. The inset shows the cell geometry. $l = q_c \cdot L$ is normalized cell thickness.

Since the boundary conditions interfere with the helicoidal ordering, characterized by the wave-vector q_c in the $smectic-C^*$ phase, the $smectic-A$ - $smectic-C^*$ phase transition will take place for finite d at a lower temperature than in the bulk. There will be some limiting thickness d_{LP}, below which the $smectic-A$ - $smectic-\overline{C}^*$ phase transition into the unwound $smectic-\overline{C}^*$ phase will occur.

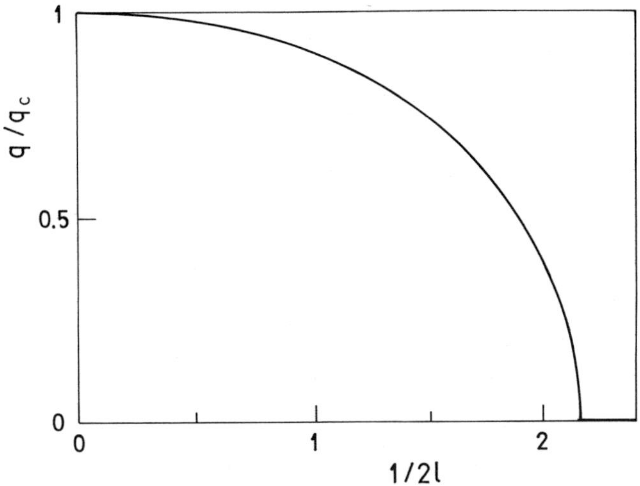

Fig.4.2.2. The magnitude of the modulation wave-vector along the $smectic-A$ - $smectic-C^*$ transition line as a function of the inverse thickness $1/2l$ of a cell. $l = q_c \cdot L$ is normalized cell thickness.

Since the boundary conditions for $d < d_{LP}$ do not interfere with the homogeneous ordering of the surface-unwound $smectic-\overline{C}^*$ phase, the phase transition boundary for $d < d_{LP}$ will be thickness-independent

$$d > d_{LP}: \quad T_c(d) < T_c(d \to \infty) \tag{4.2.6a}$$

$$d < d_{LP}: \quad T_c(d) = T_\circ = const. \tag{4.2.6b}$$

The predicted *(d,T)* phase diagram is shown schematically in Fig.4.2.3 and is very similar to the *(H,T)* phase diagram of a ferroelectric liquid crystal in a transverse magnetic field with negative diamagnetic anisotropy (see Fig.3.2.1.(b) in Section 3). The reason for this is that for $d < d_{LP}$ the structure is unwound, as favored by the wall.

If one calculates the behavior of the critical wave-vector $q_c(d)$ along the phase transition line $T_c(d)$, one can observe even more similarity with the (H,T) phase diagram. The wave-vector $q_c(d)$ goes continuously to zero at d_{LP}, which is thus a Lifshitz thickness, where the $smectic-A$, distorted $smectic-C^*$ and the unwound $smectic-\overline{C}^*$ phases meet. The value of the Lifshitz thickness depends on the surface coupling constant C_s and the liquid

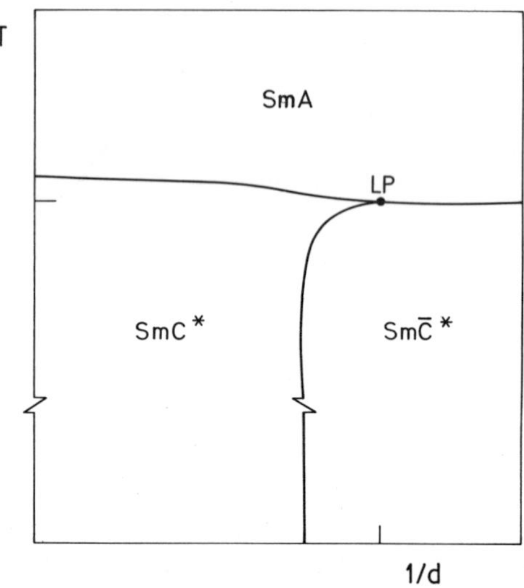

Fig.4.2.3. The *(d,T)* phase diagram of a ferroelectric liquid crystal, confined between two plan-parallel plates with quadrupolar boundary conditions.

crystalline material constants. It can be evaluated in the limit of weak and strong surface anchoring, respectively:

$$C_s \to 0, \quad d_{LP} \to 0 \tag{4.2.7a}$$

$$C_s \to \infty, \quad d_{LP} \to \frac{\sqrt{3}}{2\pi} p_o = 0.276 p_o \tag{4.2.7b}$$

Here, p_o is the period of the helix in bulk liquid crystal.

The phase boundary between the distorted $smectic-C^*$ phase and the surface unwound $smectic-\overline{C}^*$ phase can be treated analogously to the cholesteric-nematic transition (Luban *et al.*, 1974). In the constant amplitude approximation (CAA) the $smectic-C^*$-$smectic-\overline{C}^*$ transition is of second order at a critical thickness d_c. A rather nontrivial analysis gives the variation of the critical cell thickness d_c and the Lifshitz thickness d_{LP} with the anchoring strength C_s. It shows that the Lifshitz thickness d_{LP} is always smaller than the critical thickness d_c. As one can expect, they both depend on the surface coupling constant, as shown in Fig.4.2.4. For a cell spacing larger than the Lifshitz spacing there is a reentrant, distorted $smectic-C^*$ phase present just below the $smectic-A$ phase. This may have important implications for the technology of FLC devices, where the cell spacing is of the order of the helical pitch in bulk material. When cooling such a cell from the $smectic-A$ phase, one may cross the reentrant, distorted $smectic-C^*$ phase and generate disclination lines, that are always present in partially unwound ferroelectric phases (Bourdon *et al.*, 1982, Glogarova *et al.*, 1983, Glogarova and Pavel, 1984a, b).

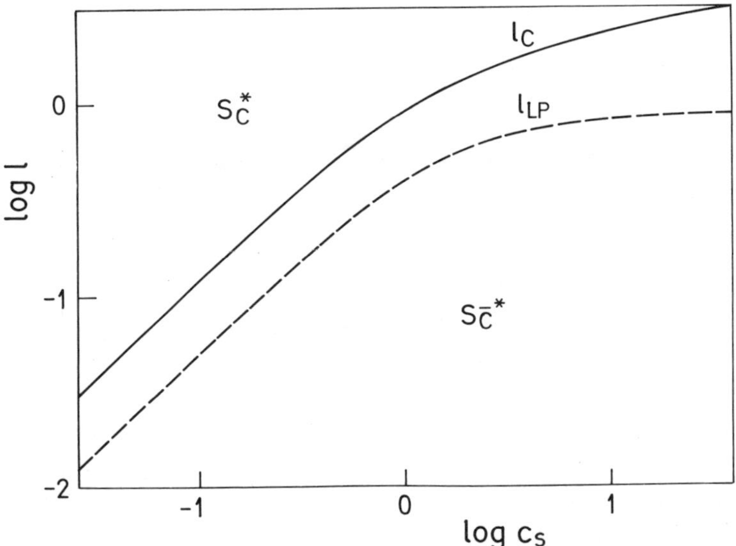

Fig.4.2.4. The dependence of the Lifshitz thickness l_{LP} and the critical thickness l_c on the surface coupling constant $c_s = C_s/(2Kq_c)$; $l = q_c \cdot L$ is normalized cell thickness.

A careful analysis of the results of the CAA approximation shows that this approximation breaks down for large coupling strengths C_s. Experimental observations of the $smectic-C^*$ structures in thin samples indeed indicate (Bourdon *et al.*, 1982, Glogarova *et al.*, 1983), that instead of a continuous director field, a system of disclination lines mediates the transition between the homogeneous orientation dictated by the surface and the helical structure in the interior of the liquid crystal. This means that we are always in the strong coupling regime. In this case, the critical thickness is (Povše *et al.*, 1993)

$$d_c = \frac{p_o}{2}\sqrt{\frac{K''}{K_3}} \tag{4.2.8}$$

Here K'' is the bend elastic constant, which is assumed to be equal to the splay elastic constant, whereas K_3 is the twist elastic constant of the $smectic - C^*$ phase.

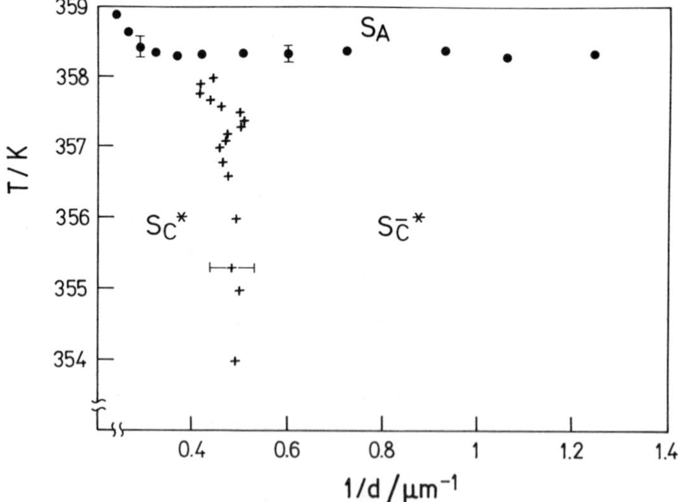

Fig.4.2.5. Phase diagram of a ferroelectric liquid crystal CE-8 confined in a cell with polyimide coated surfaces (Povše et al., 1993).

The *(d,T)* phase diagram was experimentally determined by Povše et al.(1993) in wedge type cells of the liquid crystal 4-(2'-methylbutyl)phenyl 4'-n-octylbiphenyl-4-carboxylate (CE-8), using untreated polyimide coated surfaces.

The observed phase diagram is shown in Fig.4.2.5 and is in qualitative agreement with theory.

At $T_c - T = 5K$, the critical thickness is $d_c = 2\mu m$ and increases slightly on approaching the $smectic-A$ phase. The experiment allows for the estimation of the coupling constant, $C_s > 3 \cdot 10^{-3}\, Jm^{-2}$, and indicates that this system is indeed in the strong surface coupling regime. This value of the surface anchoring energy is rather high, compared to typical surface anchoring energies for nematics, $C_s \approx 10^{-5}\, J/m^2$, but is in reasonable agreement with other observations (Cognard, 1982). Because of the experimental limitations, the critical thickness close to the $smectic-A$ phase could not be determined unambiguously, so the question of the existence of a Lifshitz point in such a system is still open.

Note4.2.1. What do polymer surfaces really look like?

If one looks closer to the surface profiles of the aligning layers that are used in the experiments, one observes rather huge irregularities on the length scale of molecular size that are certainly generators of disorder at the surface.

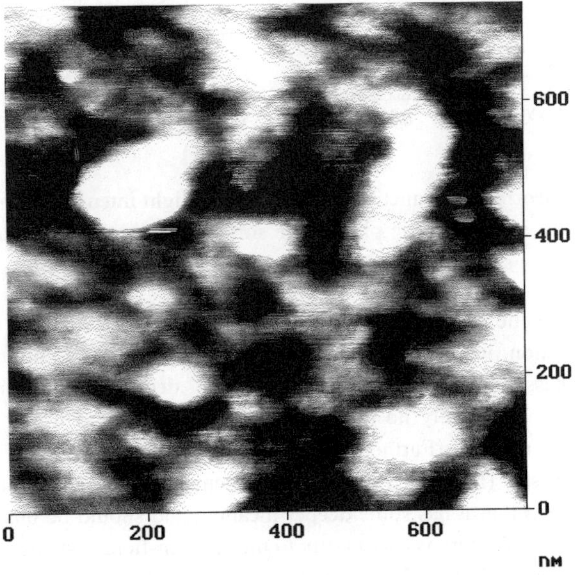

Fig.4.2.6. An Atomic Force Microscope image of a Nylon spin-coated glass surface. Total height between dark and white regions is 2.0nm

In this sense it seems likely that disordering effects of the surface dictate the thermodynamics at very low thickness. An example of a surface profile of a spin-coated nylon alignment layer as observed by the Atomic Force Microscope is shown in Fig.4.2.6.

The onset of rather complex behavior, which is due to the presence of an inhomogeneous polymer surface can, in fact, be observed already in very thin homogeneously aligned ferroelectric cells. Figure 4.2.7. shows the comparison between the autocorrelation function of the quasielastically scattered light intensity for bulk and thin layers of homogeneously aligned ferroelectric liquid crystal (Rastegar, 1996a, b).

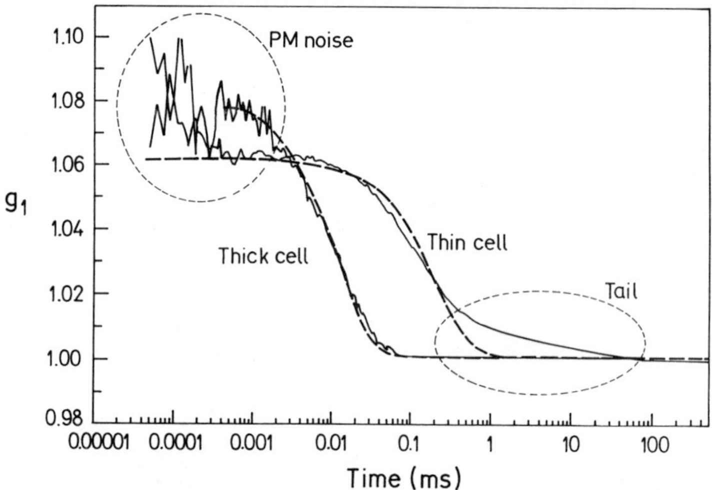

Fig.4.2.7. The autocorrelation function of the scattered light intensity in the bulk and very thin cell filled with a ferroelectric phase. The surfaces of a cell were coated with Nylon that has a topography similar to the surface shown in Fig.4.2.6.

One can clearly see the difference for long lag-times. Whereas we have in bulk a single exponential relaxation, we have in a thin ferroelectric layers an additional and unusual relaxational process, which is observed on three decades of a time scale! The phenomenon shows practically no dispersion in the q-space and it can be observed in chiral and racemic material. Furthermore it can be observed in a very thin layer of a nematic liquid crystal (Mertelj, 1998) or in a nematic confined in an aerogel. This universality suggests that it has some deeper meaning and should be observable in other systems as well. Indeed, there is a similarity to the random-field systems (Wu et al., 1992; Frisken and Cannell, 1992; Sidebottom et al., 1992). Here, in confined geometry, a disordered surface acts as an random field, and imposes a characteristic long-time dynamics.

4.3. Surface Induced Polarity: Splayed States, Surface Electroclinic Effect and Reentrance

It has been realized since the early experiments on the switching dynamics of surface stabilized ferroelectric $smectic-C^*$ phase (Handschy et al., 1983; Glogarova et al., 1984a-c), that ***surface-induced polarity*** is in fact a quite widely observed interfacial phenomenon in liquid crystals (Xue and Clark, 1990; Lee and Patel, 1990; Tripathi et al., 1991; Crandall et al., 1992; Chen et al., 1992; Moses et al., 1993). The fundamental reason for the surface polarity is not necessarily related to the surface electric properties, but is a rather general result of inversion symmetry breaking at the interface. As a consequence, the surface interaction energy of a chiral molecule depends on the orientation of the molecule and leads to polar anchoring and surface polarization. In particular, the following phenomena are mediated by surface-induced polarity:

(i) In the $smectic-A$ phase the surface polar coupling induces a small, but finite value of the spontaneous polarization close to the solid interface, as shown in Fig.4.3.1 Due to the coupling between the polarization and the tilt, a finite tilt θ_s is as well induced at the surface. This tilt is spatially inhomogeneous, i.e. it is largest at the surface and decreases as we move away from the surface into the bulk, see Fig.4.3.1.(b). This phenomenon is the so-called ***surface electroclinic effect*** and may be quite large. It can be of the order of several degrees for materials with a large bulk electroclinic coefficient (see for example Nakagawa et al., 1986a,b; Xue and Clark, 1990). The surface electroclinic effect can induce a macroscopic rotation of the smectic layers with respect to the rubbing direction, which fixes the molecular direction at the surface.

(ii) In the $smectic-C^*$ phase, the polar surface coupling can for a certain range of layer thickness lead to the so-called ***splayed polarization states***. Here, the direction of the polarization is at the upper surface forced by the surface coupling in the $+x$-direction, whereas it is forced into the $-x$-direction on the bottom surface, as shown in Fig.4.3.2. In between, the polarization field $P(x)$ will be distorted in a splayed-like fashion.

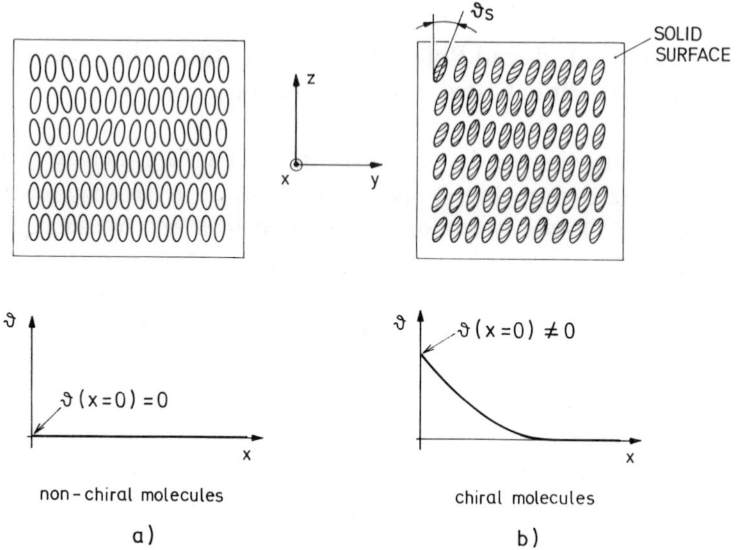

Fig.4.3.1. Surface electroclinic effect due to a polar coupling between a solid wall and the liquid crystalline molecules. Note the difference between nonchiral (a) and chiral molecules (b).

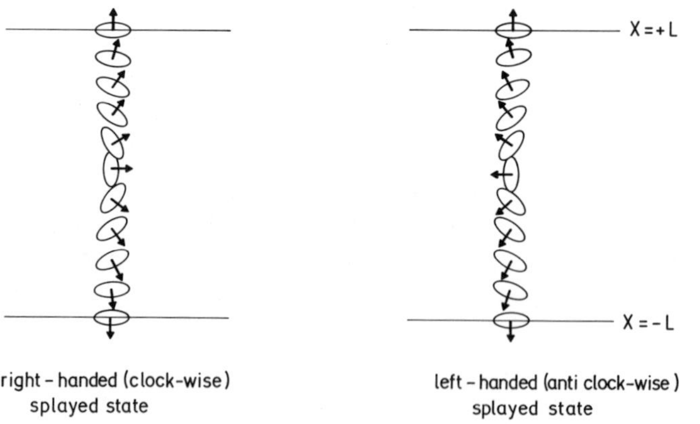

Fig.4.3.2. The splayed polarization states due to polar surface coupling. Note that there are two possible phase profiles that connect the $+\vec{P}$ and $-\vec{P}$ states at the two surfaces: one represents a clock-wise and the other an anti-clock-wise rotation of the local director, as we move from $+L$ to $-L$.

Splayed polarization states are very often found in thin ferroelectric, surface stabilized devices. When such a sample is observed between a pair of crossed polarizers, one cannot obtain a complete extinction of the light intensity, transmitted through the splayed state. This significantly reduces the contrast of surface stabilized devices. Furthermore, the contrast is diminished by the reduced apparent optical tilt angle of these splayed polarization states.

The effect of surface polarity on phase transitions in thin, homogeneously aligned ferroelectric films was discussed by Rovšek and Žekš (1995). They considered a ferroelectric liquid crystal layer, confined between two plan-parallel polar surfaces, separated by a distance $d=2L$. The polarization-renormalized free-energy density is

$$g = g_{\circ} + \tfrac{1}{2}a\left(\xi_x^2+\xi_y^2\right)+\tfrac{1}{4}b\left(\xi_x^2+\xi_y^2\right)^2 - \Lambda\left(\xi_x\frac{\partial\xi_y}{\partial z}-\xi_y\frac{\partial\xi_x}{\partial z}\right)+$$

$$+\tfrac{1}{2}K\left[\left(\frac{\partial\xi_x}{\partial x}\right)^2+\left(\frac{\partial\xi_y}{\partial x}\right)^2\right]+\tfrac{1}{2}K\left[\left(\frac{\partial\xi_x}{\partial z}\right)^2+\left(\frac{\partial\xi_y}{\partial z}\right)^2\right]-$$

$$-W\cdot\delta(x-L)\cdot\xi_y + W\cdot\delta(x+L)\cdot\xi_y \tag{4.3.1}$$

and the geometry of the problem is shown in Fig.4.3.3. For simplicity, we have used here the one elastic constant approximation and quadrupolar coupling has been omitted (see Chapter 4.2. for quadrupolar surface coupling).

The surface polarity is expressed via the last two coupling terms. A positive polar surface coupling coefficient W favors a positive equilibrium value of ξ_y at $x=L$ and a negative ξ_y at $x=-L$. Note that the electric polarization field $P(x,z)$ is related to the ξ -field

$$P_x = \varepsilon\mu\frac{d\xi_x}{dz}-\varepsilon C\xi_y \tag{4.3.2.a}$$

$$P_y = \varepsilon\mu\frac{d\xi_y}{dz}+\varepsilon C\xi_x \tag{4.3.2.b}$$

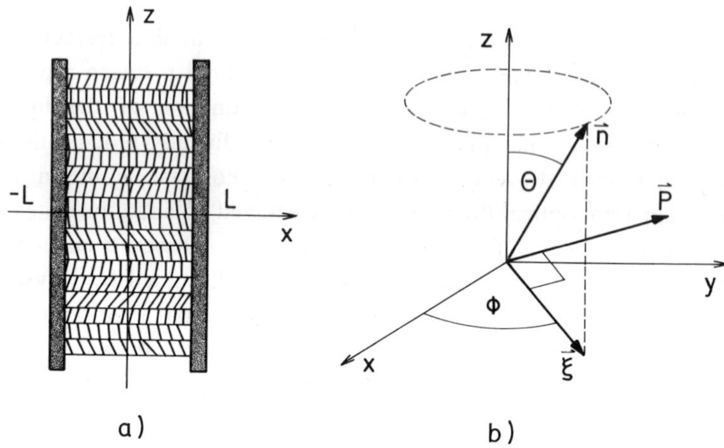

Fig.4.3.3. The ferroelectric liquid crystal is confined in between two plan-parallel plates, separated by a distance $2L$ (a). The definition of the tilt order parameter $\vec{\xi}$ and the polarization $\vec{P}$ (b). After Rovšek and Žekš, 1995.

In the $smectic-A$ phase, we consider that the system is homogeneous along the smectic layer normal. We can therefore neglect the z-dependence of the free-energy density and, after minimization, we obtain a set of equilibrium equations for the bulk

$$b\xi_x^3+\xi_x\left(a+b\xi_y^2\right)-K\frac{\partial^2\xi_x}{\partial x^2}=0 \tag{4.3.3a}$$

$$b\xi_y^3+\xi_y\left(a+b\xi_x^2\right)-K\frac{\partial^2\xi_y}{\partial x^2}=0 \tag{4.3.3b}$$

together with boundary conditions

$$\left.\frac{\partial\xi_x}{\partial x}\right|_{x=\pm L}=0 \quad \text{and} \quad \left.\frac{\partial\xi_y}{\partial x}\right|_{x=\pm L}=W \tag{4.3.4}$$

It is straightforward to see that in the $smectic-A$ phase, the solution of the above equations is $\xi_x\equiv 0$, because the surface coupling does not induce any out-of surface tilt. This simplifies the equation for $\xi_y(x)$ to:

$$b\xi_y^3 + a\xi_y - K\frac{\partial^2 \xi_y}{\partial x^2} = 0 \tag{4.3.5}$$

This equation can be solved numerically and the solutions are shown in Fig.4.3.4. for different surface coupling strengths and for different temperatures. We can see that already in the *smectic* – *A* phase there is a surface induced tilt. Due to the coupling between the tilt and polarization (Eqs.4.3.2.) a surface-induced polarization region develops in the vicinity of a solid boundary. Both the tilt and the polarization decay as we move into the interior of the cell. When the temperature is closer to the phase transition into the ferroelectric phase, the region of the surface-induced polarization is larger. The magnitude of the polarization at the surface is also larger if the surface coupling strength is larger. A finite polarization is developed throughout the cell at and below the phase transition temperature.

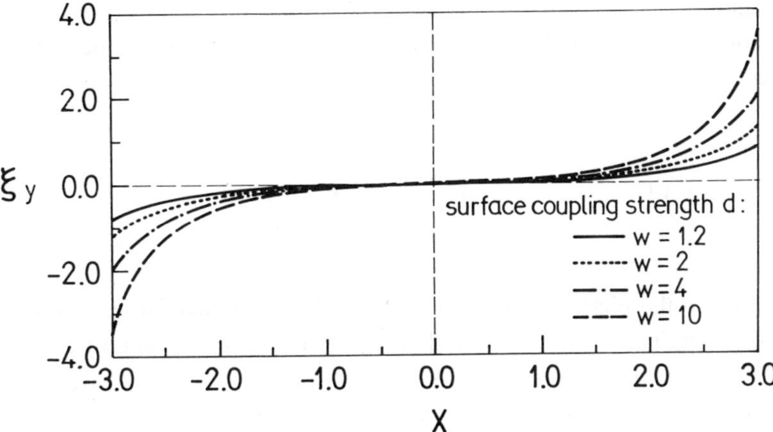

Fig.4.3.4. The profile of the tilt order parameter $\vec{\xi}$, as induced by a polar surface coupling in the *smectic* – *A* phase of a ferroelectric liquid crystal, confined in a plan-parallel cell with normalized thickness $L = 3(p_\circ/2\pi)$. The profiles are shown for different surface coupling strengths $w = W\sqrt{Kb}/\Lambda^2$ (after Rovšek and Žekš, 1995).

Further physical insight can be obtained by considering the limiting solutions of the Equation 4.3.5., which governs the magnitude and the *x*-dependence of the tilt profile in the $smectic-A$ phase. Far above the $smectic-A$ - $smectic-C^*$ phase transition temperature, the cubic term in the Equation 4.3.5. is small and for a given surface coupling strength D, the equation reduces to

$$a(T)\xi_y - K\frac{\partial^2\xi_y}{\partial x^2} = 0 \tag{4.3.6.}$$

For a single wall at $x=0$, the solution of the above equation is

$$\xi_y(x) = W\sqrt{K/a(T)}\cdot e^{-x/\lambda} \tag{4.3.7}$$

where the penetration length is $\lambda = \sqrt{K/a(T)}$. The induced tilt therefore decays exponentially, as we move from the surface into the liquid crystal. Both the penetration length λ and the magnitude of the induced tilt ξ_y° increase, as we approach the phase transition

$$\lambda(T) \approx (T-T_c)^{-1/2} \qquad \xi_y^\circ = W\sqrt{K/a} \propto (T-T_c)^{-1/2} \tag{4.3.8}$$

Very close to the phase transition, the induced tilt is rather large, and the cubic term has to be taken into account. This prevents the divergence of the induced tilt at the phase transition point. As one could intuitively expect, the surface induced tilt phenomena are more and more pronounced as we approach the phase transition temperature, because of the large susceptibility of the system.

The behavior predicted above was indeed observed in the ellipsometry studies of Xue and Clark (1990) and Moses et al. (1993) and second-harmonic and ellipsometry experiments of Chen et al. (1992). Figure 4.3.5. shows the ellipsometry results of Chen et al. (1992), which have been obtained in the $smectic-A$ phase of a ferroelectric liquid crystal on a polyimide surface.

The molecular alignment of a ferroelectric liquid crystal on a polar surface was also studied using second harmonic generation (Chen et al., 1992).

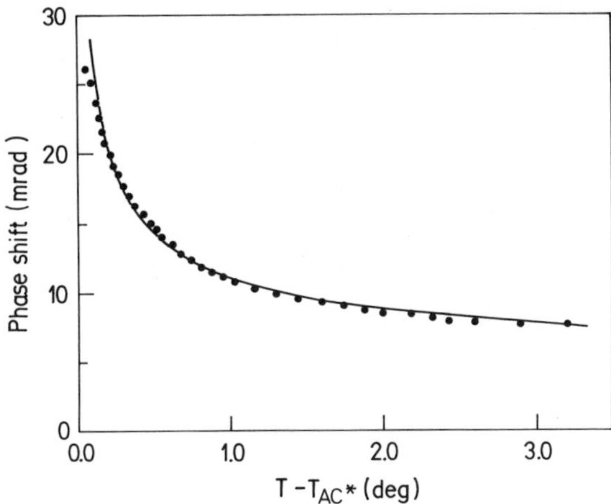

Fig.4.3.5. The phase shift in the ellipsometry experiment of Chen et al. (1992). This shift is proportional to the space-average of the birefringence and therefore to the induced molecular tilt near the surface. The solid line is the best fit to the Eq.4.3.8. Note that close to the phase transition the divergence is quenched.

Here, a very intense laser light beam is focused to the liquid crystal monolayer and the intensity of light with double frequency is measured. Using different polarizations of light, the method can extract information on the molecular ordering in a single liquid crystalline monolayer. The results of Chen et al.(1992) shown in Fig.4.3.6. clearly demonstrate polar molecular ordering within the first molecular layer in a ferroelectric liquid crystal.

We have considered so far only pretransitional effects, which are induced by polar surface coupling in the $smectic-A$ phase of a ferroelectric liquid crystal. The question arises, whether for a given thickness of a sample, the $smectic-A$ phase would transform into the helicoidally modulated $smectic-C^*$ phase or the unwound $smectic-\overline{C}^*$ phase. This question was discussed by Rovšek and Žekš (1995) and a very peculiar issue of the (d,T) phase diagram with polar boundary conditions was observed.

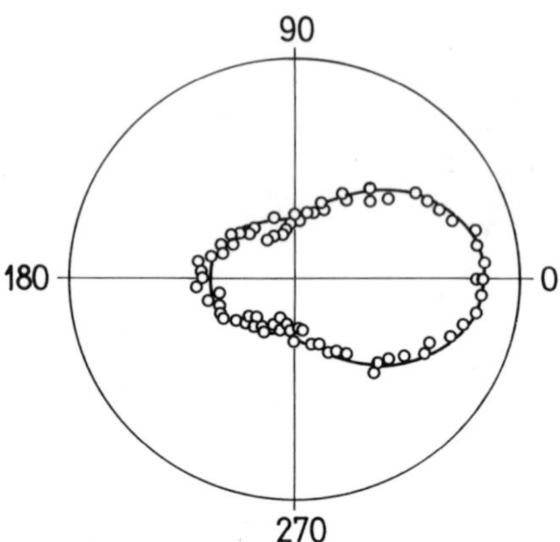

Fig.4.3.6. Polar plots of the square root of the SHG intensity, emitted from a monolayer of a ferroelectric liquid crystal on a rubbed polyimide surface (Chen et al., 1992). The asymmetry of the plot clearly reveals the polar ordering of liquid crystal molecules at the surface. The molecules are aligned along the rubbing direction, which is horizontal.

They have performed the stability analysis of the $smectic-A$ phase of a ferroelectric liquid crystal, confined to a planar cell with polar boundary conditions. The stability of the $smectic-A$ phase was tested against the appearance of small deviations of the order parameter

$$\xi_x(x,z)=\delta\xi_x(x,z) \qquad (4.3.9a)$$

$$\xi_y(x,z)=\xi_y^\circ(x)+\delta\xi_y(x,z) \qquad (4.3.9b)$$

where $\xi_y^\circ(x)$ is the equilibrium, surface-induced tilt profile of the confined $smectic-A$ phase. After expressing the deviations

$$\delta\xi_x(x,z)=\Theta_x(x)\cdot\cos qz \qquad (4.3.10a)$$

$$\delta\xi_y(x,z)=\Theta_y(x)\cdot\sin qz \qquad (4.3.10b)$$

the linear stability analysis of the $smectic-A$ phase reduces to an eigenvalue problem. The phase transition temperature from the $smectic-A$ phase to the modulated or homogeneous phase is then defined as the temperature, where the lowest of the eigenvalues becomes zero (see also Chapter 2). Furthermore, this condition also defines the critical wave-vector q_c, where the condensation of the soft mode occurs in the reciprocal space.

The analysis of the eigenvalue spectrum can be done numerically and the rather surprising results are shown in Figure 4.3.7.

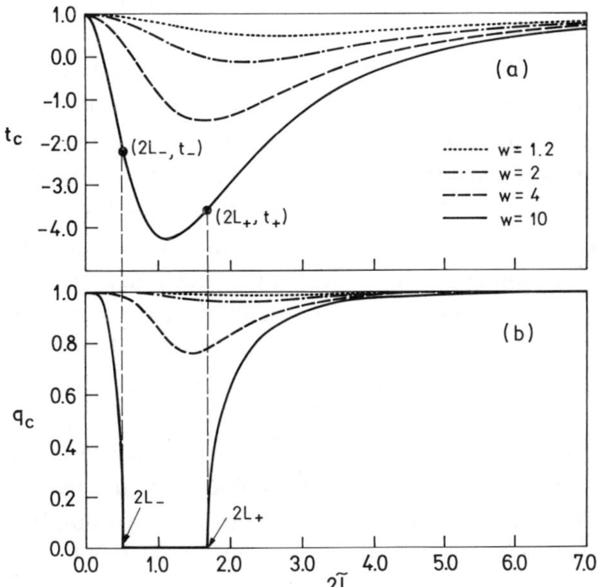

Fig.4.3.7. (a) The $smectic-A$ - $smectic-C^*$ phase transition temperature for different surface polar surface couplings $w = W\sqrt{Kb/\Lambda^2}$ in a homogeneously aligned cell of thickness $2\widetilde{L}$. (b) The critical wave-vector of the transition as a function of the cell thickness for various surface couplings. Al quantities are normalized (see Rovšek and Žekš, 1995).

Here, the $smectic-A$ - $smectic-C^*$ - $smectic-\overline{C}*$ phase transition temperature is shown as a function of normalized cell thickness $2\tilde{L}=2(2\pi L/p_\circ)$ for various strengths of the surface coupling constant *w*. For large thickness of the cell, the phase transition temperature asymptotically approaches the bulk transition temperature for a bulk chiral $smectic-C^*$ phase. By decreasing the cell thickness, the transition takes place at a lower temperature, and the phase transition line gradually decreases. This behavior is similar to the *(d,T)* phase diagram of a ferroelectric liquid crystal, confined to plan-parallel plates with quadrupolar surface coupling. However, quite surprisingly, for very low thickness, the phase transition temperature increases again and asymptotically approaches the bulk $smectic-A$ - $smectic-C^*$ transition temperature for ultrasmall cell spacing.

This behavior of the phase transition line is an indication, that there is a reentrance of modulated $smectic-C^*$ phase for very low thicknesses of the cell. In order to test this conjecture, the critical wave-vector along this phase transition line was calculated and is shown in Fig.4.3.7(b). For large thickness of the cell, the critical wave vector is equal to the bulk $smectic-A$ - $smectic-C^*$ transition wave-vector q_c. By decreasing the thickness of the cell and for large-enough surface coupling *D*, the critical wave-vector decreases continuously to zero at some critical thickness L_+. We therefore reach a triple point (d_+, T_+), where a homogeneous $smectic-A$, helicoidally modulated $smectic-C^*$ and the unwound $smectic-\overline{C}*$ phases meet. This point is therefore a Lifshitz point, if the phase transition line between the modulated $smectic-C^*$ and the homogeneous $smectic-\overline{C}*$ phases is of first order. By further decreasing the cell thickness, we notice at some critical thickness d_- the onset of helical modulation, and we enter a reentrant, helicoidally modulated $smectic-C^*$ phase. This is therefore a second triple point (d_-, T_-), which is also a good candidate for the Lifshitz point.

The surprising possibility of having a reentrant modulated phase and two Lifshitz points was explained by Rovšek and Žekš (1995). The polar boundary conditions induce in the $smectic-A$ phase a finite tilt of the molecules at confining surface, as shown in Fig.4.3.8(a). Because the tilt is in opposite directions at the two surfaces, π -splay of the director field is observed, as we move from one surface to another. This is energetically costly, because there is a transverse elastic deformation due to the splayed director profile, and the elastic energy of this deformation increases as $1/d$. For very small thickness the system will therefore try to avoid this transverse deformation. However, this

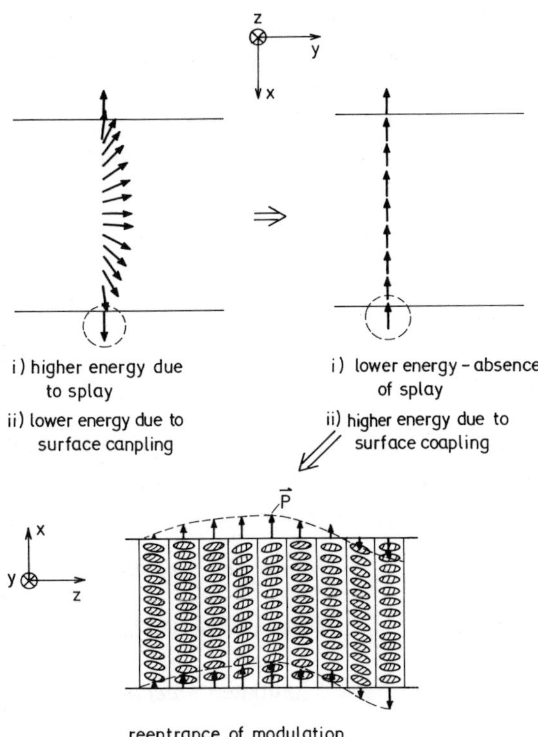

Fig.4.3.8. Scenario for the emergence of a reentrant $smectic-C^*$ phase in very thin cells with polar boundary conditions. The transverse profile of the polarization is shown. The splayed polarization state becomes energetically unfavorable and the system prefers to be homogeneous in the transverse direction. Once it is homogeneous in the transverse direction, it will spontaneously restore the helicoidal modulation, because this further decreases its free energy.

can be done only at the expense of the energy of the polar surface coupling, see Fig.4.3.8.(b).

By decreasing the cell thickness we therefore reach a situation, where the system is homogeneous in the transverse x-direction. Although this increases its surface energy, the total energy is lower because of the absence of the transverse splay. But then, it is more favorable to have also a spontaneous twist along the z-direction, because this further reduces the elastic energy (Fig.4.3.8.(c)). We see, that the polar surface coupling becomes irrelevant for the thermodynamics of very thin cells and the system behaves as a bulk sample.

In the CAA approximation, the two critical thickness d_+ and d_- can be evaluated as (Rovšek and Žekš, 1995)

$$d_{\pm} = \frac{2W}{\Lambda\theta_{\circ}^{2}}\left(1 \pm \sqrt{1 - \frac{\pi^{2}}{4} \cdot \frac{\Lambda^{2}\theta_{\circ}^{2}}{W^{2}}}\right) \tag{4.3.11}$$

The following phases are therefore stable in thin cells with polar surface coupling:

$0 < d < d_-$	Modulated $smectic - C^*$ is stable
$d_- < d < d_+$	Unwound and splayed $smectic - \overline{C}^*$ is stable
$d_+ < d$	Modulated $smectic - C^*$ is stable

As a surprise, we see that a homogeneous and unwound $smectic - \overline{C}^*$ phase ***is never stable for polar boundary conditions***. Here, only non-polar surface coupling can assure true bistability and large apparent optical tilt angles, which are necessary for good electrooptical performance of these devices. Further, one can see from the Eq.4.3.11 that for small values of the surface polar coupling constant W, the helix is never unwound and the modulated phase is stable. The predictions of the above theory have been only partially substantiated by experiments. As reported and discussed by Glogarova and Pavel (1984a, b) and by Handschy et al. (1983), only the existence of splayed polarization states was observed in thin cells with polar surface coupling, whereas the reentrance of the modulated phase for ultrathin cells has not yet been observed. It should be noted that the above analysis does not include the electrostatic energy of the polarization field, which can have a significant influence on the thermodynamics of thin ferroelectric layers. We have also neglected quadrupolar term, which can have significant effect.

Finally, Fig.4.3.9. shows the comparison of the phase diagrams of a chiral ferroelectric liquid crystal in a thin cell for quadrupolar and polar boundary conditions. For quadrupolar boundary conditions, the unwound and homogeneous $smectic-\overline{C}*$ phase is stable below some critical thickness d_c and there is a single triple point, which is a Lifshitz point. On the other hand, we have for polar boundary conditions two triple points and a reentrant helical $smectic-C^*$ phase at very small thickness. This reentrant helical phase is separated from the bulk helical phase by a region of unwound but splayed ferroelectric phase.

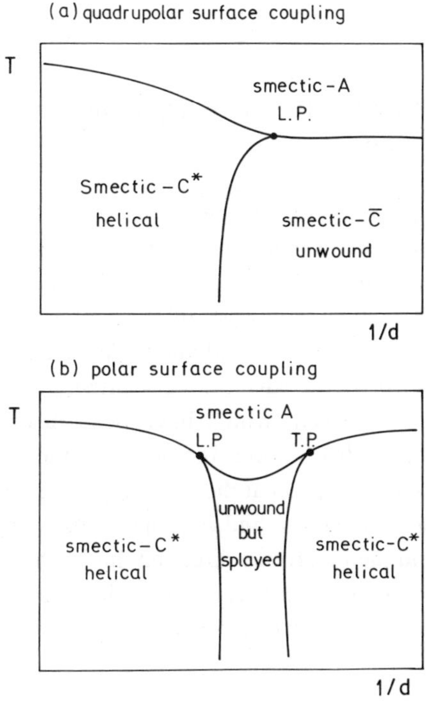

Fig.4.3.9. Comparison of the $(T, 1/d)$ phase diagrams of chiral ferroelectric liquid crystal in thin cells. (a) Quadrupolar boundary conditions result in the appearance of a Lifshitz point. (b) For polar boundary conditions, a reentrant helicoidally modulated phase appears below some critical thickness.

Note 4.3.1. Surface coupling energy of ferroelectric liquid crystals.

The polar surface coupling strengths of ferroelectric liquid crystals on polymer surfaces range from $10^{-5} - 10^{-4}\,J/m^2$. Xue et al., (1988) measured surface energy of $8\times10^{-4}\,J/m^2$ of ferroelectric liquid crystal on Indium Tin Oxide layer. Surface energy of $10^{-3}\,J/m^2$ is reported by van Haaren and Rikken (1989) for ZLI 4005 on poly(hexamethylene)terephtalamide. Panarin et al. (1995) report in-plane energy of $4\times10^{-5} - 9\times10^{-4}\,J/m^2$ for SCE-13 on rubbed PVA surfaces. Rastegar et al. (1996a) have obtained polar surface energy of $5\times10^{-5}\,J/m^2$ for SCE-9 on Nylon 6/6. This is similar to the value of $1\times10^{-4}\,J/m^2$, obtained by Škarabot et al. (1999a) for CE-8 on Nylon 6/6.

4.4. Soliton and Plane Wave Dynamics in Thin Cells with Polar Anchoring

When a ferroelectric liquid crystal is placed in between two polymer-coated surfaces, separated by a micrometer distance, one usually observes a wide variety of textures (Handschy et al., 1983; Clark and Lagerwall, 1984; Handschy and Clark, 1984). If the thickness of the cell is much smaller than the bulk helical period, the helix is unwound and we have a homogeneous ordering in the z-direction, dictated by the surface. Very often, so-called 'splayed' states of the director field are observed, which have been discovered by Handschy et al. (1983). Here, the electric dipole moments of the liquid crystalline molecules have a preferential orientation at the surface, as shown in Fig. 4.4.1. This leads to a texture, which is unwound and homogeneous along the layer normal, but is inhomogeneous in the transverse direction due to the polar surface coupling.

A very interesting consequence of the splayed polarization field $P(\vec{r})$ is the appearance of a space-charge density

$$\rho_{induced}(\vec{r}) = -div\vec{P} \tag{4.4.1}$$

This substantially influences the static and dynamic properties of a ferroelectric liquid crystal, confined between two plan-parallel plates as one has to take into account long-range electrostatic interactions of these induced charges.

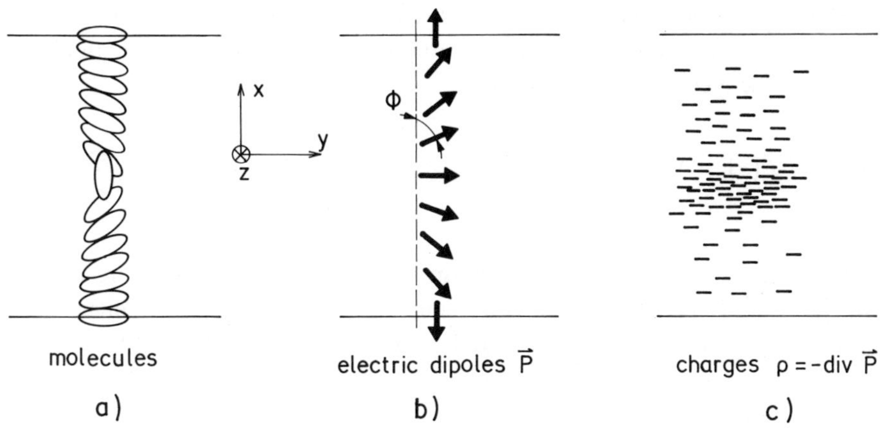

Fig.4.4.1. (a) The "splayed" states of a ferroelectric liquid crystal in very thin homogeneous layers with polar surface anchoring. The dipole moments prefer an 'inward' direction, pointing into the bounding surface. This results in a splayed state: If one moves from one surface to another, the direction of the polarization is rotated by 180° from one surface to another. (b) The distribution of the electric polarization throughout the cell. (c) The resulting space charge density due to the splay of the director field.

Let us consider the effects of the splayed polarization field on the static properties of a homogeneous ferroelectric cell. The director field $\vec{\xi}(\vec{r})$ is homogeneous in the y-z plane, and is inhomogeneous in the x-direction due to polar anchoring (see Fig.4.4.1.). In the constant amplitude and one elastic constant approximation, the free-energy density is

$$g(x) = g_{\circ} + \tfrac{1}{2} K \theta_{\circ}^{2} \left(\frac{d\Phi}{dx} \right)^{2} - \tfrac{1}{2} \vec{D}\vec{E} - \vec{P}\vec{E} \tag{4.4.2}$$

Here $\Phi(x)$ is the phase profile across the cell, $D(x) = \varepsilon\varepsilon_{\circ} E(x)$ is the electric displacement field, ε is the dielectric constant of a crystal at high frequencies and

$P = P_\circ(\cos\Phi, \sin\Phi, 0)$ is the polarization, as shown in Fig.4.3.1. The third term represents the electrostatic energy of induced dipoles and the last term represents the electrostatic energy of permanent dipoles in the local electric field. We introduce the electric potential $\varphi(x)$

$$\vec{E}(x) = -\vec{\nabla}\varphi(x) \tag{4.4.3}$$

and minimize the total free energy $G = 1/L \int g(x)dx$ with respect to $\Phi(x)$ and $\varphi(x)$. This leads to a set of coupled equations for the stationary phase profile $\Phi_\circ(x)$ and the electric potential $\varphi_\circ(x)$

$$\frac{d^2\Phi_\circ}{dx^2} + \frac{P_\circ}{K\Theta_\circ^2} \cdot \sin\Phi_\circ \frac{d\varphi_\circ}{dx} = 0 \tag{4.4.4a}$$

$$\frac{d^2\varphi_\circ}{dx^2} + \frac{P_\circ}{\varepsilon\varepsilon_\circ} \cdot \sin\Phi_\circ \frac{d\Phi_\circ}{dx} = 0 \tag{4.4.4b}$$

Here the second equation is in fact the well-known Poisson equation, which relates the charge distribution to the corresponding potential, $\varepsilon\varepsilon_\circ \nabla^2\varphi_\circ(\vec{r}) = -\rho(\vec{r}) = divP(\vec{r})$.

The above set of equations can be solved analytically in the case of zero external field and numerically for the case of finite field (Nakagawa and Akahane, 1986a, b; Akahane et al., 1989). After integration, we obtain the well known sine-Gordon equation for the stationary phase profile $\Phi_\circ(x)$

$$\frac{d^2\Phi_\circ}{dx^2} + \frac{P_\circ^2}{2\varepsilon\varepsilon_\circ K\Theta_\circ^2} \cdot \sin 2\Phi_\circ = 0 \tag{4.4.5}$$

whereas the local electric field is

$$E_x = \frac{d\varphi_\circ}{dx} = \frac{P_\circ}{\varepsilon\varepsilon_\circ} \cos\Phi_\circ \tag{4.4.6}$$

The above equations show that the presence of splayed spontaneous polarization generates an internal, local electric field, that has the same role as an external homogeneous magnetic or electric field (see, for comparison the Eqs.3.2.8 and Eq.3.3.4). This internal field has the tendency to induce a soliton-like deformation of the phase profile, and finally, for a very high polarization, to flatten the phase profile across the cell. It is easy to understand the origin of this field: any splay-like inhomogeniety of the polarization field results in the appearance of local induced charges, as shown in Fig.4.4.1(c). These tend to repel each other via the Coulomb repulsion forces and tend to make the phase profile locally homogeneous. As a result, the induced charge distribution balances itself in such a way, that the phase profile across the splayed cell obtains a soliton form. The soliton-like deformed phase profile is therefore a profile of minimum free energy. This electrostatic mechanism was considered by Galerne (1996) to explain an effective "screening" of the polarization field by the internal electric field, which is generated by the divergence of polarization.

Similarly to the magnetic or electric-field induced soliton lattice, the solution of the sine-Gordon equation, which satisfies the fixed boundary conditions $\Phi_\circ(x=-L)= \; =\Phi_\circ(x=+L)=0$ is a soliton phase profile

$$\sin\Phi_\circ = sn(u,k) \tag{4.4.7}$$

where $u = x/\xi_P k + K$ is the reduced coordinate and $\xi_P = \sqrt{\varepsilon\varepsilon_\circ K\theta_\circ^2/P_\circ^2}$ is the polarization coherence length, which is similar to the magnetic and electric coherence lengths, introduced in Chapter 3. $K(k)$ is the complete elliptic integral of the first kind and the modulus k of the Jacobi's elliptic functions is given by the transcendental equation

$$k = \frac{d}{2\xi_P}\cdot\frac{1}{K(k)} \tag{4.4.8}$$

The solutions of the above equation are sketched for two different thicknesses of the cell d in Fig.4.4.2. For a given thickness d of the cell, we obtain the following limits:

$$\text{Plane-wave-limit:} \qquad P_\circ \to 0 \quad \Rightarrow \quad k \to 0 \tag{4.4.9a}$$

$$\text{Soliton limit:} \qquad P_\circ \to \infty \quad \Rightarrow \quad k \to 1 \tag{4.4.9b}$$

There is again a crossover from the plane-wave modulation to the soliton-like modulation of the phase of the order parameter in splayed cells, as shown in Fig.4.4.3. For zero polarization, there is no internal electrostatic field to impose a distortion of the director field. On the other hand, for very high polarization, there is a very strong internal electrostatic field that expels a soliton wall to the close vicinity of the surface and sustains a rather uniform alignment in the middle of the cell, as shown in Fig.4.4.4.

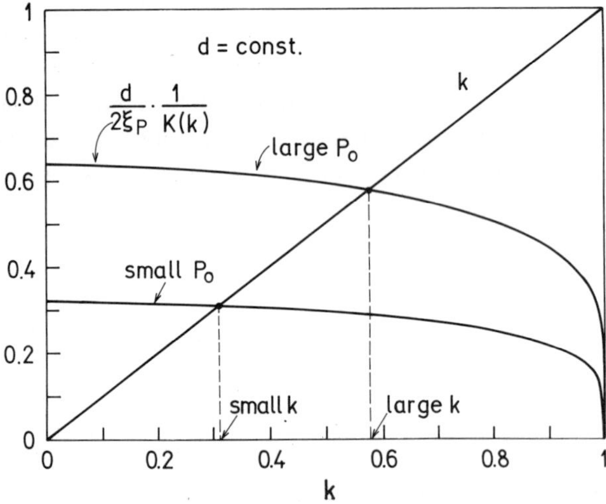

Fig.4.4.2. Sketch of the solutions of the transcendental Eq.4.4.8. For a given thickness d, we have small k for small polarization P and large k for very high polarizations.

It is easy to imagine that the total energy of the splayed state increases roughly as a square of the spontaneous polarization and eventually becomes comparable to the surface anchoring energy. We may therefore expect that for low-enough surface anchoring and very thin cells, the splayed state would become energetically unfavorable and there would be a transition to a pure bookshelf,

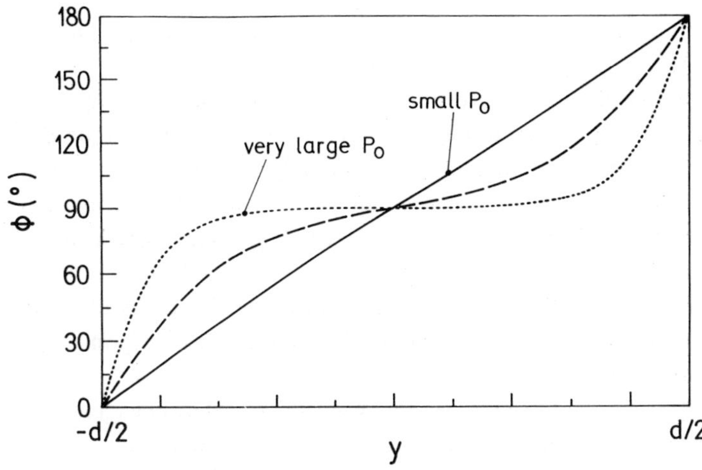

Fig.4.4.3. Crossover from the plane-wave modulation of splayed director field at small spontaneous polarization to the soliton-like modulation for very high polarization.

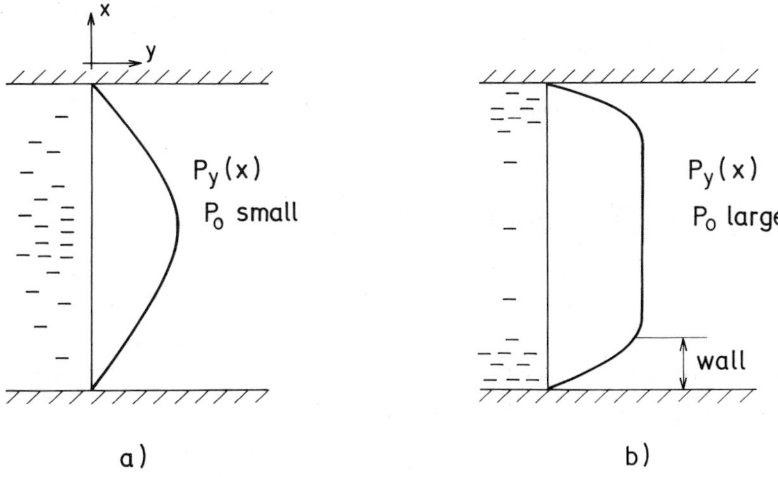

Fig.4.4.4. (a) For a very small polarization, the splayed state is nearly sinusoidally modulated. (b) For a high spontaneous polarization, the interior of the cell is nearly homogeneous, whereas close to the surface, soliton walls are present. Note the distribution of charges in both cases.

spatially homogeneous structure. This kind of phase transition has been treated by Pavel (1984), Akahane et al.(1989) and others.

One of the most interesting and also technologically important issues is the dynamics of these soliton-like splayed states. The ***nonlinear dynamics*** of these structures has been analyzed numerically by Maclennan et al. (1991), in relation to the switching of surface stabilized ferroelectric devices (Clark and Lagerwall, 1991; Abdulhalim et al., (1992); Maclennan, Clark and Handschy, 1991). In contrast to the nonlinear dynamics, which occurs at very high driving electric fields, we shall here discuss the linearized dynamics of splayed states, which is important when studying the linear dielectric response or the light-scattering spectrum.

The nonequilibrium free-energy, which determines the dynamics of soliton-splayed structures, contains both the elastic distortion energy and the electrostatic energy. We shall use here the so-called ***adiabatic approximation***, where the local electric potential $\varphi(x,t)$ is always ***in equilibrium with the instantaneous director field***. This means, that there are no other sources of the electric potential in the system, except the electric charge distribution, which arises from the divergence of the polarization field. This approximation obviously breaks down if there are ionic impurities in the sample which screen the local electric potential. The ionic motion is considered in Chapter 10.6. In the adiabatic approximation, the instantaneous value of the local electric potential $\varphi(x,t)$ is related to the instantaneous phase profile $\Phi(x,t)$ by

$$\varepsilon\varepsilon_{\circ}\frac{d^2\varphi(x,t)}{dx^2} - P_{\circ}\cos\Phi(x,t) = 0 \tag{4.4.10}$$

If we consider that the fluctuating electric field is zero at the cell boundaries (metal electrodes), the linearized Landau-Khalatnikov equation of motion appears in the well known form of Lamee's equation of order one:

$$\frac{d^2\Psi}{du^2} + \left[h - 2k^2 sn^2\Phi_{\circ}\right]\cdot\Psi = 0 \tag{4.4.11}$$

Here, Ψ is the amplitude of the phase excitation and h is the eigenvalue of the differential equation, which has been discussed in Chapter 3.4. The non-equilibrium phase profile is

$$\Phi(x,t) = \Phi_\circ(x) + \Psi \cdot e^{-t/\tau} \tag{4.4.12}$$

and the equilibrium phase profile $\Phi_\circ$ satisfies the sine-Gordon equation (4.4.5). The reduced coordinate $u = x/\xi_P k + K$ is scaled with the polarization coherence length, which was defined before. Following these considerations, the relaxation rate of the phase excitation is given by the eigenvalue h of the Lamee equation:

$$h = k^2\left(1 + \frac{\gamma\varepsilon\varepsilon_\circ}{P_\circ^2\tau}\right) \tag{4.4.13}$$

The eigenfunction, which satisfies the fixed boundary conditions is selected from the general solutions of the Lamme equation (see Chapter 3.)

$$\Psi = sn(u,k) \qquad \text{and} \qquad h = 1 + k^2 \tag{4.4.14}$$

The corresponding relaxation rate of the lowest-order mode is

$$\tau^{-1} = \frac{P_\circ^2}{\gamma\varepsilon\varepsilon_\circ k^2} = \frac{4K}{\gamma} \cdot \frac{K^2(k)}{d^2} \tag{4.4.15}$$

Let us remember that the wavelength of this excitations equals one quarter of the cell thickness d, i.e. the corresponding wave-vector is $q = \pi/d$. The above expression can be therefore understood as a "dispersion" relation for soliton-like excitations in a cell with variable thickness and is shown in Figure 4.4.5.

One of surprising features of the dispersion relation for solitary waves in splayed cells with variable thickness is the appearance of a gap in the long-wavelength limit, $q \approx 1/d^2 \to 0$. This can be easily seen by calculating a limiting value of the Eq.4.4.15:

$$G = \lim_{1/d \to 0} \tau^{-1} = \frac{P_\circ^2}{\gamma\varepsilon\varepsilon_\circ} \tag{4.4.16}$$

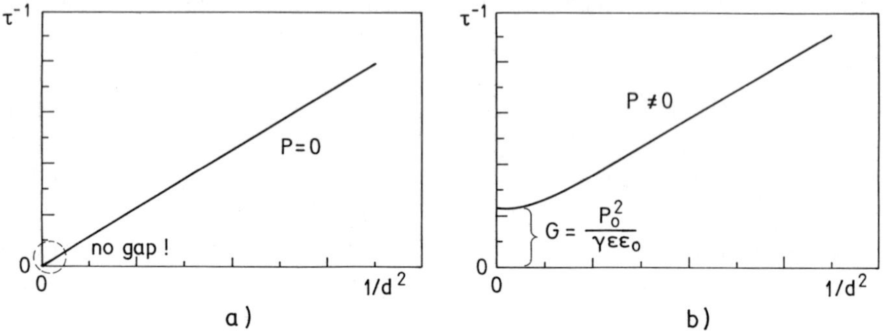

Fig.4.4.5. Dispersion relation for phase excitations in thin, splayed cell. (a) Plane wave limit $k \to 0$ and parabolic dispersion. (b) Soliton-like limit $k \to 1$ and the appearance of a gap at $q = 0$ in the dispersion relation.

Remembering the similarity to the magnetic-field and electric field-induced gaps, we can relate the appearance of this gap to the splayed geometry of the equilibrium polarization field. This field acts via induced electric charges as an internal electrostatic field, which induces a soliton-like distortion of the stationary phase profile and drives the dynamics of splayed cells into the solitary-wave regime.

The nature of this phenomenon can be further clarified by considering the form of the excitations and the limiting values of the dispersion relation (4.4.15). Fig.4.4.6. shows the effect of a soliton-like fluctuation, given by the Eq.4.4.14. One can see, that this fluctuation represents back and forth movement of a belly-like polarization profile.

Let us now look at the limiting values of the dispersion relation Eq.4.4.15. For a very small spontaneous polarization we always have the plane-wave regime and the dispersion is parabolic

$$P_\circ = 0 \Rightarrow k = 0 \quad : \qquad \tau^{-1} = \frac{K}{\gamma} \cdot \left(\frac{\pi}{d} \right)^2 \qquad (4.4.17)$$

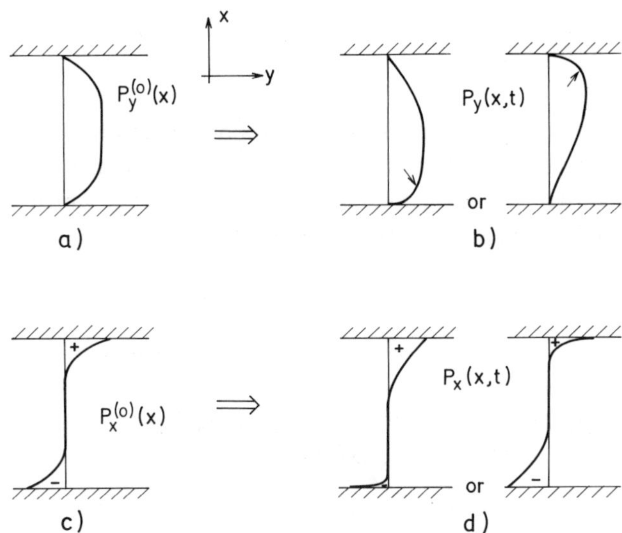

Fig.4.4.6. The lowest-order soliton-like fluctuations in a thin cell. (a) The equilibrium, soliton-like distorted polarization profile. (b) The polarization profile, where a fluctuation $\Psi = \pm sn(u,k)$ has been added. (c) and (d): The other component of the polarization. Note that in case (d), there is a nonzero average of the electric polarization along the x-direction, which means that this mode is polar. The drawings are schematic.

For a finite polarization $P_\circ$, we can have either the plane wave $(k \to 0)$ or the soliton $(k \to 1)$ regime, depending on the ratio d/ξ_P. For very small thickness we have again the plane-wave regime, but the relaxation rates depend on polarization:

$$P_\circ = const., d/\xi_P \to 0 \text{ and } k \to 0 \quad : \tau^{-1} = \frac{K}{\gamma} \cdot \left(\frac{\pi}{d}\right)^2 + \frac{K}{\gamma} \cdot \frac{1}{2\xi_P^2} \tag{4.4.18}$$

For large thickness, we have a soliton-like dynamics, which is polarization-dependent and there is a finite gap at $q = 0$:

$$P_\circ = const., K(k) \to \infty \text{ and } k \to 1 \quad : \tau^{-1}(q=0) = \frac{K}{\gamma} \cdot \frac{1}{\xi_P^2} \tag{4.4.19}$$

We see that the presence of the spontaneous polarization increases the relaxation rates of the excitations. This increase can be interpreted as a renormalization of the elastic constant for bend fluctuations of the director, which represent splay fluctuations of the polarization field (see Fig.4.4.1).

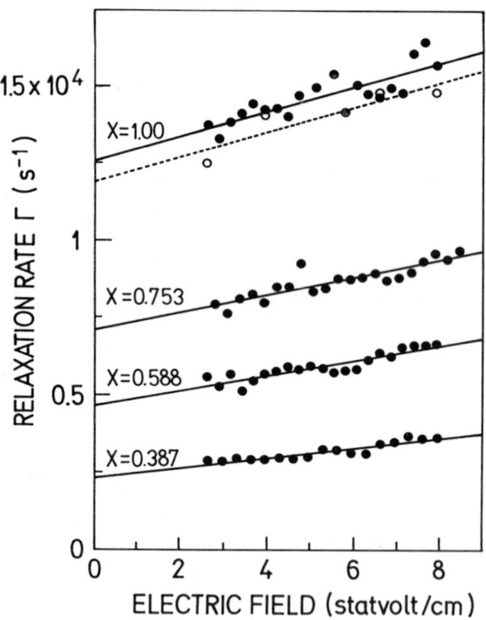

Fig.4.4.7. The relaxation rate τ^{-1} of a bend-like phase mode in a ferroelectric liquid crystal SCE-12 as a function of the applied field. The field dependence was measured at different magnitudes of the spontaneous polarization, which was determined by the fraction x of the chiral crystal in the racemate. Note the overall increase of the relaxation rates with increasing polarization, which is due to the renormalization. After Lu, Min-Hua et al.(1992).

The elastic constant increases as a square of the spontaneous polarization and this is reflected in the increase of the relaxation rate of phase excitations. The apparent increase of the elastic constant is due to the appearance of a fluctuation-induced electrostatic charge density and a consequent increase of the total free-energy of the system. This represents an additional thermodynamic force which drives the system back to equilibrium and results in an apparent increase of the liquid crystalline elastic constant.

The renormalization of liquid crystalline elastic constants due to the fluctuations of the polarization field was first considered by Palierne (1986) for the nematic phase and latter by Okano (1986) for the ferroelectric $smectic-C^*$ phase. The effect was first observed in a light-scattering experiment of Lu, Min-Hua et al.(1992), which was performed in a chiral $smectic-C^*$ material. Their results are shown in Fig.4.4.7. and Fig.4.4.8. and clearly demonstrate the polarization-renormalization of the B_1 (bend) elastic modulus.

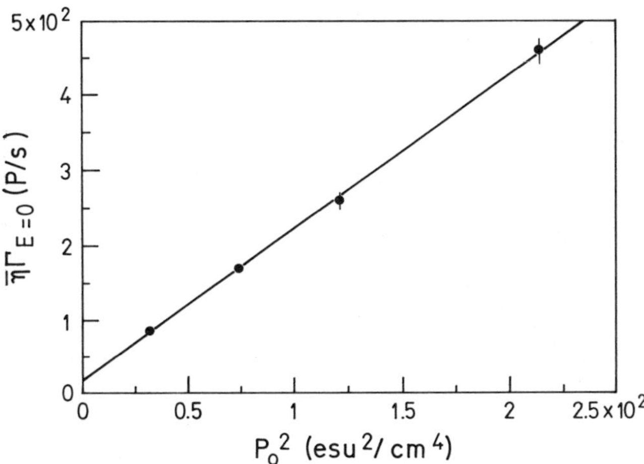

Fig.4.4.8. The bend elastic constant $(\gamma \cdot \tau^{-1} \propto B_1)$ as a function of the magnitude of the spontaneous polarization of the sample. The polarization was varied by changing the concentrations x of the chiral and racemic versions of the liquid crystal SCE-12. After Lu, Min-Hua et al.(1992). Note: $1 \cdot esu = 3.3 \times 10^{-10}\, As$.

The soliton-like dynamics in submicron ferroelectric cells was observed by Škarabot et al. (1998b) using the linear-electrooptic response technique. The relaxation rate of the phase mode, as observed in wedge-type cells of splayed ferroelectric *smectic* $-C^*$ phase is shown as a function of $1/d^2$ in Figure 4.4.9. For small thickness (i.e. large $1/d^2$), the relaxation rate follows the predicted $1/d^2$ dependence and there is apparently an overall increase of the relaxation rate which leads to a finite gap if the measurements are extrapolated to $q = 0$. There is however, an obvious crossover to the plane-wave like dynamics at larger thickness (small $1/d^2$). This can be explained by the presence of ions, which screen-out the electrostatic field. As a result of the screening, the interior electric field is diminished, which drives the order parameter dynamics into the plane-wave regime.

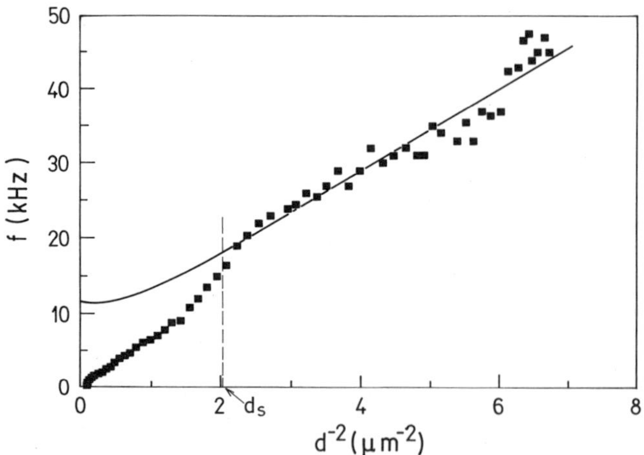

Fig.4.4.9. Relaxation rate of the phase mode in the *smectic* $-C^*$ phase of CE-8 in ultrathin, homogeneous and splayed cells. Above a thickness of approximately 0.8μm, the screening of the electric potential by ions diminishes the effect of the polarization renormalization of the elastic constant. The estimated Debye screening length is of the order of 0.5μm. The upper solid line represents a fit to Eq.4.4.15. with the known material parameters for CE-8.

Note 4.4.1. Magnitude of space charges and electric fields due to splayed polarization. We consider splayed polarization in a simple form:

$$\vec{P} = P_\circ(\cos\Phi, \sin\Phi, 0)$$

Splay induced charge density is $\rho(x) = -divP = -P_\circ q \cdot \sin\Phi(x)$, where $q = 2\pi/d$ is the wave vector of the splay deformation. By taking $P_\circ = 10\,nC/cm^2$ and $d = 1\mu m$, the maximum charge density is of the order of $\rho_{\max} \approx 10^{-24}\,As/nm^3$, which corresponds to approximately 10^{-5} elementary charges per molecule. Although this induced charge is small, the corresponding local electric field

$$E = \frac{P_\circ}{\varepsilon\varepsilon_\circ}$$

is of the order of $E \approx 2V/\mu m$. This is rather large field, which can significantly influence switching properties of splayed ferroelectric cells.

4.5. Confinement and Finite Size Effects on the Dynamics

The confinement and the effect of finite size of the sample strongly influence the long-wavelength excitations of a system. For example, in a planar sample with a finite size in the x-direction, only the discrete values are possible for the q_x component of the wave-vector of director fluctuations, as sketched in Fig.4.5.1. In the limit of strong surface anchoring, when the director is fixed at both surfaces, the allowed values for q_x are

$$q_x^n = n \cdot \frac{\pi}{d} \qquad (4.5.1.)$$

where d is the thickness of a sample. The lowest allowed wave-vector is therefore $q = \pi/d$ and corresponds to a half-wave excitation confined between two walls. If the anchoring at the surface is not infinitely strong, the director field can fluctuate at the surface. The corresponding wavelength is larger and the relaxation rates are slower until in the limit of zero anchoring the wavelength of fundamental excitation is infinite and the relaxation rates are zero.

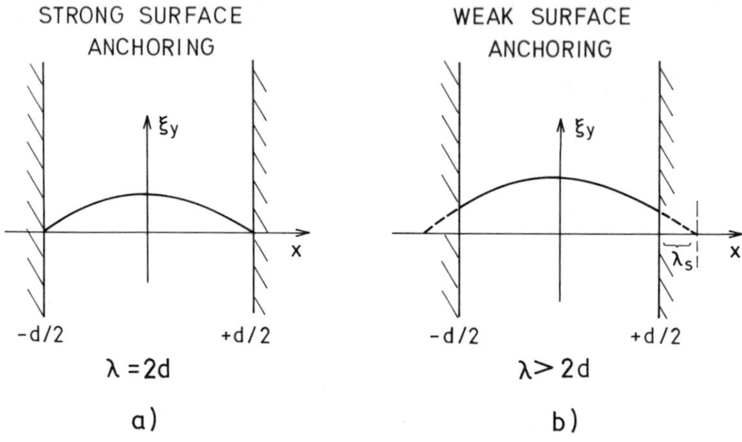

Fig.4.5.1. Fundamental excitations of a director field, confined between two flat surfaces. (a) Very strong surface anchoring. (b) Intermediate strength of the surface anchoring.

As an example of finite-size effects, let us look at the soft mode dynamics in a nonchiral $smectic-A$ phase, which is confined in a plan-parallel cell with thickness d. The free-energy density is

$$g = g_A + \tfrac{1}{2}a\left(\xi_x^2 + \xi_y^2\right) + \tfrac{1}{4}b\left(\xi_x^2 + \xi_y^2\right)^2 +$$

$$+ \tfrac{1}{2}K_3\left[\left(\frac{\partial \xi_x}{\partial z}\right)^2 + \left(\frac{\partial \xi_y}{\partial z}\right)^2\right] + \tfrac{1}{2}K'\left(\frac{\partial \xi_x}{\partial x}\right)^2 + \tfrac{1}{2}K''\left(\frac{\partial \xi_y}{\partial x}\right)^2 \quad (4.5.2)$$

Here we have assumed that the system is homogeneous in the y-direction and non-chiral $(\Lambda = 0)$. $K'' < K'$ are the elastic constants for the splay and bend-like deformation, respectively. For simplicity we take strong anchoring with fixed boundary conditions $\xi_x(x = \pm L) = \xi_y(x = \pm L) = 0$ at the confining surfaces. This implies fluctuations of the form

$$\xi_x(x,z) = e^{iq_z z} \sum_n C_n \cdot \cos(q_{nx} x) \tag{4.5.3a}$$

$$\xi_y(x,z) = e^{iq_z z} \sum_n D_n \cdot \cos(q_{nx} x) \tag{4.5.3b}$$

Here $q_{nx} = n \cdot \pi / d$ is a set of discrete wave-vectors that are allowed in the system because of boundary conditions in the *x*-direction After minimization with respect to ξ_x and ξ_y we find a phase transition into a homogeneously splayed state with $\xi_x = 0$ and $\xi_y \neq 0$ for temperatures lower than $T_c^S = T_\circ - (K''/\alpha)(q_{1x})^2$.

The finite-size effects are directly reflected in the excitation spectrum of the $smectic - A$ phase. After applying the Landau-Khalatnikov equations of motion we find in the $smectic - A$ phase a set of discrete branches of excitations, separated by forbidden frequency gaps. Each set is characterized by a discrete, *n*-th value of the transverse wave-vector and is composed of one splay- and one bend-like excitation branch. For the lowest two branches with $n = 1$ the relaxation rates of splay-like and bend-like excitations are

$$\tau_S^{-1}(\vec{q}) = \frac{\alpha}{\gamma}\left(T - T_c^S\right) + \frac{K_3}{\gamma} q_z^2 \tag{4.5.4a}$$

$$\tau_B^{-1}(\vec{q}) = \frac{\alpha}{\gamma}\left(T - T_c^{sp}\right) + \frac{K_3}{\gamma} q_z^2 + \frac{(K' - K'')}{\gamma} \cdot \left(\frac{\pi}{d}\right)^2 \tag{4.5.4b}$$

The splay mode is therefore the ***soft mode of the transition*** that softens at $\vec{q} = (\pi / d, 0, 0)$ in the reciprocal space. Due to finite-size effects, the relaxation rates of bend mode increase as $1/d^2$ and are separated from the splay modes by a thickness-dependent gap $G(d)$

$$G(d) = \frac{K'' - K'}{\gamma} \cdot \left(\frac{\pi}{d}\right)^2 \tag{4.5.5}$$

Both types of fluctuations are shown in Fig.4.5.2.

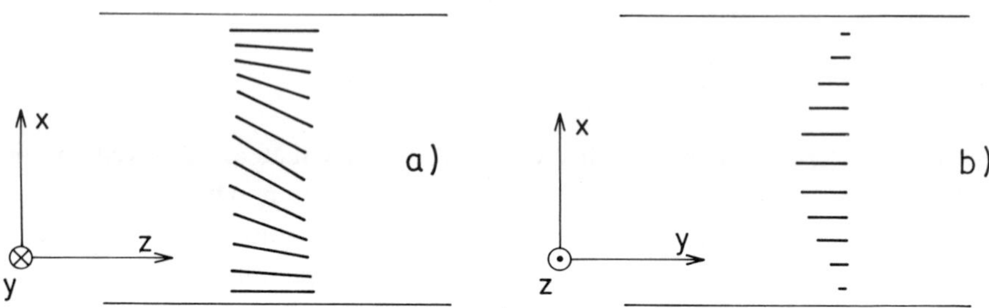

Fig.4.5.2. (a) Splay-like and (b) Bend-like fluctuations in the confined *smectic* – *A* phase. In (a), the smectic layers are perpendicular to the paper, whereas in (b), the layers are parallel to the paper.

This simple example of a phase transition in a confined geometry clearly indicates that a confinement may have similar effects on the order parameter dynamics as an external field. Whereas we have for the nonchiral *smectic* – *A* - *smectic* – *C* bulk transition a soft mode instability at $\vec{q}=0$, we have in a confined system with fixed boundary conditions a soft mode instability at $\vec{q}=(\pi/d,0,0)$ and a thickness-dependent splitting of the excitation branches. Finite-size effects in the eigen-dynamics of thin, homogeneously aligned *smectic* – $\overline{C}^*$ ferroelectric liquid crystals were observed by numerous authors (Cava et al., 1987; Panarin et al., 1994a, b; Glogarova et al., 1995). The most pronounced phenomenon is here the appearance of the so-called "thickness" mode (Panarin et al., 1994a; Glogarova et al., 1995). This is the transverse phase mode, which we have discussed in the Section 4.4. The wave-vector of this mode is discrete due to the boundary conditions and its relaxation frequency increases as $1/d^2$, when the thickness of the cell is decreased (see Fig.4.4.9.).

Finite-size effects can also be observed in the vicinity of the *smectic* – *A* - *smectic* – C^* transition. Confinement-induced increase of the relaxation rate of the soft mode at the phase transition was observed both in the quasielastic-light scattering (Rastegar et al., 1996a and b) and linear-optical response experiments (Škarabot et al., 1999) near the *smectic* – *A* - *smectic* – C^* transition in very thin cells.

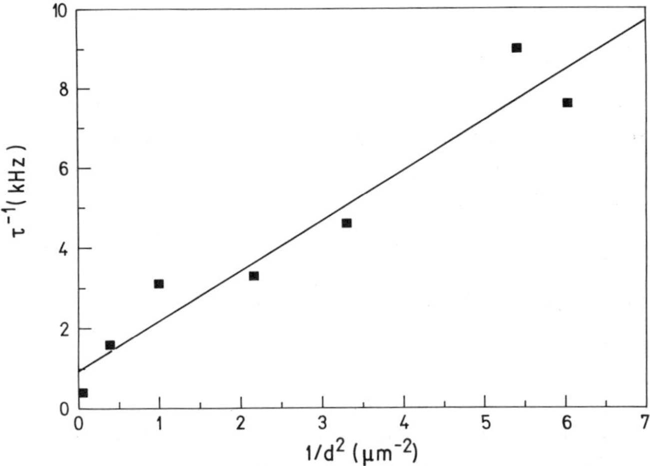

Fig.4.5.3. Relaxation rate of the soft excitation at the *smectic* $-A$ - *smectic* $-C^*$ phase transition temperature in CE-8 as a function of the inverse thickness of a homogeneously aligned sample (Škarabot, 1997).

In this case the *smectic* $-C^*$ phase is unwound by the surface anchoring and we have a phase transition of nonchiral nature, as discussed before. The linear-response measurements in these ultrathin samples show an incomplete critical slowing down. At the phase transition, there is a finite value of the measured relaxation frequency, which increases as a square of the inverse thickness of a sample, as shown in Fig.4.5.3. Similar observations were reported in a light scattering experiment on wedge-type cells (Rastegar et al., 1996b) The incomplete softening of a soft mode is here the result of confinement. It indicates that polar modes at $\vec{q}=0$, which are observed in the linear response experiment, are not critical modes of the transition in very thin ferroelectric cells. Instead, the critical slowing-down should be observed at a discrete value of the transverse wave-vector q_x, given by Eq.4.5.1.

Let us briefly discuss the case of weak anchoring (Panarin et al., 1995, Rastegar et al., 1996a). In the case of Rapini-Papoular type of anchoring (Rapini and Papoular, 1969), a surface energy term has to be added to the free-energy density (Eq.4.5.2.)

$$g_{surface} = \tfrac{1}{2}\left[\delta\left(x+L\right)+\delta\left(x-L\right)\right]\cdot W_s \xi_x^2 \tag{4.5.6.}$$

Here, ξ_x is the out-of surface component of the director and $W_s > 0$. The Landau-Khalatnikov dynamical equations lead to a diffusion equation with mixed boundary conditions. The transverse value of the wave-vector q_x^n now has to satisfy the transcendental equation

$$\frac{K}{W_s} \cdot q_x^n \cdot \tan\left(q_x^n \cdot L\right) = 0 \tag{4.5.7.}$$

Here, we have used one elastic constant K approximation. The quantity $\lambda_s = K/W_s$ is the so-called extrapolation (penetration) length (de Gennes, 1974) and is a measure for the strength of the surface coupling, as shown in Fig.4.5.4.

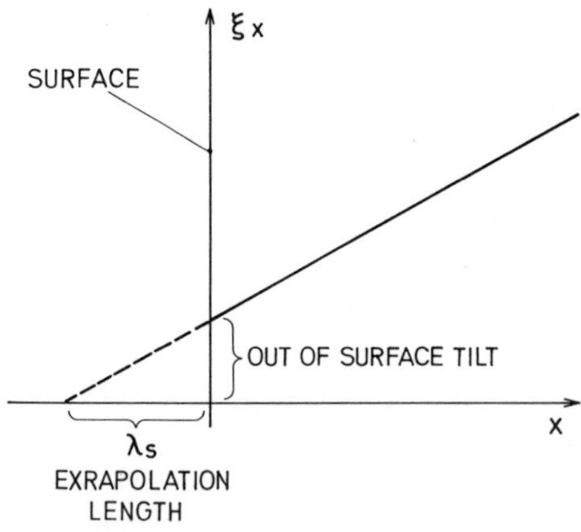

Fig.4.5.4. The out-of plane component of the tilt in the vicinity of a solid wall. Note that the surface extrapolation length λ_s determines the magnitude of the allowed fluctuation at the surface. For a large extrapolation length, the system will be virtually free and large fluctuations are allowed at a surface.

For strong couplings W_s this length is small, $\lambda_s \to 0$, and the director is nearly fixed at the boundary. This means that it cannot fluctuate at the surface and the transverse wave-length of the fundamental fluctuation wave equals $\lambda = 2d$, as shown in Fig.4.5.1.(a). In contrast, the extrapolation length is large for weak surface anchoring, and the fluctuations at a surface can be large. As a consequence, the wavelength of the fundamental wave is larger than for strong anchoring, as shown in Fig.4.5.1.(b). The system is therefore "loose" in this weak coupling regime and behaves virtually as a free, unconfined sample.

It should be noted that the surface coupling energy W_s can be determined if one measures the extrapolation length and the elastic constants are known (see, for example Rastegar et al., 1996a).

Note 4.5.1. Surface extrapolation lengths
The surface energy of ferroelectric liquid crystals on polymer surfaces is in the range of $10^{-5} - 10^{-4}\ J/m^2$. Since the elastic constants are of the order of $K \approx 10^{-11} N$, the surface extrapolation length is $\approx 1\mu m$ for weak anchoring ($W_s \approx 10^{-5}\ J/m^2$) and is much shorter, $\approx 0.1\mu m$ for strong anchoring ($W_s \approx 10^{-4}\ J/m^2$). This means that surface anchoring is important when the size of the sample is comparable to the surface extrapolation length.

4.6. Freely Suspended Ferroelectric Smectic Films

Freely suspended smectic films are formed by drawing a small amount of smectic material over a circular or rectangular hole, which is made by etching a thin glass or metal plate, as shown in Fig.4.6.1. In this way, a thin film of an integral number N of smectic layers is formed, which is stable against rupture for $N \geq 2$.

Freely suspended smectic films represent an ideal, defect-free system, where one can study phase transitions in well-defined and controllable geometry. Some of the most interesting physical phenomena that can be observed in these systems are for example free-surface freezing (Heinekamp et al., 1984; Geer et al., 1989; Swanson et al., 1989; Amador and Pershan, 1990; Geer et al., 1992; Stoebe et al., 1992b), crossover of the thermodynamics of the X-Y universality

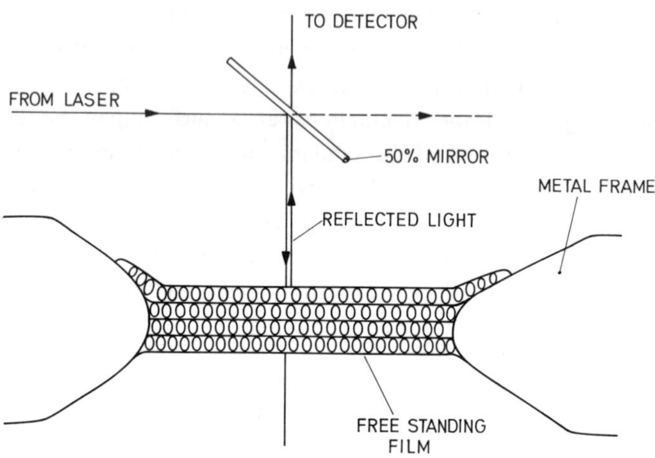

Fig.4.6.1. A free-standing smectic film is formed by spreading the liquid crystal across a sharp-edged hole. The thickness of the film is determined by measuring the intensity of a reflection of laser light.

class from three to two dimensions (Young et al., 1978; Rosenblatt et al., 1980; Heinekamp et al., 1984; van Winkle and Clark, 1984 and 1988; Amador and Pershan, 1990; Spector et al., 1993; Kraus et al., 1993; Demikhov et al., 1994), topological defects and their Brownian motion in two dimensions (Pindak et al., 1980); Hinshaw et al., 1988; Lo et al., 1990; Pang et al., 1992; Pargelis et al., 1992; Muzny and Clark, 1992; Maclennan et al., 1994; Demikhow, 1995; Demikhov et al., 1995), the onset and the dynamics of the hexatic order (Sprunt and Litster, 1987; Dierker and Pindak, 1987), phase transitions in antiferroelectric liquid crystals (Bahr and Fliegner, 1993a, b), etc. These properties were probed by very different experimental techniques, such as the quasielastic light scattering, AC microcalorimetry, surface tension technique, linear electrooptical response, ellipsometry, X-ray reflectometry and microscopic observations.

The experiment on freely suspended, ferroelectric $smectic-C^*$ films was first reported by Young et al.(1978). Using the quasielastic light scattering technique, they measured the dispersion relations for the bend and splay director modes in 2D suspended films of chiral octyloxybenzylidene-β-methyl-butyl aniline (8O.5*). Due to the presence of a finite spontaneous polarization, the bend-director mode with a wave-vector in the smectic plane represents a splay mode for the polarization. As we have already seen in the Section 4.4., this creates a time-fluctuating polarization-induced charge wave $\rho_P(\vec{r},t)$

$$\rho_P(\vec{r},t) = -div\vec{P}(\vec{r},t) \tag{4.6.1}$$

which is expected to have a significant effect on the dynamics (Young et al., 1978; Palierne, 1986; Okano, 1986; Lu et al., 1992). In particular, the long-range electrostatic interaction induces a frequency gap in the dispersion relation for the bend-director mode (Lu et al., 1992),

$$\tau^{-1}(q_x) = \frac{K_B}{\gamma} \cdot q_x^2 + \frac{P_\circ^2}{\varepsilon\varepsilon_\circ\gamma} \tag{4.6.2}$$

A finite frequency gap $G = P_\circ^2/(\varepsilon\varepsilon_\circ\gamma)$ appears, which prevents the existence of a truly gapless Goldstone mode for this branch of excitations, as shown in Fig. 4.6.2.

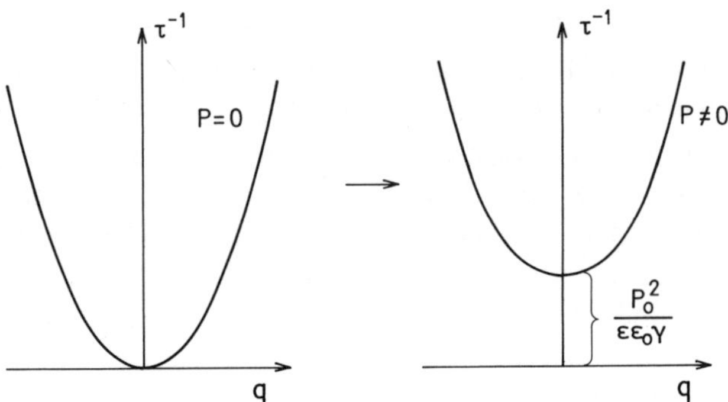

Fig.4.6.2. The crossover from the gapless excitation spectrum for zero spontaneous polarization (a) to a polarization-induced gap (b). The crossover should be observed for a pure bend-director mode with the wave-vector lying in the smectic planes.

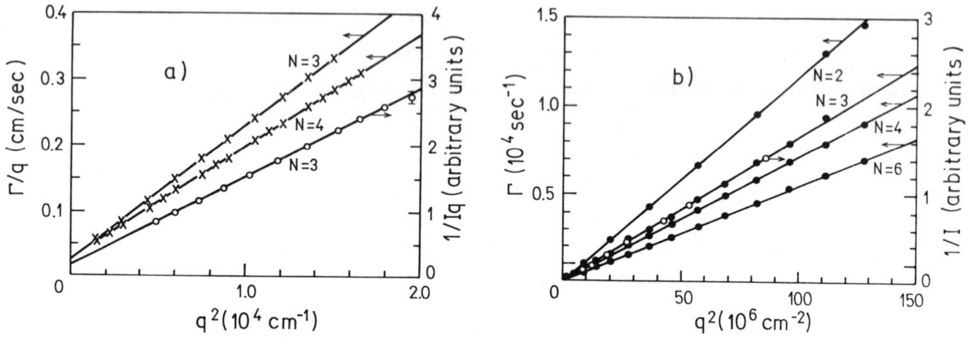

Fig.4.6.3. The dispersion relation (•,×) and the inverse intensity (o) as a function of the square of the wave-vector for the (a) bend and (b) splay director modes in freely suspended films of DOBAMBC for several film thickness (after Young et al., 1978).

The experiments of Young et al.(1978) clearly revealed the parabolic dispersion relation for the bend and splay modes in two-dimensional chiral-$smectic-C^*$ films (see Fig.4.6.3.) The experiment however failed to observe the onset of polarization-induced gap in the dispersion relation, which was probably due to very small spontaneous polarization of the sample. This polarization induced gap in the dispersion relation for a bend-mode was observed only recently in light scattering experiments of Lu et al.(1992), see Fig.4.4.7. in Section 4.4.

Later, the experiments of Pindak et al.(1978) and Rosenblatt et al.(1979, 1980) indicated an onset of a ***surface-induced order*** in freely suspended ferrolectric $smectic-C^*$ films, that was clearly demonstrated by an ellipsometry study of Heinekamp et al.(1984). The geometry of their ellipsometry experiment is shown in Fig.4.6.4.b.

Using a DC external field, they have aligned the directors in the $smectic-C^*$ phase in the *x-z* plane and used a linearly polarized laser light that was propagating along the *z*-direction, at 45° with respect to the layer normal. In this geometry, both the ordinary and the extraordinary eigenwaves are excited with equal amplitudes at the front surface of the sample.

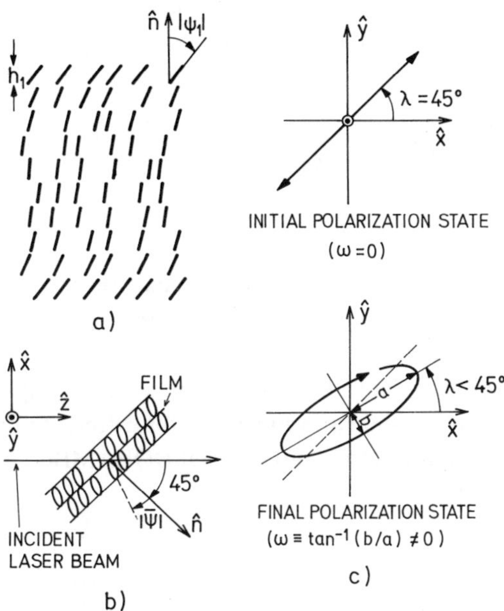

Fig.4.6.4. (a) Variation of the molecular tilt angle through a ten-layer, free standing film above the bulk transition temperature (schematic drawing). The interior layers are still *smectic* – *A* -like, whereas the two surface layers already show a finite tilt and a finite polarization, characteristic for the *smectic* – C^* phase. (b) The experimental geometry and (c), the polarization states of the incident light before and after transmission, as shown below. Note that the ordinary wave is polarized in the *y*-direction.

After passing through the sample, there is a phase difference $\Delta\Phi$ between the two eigenwaves. Because the smectic layer thickness is much smaller than the wavelength of light, the optical properties of the film are determined by the average value of the dielectric tensor across the suspended film. In this case we have an optically uniaxial film, which is composed of *N* optically uniaxial smectic layers, each with a different value of the tilt angle. The phase difference between the ordinary and extraordinary waves, propagating through such a film is proportional to the difference between the two refractive indices n_o and n_e. Whereas ordinary index n_o does not depend on tilt, the extraordinary index n_e is approximately proportional to the molecular tilt, $n_e \propto \sqrt{\varepsilon_{33}} \propto \sqrt{\vartheta^2} \propto \vartheta$. The phase difference between the two eigenwaves is in this case directly proportional to the average tilt of the molecules throughout the suspended film,

$$\Delta\Phi \approx \tfrac{1}{d}\int_0^d \left|\theta\left(u,T\right)\right| du \approx \tfrac{1}{N}\sum_{i=1}^{N} \left|\theta_i\left(T\right)\right| \tag{4.6.3}$$

Here θ_i is the tilt angle for the i-th layer. By analyzing the polarization state of the transmitted light (see Fig.4.6.4c), one can extract the space-averaged absolute value of the tilt angle.

The temperature dependence of the average tilt angle, as measured by Heinekamp et al.(1984) is shown in Fig.4.6.5. for different numbers of layers of the freely suspended ferroelectric liquid crystal DOBAMBC.

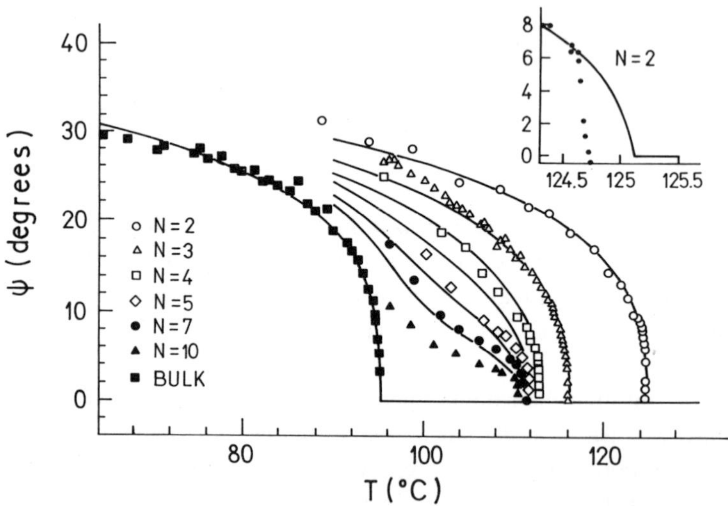

Fig.4.6.5. The variation of the average tilt angle versus temperature for different thicknesses of freely-suspended DOBAMBC. The solid curves are the best fit to a discrete mean-field theory. The inset shows data for $N = 2$, clearly demonstrating deviation from the mean-field theory.

The measurements on a two-layer film $(N = 2)$ clearly demonstrate that the two exterior layers undergo a phase transition into the ferroelectric state at much higher temperatures than the interior, bulk-like layers. This temperature difference may be as high as $30K$ for DOBAMBC and the phenomenon is similar to the surface-induced ferromagnetism in thin ferromagnetic films (Kaganov and Omelyanchouk, 1971; Binder and Hohenberg, 1972). For films with $N > 2$, the onset of the tilt angle is observed at T_N, which gradually decreases with increasing *N*. It is however obvious, that the ordering starts at the surface and propagates into the interior layers with decreasing temperature (see the set of data for $N = 10$, Fig.4.6.5.).

The observed variation of the average tilt angle

$$|\theta| = \frac{1}{N}\sum_{i=1}^{N}|\theta_i| \tag{4.6.4}$$

can be well described by a simple Landau model based on the following free-energy expansion:

$$G = G_S + G_B + G_C \tag{4.6.5}$$

Here G_S is the contribution from the two surface layers,

$$G_S = a_S(T)\left[\theta_1^2 + \theta_N^2\right] + b_S\left[\theta_1^4 + \theta_N^4\right] + c_S\left[\theta_1^6 + \theta_N^6\right] \tag{4.6.6}$$

where $a_S = \alpha_S(T - T_S)$ describes the phase transition for the two surface layers and the sixth-order term has been added to correctly describe the critical behavior (Huang and Viner, 1982). The "bulk contribution is

$$G_B = \sum_{i=2}^{N-1}\left(a\cdot\theta_i^2 + b\cdot\theta_i^4 + c\cdot\theta_i^6\right) \tag{4.6.7}$$

where $a = \alpha(T - T_B)$ and T_B is the bulk phase transition temperature. The coupling between the two neighboring layers is taken in a form

$$G_C = \sum_{i=1}^{N-1} d \cdot \left(|\theta_{i+1}| - |\theta_i|\right)^2 \tag{4.6.8}$$

It is assumed that this coupling is temperature independent and one can put $d = 1$ for convenience, as only the ratio between the magnitude of each terms in the free-energy is important. For $N = 2$, $G = G_S$ and one finds by minimization (Heinekamp et al., 1984)

$$\theta_1 = \theta_2 = \left(\frac{b_S}{3c_S}\right)^{1/2} \cdot \left[\left(1 + \frac{3t}{t_\circ}\right)^{1/2} - 1\right]^{1/2} \tag{4.6.9}$$

where $t = T - T_S$ and $t_\circ = -b_S^2/(\alpha_S c_S)$. The above expression provides an excellent fit to the $N = 2$ data except close to the transition temperature T_S, see inset to Fig.4.6.5. For an arbitrary *N*, the free-energy has to be minimized numerically and the agreement with the experiment is excellent, as can be seen in Fig.4.6.5.

The continuum version of the Landau-free energy expansion (Eq.4.6.5.) can be solved analytically (Heinekamp et al., 1984; Kaganov and Omelyanchuk, 1971) and gives further insight into the physics of the system. In particular, it allows the evaluation of the critical temperature T_N as a function of *N* from the expression (see Heinekamp et al., 1984 and Kaganov and Omelyanchuk, 1971)

$$\tanh\left\{\tfrac{1}{2}(N-2)[\alpha(T_N - T_B)]^{1/2}\right\} = \alpha'(T_S - T_B)/\alpha(T_N - T_B)^{1/2} \tag{4.6.10}$$

The relative increase of the $smectic-A$ - $smectic-C^*$ phase transition temperature in chiral ferroelectric, freely suspended smectic films with *N* smectic layers is shown in Fig.4.6.6.

It turns out that for a given *N* one can define a characteristic length, which equals the bulk mean-field correlation length ξ_B, evaluated at T_N and measured in units of the interlayer spacing:

$$\xi_B(T_N) = \left(\alpha \cdot (T_N - T_B)\right)^{-1/2} \tag{4.6.11}$$

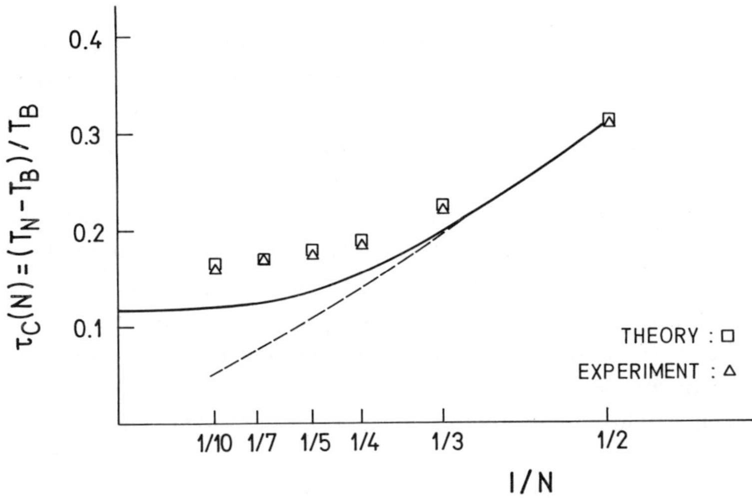

Fig.4.6.6. Temperature shift of the $smectic-A$ - $smectic-C^{*}$ phase transition temperature as a function of the inverse film thickness. The solid curve is the prediction of the continuum approximation, given by the Eq.4.6.10. Predictions of the discrete mean-field model, based on a numerical minimization of the free-energy given by Eqs.4.6.6.-4.6.8. are denoted by squares. The agreement with the experiment (triangles) is remarkable (after Heinekamp, et al.1984).

For the phase transition temperature of a N-layer film we obtain the following limits

$$N << \xi_B(T_N) \quad : \quad (T_N - T_B)/T_B \approx 1/N \qquad (4.6.12a)$$

$$N >> \xi_B(T_N) \quad : \quad (T_N - T_B)/T_B \approx const. \qquad (4.6.12b)$$

We can therefore distinguish "thick" and "thin" freely suspended $smectic-C^{*}$ films and the crossover occurs when the thickness of a film is equal to $\xi_B(T_N)$.

We have already mentioned that the onset of a finite tilt in the two surface layers at temperatures far above the bulk phase transition temperature can be considered as a result of the action of a surface field. This has led to several interesting experiments performed by Bahr and Fliegner (1992a) and later by Kraus et al.(1993) on ferroelectric compounds that show a discontinuous, first order $smectic-A$ - $smectic-C^*$ phase transition in the bulk. It was shown (Bahr and Heppke, 1990) that this phase transition line terminates at a critical point (E_c, T_c) in the (E,T) phase diagram and it was conjectured that the surface field could have a similar effect. This means that for thick freely suspended films, the $smectic-A$ - $smectic-C^*$ phase transition should be of first order. Then, by decreasing the film thickness, the conjugate field, which originates from the surface, would penetrate deep-enough into the sample to drive the phase transition close to the critical point. This was indeed observed in the ellipsometry experiment by Bahr and Fliegner (1992a) in 4-(3-methyl-2-chloropentanoyloxy)-4'-heptyloxybiphenyl (C7), as shown in Fig.4.6.7.

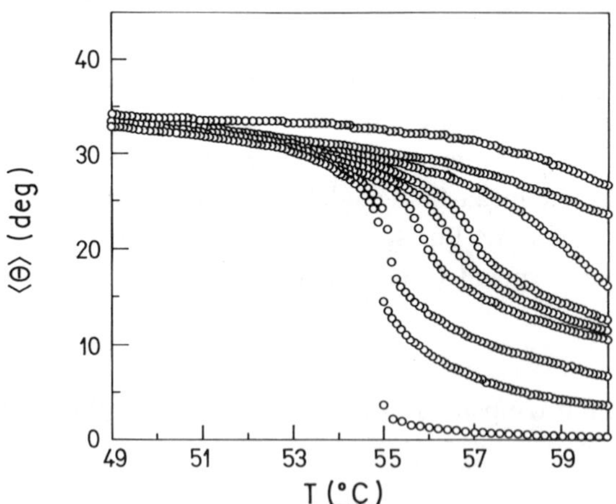

Fig.4.6.7. Temperature dependence of the average tilt angle $\langle\theta\rangle$ in various free-standing films of C7, as determined from ellipsometry (after Bahr and Fliegner, 1992a). The film thicknesses are (from above): 2, 4, 6, 8, 10, 12, 16 ± 1, 27 ± 2 and 190 ± 10 smectic layers.

Whereas a very thick sample shows a pronounced discontinuity at the phase transition temperature, this is less and less pronounced for thinner samples. For films with $N < 16$, the evolution of the tilt angle with temperature is smooth, indicating that the system is in the vicinity of a critical point in the (d, T) phase diagram. The above crossover from the first-order to a critical transition, as induced by the surface field, is well described with a simple Landau model, similar to the model of Heinekamp (1984) (see also Bahr and Fliegner, 1993b). Here the interlayer coupling is neglected and a surface field, conjugate to the order parameter is introduced in an exponentially decaying form $S_N = S \cdot e^{-s(N-1)}$.

Chapter 5

Optical Properties

5.1. Introduction

Many of the fundamental properties of ferroelectric liquid crystalline phases can be conveniently studied by different optical and electrooptical techniques, such as quasielastic light scattering, optical response measurements, selective reflection, optical interferometric techniques etc. The reason for using optical spectroscopic methods to study ferroelectric liquid crystals is in the strong coupling of the director field $\vec{n}(\vec{r})$ to the dielectric tensor field $\underline{\varepsilon}(\vec{r})$, as well as in the high sensitivity and local probing of optical methods. A good understanding of the optical properties of ferroelectric liquid crystals, which are by themselves a very attractive and interesting problems, is thus necessary. Although almost everything which is related to the problem of light propagation in these birefringent and helicoidally modulated structures seems hardly solvable analytically, there has been significant progress initiated by the work of Oldano et al.(1983;1984;1985). We shall present here some simple views and concepts, which can explain most of the basic optical phenomena in ferroelectric liquid crystals.

In the ferroelectric $smectic-C^*$ phase, each smectic layer can be considered as a weakly biaxial $smectic-C$ phase, which is described by the dielectric tensor $\underline{\varepsilon}$ with the eigenvalues ε_1, ε_2 and ε_3

$$\underline{\varepsilon} = \begin{bmatrix} \varepsilon_1 & 0 & 0 \\ 0 & \varepsilon_2 & 0 \\ 0 & 0 & \varepsilon_3 \end{bmatrix} \qquad (5.1.1)$$

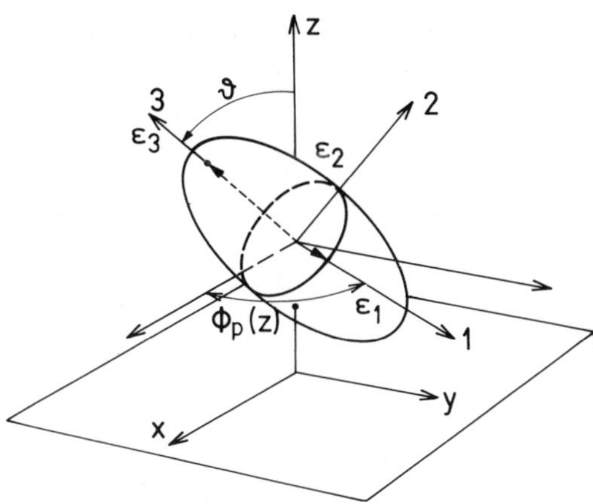

Fig.5.1.1. Local orientation of the dielectric tensor in the $smectic-C^*$ phase.

Due to symmetry arguments, one of the axis of the eigenframe (1,2,3) (e.g. $1||y||C_2$) must be parallel to the C_2 axis as shown in Fig.5.1.1. This is also the direction of the spontaneous dipole moment, see Fig.2.2.1. Furthermore, the difference between ε_1 and ε_2 is usually small, so that each layer of the $smectic-C^*$ phase can be considered as an optically uniaxial layer, with the optical axis tilted at a tilt angle θ with respect to the helical axis, as shown in Fig.5.1.1.

The dielectric tensor which defines the optical properties of a single smectic layer the $smectic-C^*$ phase is obtained by two successive rotations of the tensor $\underline{\varepsilon}$, as shown in Fig.5.1.2(a)-(c). We start from the configuration, where the eigenframe (1,2,3) coincides with the laboratory frame (x,y,z), as shown in Fig.5.1.2.(a). First the tensor $\underline{\varepsilon}$ is rotated through the tilt angle θ around the x-axis, as shown in Fig.5.1.2.(a). This is followed by a rotation through the angle $\Phi_P = \pi/2 + \Phi(z)$ around the z-axis, Fig.5.1.2(b). Note that the angle of rotation $\Phi_P(z)$ is the angle between the direction of spontaneous polarization and the x-axis, as first introduced by D.W.Berreman (1973). This angle is 90° larger than the phase angle $\Phi(z)$ between the projection of the director $\xi(z)$ and the x-axis, as shown in Fig.5.1.2(c).

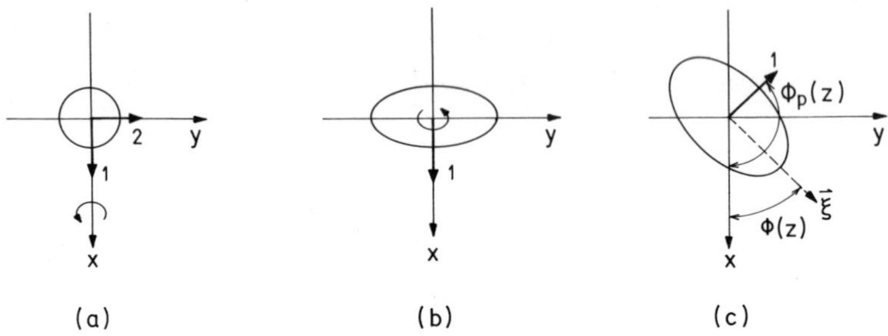

Fig.5.1.2. Successive rotations of the dielectric tensor from the upright (a) to the tilted (b) and twisted (c) position. From (a) to (b) the tensor is rotated through the tilt angle θ around the x-axis. From (b) to (c) the tensor is further rotated through the angle $\Phi_P = \pi/2 + \Phi(z)$ around the z-axis.

The resulting dielectric tensor can be written as a sum of three characteristic contributions:

$$\underline{\varepsilon}(z) = \begin{bmatrix} \frac{1}{2}\left[(\varepsilon_1+\varepsilon_2)+(\varepsilon_3-\varepsilon_2)\sin^2\theta\right] & 0 & 0 \\ 0 & \frac{1}{2}\left[(\varepsilon_1+\varepsilon_2)+(\varepsilon_3-\varepsilon_2)\sin^2\theta\right] & 0 \\ 0 & 0 & \varepsilon_3-(\varepsilon_3-\varepsilon_2)\sin^2\theta \end{bmatrix} +$$

$$+(\varepsilon_3-\varepsilon_2)\sin\theta\cos\theta\cdot\begin{bmatrix} 0 & 0 & \cos\Phi(z) \\ 0 & 0 & \sin\Phi(z) \\ \cos\Phi(z) & \sin\Phi(z) & 0 \end{bmatrix} +$$

$$+\frac{1}{2}\left[(\varepsilon_2-\varepsilon_1)+(\varepsilon_3-\varepsilon_2)\sin^2\theta\right]\cdot\begin{bmatrix} \cos 2\Phi(z) & \sin 2\Phi(z) & 0 \\ \sin 2\Phi(z) & -\cos 2\Phi(z) & 0 \\ 0 & 0 & 0 \end{bmatrix} \qquad (5.1.2.)$$

The first term is a typical dielectric tensor of a tilted, homogeneous $smectic-C$ phase, where the refractive indices change with tilt angle. The second and the third term are characteristic of the chiral, helicoidally modulated $smectic-C^*$ phase. It is important to note the ***different periods*** of these inhomogeneous terms: the second term is the so-called ***polar part of the dielectric tensor***. It repeats itself when the system is translated over a helical period $p_\circ$, i.e. it has the same translational properties as the spontaneous polarization field $P_\circ(z)$. The third term has the periodicity equal to one half of the helical period $p_\circ/2$, and is the so-called ***quadrupolar part of the dielectric tensor*** of the $smectic-C^*$ phase. Note that the structure of this term is equivalent to the dielectric tensor of a chiral nematic (de Gennes, 1974).

The problem of propagation of light waves in an infinite sample of the $smectic-C^*$ liquid crystal was first treated by D.W.Berreman (1973). He showed, that the $smectic-C^*$ phase and the chiral nematic N^* have identical properties when the light is propagating along the helical axis, whereas for oblique incidence, there are some remarkable differences. In particular, additional Bragg reflection peaks are expected in the $smectic-C^*$ phase, which are never observed in the chiral nematic phase. These, so-called ***full pitch reflection bands***, arise in the $smectic-C^*$ phase due to the double-periodic nature of the dielectric tensor (Eq.5.1.2.). The theory of light propagation along the helical axis of a $smectic-C^*$ phase was first developed by Parodi (1975) and Brunet (1975) on the basis of de Vries' theory of light propagation in chiral nematics. Whereas this problem can be solved analytically, the general form of the dispersion relation can be solved only numerically, as discussed by several authors. Among these, the discussion of Oldano (1984) gives a good insight into the physics of the problem.

The complexity of the problem of light propagation in the chiral $smectic-C^*$ phase can be seen from the structure of the wave equation, which follows from Maxwell equations for a nonmagnetic inhomogeneous medium

$$\nabla \times \nabla \times \vec{E}(\vec{r}) - \frac{\omega^2}{c_\circ^2} \cdot \underline{\varepsilon}(\vec{r}) \cdot \vec{E}(\vec{r}) = 0 \qquad (5.1.3.)$$

Here, $E(\vec{r})$ is the electric field of the light-wave, ω is the frequency of light, $c_\circ$ is the speed of light propagation in vacuum and $\underline{\varepsilon}(\vec{r})$ is the dielectric tensor of the $smectic-C^*$ phase, given by the Eq.5.1.2. This tensor is spatially periodic, which implies that the eigensolutions of the wave-equation should have the

Bloch form. A rather general discussion of the wave equation in chiral liquid crystals can be found in the work of Belyakov and Dmitrienko (1989; Taupin et al., 1978; Garoff et al., 1978).

The dielectric tensor in the wave equation (Eq.5.1.3.) can be conveniently considered as a sum of a spatially homogeneous term $\underline{\varepsilon}_\circ$ and the inhomogeneous term $\delta\underline{\varepsilon}$

$$\underline{\varepsilon}(z) = \underline{\varepsilon}_\circ + \delta\underline{\varepsilon}(z) \tag{5.1.4.}$$

The spatially inhomogeneous term in the dielectric tensor of the $smectic - C^*$ phase (i.e. the second and third term in Eq.5.1.2.) is typically by a factor $10^{-2}10^{-1}$ smaller than the principal values ε_i. This suggests that for the optical properties of the $smectic - C^*$ phase, a simple perturbative approach might be used, where spatially inhomogeneous terms play the role of a small perturbation. In this Chapter we shall describe approximations to the wave equation and compare them to the experimental results.

Note 5.1.1. Indices of refraction of ferroelectric liquid crystals

There are not many experimental data on the indices of refraction of ferroelectric liquid crystal. Here we present the indices of refraction of ferroelectric liquid crystal DOBAMBC, measured by S.Garoff (1977a).

Temperature (°C)	n_1	n_2	n_3
94.4	1.4782	1.4779	1.6998
93.8	1.4782	1.4776	1.7055
93.6	1.4782	1.4776	1.7097
93.4	1.4782	1.4775	1.7097
93.0	1.4781	1.4771	1.7129
92.0	1.4780	1.4769	1.7187
90.6	1.4779	1.4761	1.7262
89.5	1.4778	1.4761	1.7290
87.5	1.4788	1.4760	1.7359

5.2. Linear Optics of Helical Structures

Following the approach of M.A.Peterson (1983), we define a tensor-weighted inner product of the two eigenfunctions $|k,p\rangle$ and $|k',p'\rangle$, characterized by the wave-vector $\vec{k}$ and the polarization p as:

$$\langle \vec{k}', p' | \vec{k}, p\rangle = \langle \vec{E}_{k',p'} | \vec{E}_{k,p}\rangle = \frac{1}{(2\pi)^3} \int d^3\vec{r}\, \vec{E}_{\vec{k}',p'} \underline{\varepsilon}(\vec{r}) \vec{E}_{\vec{k},p} \tag{5.2.1}$$

Let $|k,p\rangle$ and $|k',p'\rangle$ represent the eigensolutions of the wave equation in the inhomogeneous medium characterized by the dielectric tensor $\underline{\varepsilon}(\vec{r}) = \underline{\varepsilon}_o + \delta\,\underline{\varepsilon}(\vec{r})$

$$\left\{ \underline{\varepsilon}_o^{-1} \nabla\times\nabla\times - \frac{\omega^2}{c_o^2} \underline{\varepsilon}_o^{-1} \delta\,\underline{\varepsilon} \right\} |\vec{k}, p\rangle = \frac{\omega^2}{c_o^2} |\vec{k}, p\rangle \tag{5.2.2}$$

where ω is the frequency and c_o is the speed of light in vacuum. The above form of the wave equation is analogous to the Schrodinger equation for the eigenvectors $|\Psi\rangle$ of the operator $\hat{A}$ in the presence of a small perturbation $\hat{V}$

$$\left(\hat{A} + \hat{V}\right) |\Psi\rangle = \lambda \cdot |\Psi\rangle \tag{5.2.3}$$

It can be shown that the operator $A \equiv \underline{\varepsilon}_o^{-1} \nabla\times\nabla\times$ is Hermitian with respect to the tensor-weighted inner product (Eq.5.2.1), so that the eigenvectors $|k,p\rangle$ are orthogonal and the eigenvalues $\omega^2(k,p)$ are always real. This allows us to use the standard perturbation techniques for the analysis of the spectrum of the eigenvalues and eigenfunctions in the presence of a small perturbation

$$\hat{V} \equiv -\frac{\omega^2}{c_o^2} \underline{\varepsilon}_o^{-1} \delta\,\underline{\varepsilon} \tag{5.2.4}$$

In the case of $\delta\underline{\varepsilon}=0$ we obtain the eigenvalue $\omega_o^2(\vec{k},p)$, describing the dispersion relation for the propagation of σ (ordinary) and π (extraordinary) linearly polarized plane waves in a birefringent, uniaxial medium. This linear dispersion relation is shown in Fig.5.2.1a and the corresponding eigenwaves are labeled $|\ k,\sigma\rangle$ and $|k,\pi\rangle$, respectively.

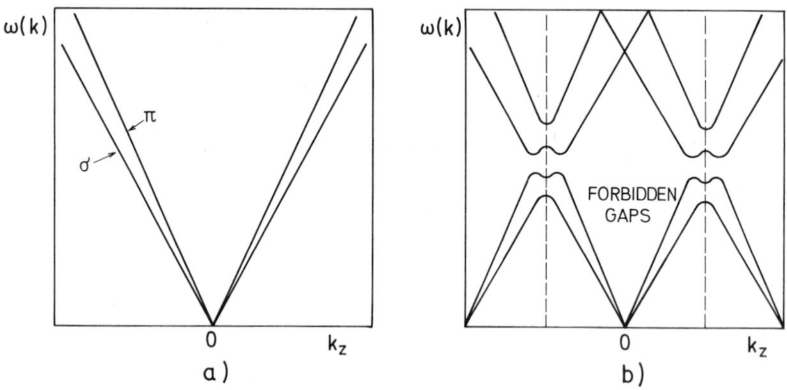

Fig.5.2.1. (a) Dispersion relation for light, propagating in a homogeneous and uniaxial medium. (b) Dispersion relation for light, propagating in a homogeneous uniaxial medium with a small periodic perturbation.

The presence of a small periodic perturbation (Eq.5.2.4) in the wave equation (Eq.5.2.2) significantly alters the situation. In particular, one has to introduce the concept of a Brillouin zone, whereas the eigensolutions should obtain the Bloch form, reflecting the appearance of a periodic perturbation term $\delta\underline{\varepsilon}(z)$. Moreover, on the basis of general considerations, one can expect that the periodic perturbation would have the strongest influence at the points of degeneration

$$\omega_o(\vec{k},p)=\omega_o(\vec{k}',p') \tag{5.2.5}$$

where we expect the appearance of forbidden frequency gaps and corresponding selective or total (Bragg) reflection.

The optical spectrum can be obtained by applying a simple perturbative approach to the wave equation (Eq.5.2.2). The perturbed eigenvalue $\omega^2(\vec{k},p)$ is given by

$$\omega^2(\vec{k},p)=\omega_o^2(\vec{k},p)\left\{1-\frac{\langle\vec{k},p|\underline{\varepsilon}_o^{-1}\delta\varepsilon|\vec{k},p\rangle}{\langle\vec{k},p|\vec{k},p\rangle}\right\}+$$

$$+\frac{\omega_o^2(\vec{k},p)}{\langle\vec{k},p|\vec{k},p\rangle}\sum_{\vec{k}',p'}\frac{\omega_o^2(\vec{k}',p')}{\omega_o^2(\vec{k},p)-\omega_o^2(\vec{k}',p')}\frac{\left|\langle\vec{k}',p'|\underline{\varepsilon}_o^{-1}\delta\varepsilon|\vec{k},p\rangle\right|^2}{\langle\vec{k}',p'|\vec{k}',p'\rangle} \tag{5.2.6}$$

whereas the approximate eigen-solution $|\vec{k},p\rangle_{cor}$ is expressed in the form of a Bloch wave

$$|\vec{k},p\rangle_{cor}\approx|\vec{k},p\rangle+$$

$$+\sum_{\vec{k}',p'}\frac{\omega_o^2(\vec{k},p)}{\omega_o^2(\vec{k}',p')-\omega_o^2(\vec{k},p)}\frac{\langle\vec{k}',p'|\underline{\varepsilon}_o^{-1}\delta\varepsilon|\vec{k},p\rangle}{\langle\vec{k}',p'|\vec{k},p\rangle}|\vec{k}',p'\rangle+\textit{higher order} \tag{5.2.7}$$

In the above expression $|k,p\rangle$ denote unperturbed eigenfunctions. We see that the perturbation $\delta\varepsilon(z)$ plays the role of a periodic potential which mixes the unperturbed eigenmodes. The strength of this mixing depends on the matrix elements

$$\langle\vec{k}',p'|\underline{\varepsilon}_o^{-1}\delta\varepsilon|\vec{k},p\rangle=\frac{1}{(2\pi)^3}\int d^3\vec{r}\,e^{i(\vec{k}-\vec{k}')\vec{r}}\,\vec{E}_{p'}^{*}\delta\,\underline{\varepsilon}(z)\vec{E}_p \tag{5.2.8}$$

which are just Fourier transform of the components of the periodic perturbation Here, the electric field is written as $E_{k,p}(\vec{r})=E_p\cdot e^{ikr}$. Because there are two different spatial periods in the perturbation (Eq.5.2.2) with wave-vectors $\vec{q}_c$ and $2\vec{q}_c$, we can expect that the perturbation would mix unperturbed eigenwaves with wave-vectors that differ by

$$\vec{k}-\vec{k}'=\vec{q}_c\ \text{ or }\ 2\vec{q}_c \tag{5.2.9}$$

where the corresponding unperturbed eigenvalues are degenerate.

Far from the degeneration points and in the limit of small tilt angles, the first order approximation to the light propagation in a periodic, birefringent crystal is valid. The approximate solutions are just σ and π linearly polarized plane waves, whereas the first order corrected eigenvalue is

$$\omega^2(\vec{k},p)=\omega_o^2(\vec{k},p)\left\{1-\frac{\langle\vec{k},p|\underline{\varepsilon}_o^{-1}\delta\varepsilon|\vec{k},p\rangle}{\langle\vec{k},p|\vec{k},p\rangle}\right\} \tag{5.2.10}$$

Here, the second term in brackets is just the space averaged value of the inhomogeneous part of the dielectric tensor. Higher order corrections to the eigenvalue $\omega^2(\vec{k},p)$ are important near the points of degeneration, i.e. near the edges of the Brillouin zone, introduced by the periodic form of the perturbation.

The first order corrected eigenvalue $\omega^2(\vec{k},p)$ in the presence of the perturbation $\delta\underline{\varepsilon}$ can be interpreted as a change in the refractive indices of the medium. This means that in this approximation the optical properties are described by the corresponding space-averaged uniaxial tensor $\langle\underline{\varepsilon}\rangle$. The first order correction preserves the form of the normal wave surface, whereas the diagonal elements of the corresponding space-averaged tensor are

$$\begin{aligned}
\langle\varepsilon\rangle_{xx} &= \tfrac{1}{2}\left[(\varepsilon_1+\varepsilon_2)+(\varepsilon_3-\varepsilon_2)\sin^2\theta\right]+\tfrac{1}{2}\left[(\varepsilon_2-\varepsilon_1)+(\varepsilon_3-\varepsilon_2)\sin^2\theta\right]\cdot\langle\cos 2\Phi(z)\rangle \\
\langle\varepsilon\rangle_{yy} &= \tfrac{1}{2}\left[(\varepsilon_1+\varepsilon_2)+(\varepsilon_3-\varepsilon_2)\sin^2\theta\right]-\tfrac{1}{2}\left[(\varepsilon_2-\varepsilon_1)+(\varepsilon_3-\varepsilon_2)\sin^2\theta\right]\cdot\langle\cos 2\Phi(z)\rangle \\
\langle\varepsilon\rangle_{zz} &= \varepsilon_3-(\varepsilon_3-\varepsilon_2)\sin^2\theta \\
\langle\varepsilon\rangle_{xy} &= \tfrac{1}{2}\left[(\varepsilon_2-\varepsilon_1)+(\varepsilon_3-\varepsilon_2)\sin^2\theta\right]\cdot\langle\sin 2\Phi(z)\rangle \\
\langle\varepsilon\rangle_{xz} &= (\varepsilon_3-\varepsilon_2)\sin\theta\cdot\cos\theta\cdot\langle\cos\Phi(z)\rangle \\
\langle\varepsilon\rangle_{yz} &= (\varepsilon_3-\varepsilon_2)\sin\theta\cdot\cos\theta\cdot\langle\sin\Phi(z)\rangle
\end{aligned} \tag{5.2.11}$$

We have considered here that the tilt angle is constant throughout the sample. In the undistorted helicoidal phase where $\Phi(z) = q_c z$, the spatial averages equal zero, i.e. $\langle \sin \Phi(z) \rangle = \langle \cos \Phi(z) \rangle = \langle \sin 2\Phi(z) \rangle = \langle \cos 2\Phi(z) \rangle = 0$, and the dielectric tensor $\langle \underline{\varepsilon} \rangle$ reduces to

$$\langle \varepsilon \rangle_{xx} = \tfrac{1}{2} \left[(\varepsilon_1 + \varepsilon_2) + (\varepsilon_3 - \varepsilon_2) \sin^2 \theta \right] \tag{5.2.12a}$$

$$\langle \varepsilon \rangle_{yy} = \tfrac{1}{2} \left[(\varepsilon_1 + \varepsilon_2) + (\varepsilon_3 - \varepsilon_2) \sin^2 \theta \right] \tag{5.2.12b}$$

$$\langle \underline{\varepsilon} \rangle_{zz} = \varepsilon_3 - (\varepsilon_3 - \varepsilon_2) \sin^2 \theta \tag{5.2.12c}$$

The components of the dielectric tensor $\langle \underline{\varepsilon} \rangle$ are here temperature dependent because of the temperature dependence of the tilt angle $\theta(T)$. In particular, "the average" ordinary index of refraction $\bar{n}_o = \sqrt{\langle \underline{\varepsilon} \rangle_{xx}}$ increases, whereas "the average" extraordinary index $\bar{n}_e = \sqrt{\langle \underline{\varepsilon} \rangle_{zz}}$ decreases with increasing tilt angle. This means that the birefringence $\Delta n = \bar{n}_e - \bar{n}_o$ should decrease with the increase of the tilt angle in the $smectic - C^*$ phase. This type of behavior was indeed observed in the measurements of the temperature dependence of birefringence in the smectic and hexatic phases of CE-8 (Muševič et al., 1996). The authors found an excellent agreement in the tilt angle, as calculated from their birefringence measurements and the X-ray (Raya et al., 1991).

One of the most spectacular manifestations of a periodic dielectric tensor $\delta \underline{\varepsilon}(z)$ is the appearance of selective or total (Bragg) reflection in highly chiral ferroelectric $smectic - C^*$ phases. Similarly to the iridescent colors that can be observed in cholesteric liquid crystals, the $smectic - C^*$ phase appears vividly colored when viewed in white-light reflection. This indicates that the $smectic - C^*$ phase strongly reflects certain wavelengths and polarizations of the incident white light. Although the phenomena was qualitatively understood for a long time, it was C.Oldano (1984) who gave a detailed explanation of the origin and number of the reflection bands in the $smectic - C^*$ phase.

In analogy to the propagation of an electron in a weak periodic potential, we note that one has to define a Brillouin zone in the presence of a weak periodic perturbation $\hat{V} \equiv -(\omega^2/c_o^2)\, \underline{\varepsilon}_o^{-1} \delta \underline{\varepsilon}$. For the $smectic - C^*$ phase, the perturbing part of the dielectric tensor has two Fourier components. The so-called ***polar part*** has a period equal to the helical period p

$$\delta\underline{\varepsilon} = (\varepsilon_3 - \varepsilon_2)\sin\theta\cos\theta \cdot \begin{bmatrix} 0 & 0 & \cos\Phi(z) \\ 0 & 0 & \sin\Phi(z) \\ \cos\Phi(z) & \sin\Phi(z) & 0 \end{bmatrix} \qquad (5.2.13)$$

whereas the ***quadrupolar part*** repeats itself every half of the helical period:

$$\delta\underline{\varepsilon} = \tfrac{1}{2}\left[(\varepsilon_2 - \varepsilon_1) + (\varepsilon_3 - \varepsilon_2)\sin^2\theta\right] \begin{bmatrix} \cos 2\Phi(z) & \sin 2\Phi(z) & 0 \\ \sin 2\Phi(z) & -\cos 2\Phi(z) & 0 \\ 0 & 0 & 0 \end{bmatrix} \qquad (5.2.14)$$

In this case, the basic vector of the reciprocal lattice is $K = 2\vec{q}_c$ and the Brillouin zone is defined as $\mathrm{BZ} \equiv [-q_c, +q_c]$. In analogy to the energy dispersion of an electron in a periodic potential, the dispersion relation for light propagation in periodic media should be periodic in the reciprocal space and we expect that at the points of degeneration of the unperturbed eigenfunctions ***band gaps*** will appear, resulting in a band-like structure of the dispersion relation. When the frequency of light, incident to the $smectic - C^*$ sample falls into this forbidden frequency regions, it will be selectively or even totally reflected. As a result, the $smectic - C^*$ sample will look colored because some of the frequencies from the white spectrum will be strongly reflected, whereas others will pass through the sample.

In order to analyze the number and the position of the forbidden frequency gaps, one has to apply perturbation theory for a set of degenerate eigenfunctions (Landau and Lifshitz, 1958). Here the matrix elements of the perturbation

$$V_{kk'} = \langle \Psi_{k'} | \hat{V} | \Psi_k \rangle \qquad (5.2.15)$$

between the unperturbed and degenerate eigenfunctions $|\Psi_k\rangle$ define the magnitude of the splitting of the eigenvalues. In our case we have to consider the matrix elements

$$\left\langle \vec{k}', p' \middle| \hat{V} \middle| \vec{k}, p \right\rangle = \frac{1}{(2\pi)^3} \int d^3\vec{r} e^{i(\vec{k}-\vec{k}')\vec{r}} \vec{E}^*_{p'} \delta \underline{\varepsilon}(z) \vec{E}_p \tag{5.2.16}$$

between the unperturbed σ and π waves. Because there are two Fourier components in Eq.5.2.16., with corresponding wave-vectors $\vec{q}_c$ and $2\vec{q}_c$ we expect two sets of forbidden bands. A straightforward calculations shows that for the ***polar part*** of the dielectric tensor only one matrix element is nonzero:

$$\left\langle \vec{k} + \vec{q}_c, \sigma \middle| \hat{V} \middle| \vec{k}, \sigma \right\rangle = 0 \tag{5.2.17a}$$

$$\left\langle \vec{k} + \vec{q}_c, \pi \middle| \hat{V} \middle| \vec{k}, \pi \right\rangle = 0 \tag{5.2.17b}$$

$$\left\langle \vec{k} + \vec{q}_c, \pi \middle| \hat{V} \middle| \vec{k}, \sigma \right\rangle \neq 0 \tag{5.2.17c}$$

whereas for the ***quadrupolar part*** all three matrix elements are finite

$$\left\langle \vec{k} + 2\vec{q}_c, \sigma \middle| \hat{V} \middle| \vec{k}, \sigma \right\rangle \neq 0 \tag{5.2.18a}$$

$$\left\langle \vec{k} + 2\vec{q}_c, \pi \middle| \hat{V} \middle| \vec{k}, \pi \right\rangle \neq 0 \tag{5.2.18b}$$

$$\left\langle \vec{k} + 2\vec{q}_c, \pi \middle| \hat{V} \middle| \vec{k}, \sigma \right\rangle \neq 0 \tag{5.2.18c}$$

The polar part of the perturbation thus removes the degeneracy only between the σ and π waves, whereas the quadrupolar part removes the degeneracy in all points. The resulting eigen-spectrum is shown in Fig.5.2.2. Here, the polar part removes the degeneracy at the crossing of σ and π waves at $\vec{q} = \vec{q}_c/2$, but for symmetry reasons it preserves the degeneracy at the crossing of two σ or two π waves that differ by $\vec{q}_c$. This results in a single reflection band denoted by C, where no light with the frequency inside the forbidden gap can propagate through the sample. On the other hand, there are three reflection bands at $\vec{q} = \vec{q}_c$ and $\vec{q} = 0$, which is the same in view of periodicity of the dispersion relation. The C-band again opens at the intersection of σ and π waves, respectively. It is a total reflection band, where no light can propagate in the medium. There are also two B-bands, B^+ and B^-, located at the intersection of two σ and two π waves, as shown in Fig.5.2.2. These are selective reflection bands, where only a certain polarization of the incident light is reflected, whereas the light of proper polarization can propagate in the liquid crystal. A similar band structure was obtained in a numerical analysis of the solutions of the wave equation in the *smectic* $-C^*$ phase by Oldano et al. (1983; 1984; 1985). The fact that this simple perturbative approach describes rather well the basic features of the optical spectrum of the phase shows that the origin of the basic optical phenomena in birefringent, modulated structures is in the form and symmetry of the dielectric tensor and not in the magnitude of its components.

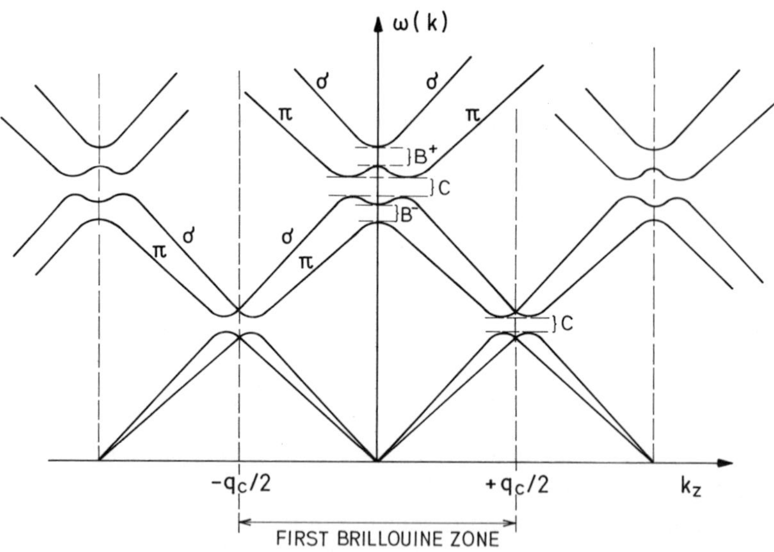

Fig. 5.2.2. Dispersion relation for light, propagating in the *smectic* $-C^*$ phase.

As it has been pointed out by Oldano (1984) this simple perturbation theory breaks down in another very interesting case of the so-called ***continuous degeneration*** of the eigenvalues, which can be met in two situations:

i) Propagation of light along the helical axis. In this case the ordinary and the extraordinary indices are equal by definition. We have therefore a situation where the eigenfrequencies of the ordinary and the extraordinary wave are equal for every $\vec{q}$. This results in a continuous degeneration of the eigenvalues as it is shown in Fig.5.2.3.

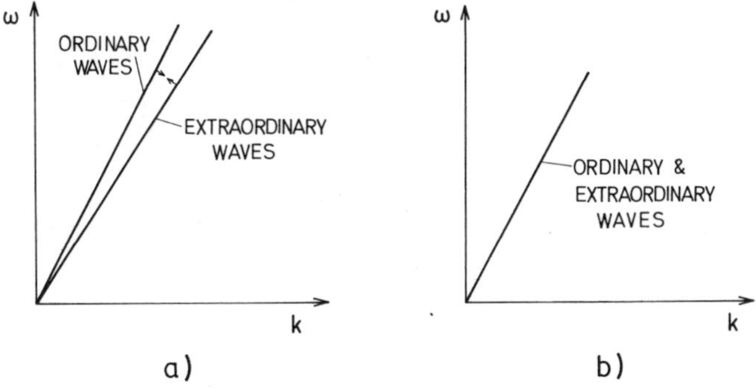

Fig.5.2.3. The dispersion relation for light propagating (a) close to the direction of the helical axis and (b) along the helical axis, i.e. the optical axis.

ii) Critical value of the tilt angle. In the limit of small θ the wave-vector surface of the $smectic-C^*$ crystal is uniaxial and the uniaxiality is positive, as shown in Fig.5.2.4. In the limit $\theta \to \pi/2$ we have a cholesteric-like structure. Here the normal velocity surface is again uniaxial, but the uniaxiality is negative, as it is shown in Fig.5.2.4b. There must be therefore a so-called critical value of the tilt angle, where the uniaxiality is zero and the normal surface of the extraordinary wave merges into the normal surface of the ordinary wave. We have in this case

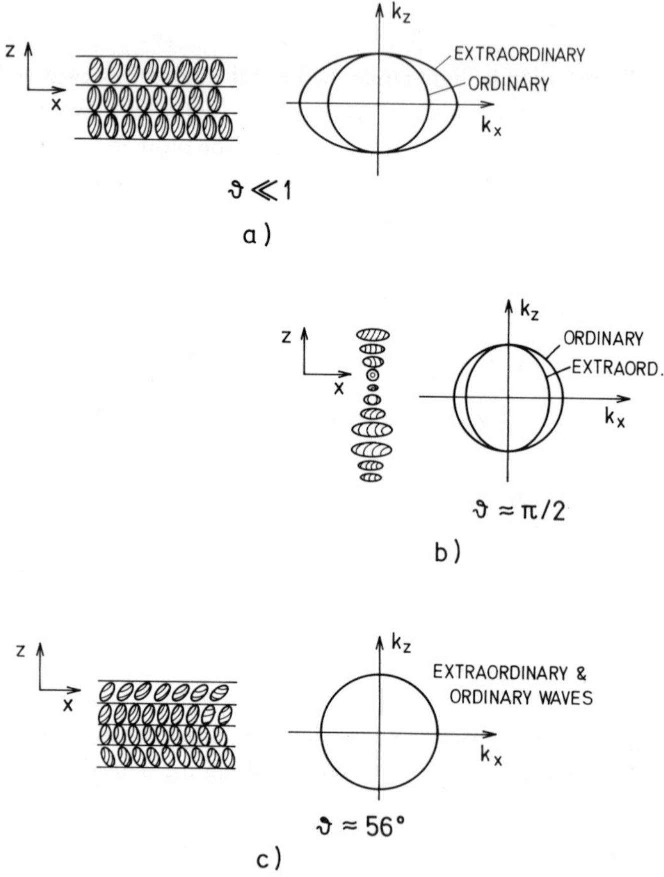

Fig.5.2.4. Structures and their wave-vector surfaces, as obtained for limiting values of the tilt angle θ. (a) In the limit of small θ we obtain *smectic – A* phase with a positive uniaxial normal velocity surface. (b) In the limit of very large tilt angle, $\theta = \pi/2$, we obtain the cholesteric structure. This phase is again uniaxial, but the uniaxiality is negative. (c) There must be therefore a certain critical value of the tilt angle, where the uniaxiality equals zero and the normal surface is spherical.

the continuous degeneracy of the eigenvalues, which is fulfilled for every direction and magnitude of the wavevector of the propagating light. This critical value of the tilt angle is rather independent of the material parameters and is close to 56 degrees, as first calculated by Oldano (1984). In both cases of degeneracy, we expect a strong mixing of the unperturbed σ and π waves, which results in a circular nature of eigen-polarizations. In this case, a perturbation analysis for degenerate eigenfunctions has to be used.

5.3. Birefringence of Helical Smectic Phases

Let us first remember that a dielectric tensor of an optically uniaxial and spatially uniform medium, such is for example the $smectic - A$ phase of a liquid crystal, is

$$\underline{\varepsilon} = \begin{bmatrix} \varepsilon_1 & 0 & 0 \\ 0 & \varepsilon_1 & 0 \\ 0 & 0 & \varepsilon_3 \end{bmatrix} \tag{5.3.1.}$$

It is well known, that the exact solutions of the wave equation (Eq.5.1.3.) in such a medium are linearly polarized electromagnetic plane-waves. There are two distinct dispersion branches of these eigenwaves, which are called ordinary (σ) and extraordinary (π) waves, respectively:

$$smectic - A: \qquad \left|\vec{k}, p\right\rangle = \begin{cases} \sigma \text{ - wave} \left|\vec{k}, \sigma\right\rangle & \omega = c_{ord} \cdot 2\pi/\lambda_\circ \\ \pi \text{ - wave} \left|\vec{\mathrm{k}}, \pi\right\rangle & \omega = c_{extr} \cdot 2\pi/\lambda_\circ \end{cases} \tag{5.3.2}$$

The polarization properties of the eigenwaves of an uniaxial crystal are shown in Fig.5.3.1. Here, the wave-vectors k_σ and k_π of the ordinary and extraordinary wave are shown as well as their electric displacements $\vec{D}_\sigma$ and $\vec{D}_\pi$.

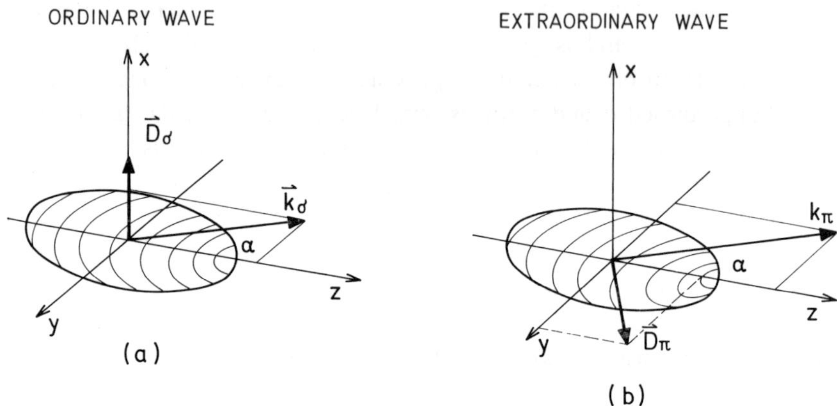

Fig.5.3.1. The polarization properties of (a) ordinary (σ) and (b) extraordinary (π) waves, propagating in a birefringent medium. The optical axis is along the z-axis and the light is propagating in the y-z plane. Note that the electric field of the ordinary wave (a) always "sees" the same dielectric constant and the speed of propagation of the ordinary wave does not depend on the angle of propagation.

The electric displacement $\vec{D}_\sigma$ of the ordinary wave is always perpendicular to the plane, containing $\vec{k}_\sigma$ and the optical axis, whereas the electric displacement $\vec{D}_\pi$ is always lying in this plane.

Due to their different nature, the ordinary and extraordinary waves travel at different speeds through the uniaxial medium. In another words, their indices of refraction are quite different and can be calculated from the Fresnel equations (see Born and Wolf, 1959). Whereas the index of refraction of the ordinary wave n_o is independent of the direction of propagation, the index of refraction of the extraordinary wave n_e is direction-dependent

$$n_o^2 = \varepsilon_1 \tag{5.3.3.a}$$

$$\frac{1}{n_e^2} = \frac{\sin^2\alpha}{\varepsilon_3} + \frac{\cos^2\alpha}{\varepsilon_1} \tag{5.3.3.b}$$

Here, α is the angle between the direction of wave propagation and the z-axis. Note that for $\alpha = 0$ both indices are equal, which means that light is propagating along the optical axis, which is in the z-direction.

The birefringence of an uniaxial crystal Δn is the difference between the extraordinary and ordinary index of refraction, i.e.

$$\Delta n = n_e - n_o \tag{5.3.4.}$$

The relations (Eq.5.3.3.) can be graphically visualized by plotting the angular dependence of the phase velocities of the ordinary and extraordinary waves,

$$c_{ord} = \frac{c_\circ}{n_o} \tag{5.3.5.a}$$

$$c_{extr} = \frac{c_\circ}{n_e} \tag{5.3.5.b}$$

or, alternatively, the angular dependence of the wave-vectors of both waves, which are shown in Fig.5.3.2. Here, $c_\circ$ is the speed of light in vacuum.

Let us come back to the problem of light propagation in a birefringent, helicaly modulated medium, such as the chiral $smectic-C^*$ phase. The question that here naturally arises is whether we can consider the $smectic-C^*$ phase as an optically uniaxial medium? What are the refractive indices in this approximation?

The answer to these questions can be given by considering the first order approximation to the wave equation. Within this approximation we start from the unperturbed ordinary and extraordinary waves of the optically uniaxial and

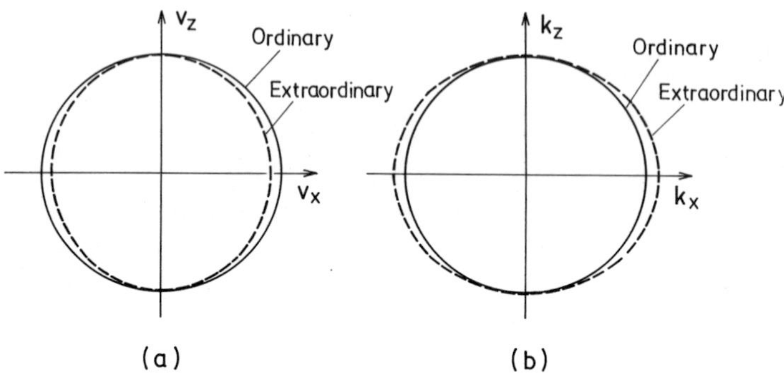

Fig.5.3.2. (a) The angular dependence of the phase velocity of an ordinary (c_o) and an extraordinary (c_e) wave in an optically uniaxial crystal. The plot is also known as the normal wave surface, see Born and Wolf (1959). (b) the angular dependence of the wave-vectors of an ordinary ($k_o = n_o \cdot k_\circ$) and the extraordinary wave ($k_e = n_e \cdot k_\circ$). Here, $k_\circ = 2\pi/\lambda_\circ$ is the wave-vector of light in vacuum.

homogeneous *smectic* $-A$ phase, and we consider the spatially inhomogeneous terms of the dielectric tensor as a small perturbation. In this approximation, we consider therefore only the first two terms in the perturbation expansion for the eigenvalues (Eq.5.2.6.),

$$\omega^2(\vec{k}, p) = \omega_\circ^2(\vec{k}, p) \cdot \left\{ 1 - \frac{\langle \vec{k}, p | \underline{\varepsilon}_\circ^{-1} \delta \underline{\varepsilon} | \vec{k}, p \rangle}{\langle \vec{k}, p | \vec{k}, p \rangle} \right\} \tag{5.3.6.}$$

and we also consider only the first term in the perturbative expansion for the eigenwaves, Eq.5.2.7.

$$\left|\vec{k},p\right\rangle_{cor} \approx \left|\vec{k},p\right\rangle = \begin{cases} \sigma \text{ - wave} \\ \pi \text{ - wave} \end{cases} \qquad (5.3.7.)$$

In the Eq.5.3.6., the second term is just the ***space-averaged*** value of the inhomogeneous part of the dielectric tensor. Higher order corrections to the eigenvalue $\omega^2(k,p)$ are important near the points of degeneration, i.e. near the edges of the Brillouin zone, introduced by the periodic form of the perturbation.

The first order corrected eigenvalue $\omega^2(\vec{k},p)$ (Eq.5.3.6.) in the presence of the perturbation $\delta\underline{\varepsilon}$ can be interpreted as a change in the refractive indices of the medium, as shown in Fig.5.3.3.

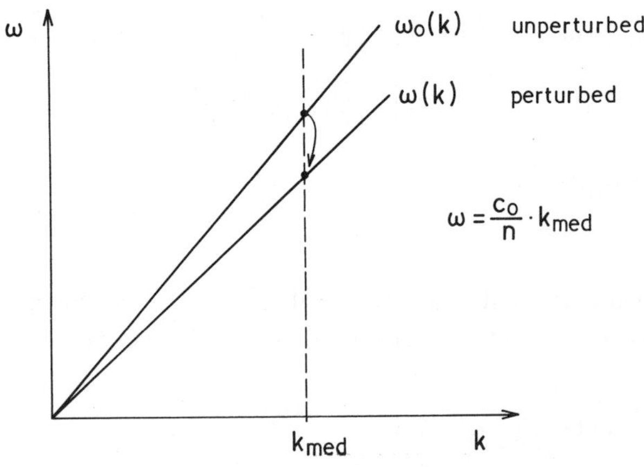

Fig.5.3.3. The interpretation of the Eq.5.3.6. For zero perturbation, the dispersion relation $\omega_\circ(k,p)$ is linear in the wave-vector $\vec{k}$. An external perturbation slightly changes the eigenvalue ω of both ordinary and extraordinary wave. This is observed as a change of the slope of the dispersion branch, which defines the index of refraction. The two indices would therefore change upon perturbation.

A straightforward calculation shows, that in the first order approximation, the optical properties of spatially inhomogeneous phases are determined by the space averaged dielectric tensor

$$\langle \underline{\varepsilon} \rangle = \begin{bmatrix} \langle \varepsilon \rangle_{xx} & 0 & 0 \\ 0 & \langle \varepsilon \rangle_{yy} & 0 \\ 0 & 0 & \langle \varepsilon \rangle_{zz} \end{bmatrix} \tag{5.3.8.}$$

which is in the case of a ferroelectric $smectic - C^*$ phase given by

$$\langle \varepsilon \rangle_{xx} = \tfrac{1}{2}\left[(\varepsilon_1 + \varepsilon_2) + (\varepsilon_3 - \varepsilon_2)\sin^2\theta\right] + \tfrac{1}{2}\left[(\varepsilon_2 - \varepsilon_1) + (\varepsilon_3 - \varepsilon_2)\sin^2\theta\right] \cdot \langle \cos 2\Phi(z) \rangle$$

$$\langle \varepsilon \rangle_{yy} = \tfrac{1}{2}\left[(\varepsilon_1 + \varepsilon_2) + (\varepsilon_3 - \varepsilon_2)\sin^2\theta\right] - \tfrac{1}{2}\left[(\varepsilon_2 - \varepsilon_1) + (\varepsilon_3 - \varepsilon_2)\sin^2\theta\right] \cdot \langle \cos 2\Phi(z) \rangle$$

$$\langle \varepsilon \rangle_{zz} = \varepsilon_3 - (\varepsilon_3 - \varepsilon_2)\sin^2\theta \tag{5.3.9.}$$

In the undisturbed helicoidal phase where $\Phi(z) = q_c \cdot z$, the spatial averages equal zero, i.e. $\langle \cos 2\Phi(z) \rangle = 0$, and the dielectric tensor $\langle \varepsilon \rangle$ reduces to

$$\langle \varepsilon \rangle_{xx} = \tfrac{1}{2}\left[(\varepsilon_1 + \varepsilon_2) + (\varepsilon_3 - \varepsilon_2)\sin^2\theta\right] \tag{5.3.10a}$$

$$\langle \varepsilon \rangle_{yy} = \tfrac{1}{2}\left[(\varepsilon_1 + \varepsilon_2) + (\varepsilon_3 - \varepsilon_2)\sin^2\theta\right] \tag{5.3.10b}$$

$$\langle \varepsilon \rangle_{zz} = \varepsilon_3 - (\varepsilon_3 - \varepsilon_2)\sin^2\theta \tag{5.3.10c}$$

The average dielectric tensor is therefore uniaxial, as expected from the helical symmetry of the unperturbed ferroelectric phase. The components of the dielectric tensor $\langle \underline{\varepsilon} \rangle$ are here temperature dependent mainly because of the temperature dependence of the tilt angle $\theta(T)$ and a slight temperature dependence of the nematic order parameter. In particular, "the average" ordinary index of refraction $n_o = \sqrt{\langle \varepsilon \rangle_{xx}}$ increases, whereas "the average" extraordinary index $n_e = \sqrt{\langle \varepsilon \rangle_{zz}}$ decreases with increasing tilt angle. This means that the birefringence $\Delta n = n_e - n_o$ of the ferroelectric $smectic-C^*$ phase is expected to decrease with the onset of the tilt angle in the $smectic-C^*$ phase. This type of behavior was indeed first observed in the measurements of the temperature dependence of birefringence in the smectic and hexatic phases of CE-8 (Muševič et al., 1996), which is shown in Fig.5.3.4. The observed temperature dependence of the tilt angle, as calculated from their birefringence measurements is shown in Fig.5.3.5. and is in good agreement with the X-ray data (Raya et al., 1991)

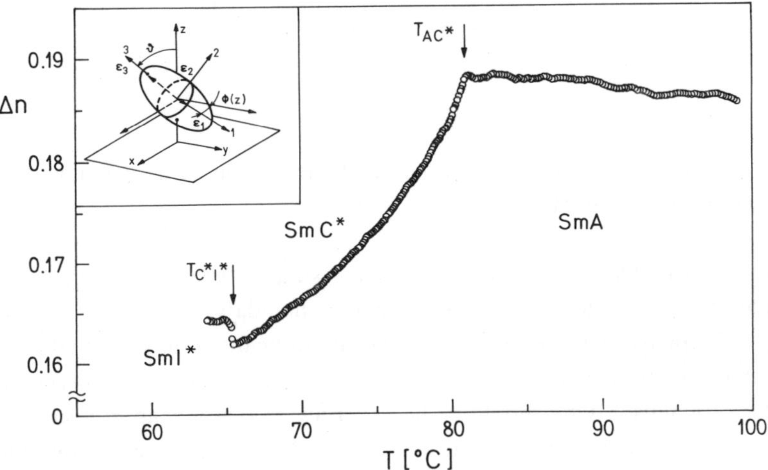

Fig.5.3.4. Temperature dependence of the birefringence in the ... *smectic–C** and *smectic–I** phases of ferroelectric liq...

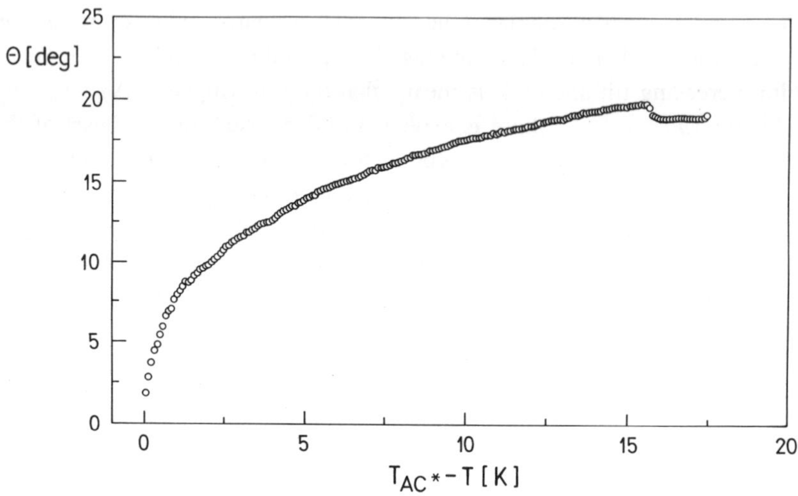

Fig.5.3.5. The temperature dependence of the tilt angle, as calculated from the birefringence measurements in CE-8.

The agreement between the predictions of the simple, first order approximation to the wave equation clearly indicates that helicoidaly modulated, unperturbed $smectic - C^*$ phase can be considered as an optically uniaxial medium with a dielectric tensor, which is in this approximation equal to the space-average of the exact dielectric tensor. One should be however aware that this approximation breaks down near the points of degeneration, which includes the propagation along the helix and at a Bragg angle.

5.4. Optical Rotation

When a linearly polarized monochromatic light is propagating along the helical axis of chiral smectic phases, the direction of the polarization of the transmitted light is rotated for an angle Ψ with respect to the direction of the polarization of the incoming light. The angle of rotation is of the order of 10 degrees for a $100\mu m$ thick $smectic-C^*$ sample with a helical period of 1μm and is much smaller than the magnitude of optical rotation in cholesteric liquid crystals. The effect of optical rotation is here a result of the helical arrangement of liquid crystalline molecules. The intrinsic optical rotation which is due to chirality of individual molecules is significantly smaller effect.

The optical rotation of polarization of light can be understood within the theory of light propagation along the helical axis of the unperturbed helical $smectic-C^*$ phase that was first treated by O.Parodi (1975) as a generalization of the theory of de Vries (1951). The wave equation, which describes the propagation of a monochromatic light wave $\vec{E}(z)e^{i\omega t}$ with frequency ω along the helical axis is

$$\vec{\nabla}\times\vec{\nabla}\times\vec{E}(z)-\underline{\varepsilon}(z)\frac{\omega^2}{c_o^2}\vec{E}(z)=0 \tag{5.4.1}$$

Here $\vec{E}(z)=(E_x,E_y,E_z)$ is the electric field of the light wave, the dielectric tensor $\underline{\varepsilon}(z)$ is given by Eq.5.1.2. and $c_\circ$ is the speed of light in vacuum. By noting that the electric displacement $\vec{D}(z)$ must be perpendicular to the wave-vector $\vec{k}=(0,0,k)$, i.e.

$$D_z=0, \tag{5.4.2}$$

the wave equation 5.4.1. is rewritten as

$$\frac{\partial^2}{\partial z^2}\begin{bmatrix}E_x\\E_y\end{bmatrix}+\frac{\omega^2}{c_o^2}\left\{\frac{\varepsilon_{//}+\varepsilon_\perp}{2}\begin{bmatrix}1&0\\0&1\end{bmatrix}+\frac{\varepsilon_{//}-\varepsilon_\perp}{2}\begin{bmatrix}\cos 2\Phi&\sin 2\Phi\\\sin 2\Phi&-\cos 2\Phi\end{bmatrix}\right\}\cdot\begin{bmatrix}E_x\\E_y\end{bmatrix}=0 \tag{5.4.3}$$

Here we have defined

$$\varepsilon_{\perp} = \varepsilon_1 \tag{5.4.4a}$$

$$\varepsilon_{//} = \frac{\varepsilon_2 \varepsilon_3}{\varepsilon_2 \sin^2\theta + \varepsilon_3 \cos^2\theta} \tag{5.4.4b}$$

where θ is the tilt angle and ε_i stands for the eigenvalues of the dielectric tensor (Eq.5.1.1.). We see that only the transverse components E_x and E_y of the electric field are important for the propagation along the helical axis. In this sense the $smectic - C^*$ phase is analogous to the chiral nematic phase and we can use the standard analysis of the eigensolutions (de Gennes, 1974). These can be found by introducing two orthogonal, left and right-handed circular polarizations E^+ and E^-

$$E^{\pm} = E_x \pm iE_y \tag{5.4.5}$$

and rewriting the wave equation as

$$\frac{\partial^2}{\partial z^2}\begin{bmatrix} E^+ \\ E^- \end{bmatrix} + \frac{\omega^2}{c_o^2}\frac{\varepsilon_{//} + \varepsilon_{\perp}}{2}\begin{bmatrix} 1 & \frac{\varepsilon_{//} - \varepsilon_{\perp}}{\varepsilon_{//} + \varepsilon_{\perp}} e^{2i\Phi} \\ \frac{\varepsilon_{//} - \varepsilon_{\perp}}{\varepsilon_{//} + \varepsilon_{\perp}} e^{-2i\Phi} & 1 \end{bmatrix}\begin{bmatrix} E^+ \\ E^- \end{bmatrix} = 0 \tag{5.4.6}$$

We note that the off-diagonal elements are temperature dependent (see Eq.5.4.4(a) and (b)) and usually very small, $\varepsilon_{//} - \varepsilon_{\perp} << \varepsilon_{//} + \varepsilon_{\perp}$. They therefore represent a periodic perturbation that couples both modes and removes their degeneracy. Note that in the absence of this periodic perturbation, both eigenvalues of the above equation are equal.

In view of the space-periodic term in the wave equation, we have to introduce the Brillouin zone in the reciprocal space and the solutions of the wave equation must obtain the Bloch structure. In the case of the unperturbed helix we have $\Phi(z) = q_c z$ and the wave-vector of the periodic term is $2q_c$. This defines the Brillouin zone as $(-q_c, +q_c)$. By introducing the Bloch structure of the eigensolutions

$$E_k^+ = e^{ikz} \sum_n a_n^+ e^{2inq_c z} \tag{5.4.7a}$$

$$E_k^- = e^{ikz} \sum_m a_m^- e^{2imq_c z} \tag{5.4.7b}$$

where a_n^{+-} are the amplitudes of circular polarizations, the wave equation (5.4.6) reduces to an infinite set of equations for the coefficients a_n^+ and a_m^-

$$\begin{bmatrix} (k+n2q_c)^2 - \frac{\omega^2}{c_o^2}\frac{\varepsilon_{//}+\varepsilon_\perp}{2} & -\frac{\omega^2}{c_o^2}\frac{\varepsilon_{//}-\varepsilon_\perp}{2} \\ -\frac{\omega^2}{c_o^2}\frac{\varepsilon_{//}-\varepsilon_\perp}{2} & (k+(n-1)2q_c)^2 - \frac{\omega^2}{c_o^2}\frac{\varepsilon_{//}+\varepsilon_\perp}{2} \end{bmatrix} \cdot \begin{bmatrix} a_n^+ \\ a_{n-1}^- \end{bmatrix} = 0 \tag{5.4.8}$$

In the above expressions, k denotes the z-component of the wave-vector that is ascribed to the Bloch function. It turns out that by redefining the wave-vector

$$K_n = k + (2n-1)q_c \tag{5.4.9}$$

we obtain an infinite number of identical pairs of linear equations. These have a solution when the corresponding determinant vanishes, i.e.

$$K^4 - 2K^2\left(q_c^2 + \frac{\omega^2}{c_o^2}\frac{\varepsilon_{//}+\varepsilon_\perp}{2}\right) + \left(q_c^2 - \frac{\omega^2}{c_o^2}\frac{\varepsilon_{//}+\varepsilon_\perp}{2}\right) - \frac{\omega^4}{c_o^4}\left(\frac{\varepsilon_{//}-\varepsilon_\perp}{2}\right)^2 = 0 \tag{5.4.10}$$

For a light wave with a given frequency ω, there are four wave-vectors $K_{(i)}$ which are the solutions of the above equation

$$K_{(i)} = \pm q_c \sqrt{1 + \frac{\omega^2}{\omega_B^2} \pm \frac{\omega}{\omega_B}\sqrt{4 + \left(\frac{\varepsilon_{//} - \varepsilon_\perp}{\varepsilon_{//} + \varepsilon_\perp}\right)^2 \frac{\omega^2}{\omega_B^2}}} \tag{5.4.11}$$

whereas the ratio of the corresponding amplitudes is

$$\frac{a^+_{(i)}}{a^-_{(i)}} = \frac{\varepsilon_{//} - \varepsilon_\perp}{\varepsilon_{//} + \varepsilon_\perp}\left(\frac{\omega}{\omega_B}\right)^2 \frac{1}{\left(1 + \frac{K_{(i)}}{q_c}\right)^2 - \left(\frac{\omega}{\omega_B}\right)^2} \tag{5.4.12}$$

Here $\omega_B = q_c c_o \sqrt{2/\varepsilon_{//} + \varepsilon_\perp}$. One should note that the Eq.5.4.11. represents the dispersion relation $K(\omega)$ (or $\omega(K)$) for the light waves, which are propagating along the helical axis. The analysis of the expressions 5.4.11 and 5.4.12 leads to the following conclusions:

(i) When light is propagating along the helical axis of a $smectic-C^*$ phase, there are four dispersion branches $\omega_i(k), i = 1...4$ as shown in Fig.5.4.1.(a) and (b). These branches can be grouped into two pairs: the first pair represents two distinct light eigenwaves propagating in the +z direction, the other two represent light waves propagating in the -z direction.

(ii) It is easy to see from the Eq.5.4.11. that the solutions $K_{(i)}$ are independent of *n*, as defined in the Eq.(5.4.9). This means that for a given frequency ω we have an infinite number of the corresponding wave-vectors $k = K - (2n-1)q_c$ (Eq.5.4.9) of the Bloch functions, which are the solutions of the wave-equation. The dispersion relation $\omega(k)$ is therefore a periodic function in the reciprocal space, as shown in Fig.5.4.2.

(iii) By inserting $K_n = k + (2n-1)q_c$ into the Eq.5.4.7., we see that the form of the eigensolutions (Eq.5.4.7) simplifies to a monochromatic plane wave, which is a trivial form of the Bloch wave,

$$E_K^+ = a^+ e^{i(K+q_c)z} \tag{5.4.13a}$$

$$E_K^- = a^- e^{i(K-q_c)z} \tag{5.4.13b}$$

The ratio of the amplitudes a_i^+/a_i^- for the i-th dispersion branch as calculated from the Eq.5.4.12, is shown in Fig.5.4.3.

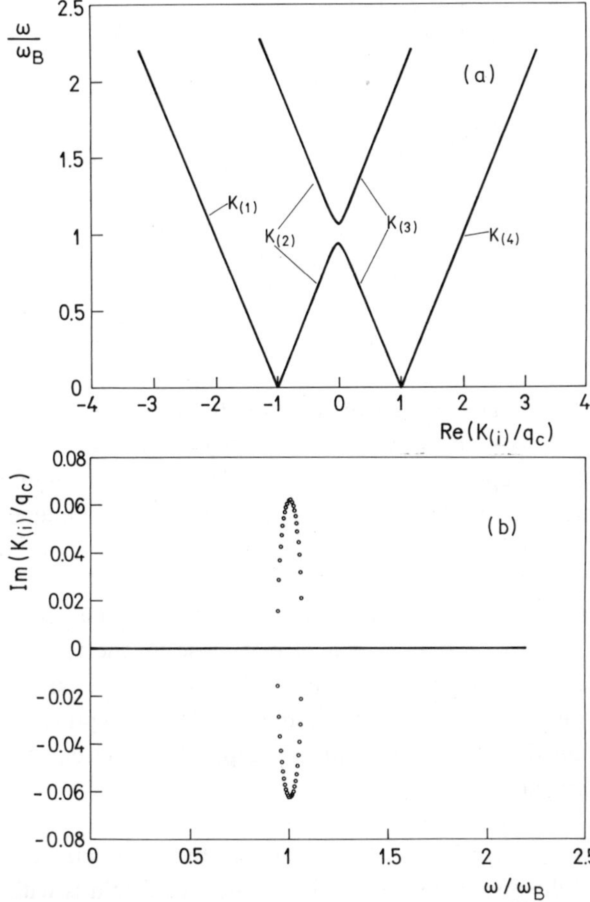

Fig.5.4.1. Dispersion relation for light propagating along the helical axis of the $smectic-C^*$ phase as calculated from the Eq.(5.4.11). Only the first Brillouin zone is shown. (a) Frequency of light, propagating along the helix, as a function of the real part of the wave-vector $K(\omega)$.(b) The imaginary part of the wave-vector versus frequency of light. Note the appearance of a finite imaginary part of the wave-vector for the branches labeled as $K_{(2)}$ and $K_{(3)}$ in (a). This leads to non-propagating (and therefore reflected) modes at the Bragg frequency.

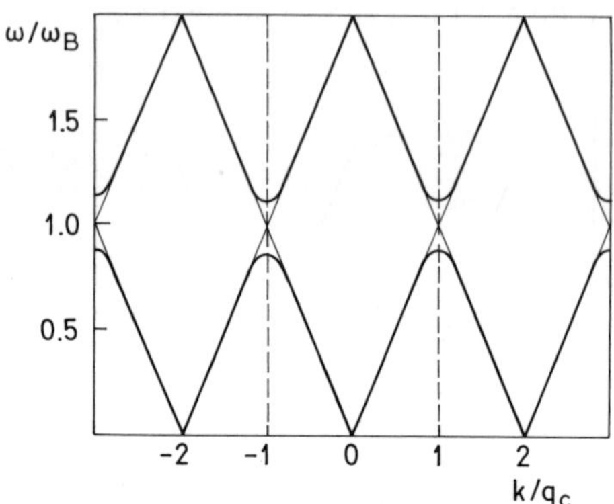

Fig.5.4.2. The dispersion relation for the light, propagating along the helical axis of a $smectic-C^*$ liquid crystal. Note that the Bragg reflection appears only for the circularly polarized light that has the same sense of rotation as the helix; the other circular polarization is transmitted.

One can clearly see that this ratio is either very large or very small, except for $\omega = \omega_B$ or $\omega \to 0$, where it is close to one for $K_{(2)}$ and $K_{(3)}$. This means that for most frequencies, the eigenmodes are either left or right-handed circularly polarized modes. The exception is here the region within the forbidden frequency gap, where the eigenmodes are linearly polarized.

As one can see from Fig.5.4.1, there is a forbidden frequency gap in one of the branches of the dispersion. When the frequency of light is within this region, the corresponding wave-vector K is purely imaginary and k is complex, $k = k' + ik''$ (Fig.5.4.1(b)). As a consequence, the Bloch functions (Eq.5.4.13) will here obtain a non-propagating form $E_k^{\pm} \propto e^{-k''z}$. A straightforward calculation shows that the width of this forbidden gap is

$$\frac{1}{\sqrt{\varepsilon_{//}}} q_c c_o = \omega_- \leq \omega \leq \omega_+ = \frac{1}{\sqrt{\varepsilon_{\perp}}} \qquad (5.4.14)$$

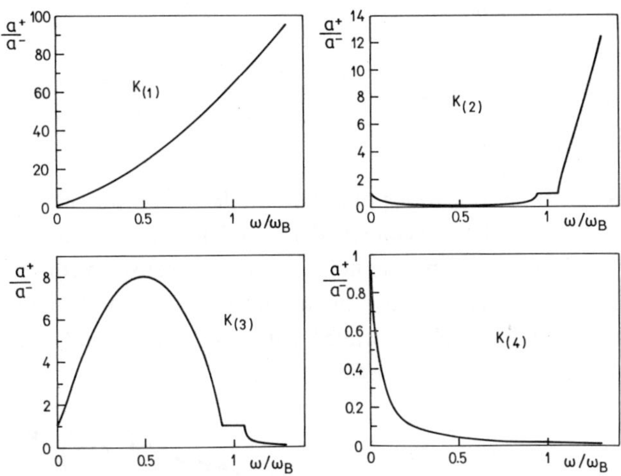

Fig.5.4.3. Polarization properties of the $\omega_i(k)$ dispersion branches. Note that the ratio a^+/a^- is very large for the dispersion branch $K_{(1)}$ and very small for the dispersion branch $K_{(4)}$. This means that $K_{(1)}$ waves are right handed circular waves, whereas $K_{(4)}$ waves are left handed circular waves. The other two branches show similar circular polarization properties except at the Bragg frequency and in the long-wavelength limit.

For the lower edge of the forbidden gap $\omega \to \omega_-$ the electric field of the light wave is

$$E_x = E_o \cos q_c z \cos \omega t$$
$$E_y = -E_o \sin q_c z \sin \omega t \qquad (5.4.15)$$

which is just a linearly polarized wave with the polarization along the projection of the long molecular axis onto the smectic layers. The mode thus "follows" the helix, because it has the same sense of rotation and the same period as the helix. For the upper edge of the forbidden gap, $\omega \to \omega_+$ the electric field is

$$E_x = -E_o \sin q_c z \sin \omega t$$
$$E_y = E_0 \cos q_c z \cos \omega t \qquad (5.4.16)$$

It is thus again a linearly polarized mode, with the electric field oriented perpendicularly to the previously discussed mode and "following" the helical structure as we move along the helix. Again the sense of rotation and the period of this mode are equal to those of the helix. Both non-propagating modes thus represent a "standing helical wave" with the helicity and period matching the helix, as it is shown in Figs.5.4.4(a) and (b).

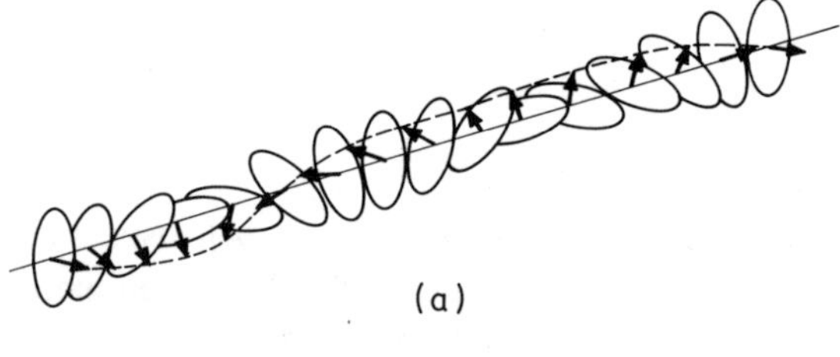

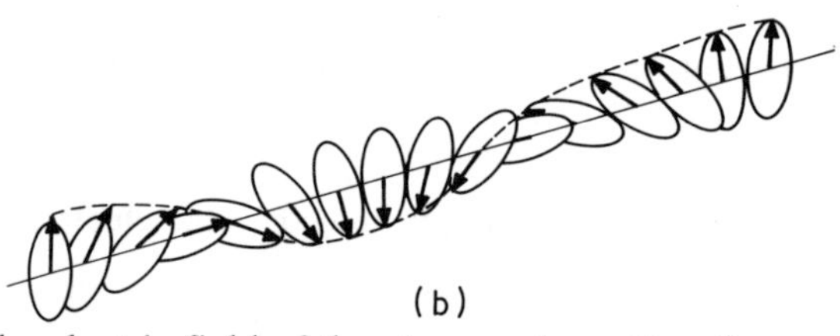

Fig.5.4.4. The electric field of the eigenmodes with a frequency at the edges of the forbidden frequency gap. (a) $\omega = \omega_+$ and (b) $\omega = \omega_-$.

The above considerations allow us to conclude, that an external circularly polarized light wave which has the frequency in the forbidden gap and the same sense of rotation as the helical $smectic - C^*$ structure, will be totally reflected for normal incidence, as it cannot propagate along the helix. On the other hand, a circular light-wave with the opposite sense of rotation will be transmitted throughout the sample. The effect is the well known selective reflection, first observed in cholesteric liquid crystals. The phenomena can be observed with bare eyes, if the $smectic - C^*$ phase is illuminated with a white light source. The sample then appears vividly colored due to selective reflection of a certain band of wavelengths.

We have seen that outside the forbidden frequency gap the eigenwaves have nearly circular character for most frequencies. This means that for a given ω we have a pair, consisting of one left-handed and one right handed circular eigenwave, which propagate in the *z*-direction, and a similar pair which propagates in the -z direction. Let us now assume that a linearly polarized light of $\omega \neq \omega_B$ is incident at $z = 0$ onto a $smectic - C^*$ sample of thickness d. At $z = 0$ this linearly polarized wave will excite both left and right-handed eigenwaves which propagate along the +z-direction. These two circular eigenwaves will travel through the sample with different phase velocities and will recombine at $z = d$ into a linearly polarized wave. However, due to different velocities of propagation, there will be a finite phase difference between the two circular waves. As a result, the direction of the polarization at $z = d$ will be rotated at an angle Ψ with respect to the incident polarization:

$$\Psi = \tfrac{1}{2}\left(K_{(1)} - K_{(3)} + 2q_c\right) d \tag{5.4.17}$$

Here $K_{(1)}$ and $K_{(3)}$ denote the wave-vectors of the eigenwaves belonging to the $\omega_1(k)$ and $\omega_3(k)$ branches. Using the equation 5.4.11 for the eigenvalues, the optical rotatory power ρ of the $smectic - C^*$ phase is

$$\rho = \frac{\Psi}{d} = -\frac{2\pi}{p}\frac{\alpha^2}{8\lambda'^2\left(1-\lambda'^2\right)} \tag{5.4.18}$$

with $\lambda' = 2\lambda_o / p(\varepsilon_{//} + \varepsilon_{\perp})$ and $\alpha = (\varepsilon_{//} - \varepsilon_{\perp})/(\varepsilon_{//} + \varepsilon_{\perp})$. Here λ_o is the vacuum wavelength of the incident light.

The validity of the above expression for the optical rotatory power of the $smectic - C^*$ phase was first tested by Seppen et al.(1988) as shown in Fig.5.4.5. They have measured the rotatory power for different wavelengths of the incident light and found a very good agreement with theory. Moreover, from the observed rotatory power they have calculated the temperature dependence of the tilt angle, which was in very good agreement with the results, obtained by other methods. This method of determining the tilt angle was also successfully used in the $smectic - C^*$ phase of a liquid crystal which showed the polarization reversal phenomenon (Muševič et al., 1989), where the standard electrooptical switching method for the determination of the tilt angle cannot be used because of vanishing electric polarization.

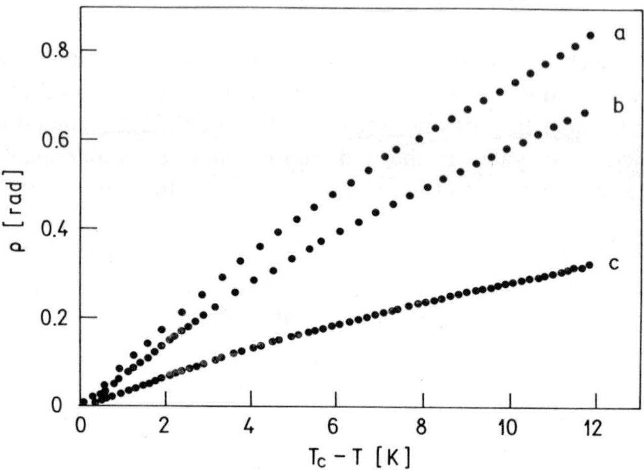

Fig.5.4.5. The optical rotatory power of $75\mu m$ thick sample of DOBAMBC for three different wavelengths. (a) $\lambda_o = 488nm$; (b) $\lambda_o = 515nm$; (c) $\lambda_o = 647nm$ (Seppen et al., 1988).

Note 5.4.1. Rotatory power of chiral smectics

Expression for the rotary power of ferroelectric liquid crystals is formally equivalent to the cholesteric liquid crystals. However, in Eq.5.4.17, the effective dielectric constants are

$$\varepsilon_{//} = \frac{\varepsilon_1 \varepsilon_3}{\varepsilon_3 \cos^2\theta + \varepsilon_1 \sin^2\theta}$$

$$\varepsilon_{\perp} = \varepsilon_2$$

For a tilt angle of $\theta \approx 20^\circ$, the corresponding difference in $\varepsilon_{//} - \varepsilon_{\perp} \approx 0.06$, which is approximately ten times smaller than in cholesterics. As the rotatory power is proportional to the square of the difference of the two dielectric constants, the rotary power of tilted smectics is typically hundred times smaller than in cholesteric liquid crystals. For DOBAMBC, rotatory power is $\approx 500 \deg/mm$ $10K$ below the phase transition into the ferroelectric phase.

5.5. Quasielastic Light Scattering

In the ferroelectric $smectic-C^*$ phase, collective excitations of the order parameter $\vec{\xi}(\vec{r},t)$ are strongly coupled to the dielectric tensor field $\underline{\varepsilon}(\vec{r},t)$ and the spontaneous polarization field $\vec{P}(\vec{r},t)$. These excitations are observable as collective, spatially correlated ***angular reorientations of the optical axis***. Similar to the fluctuations of nematic liquid crystals, the relaxation rates of the order parameter excitations in the $smectic-C^*$ phase are in the range from 10Hz to 1MHz, and can be relatively easily observed in quasielastic light scattering experiments. Whereas quasielastic light scattering allows for the determination of the dispersion relations $\tau^{-1}(\vec{q})$ for the order parameter excitations, dielectric spectroscopy measures the response of the system at the wave-vector $\vec{q}=0$, as shown in Fig.5.5.1.

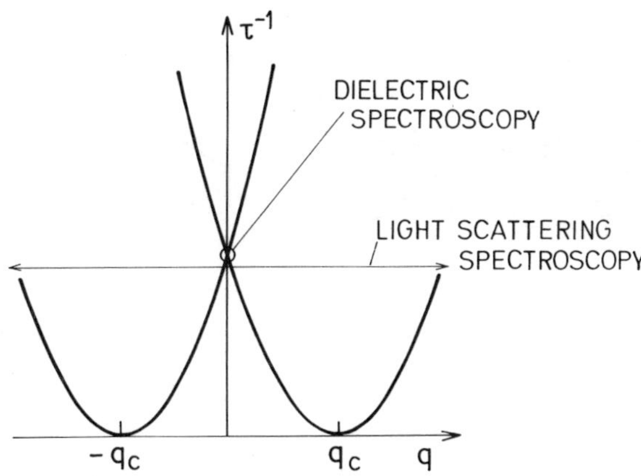

Fig.5.5.1. The dispersion relation $\tau^{-1}(q)$ of the order parameter excitations and its observability by different experimental methods. Note that dielectric spectroscopy (or other linear response spectroscopies that probe with a spatially homogeneous fields) probe the modes at $\vec{q}=0$ only. This is due to the fact that the probing field is usually spatially homogeneous and has only one Fourier component at $\vec{q}=0$. In principle, this could be extended to $\vec{q}\neq 0$ by using a spatially inhomogeneous field, like for example an optically created grating.

The physical origin of the light scattering in liquid crystals can be understood by considering the strong coupling between the director field $\vec{n}(\vec{r},t)$ and the dielectric tensor field $\underline{\varepsilon}(\vec{r},t)$ at optical frequencies (de Gennes, 1968, 1974; Litster, 1983). By remembering that an elementary excitation of the director field is an overdamped orientational wave with a wavelength $\lambda = 2\pi/q$, as shown in Fig.5.5.2.(a), we note that it represents a spatially periodic modulation of the dielectric constant of the medium, as shown in Fig.5.5.2.(b) This thermally excited orientational wave can be therefore considered as a ***phase grating***, which strongly diffracts (scatters) light when the Bragg condition is fulfilled, as shown in Fig.5.5.3. Once the phase grating is created by thermal fluctuations of the director field, it would decay in a relaxation time $\tau(q)$. As a result, the intensity of the Bragg-diffracted light would follow the decay of the phase grating, and will fluctuate in time in the same fashion.

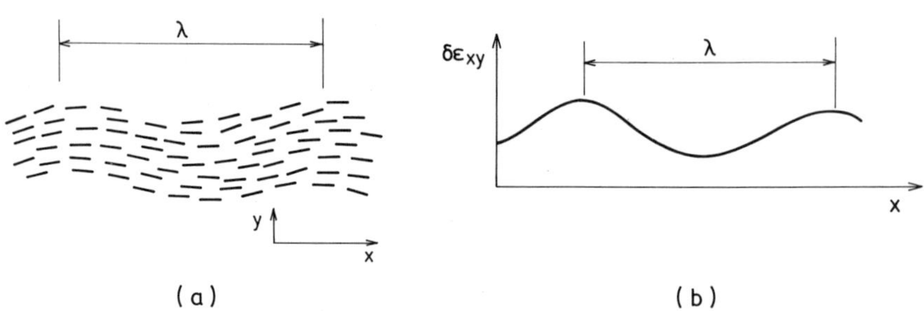

Fig.5.5.2. The origin of light scattering in liquid crystals. (a) A thermally excited orientational wave is created in the sample. A bend-like wave in a nematic liquid crystal is shown for simplicity. (b) This wave represents a modulation of the local index of refraction due to different orientations of the optical axis at different places. A phase grating is therefore created, which diffracts light.

Quasielastic light scattering, or Rayleigh scattering spectroscopy, is a well-known optical spectroscopic method, which is very useful for the study of elementary excitations in liquid crystals. Because of the extremely large optical birefringence, rather low relaxation rates and large amplitudes, the ***orientational fluctuations*** in liquid crystal scatter light strongly. Light scattering in liquid crystals can be a million times stronger than in optically isotropic liquids, where the scattering is due to thermal fluctuations of the ***magnitude of the local index of refraction***. In the ***time-resolved quasielastic light scattering experiment*** one measures the time correlation of the intensity of scattered light at different scattering wave-vectors. From the observed time correlation in the scattering intensity, one can determine the dynamics of thermally excited fluctuations of the order parameter. The quasielastic light scattering in liquid crystals thus plays a similar role as inelastic neutron scattering in the study of phonon excitations in solid crystals.

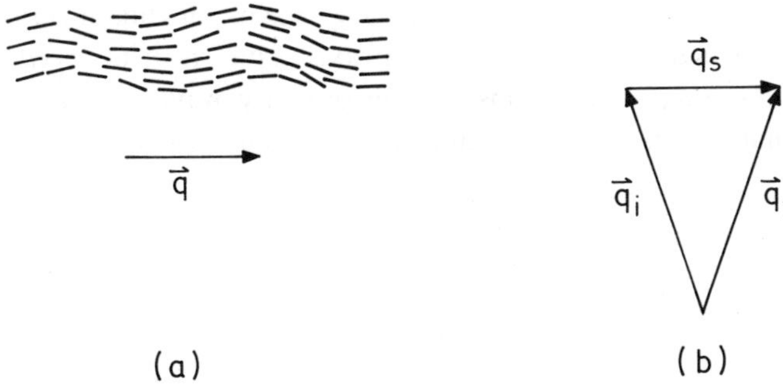

Fig.5.5.3. A bend-like excitation with a wavelength $\lambda = 2\pi/q$ is created in the nematic liquid crystal due to thermal excitation. It represents a phase grating, which strongly diffracts light when the Bragg condition for the scattered light is fulfilled. In the experiment, the magnitude and the direction of the scattering wave-vector is usually selected, and this determines (selects via Bragg condition) the fluctuations that contribute to the scattering.

The magnitude of the scattering wave-vectors which are accessible in the light scattering experiment, depend on the scattering geometry and the wavelength of the laser light. In ferroelectric liquid crystals the period of the helical modulation is in the range of the wavelength of the visible light. Using the light scattering, we can thus probe the most interesting interval in the reciprocal space (i.e. the first Brillouin Zone-BZ of the phase) by a proper selection of the scattering geometry. The magnitude of the BZ can also be adjusted by mixing right- and left-handed enantiomers to $1\text{-}10\,\mu m^{-1}$, which is close to the experimentally accessible range of laser sources.

There is however one important difference between neutron scattering in solids and light scattering in liquid crystals. This is birefringence, which is due to extremely high optical anisotropies of liquid crystals and does not occur for neutron scattering. As it is well known, liquid crystalline phases are optically uniaxial or biaxial materials with an exceptionally large birefringence of the order of $\Delta n \approx 0.1 - 0.2$. This birefringence is usually much larger than in other materials. In spatially homogeneous liquid crystalline phases such as *N*, $smectic-A$ and $smectic-C$, light can propagate as an ordinary or an extraordinary linearly polarized wave. These waves are scattered from thermally excited fluctuations of the dielectric tensor and the scattered light can again propagate as an ordinary or extraordinary wave. In liquid crystals, the orientational fluctuations of the optical axis give rise to ***predominantly depolarized scattering***. This means that an ordinary wave is scattered by the director fluctuations in the extraordinary wave and vice versa, as shown in Fig.5.5.4.

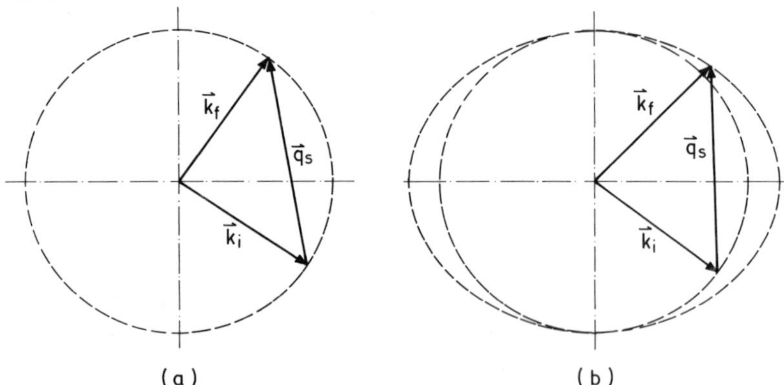

Fig.5.5.4. The difference between the light scattering in isotropic liquid (a) and in birefringent liquid crystal (b). In (a), the normal surface of an isotropic liquid is shown. There is only one branch of optical eigenmodes which can be scattered into each other. (b) Normal surface in a birefringent liquid crystal. There is a normal surface of the ordinary wave (circle) and a normal surface of the extraordinary wave. Light can be scattered between these two surfaces. Only ordinary-extraordinary scattering is shown.

Let us now discuss the light scattering by thermal fluctuations in the ferroelectric $smectic-C^*$ liquid crystal. We shall only consider the scattering of light by the order parameter modes that are characteristic for the $smectic-A$ - $smectic-C^*$ phase transition and we shall neglect the light scattering by other modes such as the smectic layer undulations etc.

In the first order approximation, which is expected to be valid far from degeneration points, the helicoidal $smectic-C^*$ phase is treated as a slightly perturbed optically uniaxial crystal. The eigenwaves of light propagation are linearly polarized ordinary $|k,\sigma\rangle$ and extraordinary $|k',\pi\rangle$ waves, with the wave-vectors $\vec{k}$ and $\vec{k}'$ and polarizations σ and π, respectively. Thermal excitations of the order parameter field $\xi(\vec{r},t)$ result in the fluctuations of the dielectric tensor field

$$\vec{\xi}(\vec{r},t)=\vec{\xi}_o(\vec{r})+\delta\vec{\xi}(\vec{r},t) \Rightarrow \underline{\varepsilon}(\vec{r},t)=\langle\varepsilon\rangle+\delta\underline{\varepsilon}(\vec{r},t) \tag{5.5.1}$$

Here $\langle\varepsilon\rangle$ denotes the time and space averaged dielectric tensor of the $smectic-C^*$ phase, which in the first order approximation (see Eq. 5.3.10.) describes the optical properties of the $smectic-C^*$ phase. Here,

$$\delta\vec{\xi}(\vec{r},t)=\left(\delta\xi_{//}(\vec{r},t),\ \ \delta\xi_{\perp}(\vec{r},t)\right) \tag{5.5.2}$$

is a small deviation of the order parameter from the equilibrium, which is created by thermal fluctuations of the system. $\delta\xi_{//}$ represents here an amplitude excitation and $\delta\xi_{\perp}$ is a phase excitation, which have been defined in the Eq.2.3.6.(a-d). We have here used the adiabatic approximation and have therefore not considered the polarization fluctuations that are in-phase with the director field.

In a scattering process, the incident light with wave-vector $\vec{k}_i$ and polarization $\vec{i}$ (which can be either σ or π) is scattered on thermal fluctuation $\delta\underline{\varepsilon}(\vec{r},t)$ into a light wave with wave-vector k_f and polarization $\vec{f}$, which can be either σ or π. In the limit of small θ , the fluctuating part of the dielectric tensor $\delta\underline{\varepsilon}(\vec{r},t)$ is derived from the Eq. 5.1.2. in terms of the nonequilibrium order parameter as

$$\delta\underline{\varepsilon}(z,t)=(\varepsilon_3-\varepsilon_1)\begin{bmatrix}0 & 0 & \cos\Phi_o z\\ 0 & 0 & \sin\Phi_o z\\ \cos\Phi_o z & \sin\Phi_o & 0\end{bmatrix}\delta\xi_\theta(z,t)+$$

$$+(\varepsilon_3-\varepsilon_1)\begin{bmatrix}0 & 0 & -\sin\Phi_o z\\ 0 & 0 & \cos\Phi_o z\\ -\sin\Phi_o z & \cos\Phi_o & 0\end{bmatrix}\delta\xi_\Phi(z,t) \quad (5.5.3)$$

Here we have used a uniaxial approximation, $\varepsilon_1=\varepsilon_2$ and we will consider only the excitations with the wave-vector along the z-direction. The generalization to other directions is straightforward. $\Phi_o(z)$ is the equilibrium phase profile, $\delta\xi_\theta=\delta\xi_{//}$ and $\delta\xi_\Phi=\delta\xi_\perp$ represent the amplitudon and the phason excitations respectively.

From the above expression we see that the excitations of the order parameter can indeed be observed as the excitations of the dielectric tensor field $\delta\underline{\varepsilon}(z,t)$. In particular,

$$\delta\varepsilon_{xz}(z,t)=(\varepsilon_3-\varepsilon_1)[\delta\xi_\theta\cos q_c z-\delta\xi_\Phi\sin q_c z]=(\varepsilon_3-\varepsilon_1)\delta\xi_x(z,t) \quad (5.5.4a)$$

$$\delta\varepsilon_{yz}(z,t)=(\varepsilon_3-\varepsilon_1)[\delta\xi_\theta\sin q_c z+\delta\xi_\Phi\cos q_c z]=(\varepsilon_3-\varepsilon_1)\delta\xi_y(z,t) \quad (5.5.4b)$$

Here we have considered for simplicity a simple plane-wave modulation of the helix in the ferroelectric $smectic-C^*$ phase, i.e. $\Phi_o=q_c z$. The fluctuations of the off-diagonal terms are therefore observable as a strong depolarized scattering between the ordinary and extraordinary waves or vice versa. Note that the magnitudes of the fluctuations of the dielectric tensor do not depend on temperature. In this respect, quasielastic light scattering can be very conveniently used to study the order parameter dynamics in the vicinity of the

$smectic - A$ - $smectic - C^*$ phase transition, where the modulation is small due to the smallness of the tilt angle.

Following the standard approach (Berne and Pecora, 1976) to the scattering of a monochromatic plane wave with polarization $\vec{i}$, wave-vector k_i, frequency ω_i and amplitude E_o

$$\vec{E}_i = \vec{i}\, E_o e^{i\vec{k}_i\vec{r} - i\omega_i t} \tag{5.5.5}$$

on the fluctuations of the dielectric tensor $\delta\underline{\varepsilon}(\vec{r},t)$, the projection of the scattered electric field amplitude $\vec{f}\vec{E}_s$ at a distance $\vec{R}$ from the sample is

$$E_s = \vec{f}\vec{E}_s(\vec{R},t) = -\frac{|\vec{k}_i|^2 E_o}{4\pi\varepsilon_o R}\delta\varepsilon_{if}(\vec{q}_s,t)e^{i\vec{k}_f\vec{R} - i\omega_i t} \tag{5.5.6}$$

Here, $\vec{q}_s$ is the scattering wave-vector, $\vec{q}_s = \vec{k}_i - \vec{k}_f$, and

$$\delta\varepsilon_{if}(\vec{q}_s,t) = \int d^3\vec{r}\left\{\vec{f}\cdot\delta\underline{\varepsilon}(z,t)\cdot\vec{i}\right\}e^{i\vec{q}_s\vec{r}} \tag{5.5.7}$$

is the Fourier transform of the fluctuating part of the dielectric tensor, projected onto the polarizations $\vec{i}$ and f of the incoming and the scattered beam, respectively. In a light scattering experiment, the time autocorrelation function

$$G^{(2)}(\tau) = \langle I(0)I(\tau)\rangle = \lim_{T\to\infty}\frac{1}{T}\int_0^T dt\, I(t)I(t+\tau) \tag{5.5.8}$$

of the scattered light intensity *I(t)* is detected. Usually, a small portion of a coherent, elastically scattered field E_{lo} of the local oscillator (i.e. the laser beam) is mixed with the quasielastically scattered field E_s, so that the experiment is in the so-called "heterodyne" regime. In the case when $|E_{lo}| >> |E_s|$, $G^{(2)}(\tau)$ is proportional to the autocorrelation function of the Fourier component of the dielectric tensor field:

$$G^{(2)}(\tau) \cong I_{lo}^2 + 2I_{lo}I_s + 2I_{lo}\,\mathrm{Re}\{\langle E_s(0)E_s^*(\tau)\rangle\} \approx \mathrm{Re}\{\langle \delta\varepsilon_{if}(\vec{q}_s,0)\delta\varepsilon_{if}^*(\vec{q}_s,\tau)\rangle\} \tag{5.5.9}$$

The time-autocorrelation function of the scattered light intensity therefore directly reflects the dynamics of the order parameter, because the fluctuations of the dielectric tensor $\delta\varepsilon_{if}(\vec{q}_s,\tau)$ are directly dependent on these fluctuations (see Eq.5.5.4.).

In the light scattering experiment, the polarizations of the incident and the scattered beam determine ("pick-up") the contributions of different modes to the light scattering. For example, in the case of scattering of ordinary polarized light, $\vec{i} = (0,1,0)$, into extraordinary polarized light, $f = (\sin\alpha, 0, \cos\alpha)$, the projection $\delta\varepsilon_{if}(z,t)$ is

$$\begin{aligned}
&\vec{f}\cdot\delta\underline{\varepsilon}\cdot\vec{i} \cong (\sin\alpha \quad 0 \quad \cos\alpha)\cdot\delta\underline{\varepsilon}\cdot\begin{pmatrix}0\\1\\0\end{pmatrix} = \\
&= \cos\alpha\cdot(\varepsilon_3-\varepsilon_1)\cdot[\cos\Phi_o\,\delta\xi_\Phi(z,t)+\sin\Phi_o\,\delta\xi_\theta(z,t)] \cong \\
&\cong \delta\xi_y(z,t)\cdot\cos\alpha
\end{aligned} \tag{5.5.10}$$

Here, α is the angle between the wave-vector of the scattered light and the x-axis. The scattering amplitude in this geometry is thus proportional to the $\delta\xi_y$ component of the order parameter, which is in turn proportional to the phase and amplitude fluctuations. The signal is most intense for $\alpha = 0$.

In the unperturbed $smectic - C^*$ phase, $\Phi_o(z) = q_c z$, and the order parameter excitations $\delta\xi_i^o(z,t) = \delta\xi_i^o \cos qz e^{-t/\tau}$ are plane waves with wave-vector $\vec{q}$. From 5.5.10 we see that the Fourier component $\delta\varepsilon_{if}(\vec{q}_s,t)$ at the scattering wave-vector $\vec{q}_s$ is finite if the wavevector of the excitation equals

$$\vec{q} = \pm\vec{q}_s \pm \vec{q}_c \tag{5.5.11}$$

The corresponding autocorrelation function of the scattered light intensity $G^{(2)}(\tau)$ in the heterodyne regime then consists of two distinct, exponentially decaying contributions from the phason and the amplitudon excitations, respectively:

$$G^{(2)}(\tau,\vec{q}_s) \cong \langle \delta\xi_\Phi(\vec{q}_s,t)\delta\xi_\Phi(\vec{q}_s,t+\tau)\rangle + \langle \delta\xi_\theta(\vec{q}_s,t)\delta\xi_\theta(\vec{q}_s,t+\tau)\rangle =$$
$$= \left\langle \left(\delta\xi_\Phi^o(\vec{q}_s,t)\right)^2 \right\rangle e^{-\tau/\tau_\Phi} + \left\langle \left(\delta\xi_\theta^o(\vec{q}_s,t)\right)^2 \right\rangle e^{-\tau/\tau_\theta}$$

(5.5.12)

Here, $\langle \delta\xi_i^2(q_s,t)\rangle$ is the mean square amplitude of the phason and the amplitudon excitation, respectively. The quasielastic light scattering experiment should thus allow the determination of both relaxation rates.

In practice, the situation is somewhat complicated in view of the large differences in the relaxation rates of the amplitude and phase modes, which result in large differences in the corresponding scattering amplitudes. The amplitudon modes, which have much higher relaxation rates, have correspondingly smaller amplitudes and are more difficult to observe.

Let us now calculate the thermal averages of the amplitude of the eigenmodes. For simplicity, we shall consider the adiabatic approximation as we did in all calculations.

The nonequilibrium value of the free-energy per unit length in a helical eigenframe is derived from the Eq.2.2.7.

$$\Delta G = G - G_\circ = \sum_q \left(\tfrac{1}{2}a(T) + \tfrac{3}{2}b\theta_\circ^2 + \tfrac{1}{2}K_3 q^2\right)\cdot\left|\delta\xi_q^{//}\right|^2 +$$
$$+ \sum_q \left(\tfrac{1}{2}a(T) + \tfrac{1}{2}b\theta_\circ^2 + \tfrac{1}{2}K_3 q^2\right)\cdot\left|\delta\xi_q^{\perp}\right|^2$$

(5.5.13)

Here, we have considered the equilibrium relation $q_c = \Lambda/K_3$ and the expansion coefficients are polarization-renormalized. Due to the $\theta_\circ^2$ terms in the expansion, we have to consider the structure of the nonequilibrium free-energy separately in the region above and below the *smectic – A - smectic – C** phase transition temperature T_c^*.

In the *smectic – A* phase, for $T > T_c^*$, the equilibrium tilt equals zero and the nonequilibrium free-energy is

$$\Delta G = G - G_\circ = \sum_q \left(\tfrac{1}{2}\alpha(T - T_c^*) + \tfrac{1}{2}K_3 q^2\right)\cdot\left(\left|\delta\xi_q^{//}\right|^2 + \left|\delta\xi_q^{\perp}\right|^2\right) \tag{5.5.14}$$

This expression is formally analogous to the energy of a system of uncoupled harmonic oscillators. We now use ***the equipartition theorem*** to evaluate the magnitude of the fluctuation amplitudes: For a classical system with the free energy quadratic in the amplitudes $|\delta\xi_q^i|$, the average free-energy per degree of freedom (per mode) equals $\frac{1}{2}k_B T$:

$$\forall q: \qquad \left(\tfrac{1}{2}\alpha(T - T_c^*) + \tfrac{1}{2}K_3 q^2\right)\cdot\left|\delta\xi_q^i\right|^2 = \tfrac{1}{2}k_B T \tag{5.5.15}$$

or

$$\left|\delta\xi_q^i\right|^2 = \frac{k_B T}{\alpha\left(T - T_c^*\right) + K_3 q^2} \qquad i = \| \text{ or } \perp \tag{5.5.16}$$

The amplitude of the soft mode with $\vec{q} = 0$ is critical and diverges at the phase transition temperature T_c^*. By remembering that the above relations have been written in the helical system, it is straightforward to see that the critical soft mode is in fact located at $q = q_c$ in the laboratory frame.

The situation is somewhat different in the $smectic - C^*$ phase, where the existence of equilibrium tilt angle breaks the symmetry of the $smectic - A$ phase and causes the splitting of the doubly degenerate soft mode. For $T < T_c^*$ the intensities of the amplitude modes are critical for $q = 0$

$$\left|\delta\xi_q^{//}\right|^2 = \frac{k_B T}{2\alpha\left(T_c^* - T\right) + K_3 q^2} \tag{5.5.17}$$

whereas the intensities of the phase modes are large for small q at all temperatures

$$\left|\delta\xi_q^{\perp}\right|^2 = \frac{k_B T}{K_3 q^2} \tag{5.5.18}$$

From the Eqs.5.5.18 and 5.5.19 we immediately see that the amplitudes of the amplitudon modes rapidly decrease below T_c^*, whereas the amplitudes of the phase modes remain finite. This is due to the term $2\alpha(T_c - T)$, which is comparable to the $K_3 q^2$ term only in the very vicinity of the phase transition and is otherwise much larger.

A similar analysis can be performed in the case of a magnetically distorted $smectic-C^*$ phase. Here the equilibrium phase equals $\sin\Phi_o(z) = sn(u,k)$ and the eigenfunctions $\delta\xi_i^o(z)$ have the form of Bloch waves (see Chapter 3.4.). This results in a slightly modified condition for the nonzero Fourier component $\delta\varepsilon_{if}(\vec{q}_s, t)$ for a given excitation

$$\vec{q} = \pm\vec{q}_s \pm m \cdot \vec{q}_c \qquad \text{for } m \text{ integer} \tag{5.4.19}$$

Whereas the condition 5.4.11 for the observation of the order parameter excitations in the unperturbed $smectic-C^*$ phase can be interpreted as the shift of the original Brillouin Zone by the reciprocal vector $\vec{q}_c$, the above condition reflects the periodicity of the dispersion relation in the reciprocal lattice of a distorted $smectic-C^*$ phase. Further discussion of the experimental set-up and technical details can be found in Chapter 10.5.

Note 5.4.1. Homodyne and heterodyne detection in practice

In practice it is quite difficult to obtain fairly well defined regime of quasielastic light scattering detection. Usually the quasielastically scattered light is mixed with the coherent light from the local oscillator by elastically scattered light on imperfections in the sample. If the number of defects in a sample is relatively high, the normalized autocorrelation function

$$g_2(\tau) = \frac{G^{(2)}(\tau)}{\langle I\rangle^2} = 1 + \frac{2I_s I_d}{\langle I_s + I_d\rangle^2} \cdot e^{-\tau/\tau s} + \frac{I_d^2}{\langle I_s + I_d\rangle^2} \cdot e^{-2\tau/\tau s}$$

will be slightly larger than 1. Here, $I = I_s + I_d$ is the total light intensity detected by the photomultiplier. In this case, the third term is negligible and the experiment is in pure heterodyne regime, where the measured autocorrelation time is equal to the correlation time of the scattering process. However, if the statically scattered light I_s is small, which happens with well oriented samples, the zero-time value of the normalized autocorrelation function increases and we get also the contribution of the third term, which is twice as fast as the second term. This is the so-called mixed regime. In practice, this regime is obtained, when the zero-delay value of the normalized autocorrelation function is above $g_2(0) \geq 1.2$. The autocorrelation function is in this case a mixture of two exponential functions and it is not possible to determine the relaxation time. The so-called homodyne detection is obtained for $g_2(0) \geq 1.8$, up to the maximum possible value of $g_2(0) = 2$. There are different possibilities of adding local-oscillator light, like, for example, using fibers to bring some of the light from the laser to the photomultiplier.

Note 5.4.2. Diffusivity coefficients in chiral ferroelectric $smectic - C^*$ phases:

In a quasielastic light scattering experiment, one usually measures the dispersion relation for the order parameter excitations. At a given wave-vector $\vec{q}$, the autocorrelation function is fitted to the expression 5.4.12., which gives us the relaxation rate $\tau^{-1}(\vec{q})$. The measured dispersion is then fitted to the quadratic function, which allows us to determine the diffusivity coefficient for a given mode. It is a ratio of the corresponding elastic constant and viscosity. For example, the phason dispersion depends on the direction of the wave-vector. For the wave-vectors along the smectic normal, the phason relaxation rate is

$$\tau^{-1}(q_z) = \frac{K_3}{\gamma} \cdot (q - q_c)^2$$

Here, K_3/γ is the phason diffusivity along the helical axis. For the wave-vector along the smectic layers, the phason dispersion can be expressed in a similar way,

$$\tau^{-1}(q_x) = \frac{K_+}{\gamma} \cdot q_x^2$$

Here, K_+/γ is the phason diffusivity along the smectic layers. In a series of experiments, these diffusivities have been determined for DOBAMBC and CE-8 and are listed in tables below.

1. CE-8:

$T - T_c [K]$	$K_+/\gamma \; [cm^2 s^{-1}]$	$K_3/\gamma \; [cm^2 s^{-1}]$
-5	3.6×10^{-5}	3.1×10^{-6}
-0.95		2.3×10^{-6}
-0.72	3.6×10^{-5}	1.6×10^{-6}
-0.70	3.3×10^{-5}	2.1×10^{-6}
-0.09		3.1×10^{-6}
-0.07		2.7×10^{-6}

-0.05		2.7×10^{-6}
-0.05	4.5×10^{-5}	0.6×10^{-5}
+0.02	3.3×10^{-5}	0.3×10^{-5}

By considering $\gamma = 0.01 kg/ms$ (Škarabot et al., 1999a), we obtain $K_3 = 3\times10^{-12} N$ and $K_+ \approx 3\times10^{-11} N$ for CE-8. The crystal is therefore ten times stiffer for elastic deformation along the smectic layers. Similar behavior can be observed in other ferroelectric smectics.

2. DOBAMBC:

$T-T_c[K]$	K_+/γ $[cm^2 s^{-1}]$	K_3/γ $[cm^2 s^{-1}]$
-0.04	1.6×10^{-5}	0.8×10^{-6}

By considering $\gamma = 0.05 kg/ms$, we obtain $K_3 = 4\times10^{-12} N$ and $7.5\times10^{-11} N$. Again, the transverse elastic constant is an order of magnitude larger than the "twist" elastic constant, $K_+ \approx 10\times K_3$.

5.6. Linear Electrooptic Response in the *smectic-A* Phase

In contrast to quasielastic light scattering spectroscopy which allows the determination of the dispersion relation of the order parameter excitations, dielectric spectroscopy is a very useful probe for determining the susceptibilities of the soft, amplitude and phase modes at $\vec{q} = 0$. However, this method suffers an inherent deficiency of probing a rather thin, book-shelf aligned sample with a large surface area. Such a restricted planar geometry can in principle by itself induce changes in the originally gapless dispersion relations via the surface-induced deformation of the helical structure.

This problem can be circumvented by an electro-optical detection technique of the linear response of the $smectic-C^*$ phase to a small external field. Originally, this technique was first used in the "electroclinic" experiments in the $smectic-A$ phase of ferroelectric liquid crystals performed by Garoff and Meyer (1977;1979). The underlying idea is the coupling between the tilt and polarization: Whenever there is an induced polarization $\delta\vec{P}$, there will also be and induced tilt $\delta\theta$, as shown in Fig.5.6.1.

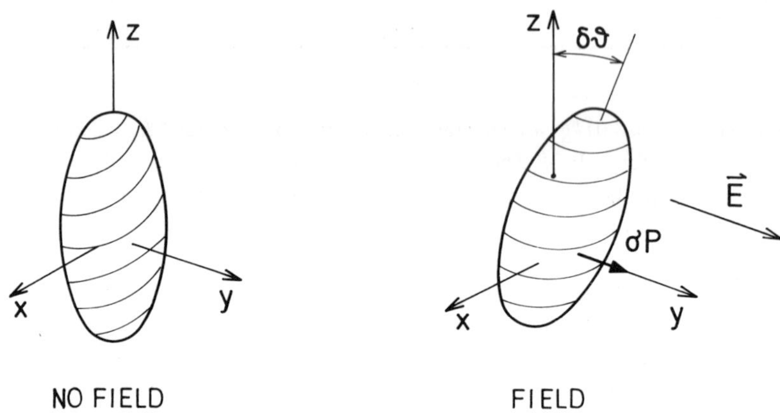

Fig.5.6.1. The electroclinic effect: An external field induces a finite electric polarization in the paraelectric *smectic* – *A* phase. Due to the linear coupling between the tilt and polarization, there will be an induced collective tilt of the molecules in the plane, perpendicular to the field direction. This induced tilt can be as large as several degrees at temperatures close to the ferroelectric phase, where the system is soft.

By applying a small electric field we induce a finite polarization in the *smectic* – *A* phase. As a result, there will be a finite collective tilt of the molecules, or, what is equivalent, the optical axis of the *smectic* – *A* phase will tilt with respect to the layer normal. The induced tilt can be measured very accurately using the optical interference techniques. Moreover, we can apply a time varying signal with a frequency ω and record the in-phase (real) and the out-of phase (imaginary) part of the signal from the photodiode detector. By sweeping the frequency of the measuring signal we can thus measure the dispersion of the real and imaginary part of the electrooptic response. As we shall see, in most cases, this is simply proportional to the dielectric dispersion. In this sense, the linear electrooptic response is a direct analog of the well-known dielectric spectroscopy.

The idea of the electroclinic (i.e. linear electrooptic) response in the *smectic* – *A* phase can be generalized to the bulk or constrained *smectic* – C^*

phase as well as other phases. For example, in the bulk $smectic-C^*$ phase, a small electric field applied perpendicular to the helix will induce a small amplitude change $\delta\theta$ and a small phase change $\delta\Phi(z)$. Both of them will result in a small rotation of the optical axis of the $smectic-C^*$. Furthermore, we will show that these ideas can be usefully generalized to the electrooptic response of thin ferroelectric cells.

Let us apply a small electric field $E = E_o e^{i\omega t} = (0, E, 0)\cdot e^{i\omega t}$ in the y-direction, i.e. perpendicularly to the layer normal of the $smectic-A$ phase. The dynamics of the resulting nonequilibrium tilt angle is calculated within the so-called adiabatic approximation (Eqs.2.4.7), where the free energy density is

$$g(z,t) = g_A + \tfrac{1}{2}a(T)(\delta\xi_{//}^2 + \delta\xi_{\perp}^2) + \tfrac{1}{2}K_3\left(\left(\frac{\partial\delta\xi_{//}}{\partial z}\right)^2 + \left(\frac{\partial\delta\xi_{\perp}}{\partial z}\right)^2\right) - \qquad (5.6.1)$$
$$-\varepsilon C(\delta\xi_{//}\cos q_c z - \delta\xi_{\perp}\sin q_c z)E_o e^{i\omega t}$$

Here, $\delta\xi_{//}$ and $\delta\xi_{\perp}$ represent small deviations of the tilt magnitude and phase, respectively. The last term is just the linear $-\vec{P}\vec{E}$ coupling term written in the helical laboratory system. After applying Landau-Khalatnikov dynamic equations for the motion of $\delta\xi_{//}$ and $\delta\xi_{\perp}$ we search for their solutions in the form

$$\delta\xi_{//} = \delta\xi_{//}^o \cdot\cos(q_c z)\cdot e^{i\omega t} \qquad (5.6.2a)$$

$$\delta\xi_{\perp} = \delta\xi_{\perp}^o \cdot\sin(q_c z)\cdot e^{i\omega t} \qquad (5.6.2b)$$

The form of the eigensolution implies that there is only a single wave-vector involved in the response. This "Ansatz" leads to the following complex form of the amplitudes of the two modes that couple to the external electric field

$$\delta\xi_{//}^o = \frac{\varepsilon C E_o}{a(T) + K_3 q_c^2}\cdot\frac{1}{1 + i\omega\tau_\theta} \qquad (5.6.3a)$$

$$\delta\xi_{\perp}^o = -\frac{\varepsilon C E_o}{a(T) + K_3 q_c^2}\cdot\frac{1}{1 + i\omega\tau_\Phi} \qquad (5.6.3b)$$

In the $smectic-A$ phase the relaxation rates of these two modes are degenerate and we have a single ***soft mode*** with a relaxation rate

$$\tau_{soft}^{-1} = \frac{\alpha(T-T_c)}{\gamma} + \frac{K_3}{\gamma}q_c^2 \tag{5.6.4.}$$

The response in the $smectic-A$ phase (Eq.5.6.3.) is therefore of the Debye type. This expresses the fact that the orientational excitations in liquid crystals are always overdamped. We furthermore observe that:

(i) Both the induced amplitude changes and the phase changes ***are degenerate*** in the $smectic-A$ phase, i.e. they have the same relaxation rates. This is consistent with the underlying general order parameter dynamics, developed in Section 2.2. By remembering that we attribute the wave-vector $\vec{q}=0$ to the external homogeneous electric field $E(t)$, we note that it can couple only to those eigenmodes in the system, that have the wave-vector $\vec{q}=0$ and which carry a finite polarization, $\langle P\rangle \neq 0$. Here $\langle\ \rangle$ denotes the space average. As we have seen before and as it is shown in Fig.5.5.1 of the previous Section 5.5., there are two observable branches of degenerate soft excitations in the $smectic-A$ phase, centered at $\vec{q}=\pm\vec{q}_c$, respectively. These two degenerate branches intersect the y-axis at the same point. The external homogeneous electric field thus couples to this polar mode at $\vec{q}=0$.

(ii) The relaxation rate of the mode at $\vec{q}=0$ is finite at the transition temperature, i.e. $\tau_\theta^{-1}(T=T_c)=(K_3/\gamma)q_c^2$. This is a result of the fact that a ***chiral system is critical at*** $\vec{q}=\vec{q}_c$ in the reciprocal space. With an external electric field we are testing the response of the chiral system at a non-critical wave-vector $\vec{q}=0$ and we observe a non-critical behavior, as shown in Fig.5.6.2. Finite frequencies of the modes at $\vec{q}=0$ thus simply reflect the mismatch of the measuring and the critical wave-vector. Note that in the limit $\vec{q}_c \to 0$ the system becomes critical at $\vec{q}=0$.

(iii) The divergence of the linear response of the $smectic-A$ phase (Eq.5.6.4.) is incomplete at T_c, i.e. we have an incomplete Curie-Weiss law, as shown in Fig.5.6.3.

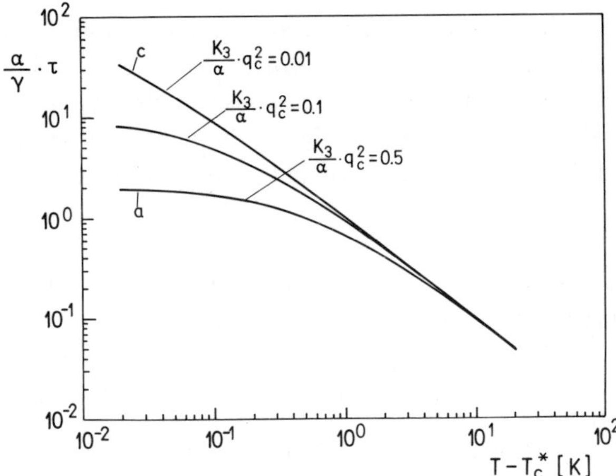

Fig.5.6.2. The incomplete softening of the soft-mode branch of excitations in an inhomogeneous system, probed by a homogeneous external field with $\vec{q} = 0$. The curve (a) corresponds to a chiral system with large q_c, whereas curve (c) corresponds to a system with small q_c.

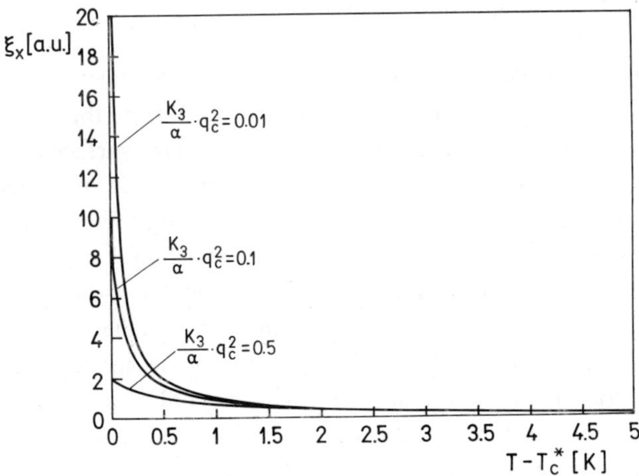

Fig.5.6.3. The incomplete Curie-Weiss law for a phase transition into an inhomogeneous phase, tested by a homogeneous external field.

At the phase transition temperature there is a finite magnitude of the linear response, $|\delta\xi_{//}| = (\varepsilon CE_o / K_3 q_c^2)$. This again expresses the fact that we are measuring the linear response of the chiral system at the non-critical wave-vector $\vec{q} = 0$. In the limit of low chirality, $\vec{q}_c \to 0$, the Curie-Weiss law is recovered and we have a true divergence of the linear response at $\vec{q} = 0$.

Finally, we calculate the induced tilt angle in the laboratory system. A straightforward calculation gives

$$\xi_x(t) = \frac{\varepsilon CE_o}{\alpha(T - T_c) + K_3 q_c^2} \cdot \frac{e^{i\omega t}}{1 + i\omega\tau} \tag{5.6.5a}$$

$$\xi_y(t) = 0 \tag{5.6.5b}$$

The molecules in the $smectic - A$ phase therefore tilt collectively in the x-direction, when an electric field is applied in the y-direction, as expected.

There are several ways of detecting this induced tilt, which depend on the geometry of the sample under study:

(i) ***Homeotropic geometry*** was used in the first electroclinic experiment performed by Garoff and Meyer (1977;1979). Here the smectic layers are aligned parallel to the chemically treated glass surface. Usually a very good, defect-free alignment can be obtained in this geometry. The transverse electric field is generated by two tiny wires, inserted into the sample cell, as shown in Fig.5.6.4(a). The liquid crystalline sample is placed in a cell that can be rotated around the z-direction, so that the optical axis of the sample is always in the x-y plane. A laser beam which is polarized at 45° with respect to the x-y plane, induces an ordinary and an extraordinary wave in the liquid crystal. These two waves than propagate at different phase velocities through the crystal and recombine at the rear glass plate. In general, the resulting polarization will be elliptical and is analyzed by an analizer, which is crossed with respect to the polarization of the laser beam. If we rotate the sample around the z-axis, we will observe an interference effect (a set of fringes) in the intensity of the output beam, as it is shown in Fig.5.6.4(b).

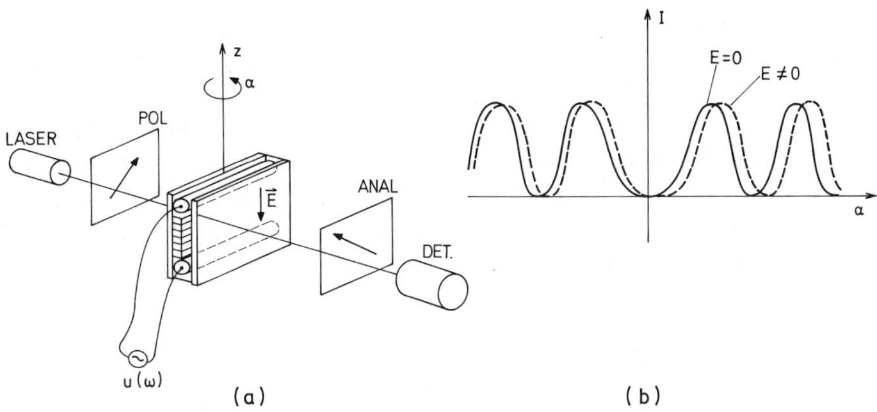

Fig.5.6.4. Detection of the electroclinic effect in the homeotropic geometry. (a) The geometry of the experiment. (b) Conoscopic fringes and the electric field induced shift of the conoscopic figure.

The interference is here a result of the coherent mixing of the ordinary and extraordinary beams and has a minimum whenever the phase difference between the two beams is a multiple of 2π :

$$\phi_e - \phi_o = n \cdot 2\pi \tag{5.6.6.}$$

Here ϕ_e and ϕ_o denote the phases of the extraordinary and ordinary waves, as they pass through the sample. Whereas ϕ_o depends only upon the thickness of the liquid crystal, ϕ_e depends also on the angle of propagation, so that the calculation of the phase difference is quite cumbersome (see for example Born and Wolf , 1959). The samples are usually very thick (100 μm) and one has to take into account different angles of refraction of both waves and other fine details. By rotating the cell around the z-axis, we can reach a half-intensity position on a selected fringe, where the intensity depends approximately linearly on the angle α of rotation of the optical axis. If we then apply a small in-plane electric field, a finite tilt of the molecules will be induced. As a consequence, a finite tilt of the direction of the optical axis will be induced, too. This is reflected as an overall shift of the conoscopic figure as a whole in a certain direction and by reversing the field, the direction of this shift will be reversed. This means that by applying a time varying electric field we will detect a time-varying intensity of the transmitted light. Using a two-channel lock-in amplifier, we can record the in-phase and the out-of-phase value of the photodiode current as a function of the measuring frequency and we obtain in the $smectic-A$ phase the Debye relaxational spectrum of the linear electrooptic response, as shown in Fig.5.6.5.

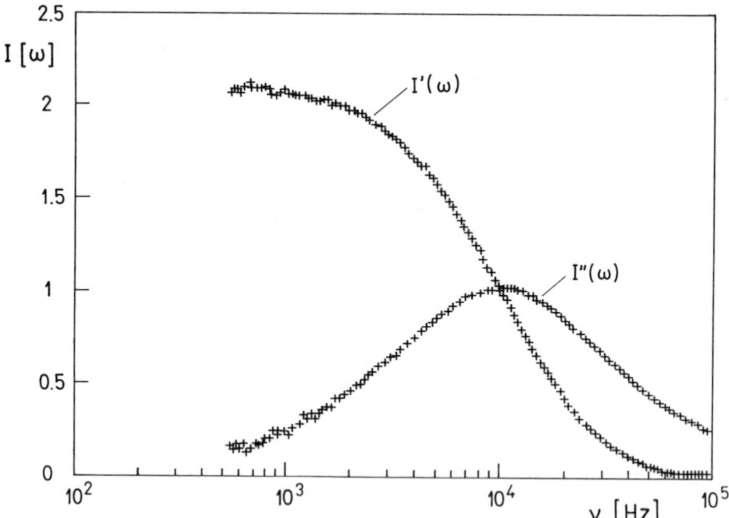

Fig.5.6.5. Real and imaginary parts of the photocurrent, induced by the linear electrooptic effect in the $smectic-A$ phase of a ferroelectric liquid crystal.

The real and the imaginary parts of the photocurrent $I = I' + iI''$ can be calculated from the complex form of the linear response amplitude (Eq.5.6.3.)

$$I' \propto \frac{\varepsilon C}{a(T)+K_3 q_c^2} \cdot \frac{1}{1+(\omega\tau)^2} \tag{5.6.7a}$$

$$I'' \propto \frac{\varepsilon C}{a(T)+K_3 q_c^2} \cdot \frac{\omega\tau}{1+(\omega\tau)^2} \tag{5.6.7b}$$

It is straightforward to determine the relaxation rate of the linear response by fitting the obtained spectra. In many cases, the relaxational frequency can be well estimated by simply reading the frequency, where the imaginary part of the photocurrent reaches a maximum value (Eq.5.6.7(b)), whereas from the Eqs.5.6.7. we can obtain only relative and not the absolute values of the amplitudes. The method thus allows very accurate and straightforward determination of the critical phenomena at $\vec{q}=0$ in a homeotropic geometry which is believed to have a negligible effects on the dynamics.

(ii) For most practical purposes, a ***homogeneous planar alignment*** of the $smectic-A$ phase is used. Here the smectic planes are perpendicular to the glass plates of the sample, which are parallel to the x-y plane and have transparent electrodes on the inner surfaces. The optical axis of the sample is again in the x-y plane and the sample can be rotated around the z-axis. The sample is therefore a birefringent plate with the birefringence $\Delta n = n_e - n_o$ and thickness *d* with the optical axis along the director $\vec{n}$, as shown in Fig.5.6.6. When this sample is placed between crossed polarizers and both polarizer and the analyzer are rotated simultaneously at an angle Ω_o around the z-direction, an interference effect between the ordinary and extraordinary wave is observed, which is similar to the one discussed previously. The transmitted intensity shows a periodic dependence on the angle of rotation Ω_o

$$I = I_o \sin^2 2\Omega_o \sin^2 \frac{\pi \Delta n d}{\lambda} \tag{5.6.8.}$$

where I_o is the intensity and λ is the vacuum wavelength of the incident light.

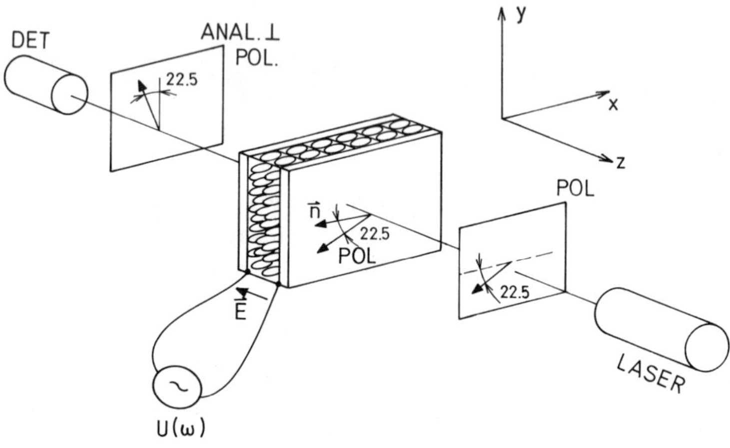

Fig.5.6.6. Linear electrooptic response in the planar geometry.

By applying a small electric field $E = E_o e^{i\omega t}$ the optical axis of the crystal is periodically rotated in the x-y plane by a small angle $\delta\theta(t)$. This is reflected in the transmitted intensity $\delta I(t)$, which is now modulated at a frequency ω

$$\delta I(t) \propto \sin(4\Omega_o) \cdot \sin^2\left(\frac{\pi \Delta n d}{\lambda}\right) \cdot \delta\theta(\omega) \cdot e^{i\omega t} \qquad (5.6.9.)$$

The intensity of the linear response has a maximum value at $\Omega_o = \pi/8 \equiv 22.5^\circ$. The complex amplitude of the response $\delta\theta(\omega)$ is given by the Eq.5.6.6. and shows the Debye relaxational behavior. There is a number of experiments which report the softening of the order parameter excitations in the $smectic - A$ phase, using linear electrooptic response (Bahr and Heppke, 1987; Qiu et al., 1988, etc). Hide et al. (1995) have used dynamic polarized infrared spectroscopy to study the dynamics of molecular conformations under an applied electric field. They report evidence for a collective rotation about the long molecular axis (i.e. the polarization mode).

5.7. Linear Electrooptics of the Bulk and Confined *smectic-C** phase

The idea of the linear elecrooptic response of the $smectic-A$ phase can be generalized to the bulk, helically modulated $smectic-C^*$ phase and other helical smectic phases. Weak external electric field induces in the chiral ferroelectric $smectic-C^*$ phase a distortion of the order parameter that is reflected in a small change of the birefringence. This linearly induced birefringence gives rise to a linear electrooptic effect.

Let us consider the dynamics of the bulk $smectic-C^*$ phase in a small external electric field $E = E_o e^{i\omega t} = (0 \quad E_\circ \quad 0)\cdot e^{i\omega t}$, which is applied in a direction perpendicular to the helix. For $T<T_c$ the nonequilibrium free-energy density is

$$g(z,t) = g_{C^*} - a(T)(\delta\xi_{//})^2 + \frac{1}{2}K_3\left[\left(\frac{\partial\delta\xi_{//}}{\partial z}\right)^2 + \left(\frac{\partial\delta\xi_{\perp}}{\partial z}\right)^2\right] - \qquad (5.7.1)$$
$$-\varepsilon C\cdot(\cos q_c z\delta\xi_{//} - \sin q_c z\delta\xi_{\perp})\cdot E_o e^{i\omega t}$$

Here g_{C^*} is the equilibrium free-energy density of the $smectic-C^*$ phase. Note that there is an asymmetry present in the amplitudes of fluctuations, which are symmetric above T_c (see Eq.5.6.1.) The solutions of the Landau-Khalatnikov equations of motion for $\delta\xi_{//}$ and $\delta\xi_{\perp}$ are taken in a form given by Eq.5.6.2. However, the complex amplitudes are now quite different below T_c:

$$\delta\xi_{//}^{\circ} = \delta\xi_{\theta}^{o} = \frac{\varepsilon C E_o}{2\alpha(T_c - T) + K_3 q_c^2}\cdot\frac{1}{1+i\omega\tau_\theta} \qquad (5.7.2a)$$

$$\delta\xi_{\perp}^{\circ} = \delta\xi_{\Phi}^{o} = -\frac{\varepsilon C E_o}{K_3 q_c^2} \cdot \frac{1}{1 + i\omega\tau_{\Phi}} \tag{5.7.2b}$$

and the degeneracy of the soft mode relaxation rates is removed below T_c:

$$\tau_{\theta}^{-1} = \frac{2\alpha(T_c - T)}{\gamma} + \frac{K_3}{\gamma} q_c^2 \tag{5.7.3a}$$

$$\tau_{\Phi}^{-1} = \frac{K_3}{\gamma} q_c^2 \tag{5.7.3b}$$

The electric field thus couples to the amplitude (Eqs.5.7.2.(a) and 5.7.3(a)) and the phase mode (Eqs.5.7.2.(b) and 5.7.3(b)) in the $smectic - C^*$ phase. Again the amplitude mode is noncritical at the phase transition temperature T_c, and the phase mode also always has a finite relaxation rate. In the linear electrooptic response experiment we should therefore observe a splitting of the soft mode branch of excitations into the amplitude and phase branch, respectively. In reality, however, the contribution of the phase mode branch is much larger than the contribution of the amplitudon mode, and we observe a large signal due to the phase mode only, except very close to T_c.

Let us now calculate the changes of the optical properties of the $smectic - C^*$ phase in a weak external electric field $E = E_o e^{i\omega t}$. For simplicity, we shall consider here only the contribution of the phase changes to the optical properties. The contribution of the amplitude changes can be calculated in a similar fashion. The reason for the changes of the optical properties is a slight distortion of the phase profile $\Phi(z,t) = \Phi_o(z) + \delta\Phi(z,t)$, where $\Phi_o(z)$ is the equilibrium phase profile. The distortion $\delta\Phi(z,t)$ is calculated from Eq.5.6.2(b):

$$\delta\Phi(z,t) = \frac{\delta\xi_{\perp}}{\theta_o} = -\frac{P_o E_o}{\theta_o^2 K_3 q_c^2} \cdot \sin q_c z \cdot \frac{e^{i\omega t}}{1 + i\omega\tau_{\Phi}} \tag{5.7.4}$$

and is a periodic function along the z-direction. The induced phase distortion $\delta\Phi(z,t)$ is reflected in the distortion of the dielectric tensor components

(Eq.5.1.2.). In the limit of small θ, where we can neglect terms of the order of θ^2 in $\delta\underline{\varepsilon}(\vec{r})$, the change of the space periodic part of the dielectric tensor due to the influence of the measuring electric field is

$$\delta\underline{\varepsilon}(z,t) \approx \begin{bmatrix} 0 & 0 & \cos(\Phi_o + \delta\Phi) \\ 0 & 0 & \sin(\Phi_o + \delta\Phi) \\ \cos(\Phi_o + \delta\Phi) & \sin(\Phi_o + \delta\Phi) & 0 \end{bmatrix} \tag{5.7.5}$$

At small tilt angles, the first order approximation is valid and the ***space averaged dielectric tensor*** $\langle\underline{\varepsilon}\rangle$ determines the optical properties of the $smectic-C^*$ phase. In the limit of a small probing field and a plane-wave modulation $\Phi_o = q_c z$, the above expression for the space averaged dielectric tensor of the $smectic-C^*$ phase can be expanded in terms of $\delta\Phi(z,t)$, resulting in

$$\langle\underline{\varepsilon}(\vec{r})\rangle = \begin{bmatrix} \varepsilon_1 & 0 & \delta \\ 0 & \varepsilon_1 & 0 \\ \delta & 0 & \varepsilon_3 \end{bmatrix} \tag{5.7.6}$$

with

$$\delta(t) = \frac{P_o E_o}{2\theta_o K_3 q_c^2} \cdot \frac{e^{i\omega t}}{1 + i\omega\tau_\Phi} \tag{5.7.7}$$

The influence of a small external field on the average dielectric tensor is shown in the δ-term, which is field-dependent and oscillates with the frequency ω of the measuring field $E(t)$. As a result, nonzero off-diagonal terms appear in the dielectric tensor, which also oscillate in time.

Let us now consider the form of the averaged dielectric tensor (Eq.5.7.6.). This form of the tensor is obtained by rotating the originally diagonal dielectric tensor with the principal values $\varepsilon_1 = \varepsilon_2$ and ε_3 around the y-axis, i.e. around the direction of the electric field. The phase distortion, as induced by the external field, therefore results in a ***small rotation of the effective dielectric tensor***, which represents a small, time periodic rotation of the optical axis around the direction of the external field, as shown in Fig.5.7.1.

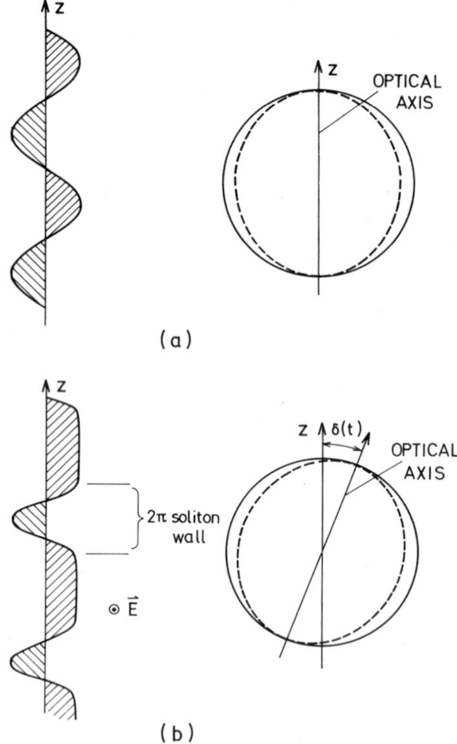

Fig.5.7.1. (a) The undistorted helical phase is optically uniaxial with the optical axis along the layer normal. (b) An applied electric field induces a small distortion and a small tilt of the optical axis.

The angle of the rotation of the optical axis is in the limit of a small distortion proportional to the external field and can easily be detected by the optical interferometric technique. From the frequency dependence of the linear electrooptic response, one can thus determine the relaxation rate of the phason mode in the unperturbed $smectic-C^*$ phase, at $\vec{q}=0$.

Similar phenomena can be observed in thin planar cells with polar boundary conditions. We consider the linear electrooptical response of such a $smectic-C^*$ phase, confined between the two plan-parallel plates (in the y-z plane) in the planar geometry with polar surface coupling. As we have seen, there is a large variety of stable director configurations, which depend on the surface anchoring, plate separation and elastic properties of liquid crystal, etc. The optical properties of these confined and birefringent phases can be conveniently studied using different techniques, as for example the Jones matrix formalism (see, for example, Berreman, D.W., 1973). We shall here use the first order approximation to the liquid crystalline optics to qualitatively explain the origin of linear electrooptic response.

When a small measuring electric field $\vec{E}=(0,E,0)e^{i\omega t}$ is applied in the x direction of the smectic layers of a thin planar $smectic-C^*$ phase, it will couple to those collective eigenmodes that have a finite space-averaged electric polarization $\langle\delta P(\vec{r},t)\rangle=(\delta P_x(t),\delta P_y(t),0)$. Due to the interconnection between the polarization and the director field, this would result in a finite value of the space averaged director field $\langle\vec{n}(\vec{r},t)\rangle=(\delta n_x(t),\delta n_y(t),0)$. Because the dielectric tensor for the optical frequencies ε_{ij} is proportional to the tensor $\vec{n}\otimes\vec{n}$, any change of $\delta\vec{n}(\vec{r},t)$ would be reflected in the dielectric tensor field. In particular, we have

$$\delta\varepsilon_{ij}\approx\delta\left(n_i n_j\right)\approx n_i^o\delta n_j+n_j^o\delta n_i \tag{5.7.8}$$

or explicitly

$$\begin{aligned}\delta\underline{\varepsilon}&\approx\begin{bmatrix}2n_x^o\delta n_x & n_x^o\delta n_y+n_y^o\delta n_x & \delta n_x\\ n_x^o\delta n_y+n_y^o\delta n_x & 2n_y^o\delta n_y & \delta n_y\\ \delta n_x & \delta n_y & 0\end{bmatrix}\approx\\ &\approx(1/\varepsilon C)^2\begin{bmatrix}2P_y^o\delta P_y & P_y^o\delta P_x+P_x^o\delta P_y & \varepsilon C\delta P_y\\ P_y^o\delta P_x+P_x^o\delta P_y & 2P_x^o\delta P_x & -\varepsilon C\delta P_x\\ \varepsilon C\delta P_y & -\varepsilon C\delta P_x & 0\end{bmatrix}\end{aligned} \tag{5.7.9}$$

In the case of polar surface coupling the structure should have mirror symmetry with respect to the y-z plane. This results in the following transformation properties of the electric polarization field

$$x \to -x: \begin{cases} P_x^o \to -P_x^o \\ P_y^o \to P_y^o \\ \delta P_x \to \delta P_x \\ \delta P_y \to -\delta P_y \end{cases} \tag{5.7.10}$$

As a result, the space averaged part of the dielectric tensor of a thin planar cell is

$$\langle \delta \underline{\varepsilon} \rangle \approx (1/\varepsilon C)^2 \begin{bmatrix} 0 & \langle P_y^o \delta P_x + P_x^o \delta P \rangle_y & \varepsilon C \langle \delta P_y \rangle \\ \langle P_y^o \delta P_x + P_x^o \delta P_y \rangle & 0 & -\varepsilon C \langle \delta P_x \rangle \\ \varepsilon C \langle \delta P_y \rangle & -\varepsilon C \langle \delta P_x \rangle & 0 \end{bmatrix} \tag{5.7.11}$$

The above form of the dielectric tensor can be interpreted as a small rotation of the originally diagonal tensor around the x, y and z-axis, respectively, which in turn represents a small rotation of the optical axis. The magnitude of this rotation is always linearly proportional to the external measuring electric field and can be detected by the optical interference techniques. By choosing proper experimental conditions (i.e. the angle of incidence and polarization), it is possible to isolate the electric field induced tilting of the optical axis in the x-y plane only. In this case we detect a change of the dielectric tensor

$$\langle \delta \underline{\varepsilon} \rangle \approx \begin{bmatrix} 0 & 0 & 0 \\ 0 & 0 & \langle \delta P_x \rangle \\ 0 & \langle \delta P_x \rangle & 0 \end{bmatrix} \tag{5.7.12.}$$

which is proportional to the dielectric susceptibility of the sample. The linear electrooptic response of the sample is thus in the first order approximation proportional to the dielectric response of the sample and can be successfully used in the study of the dynamics of confined ferroelectric liquid crystals.

Whereas it is practically impossible to measure the absolute value of the dielectric constant with this method, it is on the other hand very easy to determine the frequency dispersion of the dielectric constant just by measuring the frequency dependencies of the real and imaginary part of the linear electrooptic response. In combination with the quasi-locality of this method, we can study, for example the dependence of the order parameter dynamics on the confined geometry in wedge-type cells, where the thickness of the liquid crystal can be varied within a single cell from several tenths of a micron to several microns. Figure 5.7.2 shows the dependence of the phase mode on the cell spacing in the surface unwound $smectic-C^*$ phase of SCE9. Because of the confinement and polar surface coupling, the relaxation rate of the phase mode at $\vec{q}=0$ increases as an inverse square of the sample thickness for fixed boundary conditions.

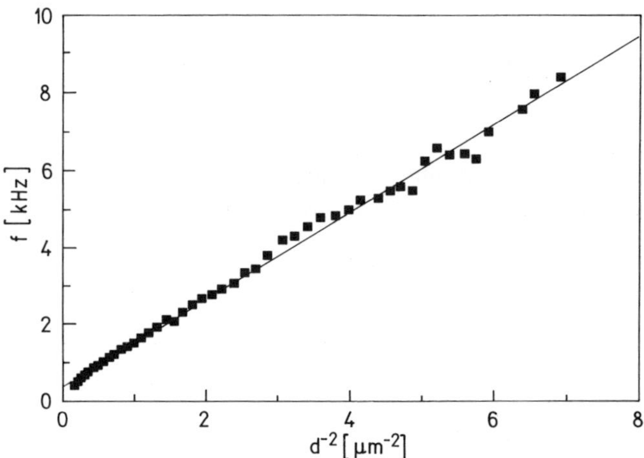

Fig.5.7.2. Dependence of the relaxation rate of the phase mode at $\vec{q}=0$ in the $smectic$–C^* phase of SCE-9 on the cell spacing, as determined from the linear electrooptic response (Škarabot, 1997).

A similar analysis can also be applied to the linear response of a π-soliton lattice (magnetically distorted helicoidal $smectic-C^*$ phase), with some particular differences. First, we know that the single mode of the unperturbed $smectic-C^*$ phase splits into an acoustic and optic-like mode, and second, in view of the periodicity of the phason dispersion relation, an infinite number of higher-frequency optic-like modes should appear at the edge of the BZ. The relaxation frequencies of these higher-frequency modes which originate from the neighboring BZ are typically one order of magnitude higher than the relaxation frequency of the two lowest modes and can be neglected in view of their small contribution to the response of the system.

The contributions of the acoustic-like and lowest lying optic-like mode to the electrooptic response of a distorted helicoidal structure can be calculated by considering the relation between the nonequilibrium polarization $\delta P(z,t)$ and the nonequilibrium tilt order parameter $\delta\xi\,(z,t)$:

$$\delta P_x = \varepsilon\mu \frac{\partial\delta\xi_1}{\partial z} - \varepsilon C\delta\xi_2 \tag{5.7.13a}$$

$$\delta P_y = \varepsilon\mu \frac{\partial\delta\xi_2}{\partial z} + \varepsilon C\delta\xi_1 \tag{5.7.13b}$$

The space-averaged nonequilibrium spontaneous polarizations $\langle\delta P\rangle$ of each mode at the edge of the BZ that are relevant for the linear response of the system at $\vec{q} = 0$ are:

The acoustic mode :

$$\langle\delta P_x\rangle \approx \varepsilon C\langle cn^2(u,k)\rangle \neq 0 \tag{5.7.14a}$$

$$\langle P_y\rangle = 0 \tag{5.7.14b}$$

The optic mode :

$$\langle P_x\rangle = 0 \tag{5.7.15a}$$

$$\langle\delta P_y\rangle \approx \varepsilon C\langle sn^2(u,k)\rangle \neq 0 \tag{5.7.15b}$$

Because the magnetic field is applied in the y-direction, we observe from the above equations that the acoustic mode represents a collective excitation with $\langle\delta P\rangle \perp H$, whereas for the optic mode $\langle\delta P\rangle // H$. The above relations thus enable a selective probing of either the optic or the acoustic-like mode via the polarization-selection rules.

5.8. Magnetic-Field Induced Biaxiality

Whereas an external DC electric field, which is applied perpendicular to the helical axis induces a ***macroscopic tilt*** of the optical axis in the plane perpendicular to the field direction, an external magnetic field is expected to induce a ***biaxiality*** in the originally uniaxial $smectic-C^*$ phase. This effect can be understood by considering the influence of both fields on the helical structure as shown for each case in Fig.5.8.1. Here, the molecular distribution is shown, as would be observed by an observer viewing the structure in a direction along the helix of a chiral smectic.

In the absence of an external field, the helical modulation is plane-wave-like and the resulting molecular distribution is isotropic when viewed along the helix (Fig.5.8.1.a). The space-averaged dielectric tensor of such a structure as a whole is therefore uniaxial, and the optical axis is in a direction along the helix. A DC electric field forces the dipoles to align along the field direction. As a result of this polar coupling, the dipoles will be preferentially aligned along the field and the directors will be preferentially aligned in a plane perpendicular to the direction of the field. This would result in a polar distribution of the director field with $\langle\cos\Phi_o(z)\rangle \neq 0$ and can be interpreted as a tilting of the optical axis of the $smectic-C^*$ phase as a whole in a plane, perpendicular to the field direction (see Eqs.5.6.5 and 5.6.6.). This was in fact the first observation of Meyer et al. (1975) that gave a clear evidence for the existence of ferroelectricity and polar coupling in liquid crystals.

On the other hand, a magnetic field does not distinguish the $-\vec{n}$ orientation of the director to the $+\vec{n}$ state. This means that it couples quadratically to the director and can therefore induce only a quadrupolar ordering, with $\langle\cos 2\Phi_o(z)\rangle \neq 0$. The diamagnetic coupling between the magnetic field and the director field is $g_m = \frac{1}{2}\Delta\chi(\vec{n}H)^2$. For a positive

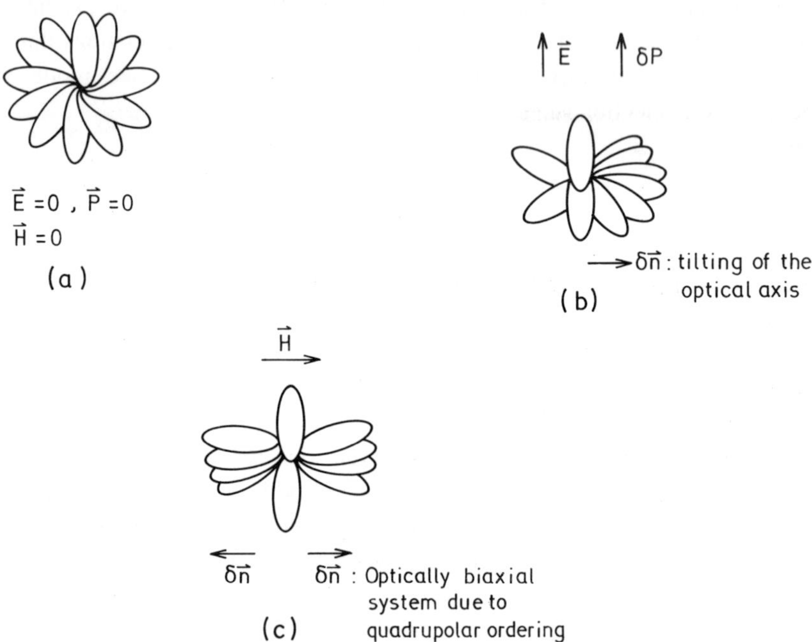

Fig.5.8.1. The distribution of molecular directors in the *smectic* $-C^*$ phase, as observed in a direction along the helix. (a) For zero external field, this distribution is isotropic in the x-y (smectic) plane and the resulting dielectric tensor is uniaxial. (b) The polar distribution of directors when a DC electric field is applied. (c) The quadrupolar distribution when a magnetic field is applied and the diamagnetic susceptibility of liquid crystal is assumed to be positive, $\Delta\chi > 0$.

diamagnetic anisotropy, the molecules will tend to align in the direction of the field, but there would be an equal probability for $+\vec{H}$ and $-\vec{H}$ directions, respectively. This would result in a quadrupolar distribution, as shown in Fig.5.8.1.(c). The distribution has two characteristic directions of the director and therefore also two optical axes. The resulting structure as a whole should be therefore biaxial. A very similar structure is obtained for negative diamagnetic susceptibility, $\Delta\chi < 0$. In this case, the magnetic field forces the molecules to align in a plane perpendicular to the direction of the magnetic field. This results in a quadrupolar distribution of the molecules that is rotated by 90° with respect to the distribution shown in Fig.5.8.1.(c).

The magnitude of this magnetic-field induced biaxiality can be calculated within the framework of the perturbational approach to the optics of chiral smectics, where the space averaged dielectric tensor in the presence of a magnetic field is

$$\langle \underline{\varepsilon} \rangle_{xx} = \tfrac{1}{2}\left[(\varepsilon_1 + \varepsilon_2) + (\varepsilon_3 - \varepsilon_2)\sin^2\theta\right] + \tfrac{1}{2}\left[(\varepsilon_2 - \varepsilon_1) + (\varepsilon_3 - \varepsilon_2)\sin^2\theta\right] \cdot \langle \cos 2\Phi_o(z) \rangle$$

$$\langle \underline{\varepsilon} \rangle_{yy} = \tfrac{1}{2}\left[(\varepsilon_1 + \varepsilon_2) + (\varepsilon_3 - \varepsilon_2)\sin^2\theta\right] - \tfrac{1}{2}\left[(\varepsilon_2 - \varepsilon_1) + (\varepsilon_3 - \varepsilon_2)\sin^2\theta\right] \cdot \langle \cos 2\Phi_o(z) \rangle$$

$$\langle \underline{\varepsilon} \rangle_{zz} = \varepsilon_3 - (\varepsilon_3 - \varepsilon_2)\sin^2\theta \tag{5.8.1}$$

In the unperturbed helical phase, where $\Phi(z) = 2\pi / p$, the spatial averages equal zero, i.e. $\langle \cos 2\Phi(z) \rangle = 0$ and the helical structure is macroscopically uniaxial with the optical axis along the z-direction.

The components $\langle \underline{\varepsilon} \rangle_{xx}$ and $\langle \varepsilon \rangle_{yy}$ depend on the magnitude of the external magnetic field $\vec{H}$, because the space averaged values of $\cos 2\Phi_o(z)$ are field-dependent (see Eq.3.2.9)

$$\langle \cos 2\Phi_o(z) \rangle = 1 - \langle sn^2(u,k) \rangle \tag{5.8.2}$$

Here, u is normalized coordinate along the layer normal and k is a measure of the distortion of the helical structure by magnetic field. The space-averaged dielectric tensor is field-dependent

$$\langle \underline{\varepsilon}(H) \rangle = \begin{bmatrix} \varepsilon - \Delta(H) & 0 & 0 \\ 0 & \varepsilon + \Delta(H) & 0 \\ 0 & 0 & \varepsilon_{zz} \end{bmatrix} \tag{5.8.3}$$

Here, $\varepsilon = \frac{1}{2}(\varepsilon_1 + \varepsilon_2) + \frac{1}{2}(\varepsilon_3 - \varepsilon_2)\sin^2\theta$ is the magnetic-field independent part, whereas the magnetic-field dependence is

$$\Delta(H) = \tfrac{1}{2}\left[(\varepsilon_2 - \varepsilon_1) + (\varepsilon_3 - \varepsilon_2)\sin^2\theta\right]\left\{2\frac{K(k) - E(k)}{k^2 K(k)} - 1\right\} \tag{5.8.4}$$

In the limit of small fields, $H \to 0, \Delta(H) \to 0$, whereas near the critical field $H \to H_c, \Delta(H) \to \frac{1}{2}\left[(\varepsilon_2 - \varepsilon_1) + (\varepsilon_3 - \varepsilon_2)\sin^2\theta\right]$. As a consequence, in the zero-field limit, $\langle \underline{\varepsilon} \rangle$ obtains the uniaxial form, whereas in the presence of a transverse magnetic field it obtains a biaxial form, which is a consequence of magnetic field distortion of the helix. For very large fields, the helix unwinds and we expect a biaxial form due to the intrinsic biaxiality of a liquid crystal. It should be stressed that the magnetic field induced biaxiality is here mainly the result of the global distortion of the phase of the order parameter and to a lesser extent the result of the intrinsic local biaxiality of the liquid crystal molecules.

The above model predicts a reduction of $\langle \underline{\varepsilon} \rangle_{xx}$, whereas $\langle \varepsilon \rangle_{yy}$ is expected to increase with the increasing magnetic field. Although the overall changes of the components of $\langle \underline{\varepsilon} \rangle$ are expected to be small, they should be observable at fields far below the critical field. In particular, magnetic field induced biaxiality in the ferroelectric phase with $\Delta\chi > 0$ should be observable as a magnetic-field induced distortion of the conoscopic figure in a plane, perpendicular to the direction of the magnetic field, i.e. the x-z plane. Figures 5.8.2 (a) and (b) show the influence of an external magnetic field on the x-z cross-sections of the normal wave surface of a ferroelectric liquid crystalline phase.

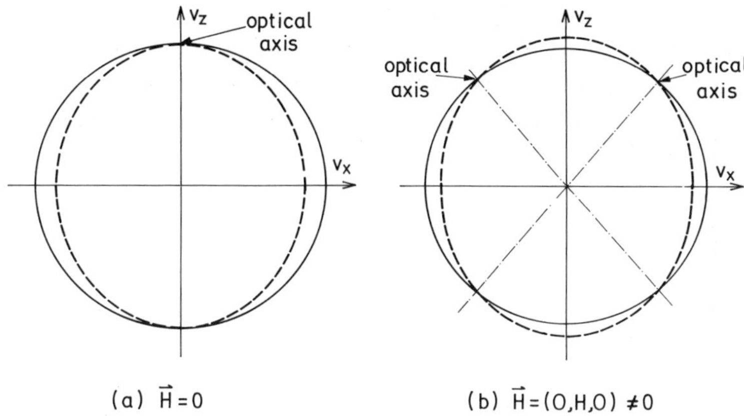

Fig.5.8.2. The normal velocity wave surface in the *smectic* $-C^*$ phase for (a) *H=0* and (b) $H \neq 0$.

In zero field, the section of a normal wave surface at *y*=0 consists of a circle and an oval, corresponding to the unperturbed ordinary (σ) and unperturbed extraordinary (π) wave, with the optical axis along *z*, see Fig.5.8.2a. The magnetic field dependence of the ordinary and extraordinary indices of refraction for the propagation of light in the x-z plane are deduced from Eqs.5.8.1 and 5.8.3

$$n_o^2 = \langle \varepsilon \rangle_{xx} (H) = \varepsilon + \Delta(H) \tag{5.8.5a}$$

$$\frac{1}{n_e^2} = \frac{\sin^2 \alpha}{\langle \varepsilon \rangle_{zz}} + \frac{\cos^2 \alpha}{\varepsilon - \Delta(H)} \tag{5.8.5b}$$

Here, α is the angle between the direction of wave propagation and the z axis. Under the influence of an external field, the circle, corresponding to the phase velocity of the ordinary wave $c = c_o / n_o$, contracts with increasing magnetic field, whereas the oval is deformed in the z-direction, as shown in Fig.5.8.2(b). The single optical axis at $H = 0$ splits into two optical axes, located symmetrically with respect to the z-axis. As a result of this splitting, the conoscopic figure which reflects the shape of the two normal wave surfaces is deformed in the x-z plane. The deformation is most pronounced near the center of the conoscopic figure, where even a small change in the index of refraction induces a strong splitting between the two optical axes. Finally, very near the critical field, the angle between the two optical axis should saturate near the value $tg(\beta/2) \approx \sqrt{\varepsilon_3/\varepsilon_1}$.

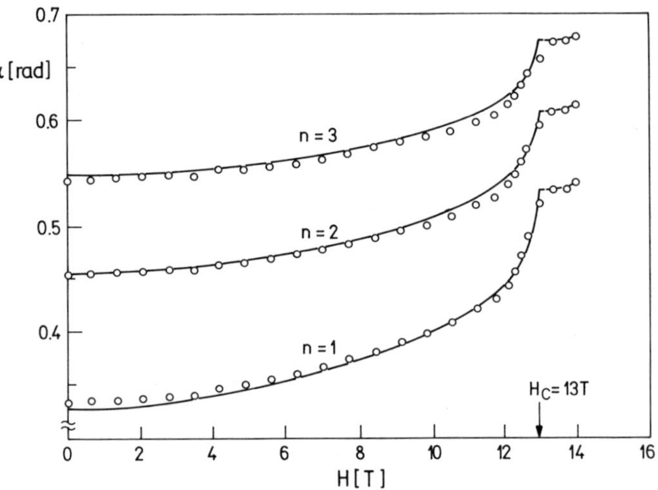

Fig.5.8.3. Magnetic field induced shift of the position of the first three intensity minima of the x-z cross section of the conoscopic figure in the antiferroelectric phase of MHPOBC. The solid line represents the best fit within the first order approximation to the optics of chiral smectics with $H_c = 13T$.

According to Eqs.5.8.4., $\Delta(H)$ depends strongly on the magnitude of the tilt angle and the difference $(\varepsilon_3 - \varepsilon_1)$. This means that the deformation of the normal surfaces is very small near T_c , where the tilt angle is small and increases upon decreasing temperature. The influence of the magnetic field on the normal surface of a helicoidal smectic phase has been observed (Muševič et al., 1993b) in an interferometric study of the magnetic field induced distortion in an antiferroelectric liquid crystal and is shown in Fig.5.8.3. An excellent agreement was found between the predicted and observed distortion of the normal velocity surface.

Chapter 6

Dielectric Dispersion

6.1. Static Response to Small Electric Fields

When a small oscillating electric field $\vec{E}_{\circ}e^{i\omega t}$ is applied in a direction perpendicular to the helical axis of the $smectic-C^{*}$ phase, it influences the liquid crystal via two different coupling mechanisms:

(i) The electric field couples linearly to the transverse dipole moments and tends to align them into the field direction. As a result, a net polarization will appear in the sample and the onset of polarization is here of orientational character. This linear effect is described by adding a linear coupling term between the field and spontaneous polarization to the free energy expression

$$g_{E} = -\vec{P}\vec{E} \tag{6.1.1}$$

(ii) The field also couples quadratically to the dielectric anisotropy $\Delta\varepsilon = \varepsilon_{//} - \varepsilon_{\perp}$ of liquid crystal and tends to align the molecules into the field direction if $\Delta\varepsilon$ is positive. The quadratic effect is invariant with respect to $+\vec{E}$ and $-\vec{E}$ and is described by the coupling term

$$g_{E^{2}} = -\tfrac{1}{2}\varepsilon_{\circ}\Delta\varepsilon\left(\vec{n}\vec{E}\right)^{2} \tag{6.1.2}$$

The quadratic orientational response of a liquid crystal to a small electric field is characterized by an induced polarization that oscillates at a second harmonic frequency of the measuring field. This second harmonic is normally small and we shall limit the discussion only to the linear response.

In the $smectic - A$ phase the external electric field can induce an electric polarization by coupling to the $q = 0$ polar eigenmode of the system. Far in the $smectic - A$ phase, this response is very small and the only contribution to the dielectric constant of the $smectic - A$ phase comes from the usual electronic contribution of the displacement of the electron clouds of liquid crystalline molecules. The response becomes large close to the $smectic - A$ - $smectic - C^*$ transition due to the softness of the system.

Let us first discuss the linear response of a chiral $smectic - A$ phase of ferroelectric liquid crystal to a small and static electric field. The total free energy density in the presence of the field is

$$g = g_A + \tfrac{1}{2} a\left(\xi_x^2 + \xi_y^2\right) + \tfrac{1}{4} b\left(\xi_x^2 + \xi_y^2\right)^2 -$$

$$-\Lambda\left(\xi_x \frac{d\xi_y}{dz} - \xi_y \frac{d\xi_x}{dz}\right) + \tfrac{1}{2} K_3\left(\left(\frac{d\xi_x}{dz}\right)^2 + \left(\frac{d\xi_y}{dz}\right)^2\right) - \varepsilon C \xi_x E_y \qquad (6.1.3)$$

In the $smectic - A$ phase the spontaneous polarization and the spontaneous tilt equal zero. An electric field $E = \begin{pmatrix} 0 & E_y & 0 \end{pmatrix}$ in the *y*-direction induces a small and spatially homogeneous polarization δP in the *y*-direction and a homogeneous tilt $\delta\xi_x$ in the *x*-direction, as shown in Fig.6.1.1.

By minimizing the total free-energy with respect to $\delta\xi_x$ the static dielectric susceptibility χ_E of the $smectic - A$ phase is

$$T \geq T_c \qquad \chi_E = \frac{\delta P_y}{E_y} = \varepsilon + \frac{\varepsilon^2 C^2}{\alpha\left(T - T_c\right) + K_3 q_c^2} \qquad (6.1.4)$$

We see that in the high-temperature limit the dielectric susceptibility equals ε, which can be interpreted as the contribution of the molecular electronic polarizibility. By lowering the temperature the linear response increases and reaches a large but finite value $\chi_E(T_c) = \varepsilon + \left(\varepsilon^2 C^2\right)/\left(K_3 q_c^2\right)$ at the phase transition point.

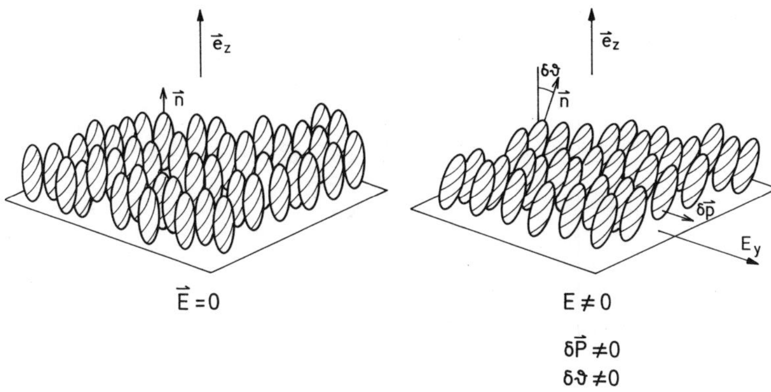

Fig.6.1.1. The electroclinic effect in the *smectic* − *A* phase of a ferroelectric liquid crystal.

In the *smectic* − C^* phase the spontaneous values of the tilt and polarization are finite and the field induces changes in both order parameters. These changes can be separated into two contributions: *(i)* the change of the magnitude of the order parameter or the amplitude change $\delta\xi_{//}$ and *(ii)* the change in the direction of the order parameter $\delta\theta_{\perp}(z)$, as shown in Fig.6.1.2.

$$\xi_x = \xi_x^\circ + \delta\xi_x = \theta_\circ \cos q_c z + \delta\xi_{//} \cos q_c z - \delta\xi_{\perp} \sin q_c z \tag{6.1.5a}$$

$$\xi_y = \xi_y^\circ + \delta\xi_y = \theta_\circ \sin q_c z + \delta\xi_{//} \sin q_c z + \delta\xi_{\perp} \cos q_c z \tag{6.1.5b}$$

Because we are looking for a linear response of the system, both changes $\delta\xi_{//}$ and $\delta\xi_{\perp}$ should be proportional to E_y. By inserting Eqs.6.1.5 into Eq.6.1.3, we obtain within the first order in E_y

$$g = g_A + b\theta_\circ^2 (\delta\xi_{//})^2 + \tfrac{1}{2} K_3 \left(\frac{d\xi_{//}}{dz} \right)^2 + \tfrac{1}{2} K_3 \left(\frac{d\xi_{\perp}}{dz} \right)^2 -$$

$$- \varepsilon C (\delta\xi_{//} \cos q_c z - \delta\xi_{\perp} \sin q_c z) E_\circ \tag{6.1.6}$$

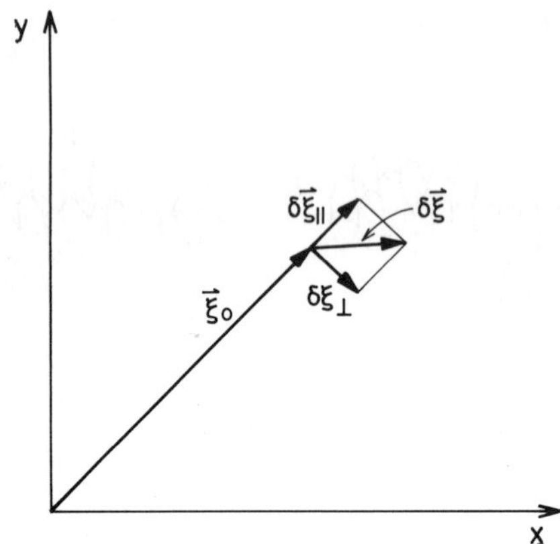

Fig.6.1.2 Change of the order parameter ξ under the influence of a small external electric field $E=(0\ E_\circ\ 0)$ applied perpendicularly to the helical axis. The change is decomposed into two orthogonal components, i.e. the amplitude change $\delta\xi_{//}$ and the phase change $\delta\xi_{\perp}$.

The electric field dependence of $\delta\xi_{//}$ and $\delta\xi_{\perp}$ can be obtained by minimizing the free energy 6.1.6. It is easy to see from the structure of the free-energy that the amplitude and phase changes are uncoupled in the harmonic approximation and represent therefore the eigenmodes of the system. We can also immediately see that a homogeneous external electric field couples to the eigenmodes with $q=0$ in the laboratory system, or, what is the same, with $q=q_c$ in the helical eigenframe:

$$\delta\xi_{//}(z)=\frac{\varepsilon C}{2\alpha(T_c-T)+K_3q_c^2}\cdot E_\circ\cdot\cos q_c z \tag{6.1.7a}$$

$$\delta\xi_{\perp}(z)=-\frac{\varepsilon C}{K_3q_c^2}\cdot E_\circ\cdot\sin q_c z \tag{6.1.7b}$$

The corresponding susceptibility of the $smectic-C^*$ is

$$\chi_E = \frac{\langle P_y \rangle}{E} = \varepsilon + \tfrac{1}{2}\varepsilon^2 C^2 \left[\frac{1}{K_3 q_c^2} + \frac{1}{2\alpha(T_c - T) + K_3 q_c^2} \right] \tag{6.1.8}$$

The dielectric susceptibility in the $smectic - C^*$ phase consists of three contributions. The first contribution equals ε and is the dielectric constant of the $smectic - A$ phase far beyond T_c. It is relatively small, $\varepsilon \approx 5 - 10$ and represents the background electronic polarizability. The two other contributions result from the coupling of the electric field to the two orientational degrees of freedom. The second contribution

$$\chi_G = \frac{1}{2} \frac{\varepsilon^2 C^2}{K_3 q_c^2} \tag{6.1.9}$$

originates from the phase fluctuations of the order parameter, i.e. it is the contribution of the phase mode with the wavevector $q = 0$ in the laboratory system. It is called the Goldstone mode contribution to the dielectric constant, or simply the Goldstone mode contribution. The expression is misleading. Let us remember that by definition the Goldstone mode is the zero-frequency, symmetry restoring mode that has a wave-vector $q = q_c$. Correspondingly, such a mode would give an infinite contribution to the linear response of system if it would be excited with an inhomogeneous field with a wave-vector $\vec{q}_c$. The true Goldstone mode cannot therefore be observed in a dielectric experiment which couples to the modes with $q = 0$ only. Instead, it can be observed in the light-scattering experiment, which can test the dynamics of the eigenexcitations of the system with finite wavevector.

Following expression 6.1.9., this "Goldstone" mode part is temperature independent, because the helical period p and therefore the wavevector of the helix $q_c = 2\pi / p$ are temperature independent, too. The third contribution

$$\chi_A = \frac{1}{2} \frac{\varepsilon^2 C^2}{2\alpha(T_c - T) + K_3 q_c^2} \tag{6.1.10}$$

originates from amplitude fluctuations and is therefore largest at the phase transition temperature, where the system becomes very soft. Nevertheless, the magnitude of the linear response remains finite at T_c. This merely expresses the fact that $q = 0$ is not the critical wave-vector of a chiral system. Instead, the true divergence and the Curie-Weiss law should be observable for the linear response of the system, measured at $q = q_c$ in the reciprocal space.

The static dielectric susceptibility in the vicinity of the $smectic - A$ - $smectic - C^*$ transition, as predicted within the simple Landau model is shown in Fig.6.1.3.

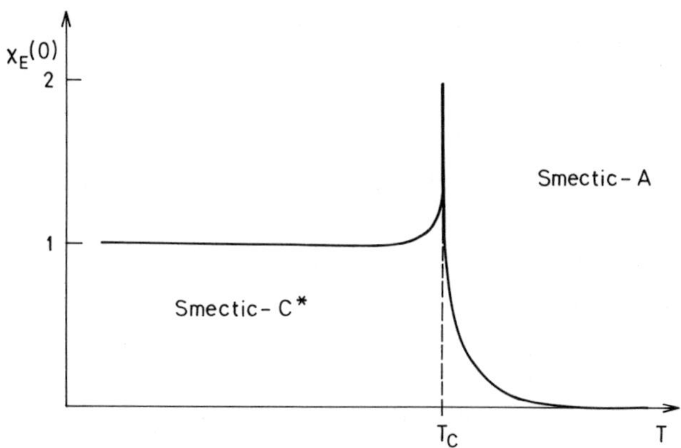

Fig.6.1.3. The linear dielectric response in the vicinity of the $smectic - A$ - $smectic - C^*$ transition within the simple Landau model. The dashed line represents the contribution of the amplitude fluctuations on the $smectic - C^*$ side of the transition.

6.2. Soft and Phase Mode Dynamics

In a dielectric experiment one measures the linear response $\chi_E(\omega)$ of a sample to a small, time-oscillating electric field $\vec{E} \cdot e^{i\omega t}$ with a measuring frequency in the range up to several GHz. This so-called dielectric dispersion spectroscopy enables us to determine the dynamical properties of the system, which are particularly interesting near the phase transitions, where the critical dynamics of a system can be observed. Dielectric spectroscopy is thus a very powerful method for studying critical phenomena of polar systems.

The linear response of a system to a time-varying electric field is formally analyzed by writing down the nonequilibrium free-energy in terms of the order parameter, where the coupling terms to the external field have to be included. On the other hand, we can get an insight into the physics by noting that an external homogeneous electric field, which is characterized by the wavevector $q = 0$, can couple to (or excite) the modes that have the same wave-vector and are polar. In other words, the field tests the polar eigenmodes of the system with the wave-vectors at the center of the reciprocal space, as shown in Fig.6.2.1.

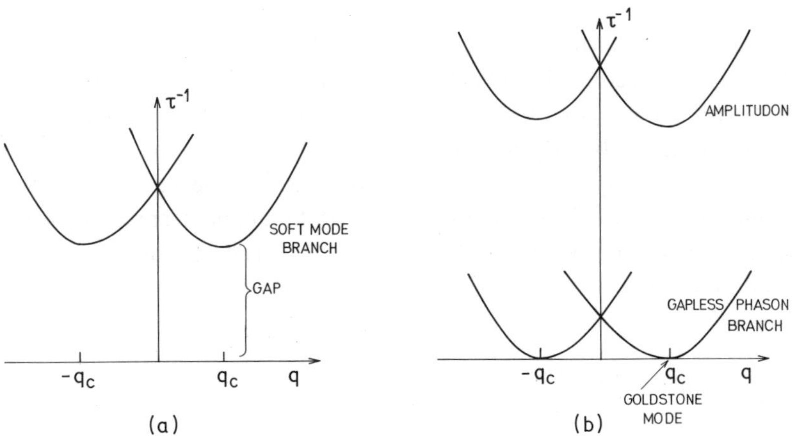

Fig.6.2.1. The dispersion of the (a) soft-mode excitations for $T \geq T_c$ and (b) amplitude and phase excitations for $T \leq T_c$. Whereas the dielectric experiment measures the linear response at $q = 0$, the quasielastic light-scattering spectroscopy measures the dynamics at finite q.

In the $smectic-A$ phase, there is only one, doubly degenerate polar mode at $q=0$ with the relaxation rate given by the Eq.2.2.30.

$$\frac{1}{\tau_S}(q=0,T)=\frac{\alpha}{\gamma_1}(T-T_c)+\frac{K_3}{\gamma_1}q_c^2 \tag{6.2.1}$$

This is a polar mode that can couple to an external electric field. Here we have neglected the high-frequency polarization modes, which do not show any critical behavior at the phase transition. The soft-mode (Eq.6.2.1.) shows a critical slowing down, which is incomplete and the relaxation rate of this mode is finite at the phase transition

$$\frac{1}{\tau_S}(q=0,T_c)=\frac{K_3}{\gamma_1}q_c^2 \tag{6.2.2}$$

Let us remember that this non-divergence of the relaxation rates expresses the fact that we are testing the dynamics of a chiral (inhomogeneous) system at $q=0$, which is a non-critical wave-vector for a chiral system close to the transition into spatially modulated phase.

In the $smectic-C^*$ phase the single soft mode at $q=0$ splits into amplitude and phase ("Goldstone") modes. Both of them are polar and their relaxational rates are given by the Eqs.2.3.31. The relaxation rate of the amplitudon mode is

$$\frac{1}{\tau_A}(q=0,T)=\frac{2\alpha}{\gamma_1}(T_c-T)+\frac{K_3}{\gamma_1}q_c^2 \tag{6.2.3}$$

and the relaxation rate of the phase mode is

$$\frac{1}{\tau_{PH}}(q=0)=\frac{K_3}{\gamma_1}q_c^2 \tag{6.2.4}$$

In the dielectric experiment we therefore expect to observe a splitting of a single relaxational mode in the $smectic-A$ phase into two relaxational modes in the $smectic-C^*$ phase, as shown in Fig.6.2.2. Whereas the relaxation rates of the amplitudon modes are expected to depend strongly on temperature, the relaxation rates of the phase modes are expected to be nearly temperature independent.

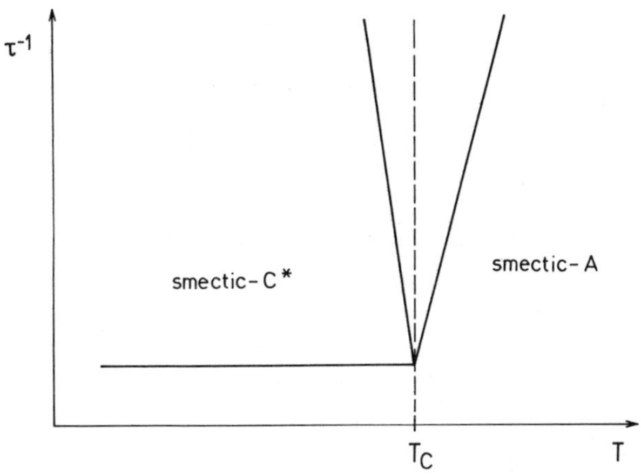

Fig.6.2.2. Splitting of the soft-mode into the amplitude and phase modes at $q = 0$.

The frequency dependence of the real and imaginary parts of the dielectric susceptibility of a ferroelectric liquid crystal can be calculated from the free-energy density (Eq.6.1.5) which includes the coupling to a small and time-varying electric field $E_y = E_{\circ} e^{i\omega t}$

$$g = g_A + \tfrac{1}{2} a\left(\xi_x^2 + \xi_y^2\right) + \tfrac{1}{4} b\left(\xi_x^2 + \xi_y^2\right)^2 -$$

$$-\Lambda\left(\xi_x \frac{d\xi_y}{dz} - \xi_y \frac{d\xi_x}{dz}\right) + \tfrac{1}{2} K_3\left(\left(\frac{d\xi_x}{dz}\right)^2 + \left(\frac{d\xi_y}{dz}\right)^2\right) - \varepsilon C \xi_x E_y \qquad (6.2.5)$$

Similar to the analysis in Sections 5.6. and 5.7. we obtain after applying the Landau-Khalatnikov equations of motion a system of two coupled equations. For $T \geq T_c$ the two eigensolutions are degenerate

$$\delta\xi_{//}(z,t) = \frac{\varepsilon CE_\circ}{\alpha(T - T_c) + K_3 q_c^2} \cdot \cos q_c z \cdot \frac{e^{i\omega t}}{1 + i\omega\tau_S} \tag{6.2.6}$$

where τ_S is the relaxation time of the soft-mode excitation with $q = 0$, given by the Eq.6.2.1. Here $\delta\xi_{//}(z,t)$ is the change of the magnitude of the order parameter, i.e. the tilt angle. Note that in the *smectic* $- A$ the phase of the order parameter cannot be distinguished from its amplitude due to the D_∞ symmetry of this phase.

For $T \leq T_c$ this axial symmetry is spontaneously broken which removes the degeneracy of the soft mode and we have two distinct eigensolutions

$$\delta\xi_{//}(z,t) = \frac{\varepsilon CE_\circ}{2\alpha(T_c - T) + K_3 q_c^2} \cdot \cos q_c z \cdot \frac{e^{i\omega t}}{1 + i\omega\tau_A} \tag{6.2.7}$$

and

$$\delta\xi_{\perp}(z,t) = -\frac{\varepsilon CE_\circ}{K_3 q_c^2} \cdot \sin q_c z \cdot \frac{e^{i\omega t}}{1 + i\omega\tau_{PH}} \tag{6.2.8}$$

Here τ_A and τ_{PH} represent the relaxation times of the amplitude and phase excitation at $q = 0$, respectively, which are given by the Eqs.6.2.3. and 6.2.4. We immediately observe that the collective motion, which is excited by the external oscillating electric field is of Debye character due to the overdamped (dissipative) nature of the elementary excitations in liquid crystal.

After a straightforward calculation we obtain the expression for the complex dielectric susceptibility of the *smectic* $- A$ phase

$$\chi_S(\omega) = \varepsilon + \frac{\varepsilon^2 C^2}{\alpha(T - T_c) + K_3 q_c^2} \cdot \frac{1}{1 + i\omega\tau_S} \tag{6.2.9}$$

where τ_s^{-1} is the relaxation rate of the soft mode, given by

$$\frac{1}{\tau_S}(q=0,T)=\frac{\alpha}{\gamma_1}(T-T_c)+\frac{K_3}{\gamma_1}q_c^2 \tag{6.2.10}$$

Note the relation between the strength of the mode and its relaxation rate, $(\chi_S(\omega=0)-\varepsilon)\cdot\tau_S^{-1}=cons\tan t$.

In the $smectic-C^*$ phase, the complex dielectric susceptibility is

$$\chi(\omega)=\chi_A+\chi_{PH}=\varepsilon+\frac{1}{2}\frac{\varepsilon^2C^2}{K_3q_c^2}\cdot\frac{1}{1+i\omega\tau_{PH}}+$$
$$+\frac{1}{2}\frac{\varepsilon^2C^2}{2\alpha(T_c-T)+K_3q_c^2}\cdot\frac{1}{1+i\omega\tau_A} \tag{6.2.11}$$

where τ_{PH}^{-1} and τ_A^{-1} are the relaxation rates of the phase and amplitude excitations, respectively, given by

$$\frac{1}{\tau_A}(q=0,T)=\frac{2\alpha}{\gamma_1}(T_c-T)+\frac{K_3}{\gamma_1}q_c^2 \tag{6.2.12}$$

and

$$\frac{1}{\tau_{PH}}(q=0)=\frac{K_3}{\gamma_1}q_c^2 \tag{6.2.13}$$

Note similar relationship between the susceptibility and relaxation rate, which means that slow modes have large susceptibilities. As one could expect, the contributions to the dielectric susceptibility at a given frequency ω of the measuring field are completely analogous to the static linear response, derived in Section 6.1. There is however a complex, frequency dependent term $1/(1+i\omega\tau_j)$. It has the following role:

(i) In the low-frequency limit, $\omega \to 0$, the response is the largest and we recover the expressions for the static dielectric susceptibility.

(ii) In the high-frequency limit, $\omega \to \infty$, the linear response becomes very small and the susceptibility of both $smectic-A$ or $smectic-C^*$ phases is equal to the electronic susceptibility ε. This expresses the fact that for very high frequencies of the measuring field the motion of the molecules cannot follow fast oscillations of the field.

(iii) At intermediate frequencies of the measuring field, there will be a phase lag between the motion of the molecules and the oscillation of the measuring field.

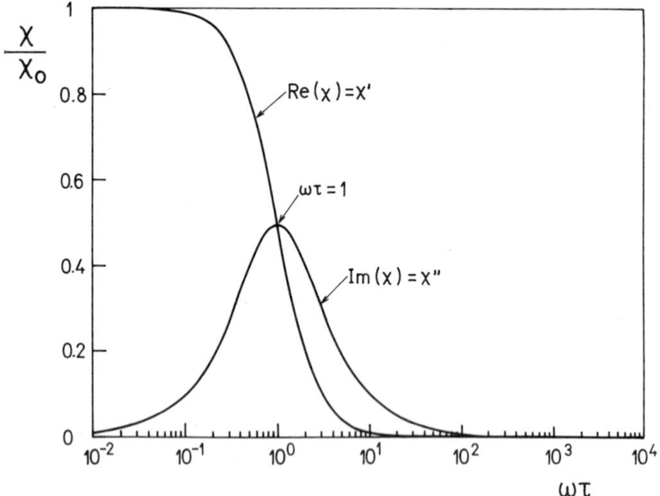

Fig.6.2.3. The real and the imaginary part of the dielectric response in the case of a single Debye relaxational process.

By writing the complex part of the response in the case of a single relaxation as a sum of a real and an imaginary part

$$\frac{1}{1+i\omega\tau_j} = \frac{1}{1+\omega^2\tau_j^2} - i\cdot\frac{\omega\tau}{1+\omega^2\tau_j^2} \tag{6.2.14}$$

we see that the phase lag between the real and imaginary parts is

$$\varphi = -arctg(\omega\tau_j) \tag{6.2.15}$$

whereas the magnitude of the response is proportional to $1/\sqrt{1+\omega^2\tau^2}$ and decreases with increasing frequency. For a single relaxation process, the phase lag equals $\varphi = -45^\circ$ when the frequency of the external field is

$$\omega = 2\pi\nu = 1/\tau_j \tag{6.2.16}$$

(i.e. $\omega\tau_j = 1$ in the Eq.6.2.15). At this value of the measuring frequency, the imaginary part of χ reaches a maximum value, as it is shown in Fig. 6.2.3. This allows us to determine the relaxation rates of the polar excitations in the system.

Sometimes, the dielectric dispersion of a system can be best visualized in an alternative way by plotting the so-called Cole-Cole diagrams, as shown in Fig.6.2.4. In the Cole-Cole plot, the real part of the measured susceptibility

$$\chi(\omega) = \chi' + i\cdot\chi'' = \chi_\infty + \frac{\Delta\chi}{1+(i\omega\tau)^\beta} \tag{6.2.17}$$

are plotted against the imaginary part for different measuring frequencies. Here, $\Delta\chi = \chi_s - \chi_\infty$, where χ_s and χ_∞ are the static and high frequency electric susceptibilities and τ is the characteristic correlation time. The parameter β measures the degree of polydispersivity of the system. For a single Debye relaxational process this represents a circle that has its highest point at the relaxational frequency of the system. From the Cole-Cole plot we can therefore determine the relaxation rate of the relaxational process in the sample in a straightforward way. In addition, any deviation from a single relaxational dynamics is also easily visually detected from the shape of the Cole-Cole plot.

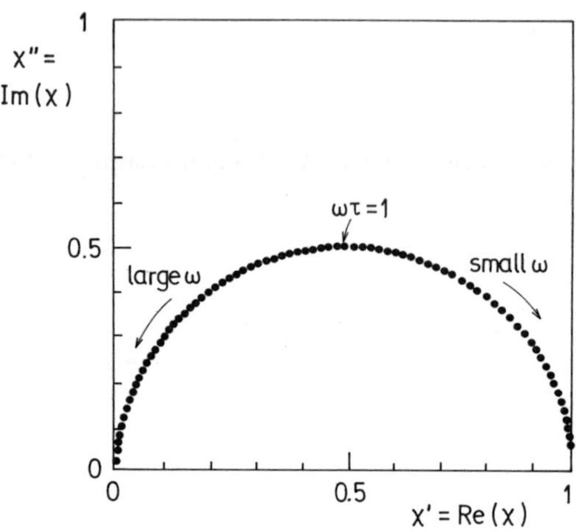

Fig.6.2.4. Cole-Cole plot for a single Debye relaxational process.

There is a very large number of dielectric measurements in the ferroelectric $smectic-C^*$ phase reported in the literature (Ostrovski et al., 1978; Žekš et al., 1979; Martinot-Lagarde and Durand, 1980; Martinot-Lagarde and Durand, 1981; Benguigui, 1982; Levstik et al., 1987; Pavel et al., 1987; Bahr et al., 1987; Filipič et al., 1988; Ezcurra et al., 1989; Biradar et al., 1989; Wrobel et al., 1989; Levstik et al., 1990; Kutnjak et al., 1990; Carlsson et al., 1990; Levstik et al., 1991a; Levstik et al., 1991b; Bahr et al., 1992; Gouda et al., 1992; Krishna Prasad et al., 1993). Figure 6.2.5. shows a series of Cole-Cole plots, as obtained from the dielectric dispersion measurements in a ternary ferroelectric mixture (Filipič et al., 1988).

The temperature dependence of the relaxation rates, as determined from the Cole-Cole plots shows a critical slowing-down. The divergence of the soft mode relaxation times is incomplete, as expected from the noncritical wave-vector in the experiment. At the phase transition, see Fig.6.2.6., the soft mode relaxation rate is usually finite, typically in the range from 100Hz to few kHz. In the $smectic-C^*$ phase, the intensity of the linear response is almost completely due to the orientational, phase mode susceptibility. The corresponding relaxation rate is related to the wave-vector of the helix, and in many cases shows a strong

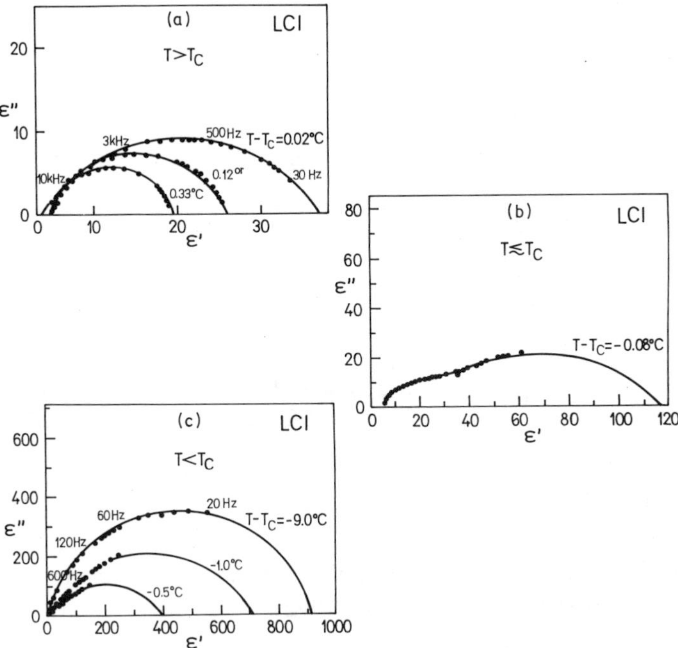

Fig.6.2.5. Cole-Cole plots of the linear response in the vicinity of the $smectic-A$ - $smectic-C^*$ phase transition (Filipič et al., 1988). (a) The $smectic-A$ phase. (b) In the $smectic-C^*$ phase, close to the transition. Note the elongated shape of the Cole-Cole plot, which indicates the presence of two relaxation contributions. (c) The $smectic-C^*$ phase, far below transition.

temperature dependence, because the period of the helix is temperature dependent. In linear response measurements, such as the dielectric experiment, one has to assure a reasonably low value of the measuring voltage. Higher electric fields may drive the experiment into a nonlinear regime, which has been discussed by Kutnjak-Urbanc et al.(1991), Orihara and Ishibashi (1993) and Orihara et al.(1995). One of the indications of the nonlinear regime is the appearance of higher-harmonics in the susceptibility and a voltage-dependence of the order parameter relaxation rates.

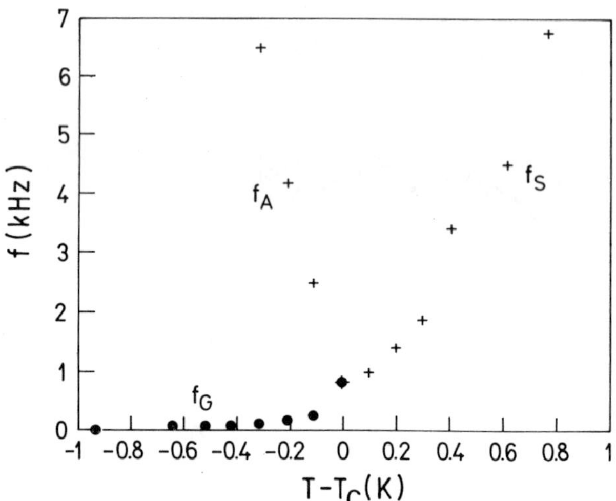

Fig.6.2.6. Relaxation rates of the soft and phase mode in a ferroelectric liquid crystal SCE-9 (Levstik et al., 1992).

Note 6.2.1. Dielectric constant of ferroelectric liquid crystals:
The dielectric susceptibility of a ferroelectric liquid crystal (Eq.6.2.11) can be written as

$$\chi(\omega) = \varepsilon + \frac{\Delta\varepsilon_{PH}}{1 + i\omega\tau_{PH}} + \frac{\Delta\varepsilon_A}{1 + i\omega\tau_A}$$

Here, $\Delta\varepsilon_{PH}$ and $\Delta\varepsilon_A$ are the dielectric strengths of the phase and amplitude mode, respectively. These are equal to the static dielectric constants of modes. The dielectric strengths depend on the chemical properties of the material and are usually larger for materials with large polarization and large helical period. For example, $\Delta\varepsilon_{PH} \approx 6$ for DOBAMBC and is much smaller than $\Delta\varepsilon_{PH} \approx 120$, which is measured in DOBA-1-MPC (Levstik et al., 1990; Kutnjak, 1991). Note that the dielectric constant in ferroelectric liquid crystals exhibits strong frequency dependence, because the relaxation rates are of the order of $\approx 1kHz$ or less.

Note 6.2.2. Rotational viscosity in ferroelectric liquid crystals:
Dielectric spectroscopy can be used to determine the rotational viscosity of ferroelectric liquid crystals (Kutnjak, 1990; Levstik et al., 1990, 1991b,1992). Typical rotational

viscosities in the ferroelectric $smectic-C^*$ phase are of the order of $0.01-0.1kg/ms$. For example, the rotational viscosity of DOBA-1-MPC is $\gamma_\Phi \approx 0.06kg/ms$ close to the phase transition into the $smectic-A$ phase and increases to $\gamma_\Phi \approx 0.09kg/ms$ 14K below the transition. The temperature dependence is Arrhenius, $\gamma_\Phi \approx \exp(\mu/k_BT)$, with activation energy $\mu = 0.33eV$ for DOBA-1-MPC and $\mu = 0.51eV$ for DOBAMBC. In the $smectic-A$ phase, the viscosity of the soft mode is smaller and also shows Arrhenius behavior. The viscosity coefficient of the soft mode splits into two viscosity coefficients at the phase transition.

6.3. Dielectric Response of a Multisoliton Lattice

In contrast to the unperturbed $smectic-C^*$ helical structure, which is a state of continuous helical symmetry, the multisoliton lattice in a ferroelectric liquid crystal is characterized by a discrete translational lattice symmetry (see Chapter 3). This reduction of a lattice symmetry is a result of the action of either an external electric or magnetic field and should be distinguished from the spontaneous symmetry breaking, which appears when the temperature of a system is lowered. The reason why these two phenomena should be distinguished is in their consequences. Whereas the spontaneous symmetry breaking of a continuous symmetry leads to the appearance of gapless Goldstone excitations, external fields produce a soliton structure and a periodic potential which leads to the band-structure of the order parameter spectrum and induce forbidden frequency gaps (see Chapters 2 and 3).

The appearance of the band-structure in the order parameter spectrum in the multisoliton lattice has a direct consequence for the linear response to a homogeneous external field: Whereas in the unperturbed helical $smectic-C^*$ phase there is a single amplitude and a single phase mode at $q=0$ (see Fig.6.3.1a), an infinite set of the phase and amplitude modes appears in the spectrum of a multisoliton lattice (see Fig.6.3.1b, where only phase modes are shown). This means that one should observe a polydisperse linear response of the $smectic-C^*$ under an electric or magnetic field, and, eventually, separate the individual modes.

In Chapter 3. we have calculated the phase excitation spectrum of a π-***soliton*** lattice, which is obtained by applying a magnetic field $H=(0,H,0)$ perpendicular to the helix of the $smectic-C^*$ phase with positive diamagnetic anisotropy $\Delta\chi > 0$. At $q=0$, the two lowest eigensolutions Ψ_i and

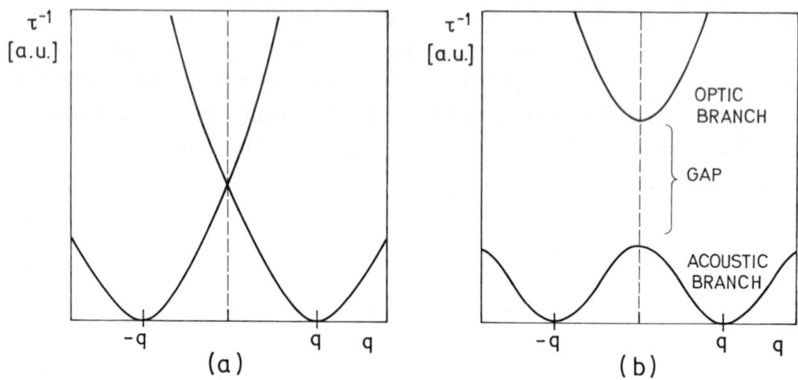

Fig.6.3.1. (a) The spectrum of the order parameter excitations in unperturbed *smectic* $-C^{*}$ phase. (b) π-soliton-distorted *smectic* $-C^{*}$ structure.

corresponding eigenvalues can be expressed analytically (see Eqs.3.4.15. and 3.4.16.)

$$\Psi_1 = sn(u,k) \qquad \tau_{+}^{-1}(H) = \frac{\Delta\chi}{\gamma}\cdot\frac{1}{k^2}\cdot H^2 \tag{6.3.1a}$$

$$\Psi_2 = cn(u,k) \qquad \tau_{-}^{-1}(H) = \frac{\Delta\chi}{\gamma}\cdot\frac{1-k^2}{k^2}\cdot H^2 \tag{6.3.1b}$$

Here, u and k are the reduced coordinate and the modulus of the elliptic functions, respectively and Ψ_i represents the amplitude of the fluctuations of the phase of the order parameter (see Section 3.4.).

As we have shown in Chapter 3, the relaxation rate of the first mode (the so-called optic mode) increases with the field, whereas the relaxation rate of the second mode (the so-called acoustic mode) decreases with the field, which leads to a field induced gap at the center of the Brillouin zone at $q = 0$. Here, optic and acoustic modes represent higher frequency and lower frequency modes, similar to the phonon modes in solids.

The polarization properties of the above modes can be calculated from the relation between the spontaneous polarization and the tilt (Eqs.2.1.19.)

$$\delta P_x = \varepsilon\mu \frac{d\delta\xi_x}{dz} - \varepsilon C\delta\xi_y \tag{6.3.2a}$$

$$\delta P_y = \varepsilon\mu \frac{d\delta\xi_y}{dz} + \varepsilon C\delta\xi_x \tag{6.3.2b}$$

Here $\delta P(z)$ and $\delta\xi(z)$ represent the amplitudes of thermally excited fluctuations of polarization and tilt, respectively. For $q = 0$ modes, we obtain the following space-averaged amplitudes of the polarization fluctuations:

optic mode Ψ_1:

$$\langle \delta P_x \rangle = 0 \tag{6.3.3a}$$

$$\langle \delta P_y \rangle \approx -\varepsilon C \int_0^{4K} sn^2(u,k) \cdot du \neq 0 \tag{6.3.3b}$$

acoustic mode Ψ_2:

$$\langle \delta P_x \rangle \approx -\varepsilon C \int_0^{4K} cn^2(u.k) \cdot du \neq 0 \tag{6.3.4a}$$

$$\langle \delta P_y \rangle = 0 \tag{6.3.4b}$$

We see that both modes are polar and represent spatially homogeneous fluctuations of macroscopic polarization. They can therefore couple to an external measuring electric field and contribute to the linear susceptibility of the system. The polarizations of both modes are mutually orthogonal: Whereas the optic-like mode carries a dipole moment in the direction of the magnetic field $\vec{H} = (0, H, 0)$, the acoustic-like mode represents polarization fluctuation in a perpendicular direction.

The nature of these two polar modes can be understood by considering the effect of a small, measuring electric field, applied perpendicular to the soliton-like distorted helix. As shown in Fig.6.3.2., the magnetic field creates homogeneously aligned domains, separated by the π-domain walls. Within the domains, the electric dipoles are also nearly homogeneously aligned. The polarization in one half of the domains points, for example, out of the paper, whereas in the other half it points into the paper. The overall soliton structure is therefore non-polar, as expected from the nonpolar diamagnetic coupling, which creates this soliton structure. If a small electric field is applied perpendicularly to the direction of the magnetic field, the polarization in one half of the domains will already have properly aligned dipoles into the field direction, whereas the other half will try to minimize the electrostatic energy by adjusting the global phase profile. This can be done in the most efficient way by slightly shifting the centers of the π-domain walls in such a manner that the domains of properly aligned dipoles will expand, whereas the neighboring domains will shrink, as illustrated in Fig.6.3.2a. We see that a small oscillating electric field induces oscillations of the position of the domain walls. It is easy to see from the Eqs.6.3.1 that this kind of motion actually represents the acoustic-like phase mode at $q = 0$, i.e. it is an out-of-phase motion of the neighboring domain walls.

When the magnetic field is increased towards the critical value, the distance between the domain walls increases, i.e. the period of the soliton lattice increases and finally diverges at the critical field. Because of the increase of the intersoliton distance, the soliton domain walls are less and less coupled to each other, and the soliton lattice becomes softer and softer as we approach the critical field. Finally, at the critical field, the solitons are nearly free. There is no restoring force for their movement and the corresponding relaxation rate goes to zero. This is in fact exactly what is predicted by the theory (Eq.6.1.3b), i.e., at the critical field, the relaxation rate of the acoustic mode equals zero.

If the electric field is applied into the direction of the magnetic field, both types of domains have dipoles aligned perpendicular to the field direction and will try to rotate dipoles into the field (Fig.6.3.2b). This is always accompanied with an increase of the diamagnetic energy and the corresponding relaxation rate will increase monotonously with the magnetic field, as predicted for the optic-like phase mode at $q = 0$.

The external magnetic field therefore influences the dynamics of the $smectic - C^*$ structure. In view of the interconnection between the relaxation rates of the eigenmodes and their intensities (i.e. slower modes have large intensities, see Eqs.6.2.9-13. and the discussion therein), there should be a large effect to the intensity of the lowest, acoustic-like mode, which has a vanishing relaxation rate at the critical field. It can be shown (see R.Blinc et al., 1991) that

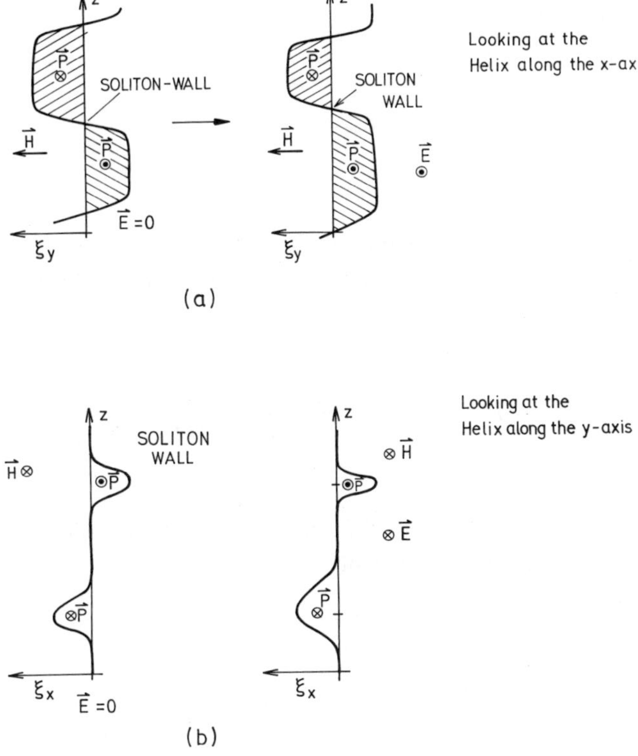

Fig.6.3.2. (a) The effect of a small electric field, applied perpendicular to the helix and perpendicular to the external magnetic field. (b) Field is applied parallel to the external magnetic field.

the analysis of the magnitude of the linear response of a π-soliton $smectic-C^*$ structure is formally equivalent to the commensurate-incommensurate transition in solids (Levstik et al., 1982; Prelovšek 1982).

In the constant amplitude approximation the static, linear dielectric susceptibility $\chi(\omega=0)$ of the acoustic-like phase mode at $q=0$ is

$$\chi_{ac}=\varepsilon+\frac{\varepsilon^2C^2}{\Delta\chi H^2}\left[\frac{E(k)}{(1-k^2)K(k)}-1\right] \tag{6.3.5}$$

For $H \to 0$ we recover the familiar phase (Goldstone) mode contribution to the dielectric susceptibility

$$H \to 0 \ : \qquad \chi_{ac} = \varepsilon + \tfrac{1}{2}\frac{\varepsilon^2 C^2}{2K_3 q_c^2} \tag{6.3.6}$$

On the other hand, the inter-soliton distance diverges for $H \to H_c$ and the divergent part of the static dielectric susceptibility shows a logarithmic-corrected divergence

$$H \to H_c \ : \qquad \chi_{ac} = \varepsilon + \frac{C_H}{(H_c - H)|\ln(H_c - H)/H_c|} \tag{6.3.7}$$

This is therefore a 'magnetic' Curie-Weiss law with a logarithmic correction. The magnetic Curie-Weiss constant

$$C_H = \frac{\varepsilon^2 C^2}{\Delta\chi H_c} \approx 1.6 H_c \tag{6.3.8}$$

is rather high and the magnetic-induced soliton lattice should thus lead to a large anomaly in the static dielectric behavior in $\chi_{ac} = \partial\langle P_x\rangle/\partial E_x$. However, it should be pointed that the divergence in the static susceptibility is accompanied by the tremendous increase of the phason relaxation time, as we approach the critical field $H \to H_c$. Because the dielectric response is always measured at a finite frequency ω, the expected increase of the dielectric constant is mediated by the dielectric relaxation

$$\chi'(\omega) = \frac{\chi}{1 + \omega^2 \tau_{ac}^2} \tag{6.3.9}$$

and eventually becomes very small and unobservable. Here, τ_{ac}^{-1} is the relaxation rate of the acoustic mode, which vanishes at the critical field (see Eq.6.3.1b).

Dielectric response of the soliton-like distorted $smectic-C^*$ phase was reported some time ago by Muševič et al.(1983). They failed to observe a Curie-Weiss magnetic behavior in DOBAMBC. Instead, the dielectric constant showed a rather constant value up to the critical field and then a sudden drop to a very low value. There are several possible explanations for such a behavior (see Blinc et al., 1991), but in view of recent observations, it seems that the experiment was not performed in an equilibrium state. The reason for this is the fact that it takes a very long time (tens of minutes) for a $smectic-C^*$ phase to reach an equilibrium after the magnetic field is slightly changed. This reflects the fact that the soliton structures in ferroelectric liquid crystals are built-up as a delicate balance between the external and elastic forces and the soliton structure needs a long time to reach its equilibrium.

A very similar analysis can be done for the 2π-soliton structure, which is induced by the application of an external DC electric field in a direction perpendicular to the helix (see Kutnjak-Urbanc and Žekš, 1995). Besides the general features, which are identical to the π-soliton lattice, the excitation spectrum is here slightly different than in the case of a magnetic field. As can be seen from Fig.6.3.3b., the forbidden gaps appear off-center of the reciprocal space, which is due to the symmetry of the coupling term between the field and polarization.

The forbidden gap is therefore not expected to be observed in a dielectric experiment. Instead, first a monotonous decrease of the relaxation rates of the modes with $q=0$ is expected and then a linear increase with the magnitude of the electric field, as shown in Fig.6.3.3d. Although there is a number of dielectric experiments reported for the $smectic-C^*$ phase in an external electric field, none has observed the predicted behavior. An exception is here the linear-electrooptic experiment by Pavel and Glogarova (1991) and a quasielastic-light scattering experiment by Drevenšek et al. (1991).

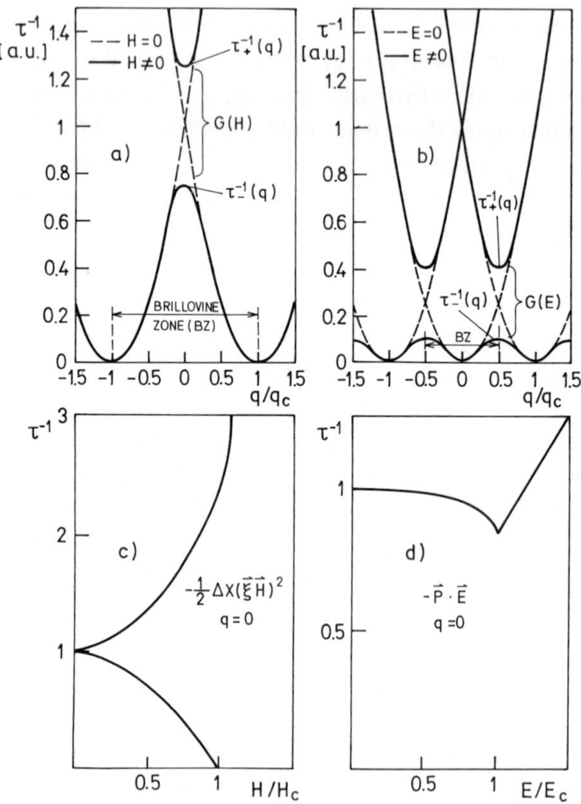

Fig.6.3.3. (a) The phason excitation spectrum in a π-soliton lattice. (b) The phason spectrum in 2π -soliton lattice. (c) Field dependence of the phason relaxation rate at $q = 0$ for the magnetic-field induced soliton lattice and electric-field induced soliton lattice, shown in (d).

6.4. Polarization Noise in a Ferroelectric Liquid Crystal

Macroscopic manifestations of spontaneous fluctuations of natural systems have always attracted a great interest of scientists since the pioneering work of J.B.Johnson (1928) and H.Nyquist (1928) on the electric noise in resistors. Since that time, a number of analogous phenomena have been observed in very diverse condensed matter systems, like magnetic materials, biological systems and membranes and insulators (for a review, see, for example Ocio et al., 1985, and Bouchiat and Ocio, 1988). In dielectrics and solid ferroelectrics, it is rather difficult to observe these phenomena due to the smallness of the fluctuating electric current (Israeloff, 1996; Bittel et al., 1965). In ferroelectric liquid crystals, the observation of the voltage noise due to the fluctuations of the polarization field was first reported by Maruyama in 1980 and then analyzed extensively by Muševič et al. in 1997.

The Johnson noise in solid conductors is due to the random movement of the charge carriers, for example free electrons. This gives rise to a fluctuating current $I(t)$ through the resistor R and induces a fluctuating voltage $U(t)$ across the resistor. Whereas both the time-averaged current and voltage equal zero, $I(t)=0$ and $U(t)=0$, the mean-square values of the fluctuating current and voltage are finite,

$$\overline{I^2(t)} \neq 0 \tag{6.4.1a}$$

$$\overline{U^2(t)} \neq 0 \tag{6.4.1b}$$

This means that the fluctuating electric current can dissipate power to an external detector, as for example an ampere-meter or volt-meter and can therefore be observed via this dissipation.

The magnitude of the fluctuating current or voltage was first calculated by Johnson (1928) and Nyquist (1928). The theory was later generalized to other linear dissipative systems by Callen et al. (1951) and R.Kubo (1957), who introduced the concept of generalized forces and dissipation. The spectral power density of the Johnson fluctuating current $S_I(\nu)$ is

$$S_I(\nu)=\frac{\overline{dI^2(t)}}{d\nu}=\frac{4k_BT}{R} \tag{6.4.2}$$

whereas the corresponding spectral power density of the Johnson fluctuating voltage $S_U(\nu)$ is

$$S_U(\nu)=\frac{\overline{dU^2(t)}}{d\nu}=4k_BT\cdot R \tag{6.4.3}$$

Here, k_B is the Boltzman constant and R is the resistance of the resistor. It is important to note that the spectrum is white up to the characteristic frequency, which corresponds to the time between the electron collisions. Furthermore, the magnitudes of the fluctuating current or voltage depend only on the magnitude of the resistance of the resistor and not on the nature of electric conductivity.

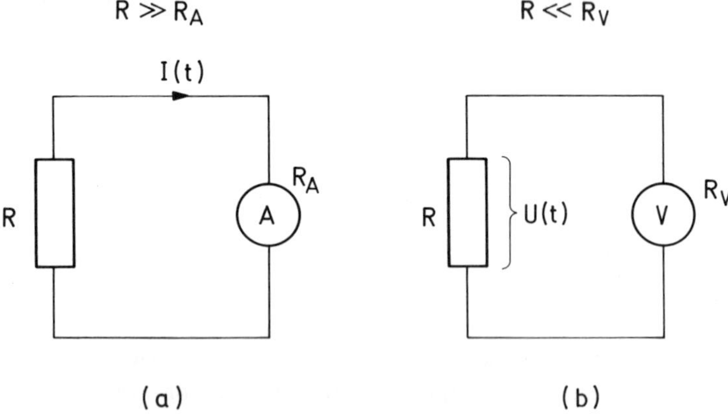

Fig.6.4.1. The experimental geometry for measuring the (a) current power spectral density of noise and (b) voltage power spectral density. In (a), a low-impedance ampere-meter, with a measuring frequency window $d\nu$ is connected to the noise-generating element R. This is therefore a short-circuit electrical scheme, in contrast to (b), where a high-impedance voltmeter is connected to R, thus measuring an open-circuit noise voltage.

This means that the Johnson noise of a $100k\Omega$ carbon resistor is equal to the Johnson noise of a $100k\Omega$ metal-film resistor. At room temperature, the *RMS* value of the Johnson voltage is of the order of $\approx 100nV$ in a frequency interval of $d\nu = 1Hz$ and can be easily observed either in a "current" or "voltage" sensing experimental geometry, as shown in Fig.6.4.1. It should be noted, that there are other sources of noise in resistors and electric circuits (see, for example Van der Ziel, 1959).

Whenever there is a long range order in a system, there are spontaneous, thermally excited fluctuations present in a system, which tend to restore the symmetry that was broken by the onset of a long range order. In solid ferroelectrics and also in liquid crystalline ferroelectrics, the spontaneous fluctuations of the polarization field $P(\vec{r},t)$ generate the polarization current density $j = \partial P / \partial t$, as shown in Fig.6.4.2.

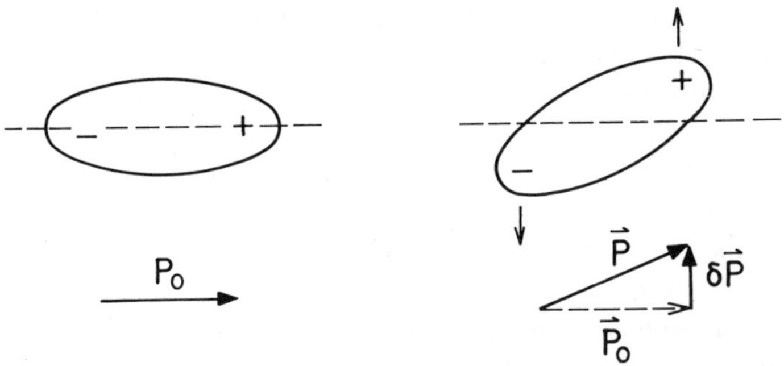

Fig.6.4.2. The polarization current due to a small rotation of a permanent dipole moment $P(\vec{r},t)$. Note that positive and negative ends of the dipole are displaced in opposite directions, which gives rise to a finite electrical current density.

The phenomenon is described by the well known fluctuation-dissipation theorem (Callen and Welton, 1951; R.Kubo, 1957), which states a general relationship between the response of a given system to external fields and the spontaneous fluctuations in the absence of external disturbances. For dielectrics, this leads to a spectral power density $\vec{q} = 0$ of the polarization-fluctuation generated short-circuit electrical current

$$S_I(\omega) = 4k_B T \cdot \omega \cdot C(\omega) \tag{6.4.4}$$

Here, $C(\omega) = C' - iC''$ is the capacitance of a capacitor, which is filled with a material with a complex dielectric constant $\varepsilon = \varepsilon' - i\varepsilon''$. The meaning of the Eq.6.4.4. can easily be understood. Whenever a spontaneous electrical current is created in a capacitor, it builds-up an electric field across the electrodes. However, power can only be dissipated by the resistive component of the current, that is in-phase with the voltage. Such a displacement current can only be generated by the lossy part of the dielectric susceptibility and the available noise power $k_B T d\nu$ is dissipated on the resistance $1/\omega C''$. The polarization noise is therefore the Johnson noise of this dissipation. As one can see from the Eq.6.4.4., the spectral power density of the polarization-generated current is proportional to the imaginary part $\varepsilon''(\omega)$ of the dielectric constant. This quantity is usually peaked around the characteristic relaxational or resonant frequency of the system, where the power dissipation is maximal. In view of the experimental limitations, it is therefore easier to observe polarization noise in systems with a slow dynamics , i.e. with long characteristic relaxation times, then in systems with a fast dynamics.

Ferroelectric smectic liquid crystals are dissipative systems with a non-zero spontaneous electric polarization and low frequency order parameter relaxation modes. In comparison to solid ferroelectrics, the characteristic relaxation rates of the normal modes are much lower and are in the kilohertz region. The magnitude of the *RMS* current through the capacitor, filled with a ferroelectric liquid crystal, will for a given geometry depend on the imaginary part of the dielectric constant at $\vec{q} = 0$. This can be easily understood, because only spatially homogeneous fluctuations (with $\vec{q} = 0$) can give rise to a spatially homogeneous electric current density. If we consider a simple situation with a single-mode dynamics, the corresponding complex dielectric constant can always be written as

$$\varepsilon(\omega) = \varepsilon(\infty) + \frac{\Delta\varepsilon}{1 + i\omega\tau} \tag{6.4.5}$$

Here, $\varepsilon(\infty)$ is the dielectric constant in the high-frequency limit, $\Delta\varepsilon = \varepsilon(0) - \varepsilon(\infty)$ is the dielectric strength, i.e. the difference between the static and the high-frequency dielectric constant and τ is the relaxation time of the mode, which gives rise to a linear response at $\vec{q}=0$. A straightforward calculation shows that the spectral power density of the polarization-fluctuations-induced electric current through a short-circuit capacitor filled with a ferroelectric liquid crystal is

$$S_I(\nu)=\frac{\overline{dI^2(t)}}{d\nu}=4k_BT\cdot C_\circ\cdot\frac{\varepsilon(0)}{\tau}\cdot\frac{\omega^2\tau^2}{1+\omega^2\tau^2} \qquad (6.4.6)$$

The frequency dispersion of $S_I(\nu)$ is shown in Fig.6.4.3. Interestingly, the curve has a "high-frequency shoulder" that appears exactly at the relaxation frequency τ^{-1} of the system. In the limit of high frequencies, $\nu\to\infty$, the spectra obtains a finite value

$$\nu\to\infty \qquad S_I(\nu)\to 4k_BT\cdot C_\circ\cdot\frac{\varepsilon(0)}{\tau} \qquad (6.4.7)$$

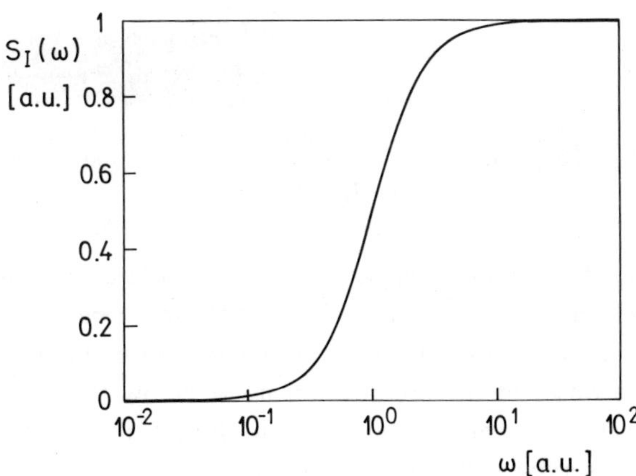

Fig.6.4.3. The short-circuit current spectral power density, calculated from the Eq.6.4.6.

It is important to note the high-frequency limit of the short-circuit power spectra depends only on the ratio $\varepsilon(0)/\tau$. This has a consequence, that high frequency modes are enhanced. These modes usually have small dielectric strengths and their contribution to the static dielectric constant is small. However, their relaxation time is also small, which can results in large value of the electric current. A rough estimate shows that the maximum *RMS* electric current in a frequency window of $d\nu \approx 10Hz$ is of the order of $1pA$ for a material with a static dielectric constant $\varepsilon(0)=20$, relaxation time $\tau \approx 1ms$ and capacitance of an empty cell of $\approx 200pF$. This is within the experimentally accessible range of electrical current noise measuring equipment.

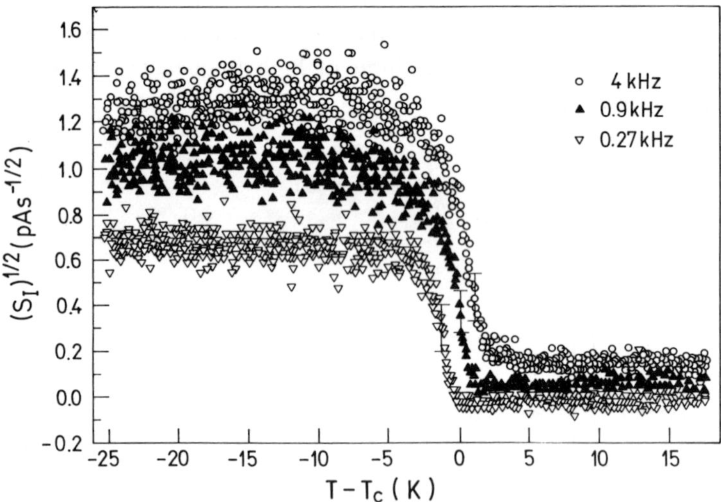

Fig.6.4.4. The temperature dependence of the polarization noise on going from the paraelectric *smectic – A* to the ferroelectric *smectic – C** phase in a $1\mu m$ thick sample of SCE-9. These are raw data and no background has been subtracted. The onset of the polarization noise below the phase transition is clearly visible. The data are scattered due to the intrinsic noise of the lock-in amplifier, which served as the current-meter.

Although the first experiment on the electrical noise in ferroelectric liquid crystals was reported by Maruyama (1980), the electrical noise spectra were first extensively analyzed by Muševič et al. (1997). The authors reported on the observation of the polarization current noise in thin cells of the commercially available ferroelectric liquid crystal SCE-9. The measurements were performed in a short-circuit geometry and the onset of consirable noise at the phase transition from the paraelectric $smectic-A$ to the ferroelectric $smectic-C^*$ phase of $1\mu m$ sample of SCE-9 is shown in Fig.6.4.4. for three different measuring frequencies. Whereas there is only a small background noise level of $\approx 0.1pA$ in the paraelectric phase, the noise level significantly increases by entering the ferroelectric phase. The *RMS* short-circuit electric current is in the ferroelectric phase within the theoretically estimated value of $\approx 1pA$, expected for the polarization noise current. One can also note that at higher measuring frequencies, the current noise starts to increase slightly above the phase transition temperature T_c, which is an indication of a mode, coming into the spectra from the high-frequency side. This was explained by measuring the spectral power density of the short-circuit electric current, which is shown in Fig.6.4.5. for different temperatures in the vicinity of the phase transition.

One can clearly see that deep in the paraelectric $smectic-A$ phase, there is practically no detectable electrical noise, except for the high-frequency shoulder that shifts to the low-frequency side of the spectrum by decreasing the temperature. Several degrees below the temperature, where the authors first observed this shoulder in the current noise spectrum, the spectra stabilized and did not change by further cooling the sample. This temperature dependence was explained by the onset of the soft mode in the paraelectric $smectic-A$ phase and a critical slowing down of this collective mode. There is a slight deviation of the observed spectra from the single-mode dynamics in the vicinity of the phase transition, which could be an indication of the observation of the amplitude mode in the ferroelectric phase. The splitting is however difficult to resolve in this data.

One of the important features of the observed short-circuit power spectra is the fact that on the high-frequency side, the spectra saturate at nearly the same level, regardless of temperature. This can be explained by considering the Equation 6.4.7. and remembering the temperature dependence of the static dielectric constant and the soft mode relaxation times in the vicinity of the transition. The static dielectric constant is

$$\varepsilon(0) \approx \frac{1}{T - T_c} \tag{6.4.8}$$

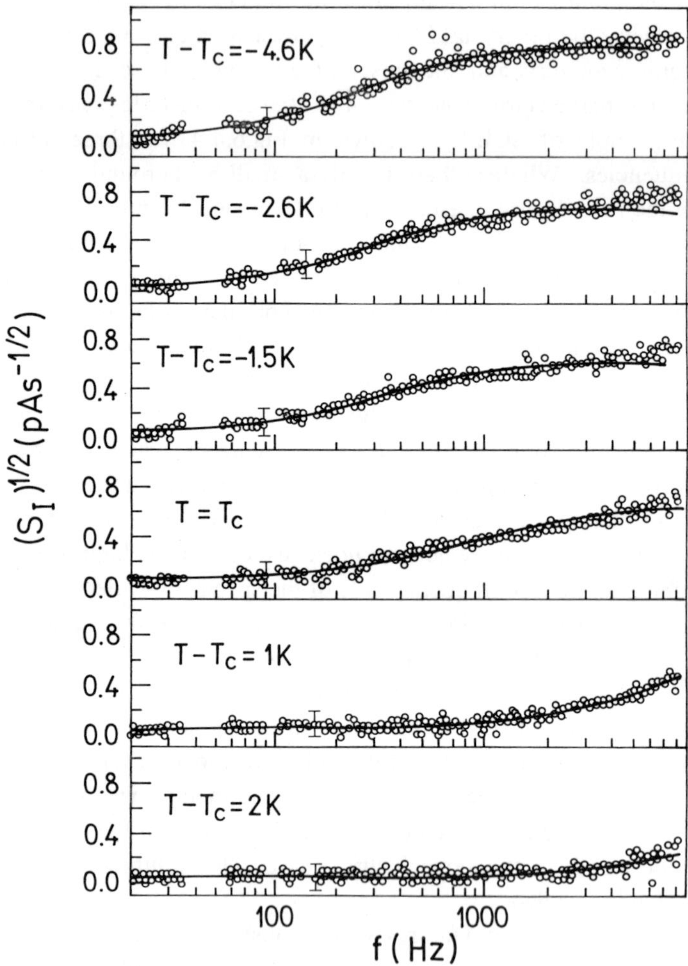

Fig.6.4.5. Square root of the short-circuit power spectra of the electric current, fluctuating through a $1\mu m$ thick layer of a ferroelectric liquid crystal SCE-9. The solid lines are the best fit to the Eq.6.4.6. and allow for the determination of the temperature dependence of the relaxation time.

The soft mode relaxation time diverges as

$$\tau \approx \frac{1}{T - T_c} \tag{6.4.9}$$

on the higher temperature side of the transition. The ratio $\varepsilon(0)/\tau$ is temperature independent, which means that the higher-relaxation modes are enhanced in the short-circuit power spectrum.

The relaxation times, which have been calculated from the observed current spectra have been compared to the linear electrooptic response and capacitance bridge measurements, as shown in Fig.6.4.6. A very good agreement was found between the noise and linear response data, clearly indicating the spontaneous polarization-fluctuation origin of the observed noise spectra.

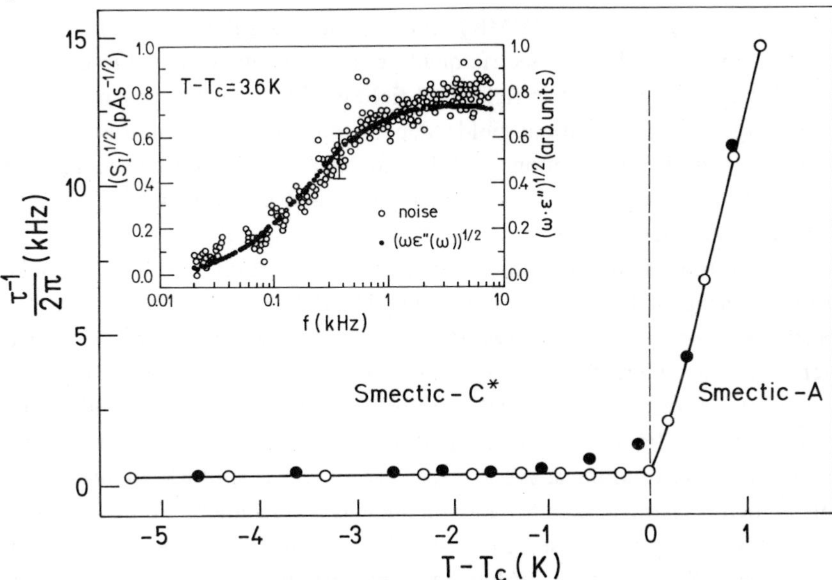

Fig.6.4.6. The temperature dependence of the order parameter relaxation rates τ^{-1}, as determined from the spectral power density measurements of the short-circuit current (full dots) and by the linear response (empty dots) in a $1\mu m$ sample of SCE-9. The inset shows a comparison of the short-circuit power density (empty dots) with the product $\omega \cdot \varepsilon''(\omega)$, determined by the capacitance bridge. After Muševič et al., 1997.

Chapter 7

NMR in Ferroelectric and Antiferroelectric Liquid Crystals

7.1. Introduction

Nuclear magnetic resonance (NMR) and nuclear quadrupole resonance (NQR) use the magnetic dipole and electric quadrupole moments of nuclei to determine the local magnetic and electric fields and field gradients at the nuclear sites. The time averaged values of these fields depend on the details of the molecular structure and molecular motions. NMR and NQR can be thus used to obtain microscopic information about molecular order and structure which cannot be obtained by other techniques.

The most important aspect of NMR for the study of liquid crystals is that the ***splitting*** of the proton or deuteron NMR line is directly related to the ***nematic orientational order parameter*** S and to the ***orientation of the nematic director*** $\vec{n}$. The interaction of the nuclear magnetic moment with the external magnetic field results in several equidistant Zeeman energy levels. These Zeeman levels in liquid crystals are perturbed by the interaction of the nuclear spins with the surroundings. As a result, we obtain a splitting of the Zeeman levels, which is characteristic for a given nucleus and its interaction.

For example, in case of protons with a spin $I = 1/2$, ***magnetic dipole-dipole interactions with neighboring spins*** are the main perturbation. For deuterons with a spin $I = 1$, the ***electric quadrupole interaction with the electric field gradient (EFG) tensor*** at the nuclear site is dominant. The same is true for ^{14}N nuclei ($I = 1$). For ^{13}C nuclei ($I = \frac{1}{2}$) the chemical shift tensor $\underline{\sigma}$ interaction is important. The chemical shift tensor $\underline{\sigma}$ measures the screening of the nucleus by the surrounding electrons and relates the external magnetic field $\vec{H}_\circ$ to the actual field $\vec{H}$ "seen" by the nucleus,

$$\vec{H} = \vec{H}_\circ(\underline{1} - \underline{\sigma}) \tag{7.1.1}$$

For an isolated and rigid pair of ***protons***, ***dipole-dipole interactions*** split the NMR line into a doublet, which is separated by:

$$\Delta\nu_d = \frac{3}{2\pi} \cdot \frac{\gamma^2 h}{r^3} \cdot \frac{1}{2}\left(3\cos^2\theta_\circ - 1\right) \tag{7.1.2}$$

where γ denotes the proton gyromagnetic ratio, r is the inter-proton distance and $\theta_\circ$ the angle between the inter-proton vector and the external magnetic field $B_\circ$. For rigid ***deuterons*** we find a ***quadrupolar splitting*** of the deuteron NMR line

$$\Delta\nu_q = \frac{3}{2} \cdot \frac{e^2 qQ}{h} \cdot \frac{1}{2}\left(3\cos^2\theta_\circ - 1\right) \tag{7.1.3}$$

Here $e^2 qQ / h$ stands for the static deuteron quadrupole coupling constant and $\theta_\circ$ is the angle between the largest principal axis of the EFG tensor (usually parallel to the C-D bond) and the direction of the external magnetic field.

In a liquid crystal, the angular factor $\frac{1}{2}(3\cos^2\theta_\circ - 1)$ is time-averaged over all molecular motions which are faster than $(2\pi\Delta\nu_d)^{-1}$ or $(2\pi\Delta\nu_q)^{-1}$ respectively. It is thus averaged over conformational changes of molecules, molecular rotations around the long axis and fluctuations of the long axis around the preferred direction $\vec{n}$. For an uniaxial phase we thus find a dipolar splitting of the NMR line for a proton pair (Doane 1979)

$$\Delta\nu_d = \frac{3}{2\pi} \cdot \frac{\gamma h}{r^3} \cdot \frac{1}{2} S \cdot \left(3\cos^2\theta_B - 1\right) \cdot \left\langle \frac{1}{2}\left(3\cos^2\beta - 1\right) \right\rangle \tag{7.1.4}$$

and a quadrupolar splitting for a deuteron pair

$$\Delta\nu_q = \frac{3}{2} \cdot \frac{e^2 qQ}{h} \cdot \frac{1}{2} S \cdot \left(3\cos^2\theta_B - 1\right) \cdot \left\langle \frac{1}{2}\left(3\cos^2\beta - 1\right) \right\rangle \tag{7.1.5}$$

Here θ_B describe the orientation of the nematic director with respect to the magnetic field direction, β describes the angle between the inter-proton vector (or the largest principal axis of the EFG tensor) and the long molecular axis and S represents the ***nematic orientational order parameter***

$$S = \left\langle \frac{1}{2}\left(3\cos^2\vartheta - 1\right)\right\rangle \tag{7.1.6}$$

where ϑ stands for the angle between the instantaneous direction of the long molecular axis and the local director $\vec{n}$. $\langle\ \rangle$ stands for the ensemble average. It should be noted that $S = S_{zz}$ and that in deriving Eq.7.1.4 and 7.1.5 the biaxiality term $\frac{1}{2}\left(S_{xx} - S_{yy}\right)\left(\sin^2\beta\cos 2\alpha\right)\left(3e^2qQ\right)/(4h)\left(3\cos^2\theta_B - 1\right)$ has been neglected. Here α would describe the azimuthal angle of the inter-proton vector (or EFG direction) with the long molecular axis.

In the isotropic phase, $S = 0$, and the splitting of the NMR line vanishes (Fig.7.1.1). Here the spin interactions are averaged out by fast and isotropic molecular reorientations.

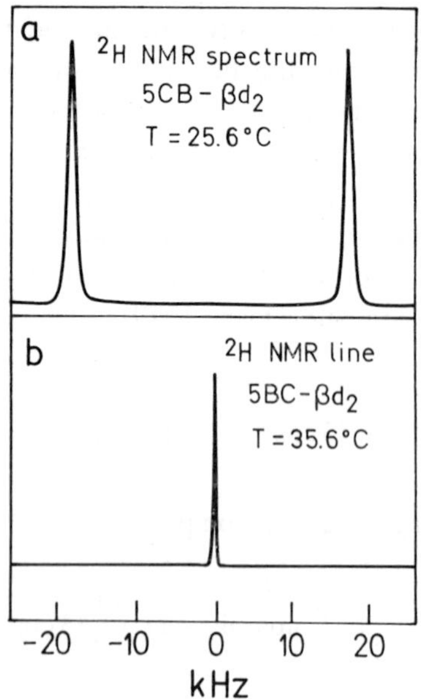

Fig.7.1.1. Deuteron NMR spectrum of 5CB-β-D_2 in the (a) nematic and, (b) isotropic phase.

In a bulk nematic, the angle θ_B between the nematic director and the external magnetic field is spatially uniform. For compounds with a positive diamagnetic anisotropy (such as benzene-based liquid crystals), $\Delta\chi > 0$, $\theta_B = 0$ and the director is aligned along the direction of the magnetic field. In case of negative magnetic anisotropy, which is characteristic of cyclohexane-based compounds, the molecules tend to align in a plane that is perpendicular to the magnetic field.

In a spatially confined or ***chiral liquid crystal***, on the other hand, the director field is space dependent, $\vec{n} = \vec{n}(\vec{r})$ (see, for example, Vilfan and Vrbančič, 1996 and references therein). In some confined systems the magnitude of the nematic order parameter may as well vary over the sample $S = S(\vec{r})$. The deuteron NMR line splitting thus becomes

$$\Delta\nu_q(\vec{r}) = \Delta\nu_{q^\circ} \cdot \frac{1}{2}\left[3\cos^2\theta_B(\vec{r}) - 1\right] S(\vec{r}) \tag{7.1.7}$$

where $\Delta\nu_{q0}$ denotes the splitting in the perfectly oriented nematic phase ($S = 1$) with the director oriented along the magnetic field $(\theta_B = 0)$. In view of the spatial dependence of $\Delta\nu(\vec{r})$ we have here, similarly as in chiral systems, a frequency distribution rather than a line splitting. In case of fast translational diffusion, $\vec{n}(\vec{r})$ is also time dependent and additional motional averaging may occur.

In this Chapter we shall describe the use of deuteron NMR, ^{14}N NQR and ^{13}C NMR for the study of the local structure of ferroelectric and antiferroelectric liquid crystalline phases. It will be shown that the angular dependence of the deuteron NMR spectra provides microscopic evidence for the alternating tilt model of the antiferroelectric $smectic-C_A^*$ phase and the presence of a soliton lattice, which is induced in helical phases due to the coupling to the external magnetic field. We shall also show that ^{14}N NQR and ^{13}C NMR on the other hand allow for a measurement of the ***tilt induced polar and quadrupolar biasing of the rotation*** of the transverse molecular dipoles.

7.2. Deuteron NMR in Ferroelectric Liquid Crystals

Deuteron NMR in partially deuterated ferroelectric liquid crystals represents an extremely powerful tool for the determination of the local order at the molecular level as well as for a determination of the nature of the helicoidal modulation wave and its distortion in the presence of external fields (Wu and Doane, 1987, Zalar et al., 1998) and confining walls (Golemme et al., 1988, Vilfan et al., 1989, Crawford et al., 1991, Crawford and Žumer, 1996). The full power of deuteron NMR can be however exploited only if we are able to measure the angular dependence of the deuteron NMR spectra by varying the angle between the external magnetic field and the helicoidal axis which coincides with the normal to the smectic planes. In this way the soliton density and the anisotropy of the critical magnetic field for the unwinding of the $smectic-C^*$ helix have been determined (Zalar et al., 1998). One can also determine the point group symmetry of a given phase and - if it is large enough - the amount of the polar and quadrupolar biasing of the rotation around the long molecular axis.

The deuteron has a nuclear spin $I=1$ and hence a non-zero electric quadrupole moment in addition to the magnetic dipole moment. The Hamiltonian H of our problem is therefore the sum of a nuclear Zeeman part H_Z and a quadrupolar part H_Q:

$$H = H_Z + H_Q \tag{7.2.1}$$

The Zeeman part measures the interaction of the magnetic moment of the deuteron with the external magnetic field $\vec{H}$

$$H_Z = -\gamma\hbar \vec{I}\,\vec{H} = -\gamma\hbar I_z H_z \tag{7.2.2}$$

The quadrupolar part H_Q, on the other hand, measures the interaction of the electric quadrupole moment tensor $Q_{\alpha\rho}$ with the local electric field gradient (EFG) tensor $V_{\alpha\beta}$ at the deuteron site

$$H_Q = \tfrac{1}{6}\sum_{\alpha,\beta} V_{\alpha\beta}\, Q_{\alpha\beta} \tag{7.2.3}$$

This quadrupolar part is sensitive to the local molecular order at a given nuclear site. The EFG tensor of a C-D deuteron is approximately axially symmetric with its largest principal axis (V_{ZZ}) pointing along the C-D (i.e. carbon-deuteron) bond. As the molecular rotation of the ··C-D bond around the long molecular axis is fast compared with the NMR frequencies, we deal with a time averaged deuteron EFG tensor of cylindrical symmetry with its largest principal axis pointing along the average direction of the long molecular axis. Here we neglected the small polar biasing of the molecular rotation, which will be discussed latter. Assuming that $\mathrm{H}_Z >> \mathrm{H}_Q$ we can rewrite Eq.7.2.1-3

$$H_Q = h\nu_Q\left(I_z^2 - \frac{I(I+1)}{3}\right) \tag{7.2.4}$$

Here

$$\nu_Q = \frac{3e^2qQ}{4I(2I-1)h} \tag{7.2.5}$$

and

$$eq = V_{ZZ} \tag{7.2.6}$$

is the largest eigenvalue of $V_{\alpha\beta}$. We assumed axial symmetry of $V_{\alpha\beta}$, i.e. that the two eigenvalues $V_{XX} = V_{YY}$ are equal, so that the asymmetry parameter η is zero:

$$\eta = \frac{V_{XX} - V_{YY}}{V_{ZZ}} = 0 \tag{7.2.7}$$

e^2qQ/h is the deuteron quadrupole coupling constant with Q describing the (scalar) deuteron quadrupole moment. e^2qQ/h is of the order of 165 kHz for aliphatic deuterons and of the order of 185 kHz for aromatic deuterons.

The energy levels of H_Z as well as of $H_Z + H_Q$ are illustrated in Fig.7.2.1. In the absence of quadrupolar coupling, $\omega_Q = 0$, the nuclear spin energy levels are equidistant resulting in a single deuteron NMR line at $\nu_L = \omega_L / 2\pi$. For $\omega_Q \neq 0$, on the other hand, the nuclear spin energy levels become non-equidistant and we get two deuteron NMR lines corresponding to the $m = +1 \leftrightarrow 0$ and $m = 0 \leftrightarrow -1$ transitions instead of a single line.

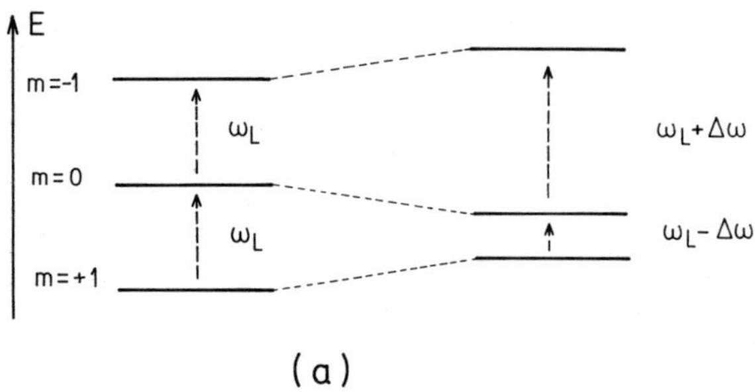

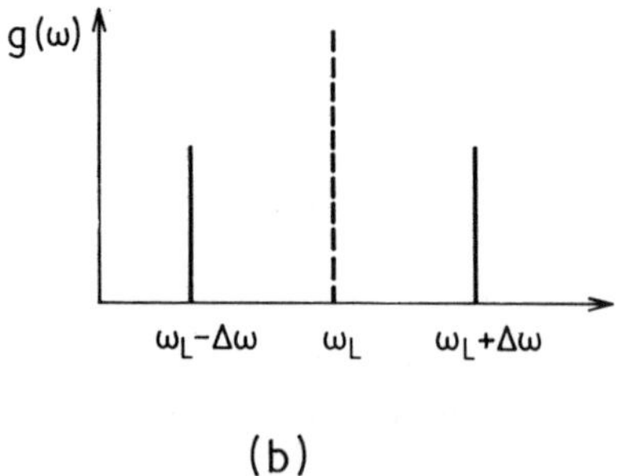

Fig.7.2.1. (a) Schematic presentation of the quadrupole perturbed Zeeman energy levels for $I = 1$. (b) Schematic presentation of the splitting of the deuteron NMR line in the presence of quadrupole interactions.

The frequencies of the two deuteron NMR lines, which are symmetrically displaced from ν_L are now obtained for a general orientation of the sample in the magnetic field (Zalar et al., 1998)

$$\nu = \nu_L \pm \frac{3}{8}\nu_Q . S . \left[3(\vec{n} . \vec{H} / H)^2 - 1\right] = \\ = \nu_L \pm \frac{3}{8}\nu_Q . S \cdot \left[(3\sin\theta_B . \sin\theta \ \sin\Phi + \cos\theta_B \cos\theta)^2 - 1\right] \tag{7.2.8}$$

Here ν_L is the deuteron Larmor frequency in the absence of quadrupole interactions, $\nu_Q = e^2\overline{q}Q/h$ is the deuteron quadrupole coupling constant averaged over the molecular motion and S is the nematic order parameter, $0 \le S \le 1$. Here, $\overline{q} = q \cdot \frac{1}{2} \cdot (3\cos^2\beta - 1)$, see also page 303.

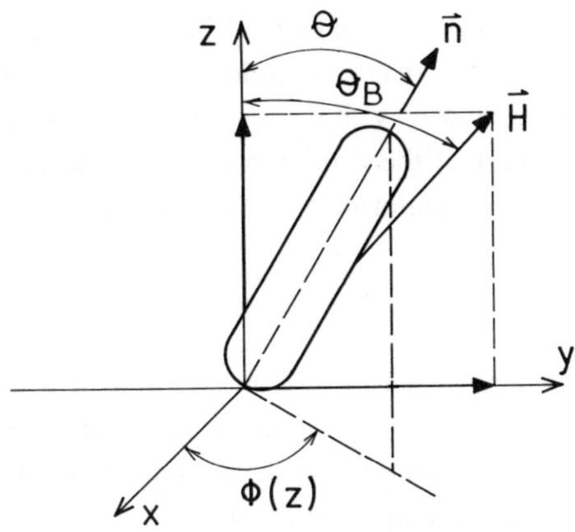

Fig.7.2.2. The orientation of the external magnetic field and the director of tilted smectic phase of liquid crystals.

The unit vector $\vec{n}$ represents the average direction of the long molecular axis, i.e. the molecular director. In the molecular fixed frame with the z-axis normal to the smectic layers one has

$$\vec{n} = (\sin\theta \cos\Phi, \ \sin\theta \sin\Phi, \ \cos\theta) \tag{7.2.9}$$

and

$$\vec{H} = (0, H\sin\theta_B, H\cos\theta_B) \tag{7.2.10}$$

Here θ and Φ are the tilt and azimuth angles of the molecular director and θ_B denotes the angle between the magnetic field $\vec{H}$ and the normal $\vec{z}$ to the smectic layers, as shown in Fig.7.2.2.

In the $smectic-A$ phase the normals to the smectic layers $\vec{z}$ and the molecular directors are aligned parallel to the direction of the external magnetic field: $\vec{z} \parallel \vec{n} \parallel H$ for a liquid crystal with a positive diamagnetic anisotropy. Experiments show that when we cool a chiral smectic liquid crystal from the $smectic-A$ into the $smectic-C^*$ phase, the smectic layers do not move or break, but we get a non-zero tilt $(\theta \neq 0)$ of the molecular directors away from the normals to the smectic layers: $\angle(\vec{z},\vec{n}) \neq 0$. If we now change the orientation of the sample with respect to the magnetic field, so that $\theta_B \neq 0$, the smectic layers will slowly reorient in such a way that after some time the normal to the layers will again become parallel to the direction of the external magnetic field. The angular dependence of the deuteron NMR spectra can therefore be measured only if:

(i) We change the orientation of the sample in the magnetic field and measure the spectra so fast that the acquisition of the NMR spectra is finished before the smectic layers can adjust to the new situation and reorient, so that we have again $\theta_B = 0$.

(ii) We use a large number of thin sample cells so that the arrangement of the smectic layers is determined not by the magnetic field but by the interaction with the surface of the walls. The same techniques can be of course used to measure the orientation dependence of the NMR spectra in the $smectic-A$ and the $smectic-C^*$ phases.

The experiment *(i)* therefore requires fine synchronization of the rotation of the sample and the NMR pulse sequence. Figure 7.2.3. shows an example of the goniometer-NMR spectrometer synchronization, which can be used to measure the angular dependence of deuteron NMR in chiral smectics.

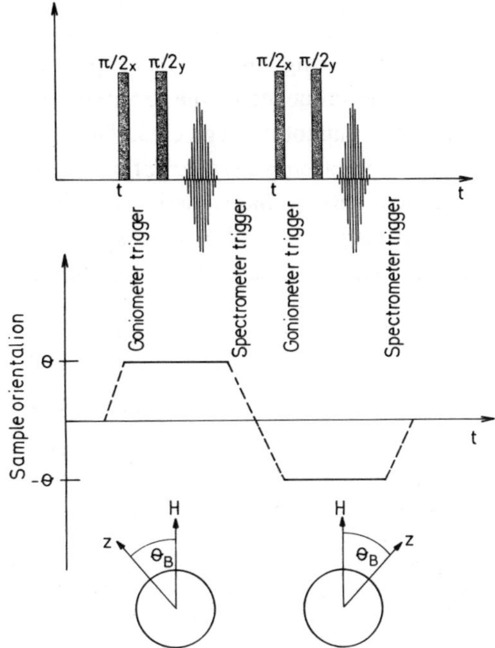

Fig.7.2.3. The synchronization of the NMR measuring pulse sequence and the rotation of the sample, which is performed with a stepper motor and a goniometer.

7.3. Deuteron NMR Spectra in the Smectic-A, Smectic-C* and Smectic-C Phases

Equation 7.2.8. adequately describes the deuteron NMR spectrum in the $smectic-A$ phase where the tilt angle is zero, ($\theta = 0$), as well as in the non-chiral $smectic-C$ phase, where the tilt angle is non-zero ($\theta \neq 0$) and the phase angle Φ does not vary with the spatial coordinate z along the normal to the smectic layers. On the other hand, in the modulated $smectic-C^*$ phase, the phase angle $\Phi(z)$, which describes the position of the molecules on the surface of the tilt cone of opening 2θ, varies with z, as the average direction of the long molecular axis precesses when we move from one smectic layer to another.

When we are considering the NMR experiments in helicoidally modulated smectic phases, the influence of the external magnetic field which is used in the NMR experiment, cannot be neglected. These magnetic fields are in the range of $1-10T$ and have a considerable influence on the director phase profile $\Phi(z)$ due to rather strong diamagnetic coupling. We have seen in Chapter 3. that an external magnetic field, which is applied ***perpendicularly to the helical axis*** of a chiral ferroelectric liquid crystal, ***induces a π-soliton lattice***, as shown in Fig.7.3.1.

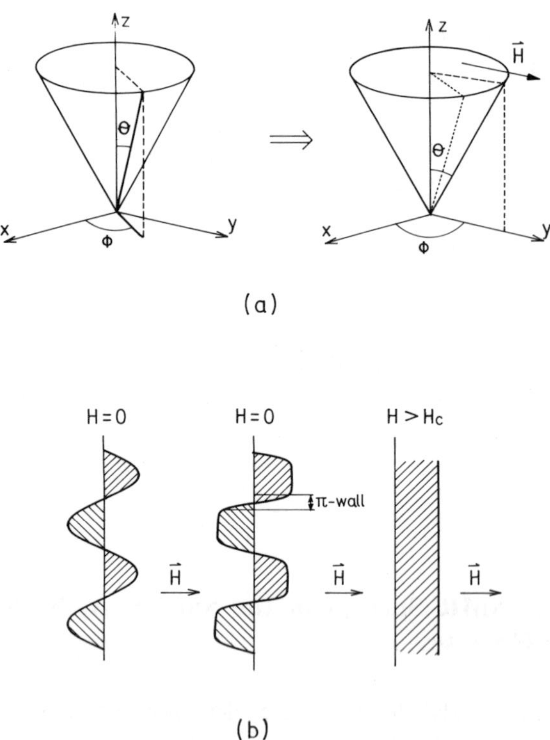

Fig.7.3.1. Soliton-like deformation of the originally "smooth" helix of a ferroelectric liquid crystal with a positive diamagnetic anisotropy in an external magnetic field, applied perpendicularly to the helix.

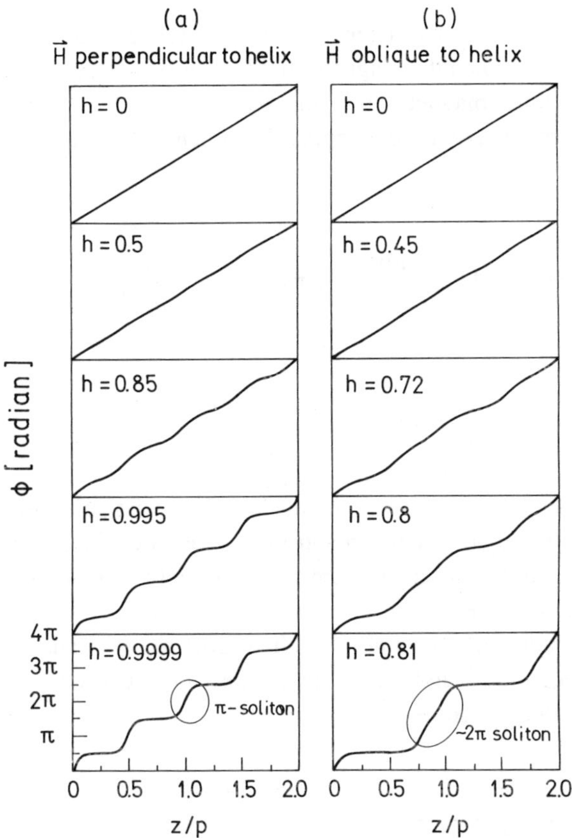

Fig.7.3.2. The distortion of the phase profile $\Phi(z)$ of a ferroelectric liquid crystal with positive diamagnetic anisotropy in the external magnetic field. (a) The magnetic field is applied perpendicular to the helix. (b) The magnetic field is applied in a direction which is only 1° away from the perpendicular direction. Note the difference in the shape of the soliton walls. The tilt angle is $\theta = 10^{\circ}$ and $h = H/H_c$ is normalized magnetic field. The phase profile was calculated by minimizing free energy density given by Eq.7.3.1.

It has been shown that the distortion of the phase profile is considerably different, when the magnetic field is applied at an oblique angle with respect to the helical axis (Zalar et al., 1998). Instead of a π-soliton lattice, we have for an oblique magnetic field ***2π solitons***, as can be seen from the phase profiles in

Fig.7.3.2. Furthermore, it has been shown, that the critical magnetic field is significantly reduced when the magnetic field is applied at an oblique angle.

The relevant magnetic free energy density of a ferroelectric liquid crystal in an external magnetic field applied at a general angle Φ with respect to the helical axis is

$$g_M = \sin^2\theta \cdot \left[\tfrac{1}{2} K_\theta \left(\frac{d\Phi}{dz} \right)^2 - 2\pi \left(K_{22} / p_0 \right) \frac{d\Phi}{dz} \right] -$$

$$- \sin^2\theta \cdot \tfrac{1}{2} \Delta\chi \, H^2 \left(\sin\theta_B \sin\Phi + \cos\theta \cos\theta_B \right)^2 \qquad (7.3.1)$$

Here $K_\theta = K_{22} \sin^2\theta + K_{33} \cos^2\theta$ is an effective twist splay elastic constant, $\Delta\chi = \chi_\parallel - \chi_\perp$ is the diamagnetic anisotropy, $p_\circ$ denotes the period of the helix $\Phi(z) = (2\pi / p_\circ)z$ in the absence of the magnetic field and we assumed for sake of simplicity that $K_{22} \approx K_{33}$.

Minimization of $G = \int g dV$ with respect to Φ leads to a non-linear Sine-Gordon equation and a soliton-like deformation of the helix $\Phi = \Phi(z, H/H_c, \theta, \theta_B)$. Here

$$H_C = \frac{\pi^2}{p_\circ} \sqrt{K_{33} / \Delta\chi} \qquad (7.3.2)$$

is the critical field for the unwinding of the helix (Fig. 7.3.1.) at $\theta_B = 90^\circ$. The multi-soliton lattice which represents the deformation of the helix has for a general direction of the field the form of a 2π - soliton lattice. An exception is when the magnetic field is applied parallel to the smectic layers and we have a π - type soliton lattice.

The magnetic field induced soliton structure substantially influences the shape of the NMR spectra of ferroelectric liquid crystals. For a general orientation of the magnetic field with respect to the helix of the ferroelectric liquid crystal we shall have a frequency distribution $f(\nu)$ since different molecules with different phase angles in different smectic layers will have different quadrupole perturbed NMR frequencies (see Fig.7.3.3).

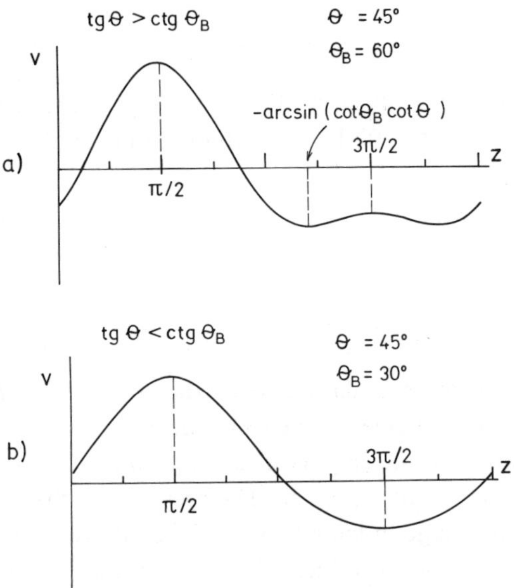

Fig.7.3.3. (a) Spatial variation of the deuteron NMR frequency in the ferroelectric $smectic-C^*$ phase for $tg\theta > ctg\theta_B$ and, (b) for $tg\theta < ctg\theta_B$.

We have

$$f(\nu)d\nu = N(z)dz \tag{7.3.3}$$

where $N(z)$ is the number of molecules in a range between z and $z+dz$ and $f(\nu)$ is the number of molecules which are resonating between ν and $\nu + d\nu$. Since $N(z) = const$, we have:

$$f(\nu) = \frac{const}{d\nu / dz} = \frac{const}{\left|\frac{d\nu}{d\Phi}\right| \cdot \left|\frac{d\Phi}{dz}\right|} \tag{7.3.4}$$

In the plane wave limit, Φ is a linear function of z so that

$$\frac{d\Phi}{dz} = const \tag{7.3.5}$$

and $f(\nu)$ is determined by $|d\nu / d\Phi|$ using Eq.7.2.8. Singularities will appear in the spectra whenever $|d\nu / d\Phi| = 0$. In the soliton limit, Φ is an non-linear function of z so that

$$\frac{d\Phi}{dz} \neq const. \tag{7.3.6}$$

and we have in addition to the singularities occurring when $|d\nu / d\Phi| = 0$ also "soliton lines" or "commensurate lines" when $d\Phi / dz = 0$.

As will be shown later, the singularities due to $d\Phi / dz = 0$ occur at the same phase angles as those for $|d\nu / d\Phi| = 0$. The presence of the soliton lattice is thus here detected via a change in the intensity of the different singularities in $f(\nu)$, rather than via the presence of new lines as in solid incommensurate crystals. Instead of looking at the soliton like deformation of the helix $\Phi(z)$, we can look at the magnetic field induced change of the distribution of the molecules on the surface of the cone of opening 2θ (see Fig.7.3.4). Here we have $N(\Phi)d\Phi = f(\nu)d\nu$ where $N(\Phi)$ is number of molecules on the cone between Φ and $\Phi + d\Phi$. The frequency distribution is now obtained as

$$f(\nu) = \frac{N(\Phi)}{|d\nu / d\Phi|} \tag{7.3.7}$$

Here $N(\Phi)$ describes the distribution of the molecules on the surface of the tilt cone, whereas $d\nu / d\Phi$ is evaluated from relation (7.2.8), which describes the dependence of the NMR frequency on the azimuth angle Φ.

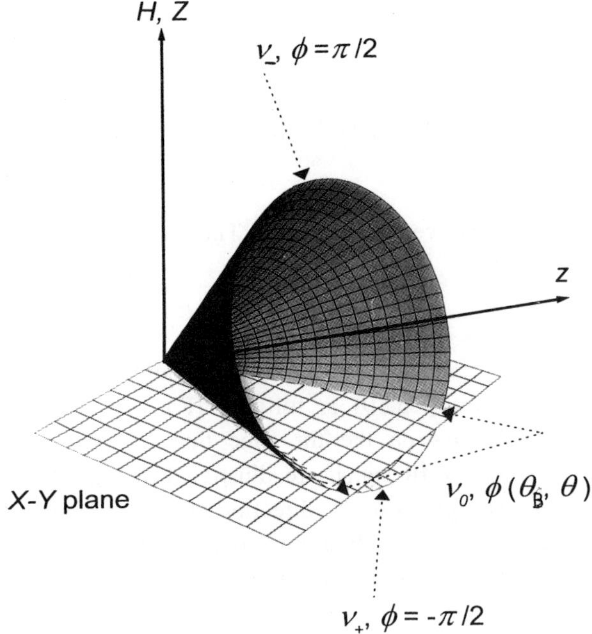

Fig.7.3.4. Singularities in the deuteron NMR frequency distribution of a helicoidal $smectic-C^*$ ferroelectric liquid crystal.

As already mentioned we have in the $smectic-C^*$ phase a distribution of azimuth angles $\Phi = \Phi(z; H, \theta, \theta_B)$ which result in a frequency distribution $f(\nu)$. This distribution has intensity singularities whenever $|d\nu / d\Phi| = 0$, i.e. at phase angles

$$\Phi_{\circ} = -\arcsin(\cot\theta_B . \cot\theta) \tag{7.3.8a}$$

$$\Phi_{-} = \pi / 2 \tag{7.3.8b}$$

$$\Phi_{+} = 3\pi / 2 \tag{7.3.8c}$$

These angles correspond to frequencies

$$\nu_0 = \pm\frac{3}{8}\nu_Q . S \tag{7.3.9a}$$

$$\nu_- = \pm\frac{3}{8}\nu_Q.S\left[3\cos^2(\theta_B - \theta) - 1\right] \qquad (7.3.9b)$$

$$\nu_+ = \pm\frac{3}{8}\nu_Q.S\left[3\cos^2(\theta_B + \theta) - 1\right] \qquad (7.3.9c)$$

Here we put for sake of convenience $\nu_L = 0$. The singularity at ν_0 occurs only if

$$\cot\theta_B < \tan\theta \qquad (7.3.10)$$

i.e. at large enough θ_B and is due to the non-sinusoidal form of the $\nu = \nu(\Phi)$ relation if the inequality (7.3.10) is fulfilled (Fig. 7.3.3(a) and (b)). The positions of the molecules on the tilt cone that contribute to singularities (Eq.7.3.9a-c) and to the NMR signal are illustrated in Fig.7.3.4. Note that the number of singularities depends on the direction of the magnetic field.

For the non-distorted modulation wave, $h = 0$, $\Phi(z)$ linearly increases with z so that the distribution of the molecules on the cone is uniform (Fig. 7.3.2)

$$N(\Phi) = const, \quad \frac{d\Phi}{dz} = const. \qquad (7.3.11)$$

In a finite magnetic field, $\Phi(z)$ becomes a non-linear function of z (Fig. 7.3.2) so that

$$N(\Phi) \neq const, \quad \frac{d\Phi}{dz} \neq const. \qquad (7.3.12)$$

The relative intensity of the spectrum increases at those phase angles, where $d\Phi / dz$ reaches a minimum value. In the presence of a perpendicular magnetic field this happens at $\Phi_- = \pi / 2$ and $\Phi_+ = 3\pi / 2$ and we have an additional enhancement of both "singularities" at ν_- and ν_+ without changing their relative positions or intensity. On the other hand, for a general orientation of the helical axis with respect to the magnetic field, the soliton singularities appear at $\Phi = \pi / 2$ and $\Phi = 5\pi / 2$ but not at $\Phi = 3\pi / 2$. Most of the molecules are thus for $\Delta\chi > 0$ found at the angle $\Phi_- = \pi / 2$ which is energetically more favorable than $\Phi_+ = 3\pi / 2$. Consequently, the peak at ν_- has a higher intensity than the one at ν_+:

$$\frac{I(\nu_-)}{I(\nu_+)} > 1 \qquad (7.3.13)$$

The soliton density can be thus determined from the intensity ratio $I(\nu_-)/I(\nu_+)$. In the $smectic-\overline{C}$ phase, where the helix is completely unwound, the singularity at ν_+ is completely suppressed. Here only one sharp resonance line on each side of ν_L is expected.

The deuteron NMR spectra of α - CD_2 deuterated CE-8 (see Fig. 7.3.5) have been measured in the $smectic-A$, $smectic-C^*$ and $smectic-\overline{C}$ phases. The $smectic-A$ - $smectic-C^*$ transition takes place here at $T_c = 354K$ with $p_\circ = 1.2\mu m$. (see Figs. 7.3.6 to 7.3.9).

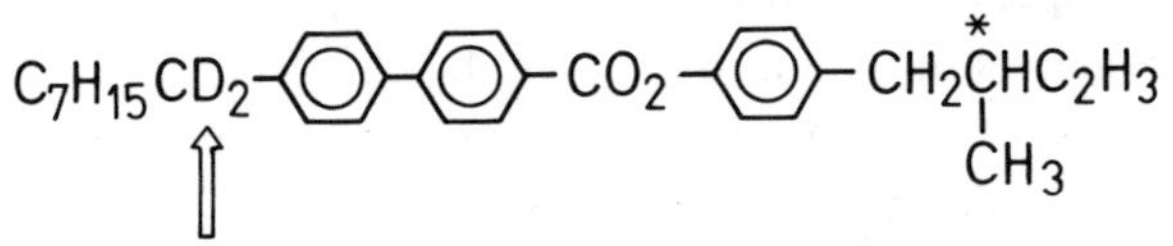

$C_7H_{15}CD_2$–⟨⟩–⟨⟩–CO_2–⟨⟩–$CH_2\overset{*}{C}HC_2H_3$ (CH_3 on the starred carbon; arrow under CD_2)

Fig.7.3.5. Molecular formula of α -CD_2 deuterated CE8. Note the α-position.

In the $smectic-A$ phase only two sharp satellite lines have been seen at all θ_B. This shows that both C-D groups are equivalent because of fast rotation around the long molecular axis. In the $smectic-A$ phase the splitting increases with decreasing T according to Eq.7.2.8 in view of the increase of the nematic order parameter S (see Fig. 7.3.6). At the $smectic-A \rightarrow smectic-C^*$ transition the splitting decreases at $\theta_B = 0^\circ$ in view of the tilting of the molecules away from the direction of the magnetic field:

$$\frac{\Delta\nu}{\Delta\nu_{SMA}} = \frac{1}{2}\left(3\cos^2\theta - 1\right),\ \theta_B = 0^0 \qquad (7.3.14)$$

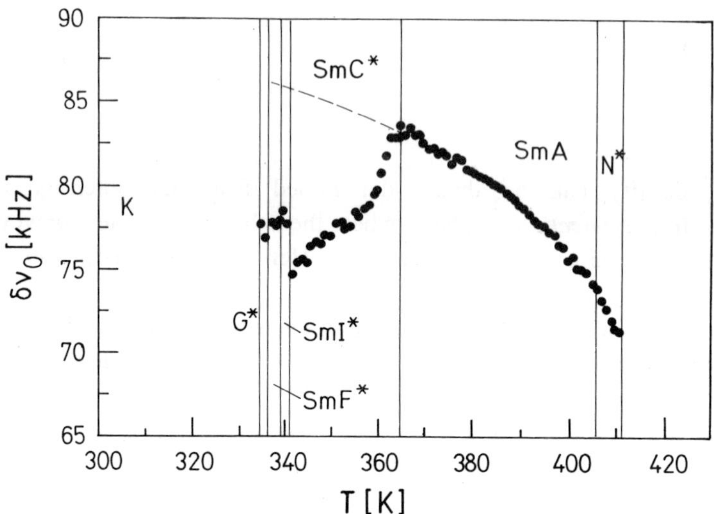

Fig.7.3.6. Temperature dependence of the quadrupole splitting of the deuteron NMR spectra of CE8 at $\theta_B = 0$ in the chiral nematic, *smectic* $-$ *A* and *smectic* $- C^*$ phases. The dotted line shows the extrapolated splitting in the absence of the molecular tilt (Abramič, 1995).

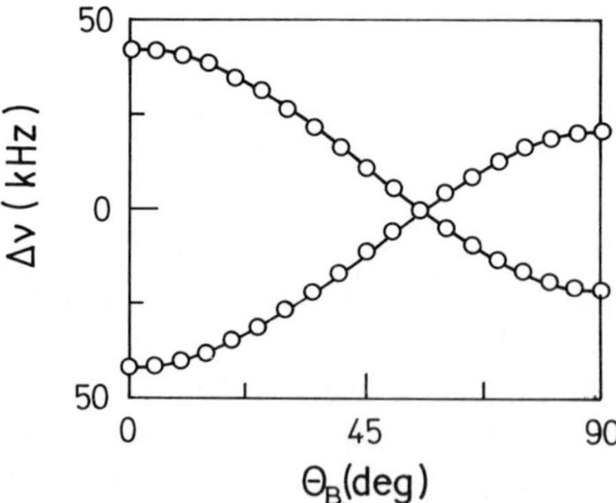

Fig.7.3.7. Angular dependence of the positions of the deuteron NMR satellite lines relative to the Larmor frequency in the *smectic* $-$ *A* phase of CE-8 (Zalar et al., 1998).

The angular dependence of the positions of these two lines with respect to the Larmor frequency $\nu = \nu(\theta_B)$ is shown in Fig.7.3.7. for the $smectic-A$ phase at $\nu_L = 58.3 MHz$ and $T = 360K$. The solid line represents a fit to expression (7.2.8) yielding zero tilt angle, $\theta = 0$, as expected for the untilted $smectic-A$ phase.

The angular dependencies $\nu(\theta_B)$ of the relative positions of the singularities in the $smectic-C^*$ phase at $T = 342K$ are shown Fig.7.3.8. We have now two ν_+ and two ν_- lines for a general θ_B. The solid lines represent the best fit to Eq.7.3.9. with $\theta = 14^\circ$. The two $\alpha - CD_2$ deuterons are again equivalent, which demonstrates that the molecular rotation is still fast on the NMR time scale and the polar biasing at the $\alpha - CD_2$ position is small.

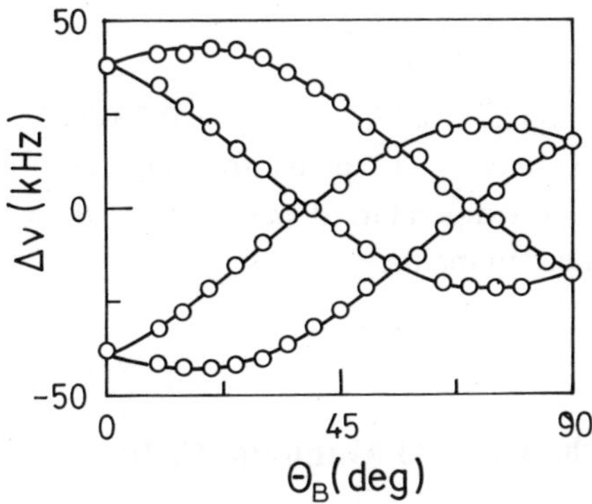

Fig.7.3.8. Angular dependence of the positions of the singularities in the deuteron NMR spectrum of CE-8 in the $smectic-C^*$ phase (Zalar et al., 1998).

The temperature dependence of the molecular tilt angle $\theta(T)$, as determined from deuteron NMR is shown in Fig.7.3.9. The solid line represents the best fit to a power law

$$\theta = \theta_\circ [(T_C - T)/T_C]^\beta \tag{7.3.15}$$

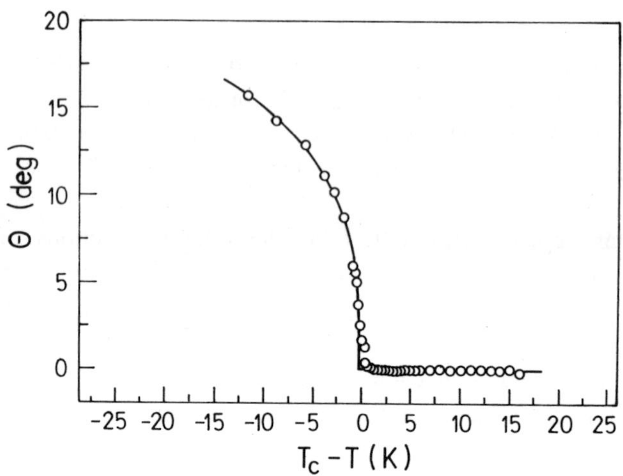

Fig.7.3.9. Molecular tilt $\theta(T)$ in CE-8 determined via deuteron NMR. (after Zalar et al., 1998).

with $\theta_\circ = 43^\circ \pm 4^\circ$, $T_C = 353.7 \pm 0.3\,K$ and $\beta = 0.3 \pm 0.03$. Sharp spectra are obtained in the unwound $smectic - \overline{C}$ phase for $H > H_C$, similar to those of the $smectic - A$ phase. The temperature dependence of the tilt angle is in good agreement with optical measurements.

7.4. Anisotropy of the Critical Magnetic Field

We have already mentioned in the previous Section 7.3. that the shape of the magnetic field distorted helical structure of a ferroelectric liquid crystal depends significantly on the direction of the magnetic field with respect to the helix. This should be reflected in the NMR spectra, which are very sensitive to phase profile.

Recently, a deuteron NMR experiment has been performed in the $smectic - C^*$ phase of α-CD_2 deuterated CE-8 by Zalar et al. (1998). In this experiment, the NMR spectrum was measured at different angles of the magnetic field with respect to the helical axis and the measured deuteron NMR spectral lineshapes are shown in Fig.7.4.1 for $T_c - T = 1K$ and in Fig.7.4.2 for $T_C - T = 3\,K$.

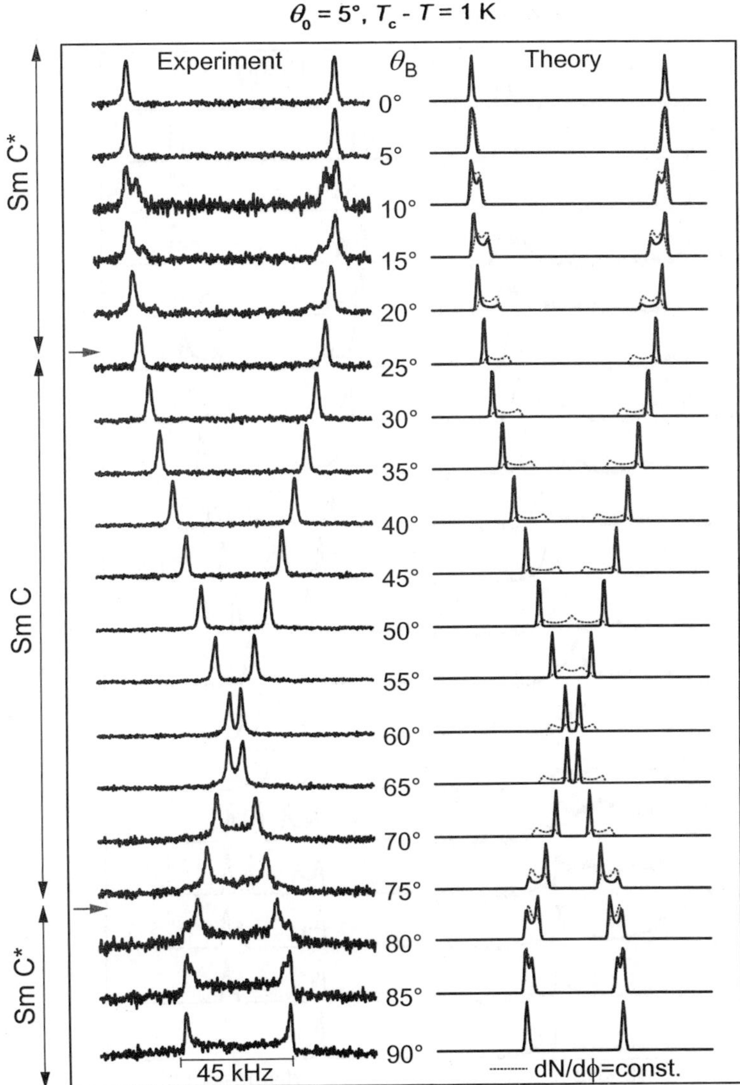

Fig.7.4.1. The angular dependence of the deuteron NMR line shape in the $smectic-C^*$ phase of CE-8 - αD_2 at $T_C - T = 1K$. A transition into the unwound $smectic-\overline{C}^*$ phase takes place between $25^\circ \leq \theta_B \leq 70^\circ$. The agreement with theory is rather good (Zalar et al., 1998).

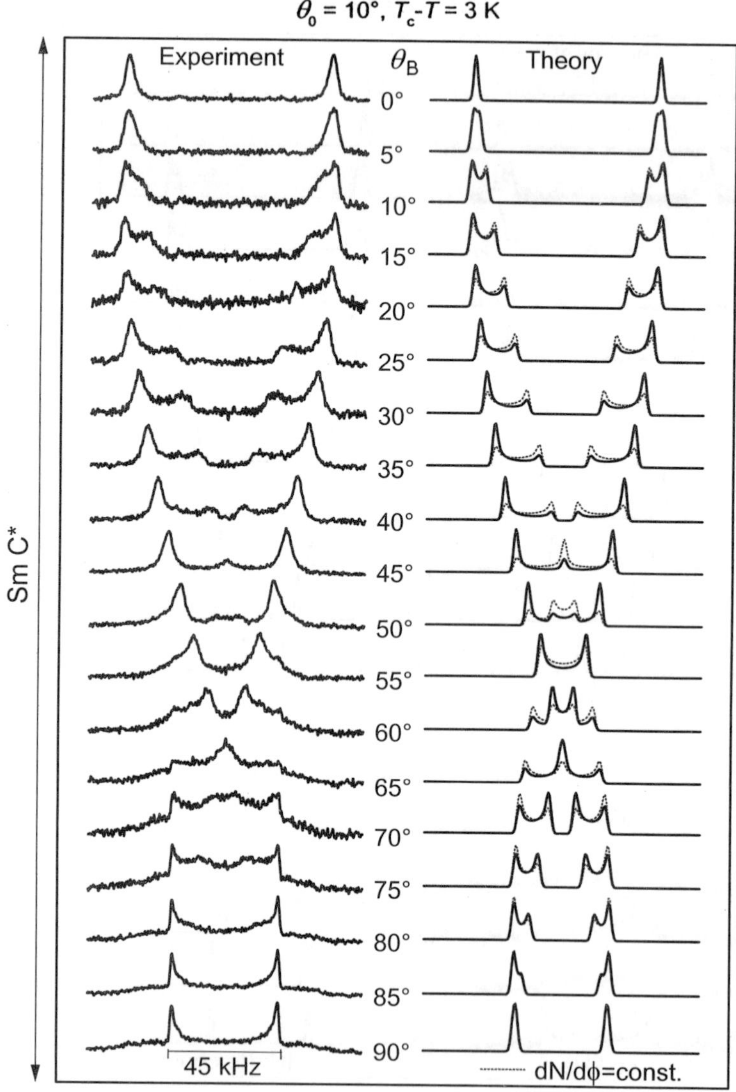

Fig.7.4.2. Angular dependence of the deuteron NMR line shape in CE-8 - $\alpha - CD_2$ at $T_C - T = 3K$. Note that there is no transition into the unwound $smectic - \overline{C}$ phase (Zalar et al., 1998).

When the direction of the magnetic field is between $\theta_B = 0$ and $\theta_B = 20^\circ$ with respect to the helical axis, the spectra show the presence of an inhomogeneous frequency distribution $f(\nu)$ as predicted by Eq.7.3.9. for the helicoidal $smectic-C^*$ phase. The agreement between the experimental and theoretical lineshapes is good. On the other hand, in the range of $25^\circ < \theta_B < 70^\circ$, the line broadening clearly disappears and we get two sharp satellite lines similar to the $smectic-A$ phase. This shape of spectra can only be explained by a transition from the magnetic-field distorted helical $smectic-C^*$ phase into the unwound $smectic-\overline{C}$ phase, which is induced above a certain value of θ_B. This is possible only if the critical field for the unwinding of the $smectic-C^*$ helix is very anisotropic, i.e. it depends strongly on the orientation of the magnetic field with respect to the helical axis. This conjecture is supported by the observation that the modulated $smectic-C^*$ phase reenters at $\theta_B = 80^\circ$ (see Fig.7.4.1), where the inner singularity reappears and the critical field is again higher than the applied field.

The conjecture of the strong anisotropy of the critical magnetic field is supported by NMR measurements at a slightly lower temperature, as shown in Fig.7.4.2. The spectra clearly demonstrate that at this temperature the critical field H_c is always higher than the applied field H and the helix is never unwound. The decrease of the critical magnetic field at intermediate sample orientations is thus pronounced only in the close vicinity of T_c, i.e. at small values of the tilt angle.

The observed anisotropy of the critical field H_c can be well explained by considering the free-energy density (Eq.7.3.1). Minimization of this free energy density with respect to the phase angle $\Phi(z)$ leads to the stationary soliton-like phase profile. The period of this soliton profile diverges at a critical field h_c

$$h_c(\theta,\theta_B) = \frac{H_c(\theta)}{H_c} = \sin^{-1}(\theta_B)\left[\sqrt{1+\cot\theta_B.\cot\theta} + \cot\theta_B.\cot\theta \cdot \ln\frac{1+\sqrt{1+\cot\theta_B.\cot\theta}}{\sqrt{\cot\theta_B.\cot\theta}}\right]^{-1} \tag{7.4.1}$$

Here, H_c is the critical magnetic field for the orientation perpendicular to the helix, which is given by Eq.7.3.2.

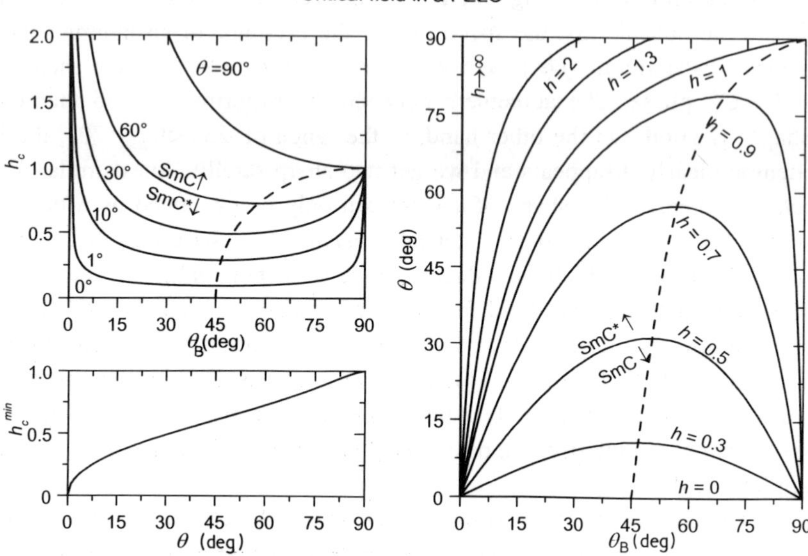

Fig. 7.4.3. Critical field $h_c = H_c(\theta)/H_c$ as a function of the angle θ_B of the magnetic field with respect to the helical axis. A large decrease of the critical field is observed at intermediate orientations for temperatures close to T_c (after Zalar et al., 1998). The dashed line shows the tilt angle at which the minimal critical field is reached for a given θ_B. θ_B is the angle between the normal to the smectic layers and the direction of the external magnetic field, θ is the tilt angle.

To a great surprise, the critical magnetic field for the unwinding of the helix is not a monotonous function of the orientation of the magnetic field with respect to the helix. Instead, as it is shown in Fig.7.4.3., this field exhibits a minimum at some intermediate orientations. As a consequence, we obtain for a given value of the tilt angle θ and a given value of the external magnetic field a phase sequence: modulated $smectic-C^*$-unwound $smectic-\overline{C}$-modulated $smectic-C^*$, as illustrated by the dashed horizontal line in Fig.7.4.3. We therefore get a reentrant $smectic-C^*$ as a function of the orientation of the magnetic field with respect to the helix. The width of this reentrant phase depends on the magnitude of the tilt angle, as can be seen from Fig.7.4.3.

The comparison between the experimental and theoretical deuteron NMR spectra in the $smectic-C^*$ phase at $T_c - T = 3K$ and different orientations (Fig.7.4.2) of the sample with respect to the applied magnetic field clearly shows the magnetic field induced distortion of the helix and the 2π -soliton nature of the $\Phi(z)$ distribution for $\theta_B \neq 0^\circ$ and $\theta_B \neq 90^\circ$. The best fit to the theoretical model yields $H_c = 30 \pm 6T$. This agrees rather well with the value $H_c \approx 35T$ estimated from the relation $H_c \propto 1/p_0$ and magnetic-field experiments of Muševič et al. (1992, 1994).

7.5. Polar and Quadrupolar Biasing of the Molecular Rotation

The appearance of a spontaneous polarization $P = (P_x, P_y, 0)$ in a direction perpendicular to the tilt plane of the $smectic-C^*$ phase is necessarily connected with a biasing of the rotation of the transverse molecular dipoles around the long molecular axis. The potential field for the rotation of the molecule around its long axis can be expressed as:

$$V(\Psi) = -a_1\theta \cos\Psi - a_2\theta^2 \cos 2\Psi \tag{7.5.1}$$

Here Ψ measures the deviation of the transverse dipole orientation from the direction perpendicular to the tilt. The first term in Eq.7.5.1. is linear in the tilt angle. It is non-zero only for chiral molecules. If $a_1 > 0$ this term induces polar order in the direction $\Psi = 0$.

The second term is quadratic in the tilt and is non-zero also for achiral molecules. For $a_2 > 0$ it induces a quadrupolar order in a direction perpendicular to the tilt. The polar and quadrupolar rotational order parameters are now $\langle\cos\Psi\rangle$ and $\langle\cos 2\Psi\rangle$. Since the polar or quadrupolar biasing of the rotation around the long molecular axis affects the motional averaging of the $-C-D$ EFG tensor in different ways, it shows up in a change of the deuteron NMR frequency or intensity singularities in the frequency distribution.

When the magnetic field is parallel to the normal to the smectic layers, i.e. for $\theta_B = 0$, we get just two deuteron NMR lines at

$$\nu_{\theta_B=0} = \nu_L \pm \frac{3e^2qQ}{4h} S\left[A + B\langle\cos\Psi\rangle + C\langle\cos 2\Psi\rangle\right] \tag{7.5.2}$$

where

$$A = \frac{1}{4}\cos^2\gamma\left(3\cos^2\theta - 1\right) \tag{7.5.3a}$$

$$B = \frac{1}{4}\sin 2\gamma \,.\, \sin 2\theta \tag{7.5.3b}$$

$$C = \frac{1}{4}\left(\sin^2\gamma + 1\right)\sin^2\theta \tag{7.5.3c}$$

Here γ describes the angle between the normal to the D-C-D plane and the molecular director (Abramič, 1995). It is of the order of $\gamma \approx 35.5^\circ$.

When the magnetic field is applied parallel to the smectic layers, i.e. for $\theta_B = \pi/2$, we get four intensity singularities in the deuteron NMR frequency distribution:

$$\begin{aligned}\nu_{\theta_B=\frac{\pi}{2}} &= \nu_L \pm \frac{3e^2qQ}{4h} S\left[A_1\langle\cos 2\Psi\rangle + D_1\langle\cos\Psi\rangle + G_1\right] \pm \\ &\pm \sqrt{\left[B_1\langle\cos 2\Psi\rangle + E_1\langle\cos\Psi\rangle + H_1\right]^2 + \left[C_1\langle\cos 2\Psi\rangle + F_1\langle\cos\Psi\rangle\right]^2}\end{aligned} \tag{7.5.4}$$

The frequency distributions for $\theta_B = 0^\circ, 15^\circ, 30^\circ, 45^\circ, 60^\circ, 75^\circ$, and 90° are shown in Fig.7.5.1 for (a) $\langle\cos\Psi\rangle = 1$, $\langle\cos 2\Psi\rangle = 0$, (b) $\langle\cos\Psi\rangle = 0$, $\langle\cos 2\Psi\rangle = 0$ and (c) $\langle\cos\Psi\rangle = 0$, $\langle\cos 2\Psi\rangle = 1$ (D. Abramič, 1995). The differences between the spectra are large enough to allow for a reliable determination of $\langle\cos\Psi\rangle$ and $\langle\cos 2\Psi\rangle$ provided the polar and quadrupolar order parameter are larger than 10^{-2}. This condition could be fulfilled only if the deuterated $-CD_2$ group is close enough to the chiral group. For example, in $\alpha - CD_2$ - deuterated CE-8 this is not the case. Such an experiment has been recently performed (Yoshida et al., 1999) in the antiferroelectric $smectic - C_A^*$ phase of MHPOBC deuterated at the methylene sites $j = 24$ and $j = 25$ next to the chiral group. Here two different quadrupolar splittings are seen for each methylene unit. This means that the molecular rotation is asymmetrically hindered. For the two deuterons attached to

$C(24)$, for example, rotation about the $C(23)-C(24)$ bond must be extremely asymmetric so that the configuration probabilities for the gauche$^+$ and gauche$^-$ conformations are remarkably different and the inversion symmetry at the $C(24)$ position is broken. This accounts for the relatively large in-plane polarizations of antiferroelectric liquid crystals. A more quantitative analysis should yield values of $\langle\cos\Psi\rangle$ and $\langle\cos 2\Psi\rangle$. No such splitting is seen at the achiral chain.

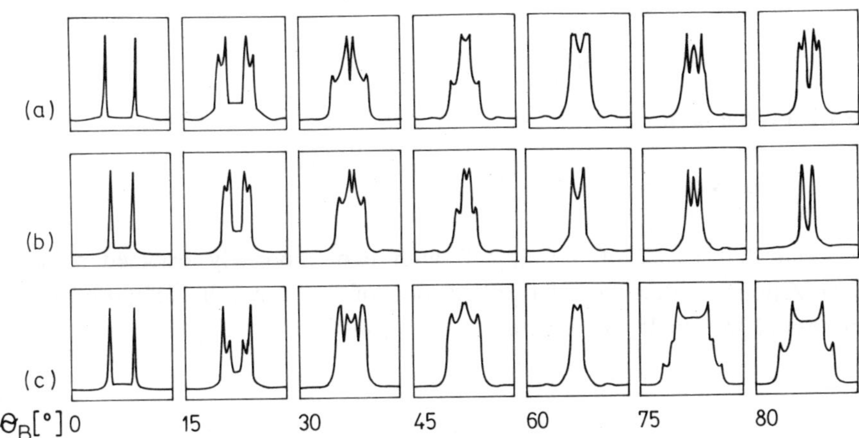

Fig.7.5.1. Calculated deuteron NMR frequency distributions for different values of the angle θ_B between the normal to the smectic layers and the direction of external magnetic field. The calculations are made for the $smectic-C^*$ phase and for different values of the polar and quadrupolar order parameters. (a) $\langle\cos\Psi\rangle=1$, $\langle\cos 2\Psi\rangle=0$, (b) $\langle\cos\Psi\rangle=0$, $\langle\cos 2\Psi\rangle=0$ and (c) $\langle\cos\Psi\rangle=0$, $\langle\cos 2\Psi\rangle=1$. The angle γ was taken as $25°$ (D. Abramič, 1995).

Again the constants $A_1, B_1, C_1, D_1, E_1, H_1, F_1$, and G_1 are known functions of γ and the tilt angle θ (D. Abramič, 1995). By combining Equation 7.5.2. and 7.5.4 we can obtain $\langle\cos\Psi\rangle$ and $\langle\cos 2\Psi\rangle$ from the spectra at $\theta_B=0$ and $\theta_B=\pi/2$. Still another possibility is to compare the experimental and theoretical frequency distributions at different θ_B.

7.6. Deuteron NMR Spectra in the Antiferroelectric Phase of Liquid Crystals

In ferroelectric liquid crystals the directions of both the in-plane polarization and the average molecular tilt vary slowly in space as the period of the ferroelectric helix is large compared to the distance between the smectic layers. In the antiferroelectric $smectic-C_A^*$ phase, on the other hand, the directions of the in-plane polarization as well as of the average molecular tilt alternate as one goes from one smectic layer to the next:

$$\theta \to -\theta, \ \vec{P} \to -\vec{P} \qquad (7.6.1a)$$

or what is the same:

$$\Phi \to \Phi + \pi \qquad (7.6.1b)$$

We can thus speak of a unit cell with two oppositely oriented sublattice polarizations so that the total polarization of the unit cell, composed of two successive smectic layers, is zero.

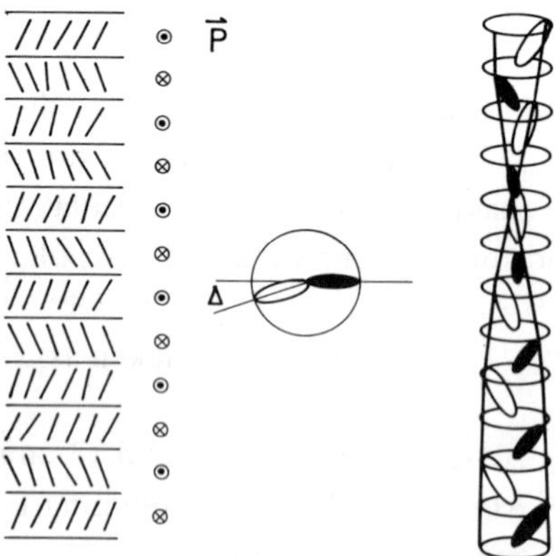

Fig.7.6.1. Schematic presentation of the structure of the antiferroelectric (AFE) $smectic-C_A^*$ phase.

Superimposed on this rapid alternation of the sign of the polarization and the tilt is a chirality induced slow helicoidal precession of each of the two sublattice polarizations and tilts resulting in a double helicoidal structure, as shown in Fig.7.6.1. The period of the two helices is comparable to that found in the $smectic-C^*$ phase of ferroelectric liquid crystals.

Similarly as in the $smectic-C^*$ phase we may assume also in the $smectic-C_A^*$ phase the existence of a fast and only weakly biased rotation of the molecules around their long axes. Expression 7.2.8. should thus in principle also describe the deuteron NMR spectra of antiferroelectric liquid crystals and not only of ferroelectric liquid crystals. There are however two peculiarities which are introduced by the alternation of the average orientation of the molecules in two successive layers, $\Phi \rightarrow \Phi + \pi$:

(i) For a general (θ_B) orientation of the magnetic field with respect to the helical axis, the magnetic field induced soliton lattice is always of the π - symmetry , which is in contrast to the $smectic-C^*$ ferroelectric phase.

(ii) Molecular diffusion between smectic layers, if present, can result in a motional averaging of the deuteron EFG tensor of two successive layers. This effect should not be important in the ferroelectric $smectic-C^*$ phase where the molecular orientation varies only slowly in space and the averaged deuteron EFG tensor of two successive layers is practically identical to the one of a single layer.

The molecular structure of chiral $\alpha-CD_2$-deuterated antiferroelectric liquid crystal MHPOBC is shown in Fig.7.6.2.

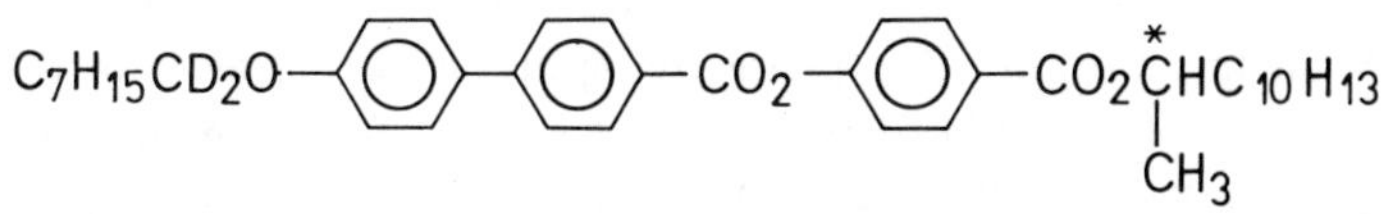

$$C_7H_{15}CD_2O-\langle\bigcirc\rangle-\langle\bigcirc\rangle-CO_2-\langle\bigcirc\rangle-CO_2\overset{*}{C}H(CH_3)C_{10}H_{13}$$

Fig.7.6.2. Molecular structure of α -CD_2 deuterated MHPOBC

On cooling from the isotropic phase the system makes first a transition into the $smectic-A$ phase at 423 K and then into the antiferroelectric $smectic-C_A^*$ phase at 390.3 K. Between the $smectic-A$ and $smectic-C_A^*$ phases there are three narrow phases $smectic-C_\alpha^*$, $smectic-C^*$ and $smectic-C_\gamma^*$.

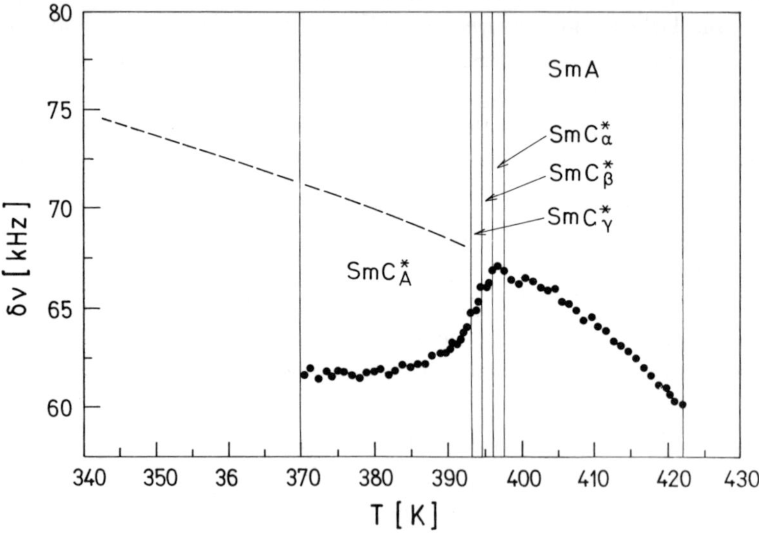

Fig.7.6.3. Temperature dependence of the splitting of the deuteron NMR spectra of MHPOBC at $\theta_B = 0$ in the $smectic-A$, $smectic-C_\alpha^*$, $smectic-C^*$, $smectic-C_\gamma^*$ and $smectic-C_A^*$ phases. The dotted line shows the extrapolated splitting in the absence of a molecular tilt (Abramič, 1995).

The temperature dependence of the quadrupole perturbed deuteron NMR line splitting at $\theta_B = 0$ is shown in Fig.7.6.3. for the $smectic-A$ and $smectic-C_A^*$ phases. In the $smectic-A$ phase the splitting increases with decreasing temperature in view of the increase in the nematic order parameter S according to Eq.7.2.8. In the $smectic-C_A^*$ phase the splitting decreases with decreasing temperature due to the tilting of the molecules away from the direction of the external magnetic field similarly as in the ferroelectric $smectic-C^*$ phase.

$$\frac{\Delta\nu}{\Delta\nu_{SmA}} = \left(3\cos^2\theta - 1\right), \quad \theta_B = 0^\circ \tag{7.6.2}$$

The angular (θ_B) dependence of deuteron quadrupole splitting in the $smectic-A$ phase of antiferroelectric MHPOBC is similar as in the $smectic-A$ phase of ferroelectric CE-8 (see Fig.7.3.7). The splitting between the two lines first decreases with increasing θ_B, becomes zero at the magic angle $\theta_B = 54^\circ$ and then increases as $\theta_B \to 90^\circ$. The fit to expression 7.2.8. yields a zero tilt angle, $\theta = 0$, as expected.

The surprise comes in the antiferroelectric $smectic-C_A^*$ phase. Deuteron NMR spectra in the antiferroelectric $smectic-C_A^*$ phase are shown in Fig.7.6.4. for different angles between the magnetic field and the helical axis. The corresponding angular (θ_B) dependence of the quadrupole splitting in the antiferroelectric $smectic-C_A^*$ phase of MHPOBC is shown in Fig.7.6.5 and can be fitted with a zero tilt angle $(\theta = 0)$. This means that from the point of view of the deuteron NMR experiment, the antiferroelectric herringbone structure is similar to the non-tilted $smectic-A$ phase. The largest principal axis (V_{ZZ}) of the effective deuteron EFG tensor is normal to the smectic layers and not tilted away from this direction as expected in the basis of the structure of the antiferroelectric $smectic-C_A^*$ phase and the observed decrease in the splitting (Fig.7.6.4.) at the $smectic-A \to smectic-C_A^*$ transition. This can be understood by the presence of rapid molecular diffusion between adjacent molecular layers followed by a molecular reorientation $\Phi_i \to \Phi_i + \pi$ which averages out the $+\theta$ tilt in one layer and the $-\theta$ tilt in the other layer (see Blinc et al., 1974). As a consequence of this averaging, the antiferroelectric $smectic-C_A^*$ structure appears identical to the $smectic-A$ structure. The condition for motional averaging is that the molecules diffuse across a distance ℓ where the resonance frequencies differ by $\delta\nu$ in a time τ which is short compared to the inverse value of the difference of the resonance frequencies:

$$\frac{\delta\nu . \ell^2}{2D} << 1 \tag{7.6.3}$$

Here D is the translational diffusion constant. In the $smectic-A$ phase, where $\theta = 0$ and the magnetic field is normal to the smectic layers $(H \parallel z)$, the $\alpha - CD_2$ deuteron EFG tensor can be expressed in the x, y, z eigenframe as:

$$\underline{V} = eq \begin{pmatrix} -1/2 & 0 & 0 \\ 0 & -1/2 & 0 \\ 0 & 0 & 1 \end{pmatrix} \tag{7.6.4}$$

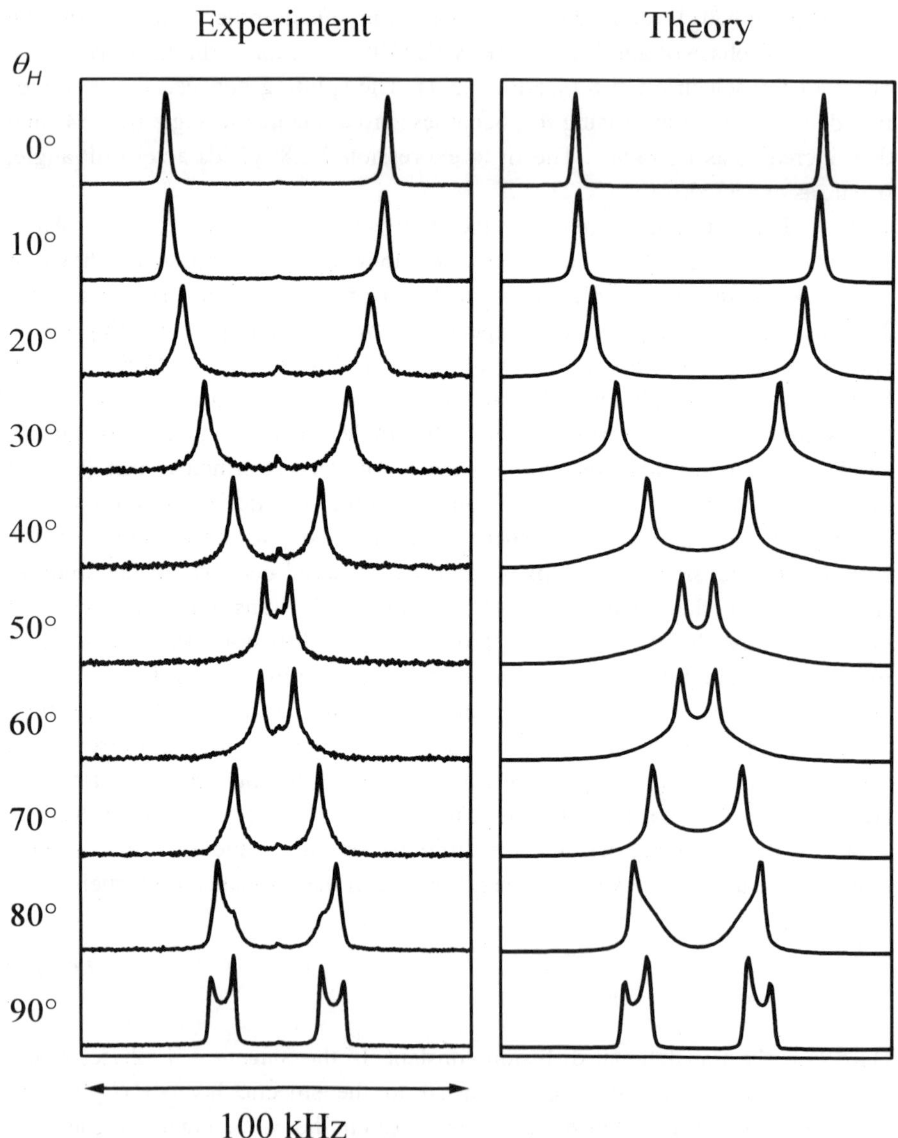

Fig.7.6.4. Deuteron NMR spectra in the antiferroelectric *smectic* $-C_A^*$ phase of MHPOBC, measured at T=380K for different orientations of the magnetic field with respect to the helical axis. (Zalar et al., unpublished).

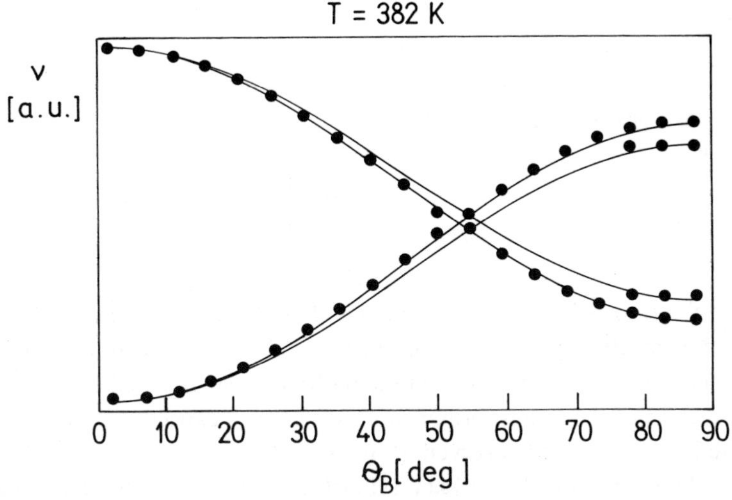

Fig.7.6.5. Comparison between the experimental and theoretical (full line) angular dependencies of the positions of the singularities in the deuteron NMR spectra of $\alpha - CD_z - MHPOBC$ in the antiferroelectric $smectic - C_A^*$ phase. The calculations are made for the tilt angle $\theta = 16^\circ$ (Zalar et al., unpublished). The solid line is the fit to Eqs.7.6.19 and 7.6.20.

Here we assume that in view of the fast molecular rotation around the long molecular axis, the largest principal axis (V_{ZZ}) is parallel to the molecular director: $V_{ZZ} \parallel z$.

Let us now tilt the molecules for $\pm\theta$ around the y - axis. The resulting EFG tensor is of the form:

$$\underline{V}_{\pm\theta} = eq\begin{vmatrix} \sin^2\theta - \frac{1}{2}\cos^2\theta & 0 & \pm\frac{3}{2}\sin\theta\cos\theta \\ 0 & 1/2 & 0 \\ \pm\frac{3}{2}\sin\theta\cos\theta & 0 & \cos^2\theta - \frac{1}{2}\sin^2\theta \end{vmatrix} \tag{7.6.5}$$

The time - averaged EFG tensor

$$\langle V\rangle = \frac{1}{2}(V_+ + V_-) \tag{7.6.6}$$

is again diagonal

$$\langle V \rangle = eq \begin{vmatrix} \sin^2\theta - \frac{1}{2}\cos^2\theta & 0 & 0 \\ 0 & -1/2 & 0 \\ 0 & 0 & \cos^2\theta - \frac{1}{2}\sin^2\theta \end{vmatrix} \tag{7.6.7}$$

As long as the tilt angle is smaller than $\theta \leq 35.26^\circ$, the largest principal axis is normal to the smectic layers (i.e. V_{ZZ}). For $\theta > 35.26^\circ$, $\langle V_{YY} \rangle$ will become larger than V_{ZZ} .The experimentally observed tilt angles are usually smaller than 35 $^\circ$ so that the direction of the largest principal axis is always expected to be normal to the smectic layers as indeed observed. (Fig.7.6.4.)

It should be noted that expression (7.6.7) also explains the observed decrease in the quadrupole splitting for $\theta_B = 0$ with decreasing temperature in the $smectic - C_A^*$ phase, which is shown in Fig.7.6.3. We namely have:

$$V_{ZZ} = eq\left(\cos^2\theta - \frac{1}{2}\sin^2\theta\right) = eq\frac{1}{2}\left(3\cos^2\theta - 1\right) \tag{7.6.8}$$

so that V_{ZZ} decreases with increasing tilt angle in agreement with expression (7.6.2).

For the sake of simplicity, expressions (7.6.5 - 7.6.7) have been derived for the phase angle $\Phi = 0$ and $H \parallel z$. For a general Φ one should not only rotate the EFG tensor (Eq.7.6.4) around the y axis for $\pm\theta$ but also perform a rotation for an angle Φ around the z-axis.

In the general case one should also make a transformation to the frame where $\vec{H}$ makes an angle θ_B with the z-direction. In this general case the time averaged value of the EFG tensor component $\langle V_{ZZ} \rangle$ is

$$\langle V_{ZZ} \rangle = e\langle q \rangle \left[\frac{1}{2}\left(3\cos^2\theta_B - 1\right) - \frac{1}{2}\eta\sin^2\theta_B\cos 2\Phi\right] \tag{7.6.9}$$

where

$$\langle q \rangle = \frac{1}{2}q\left(3\cos^2\theta - 1\right) \tag{7.6.10}$$

and the asymmetry parameter η is

$$\eta = \frac{3\sin^2\theta}{3\cos^2\theta - 1} \tag{7.6.11}$$

For a general orientation of the antiferroelectric $smectic - C_A^*$ sample in the magnetic field, the time averaged frequencies of the two deuteron quadrupole-perturbed NMR lines, which are symmetrically displaced from the Larmor frequency ν_L are

$$\nu_\pm = \nu_L \pm \frac{3}{4} \cdot S \cdot \frac{e^2 \langle q \rangle Q}{\hbar} \left[\frac{1}{2} \left(3\cos^2\theta_B - 1 \right) - \frac{1}{2} \eta \sin^2\theta_B \cos 2\Phi \right] \tag{7.6.12}$$

The fluctuations of the long molecular axis around the nematic director are taken into account by introducing the nematic order parameter S. One should note that the above expression can also be directly derived from Eq.7.2.8., obtained for the ferroelectric $smectic - C^*$ phase, if one performs the time average of expression 7.2.8. over the azimuthal angle $\Phi(t)$:

$$\Phi \to \Phi + \pi \to \sin\Phi \to -\sin\Phi \tag{7.6.13}$$

Here we assumed that the length of the antiferroelectric helical pitch is much longer than the smectic interlayer distance. In such a case the chirality induced additional small change in Φ can be neglected in the averaging process on the scale of a few interlayer distances where the condition (Eq.7.6.3) for fast diffusion is fulfilled.

In view of the chiral helicoidal modulation, we shall have also in the antiferroelectric $smectic - C_A^*$ phase an inhomogeneous frequency distribution $f(\nu)$. In view of the azimuthal dependence of the NMR frequency in the helicoidally modulated bulk sample we again have

$$f(\nu)d\nu = N(z)dz \tag{7.6.14}$$

where $N(z) = const.$ and

$$f(\nu) = \frac{const}{\left|\frac{d\nu}{d\Phi}\right| \cdot \left|\frac{d\Phi}{dz}\right|} \tag{7.6.15}$$

In the plane wave limit $d\Phi/dz = const \neq 0$. Singularities in the spectral frequency distribution $f(\nu)$ are here obtained when

$$\frac{d\nu}{d\Phi} = 0 \Rightarrow \sin(2\Phi)\sin^2\theta_B.\sin^2\theta = 0 \tag{7.6.16}$$

The corresponding solutions for Φ are

$$\Phi_I = 0, \pi .. \tag{7.6.17}$$

and

$$\Phi_{II} = \frac{\pi}{2}, \frac{3\pi}{2} .. \tag{7.6.18}$$

The frequencies of the two singularities ν_I and ν_{II} are thus:

$$\nu_I(\Phi_I = 0) = \pm\frac{3}{8}\nu_Q S\left[3\cos^2\theta_B \cos^2\theta - 1\right] \tag{7.6.19}$$

and

$$\nu_{II}\left(\Phi_{II} = \frac{\pi}{2}\right) = \pm\frac{3}{8}\nu_Q \, S\left[3(\cos^2\theta_B \cos^2\theta + \sin^2\theta_B \sin^2\theta) - 1\right] \tag{7.6.20}$$

Here we put for sake of convenience $\nu_L = 0$.

For small θ_B or small tilt angle θ, the two singularities practically coincide. They will be clearly separated only at large θ_B values, i.e. close to 90° and at large enough tilt angles. This is indeed seen in the experiments, as shown in Fig.7.6.5. which agree rather well with the above theory. Deuteron NMR thus provides for an independent microscopic verification of the alternating tilt model of the antiferroelectric $smectic-C_A^*$ phase.

It should be noted that expressions (7.6.19-20) are rather different from the expression for the singularities in the ferroelectric $smectic-C^*$ phase. In particular, in the angular (θ_B) dependence of the positions of the two singularities, there is no phase angle Φ in the expression $[3\cos^2(\theta_B \pm \theta)-1]$ which would distinguish the θ_B dependence in the $smectic-A$ and $smectic-C_A^*$ phases. This is of course a consequence of the fact that in the antiferroelectric $smectic-C_A^*$ phase, the effective direction of the largest principal axis of the motionally averaged EFG tensor is normal to the smectic layers, i.e. it points in the same direction as in the $smectic-A$ phase.

Another important difference between the antiferroelectric $smectic-C_A^*$ and ferroelectric $smectic-C^*$ phases is that the third singularity at $\nu_0 = \pm\frac{3}{8}\nu_Q S$, which appears in the $smectic-C^*$ phase for $tg\theta > ctg\theta_B$ is absent in the $smectic-C_A^*$ phase if motional averaging takes place. It should also be noted that the angular dependence in Fig.7.6.5. would be completely different if the angle Φ between the tilt directors in successive layers would not be 180°. This allows for an independent and direct test of the structure of the antiferroelectric $smectic-C_A^*$, and also $smectic-C_\alpha^*$ and $smectic-C_\gamma^*$ phases.

Finally it should be mentioned that a comparison between experimental and theoretical NMR lineshapes in the $smectic-C_A^*$ phase (Fig.7.6.4a-b) also allows for a determination of the inter-layer component of the mass self-diffusion tensor. This component amounts in the $smectic-C_A^*$ phase to $D_\perp \approx 4\times10^{-9} cm^2 s^{-1}$ and is by nearly two orders of magnitude smaller than in the ferroelectric $smectic-C^*$ phase, where all layers are tilted in the same direction. This result disagrees with forced Rayleigh scattering studies (Moriyama et al., 1995) of dye diffusion in the $smectic-C^*$ and racemic $smectic-C_A$ phases. It shows that the diffusion of liquid crystal molecules in the $smectic-C_A$ phase is much slower than that of dye molecules.

7.7. NMR in the Soliton-like deformed Antiferroelectric Phase

In order to determine the spatial variation of the long pitch chiral phase profile $\Phi(z)$ in the antiferroelectric $smectic-C_A^*$ phase in the presence of a magnetic field, let us first evaluate the magnetic free energy density averaged over the two $smectic-C_A^*$ layers:

$$\langle g_H \rangle = -\tfrac{1}{2}\Delta\chi < (\overline{H}\overline{n})^2 > = -\frac{1}{2}\Delta\chi H^2 \sin^2\theta\left(\sin^2\theta_B \sin^2\Phi + \cos^2\theta_B . ctg^2\theta\right) \tag{7.7.1}$$

The second term in this expression $\cos^2\theta_B . ctg^2\theta$ does not depend on Φ and can be therefore disregarded when discussing the effects of the soliton structure. The relevant part of $\langle g_H \rangle$ for a general θ_B is thus

$$\langle g_H \rangle^{SmC_A^*} = -\frac{1}{2}\Delta\chi H^2 \sin^2\theta_B \sin\theta \sin^2\Phi \tag{7.7.2}$$

Let us now compare the above expression with the corresponding magnetic free energy density of a ferroelectric $smectic-C^*$ layer:

$$\langle g_H \rangle^{SmC^*} = -\frac{1}{2}\Delta\chi H^2 \sin^2\theta\left(\sin^2\theta_B \sin^2\Phi + 2\sin\theta_B \cos\theta_B \sin\Phi . ctg\theta\right) \tag{7.7.3}$$

For $\theta_B = 90^\circ$, i.e. for the magnetic field applied parallel to the smectic layers, Equation 7.7.3 becomes

$$\langle g_H \rangle^{SmC^*} = -\frac{1}{2}\Delta\chi H^2 \sin^2\theta \sin^2\Phi \tag{7.7.4}$$

This is identical to $\langle g_H \rangle^{SmC_A^*}$, Eq.7.7.2. if we introduce an effective magnetic field

$$H_{eff} = H \sin\theta_B \tag{7.7.5}$$

One can thus see that the soliton structure $\Phi = \Phi(z)$ of the antiferroelectric $smectic - C_A^*$ phase for a general (θ_B) orientation of the magnetic field with respect to the helix resembles that of the ferroelectric $smectic - C^*$ phase for $\theta_B = 90^\circ$, if the applied magnetic field H in the $smectic - C_A^*$ phase is replaced by an effective magnetic field $H_{eff} = H \sin\theta_B$. This is true as

$$\langle g_H \rangle^{SmC_A^*}(H, \theta, \theta_B, \Phi) = \langle g_H \rangle^{SmC^*}(H \sin\theta_B, \theta, \pi/2, \Phi) \tag{7.7.6}$$

In contrast to the ferroelectric $smectic - C^*$ phase where the soliton lattice has 2π - symmetry for a general $\theta_B \neq 90^\circ$, it always has a π - symmetry in the antiferroelectric $smectic - C_A^*$ phase, regardless of the direction of the field. We therefore always have π -soliton lattice, irrespective of the orientation of the magnetic field with respect to the smectic layers. It should be noted that the above considerations assume that the basic alternating tilt structure of the $smectic - C_A^*$ phase is not affected by the external magnetic field. It is only the long pitch chiral modulation which becomes distorted in the presence of the external magnetic field.

The soliton structure does not introduce any additional singularities into the deuteron NMR spectrum of the $smectic - C_A^*$ phase as $d\Phi / dz = 0$ at $\Phi = \pi/2$, $3\pi/2$,..., i.e. at the same azimuthal angles where already $d\nu / d\Phi = 0$. The soliton like structure of the $smectic - C_A^*$ phase in an external magnetic field, applied at an arbitrary direction with respect to the helical axis is presented in Fig.7.7.1.

In conclusion, the angular dependence of the critical field for the unwinding of the antiferroelectric $smectic - C_A^*$ helix can be thus obtained from the calculations that have been performed for the ferroelectric $smectic - C^*$ phase by replacing H with $H \sin\theta_B$.

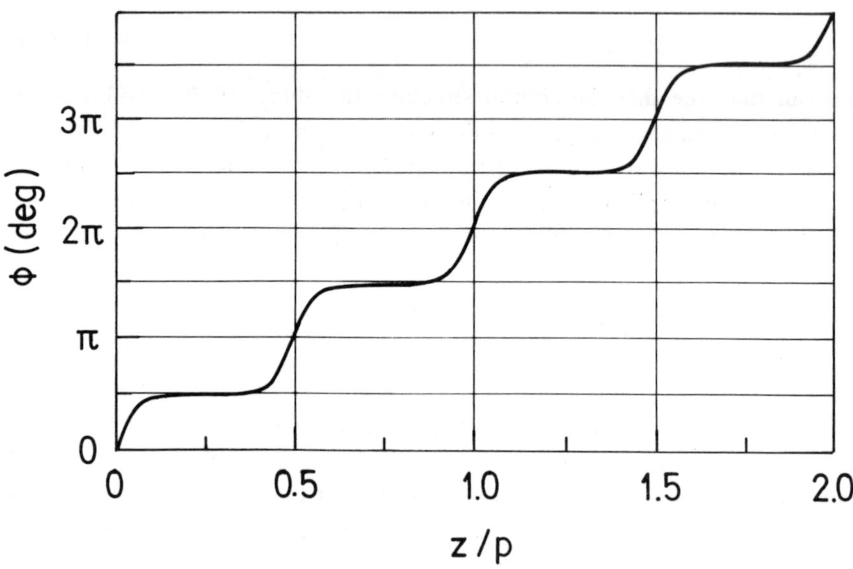

Fig.7.7.1. Soliton like distortion of the long pitch chiral phase profile of an antiferroelectric $smectic-C_A^*$ liquid crystal in an external magnetic field for an arbitrary orientation of the magnetic field with respect to the helical axis. The magnetic field is close to the critical magnetic field for the unwinding of the helix, $H/H_c = 0.99$. Note the π-soliton like structure.

7.8. ^{14}N Nuclear Quadrupole Resonance in Tilted Smectic Phases

The two questions we wish to answer by ^{14}N (I=1) Nuclear Quadrupole Resonance (NQR) in smectic phases are:

> *(i)* Is there an ordering of the transverse molecular axes in the *smectic − C* and other tilted smectic phases?
>
> *(ii)* What is the nature of this ordering of the short molecular axes, if it exists at all? How large is the quadrupolar order and how large is the polar ordering?

The possible biasing of the rotation of the outboard transverse molecular dipoles (i.e. the C-N=CH-C groups) around the long molecular axis in achiral TBBA and chiral TBACA has been studied via a determination of the asymmetry parameter η

$$\eta = \frac{V_{XX} - V_{YY}}{V_{ZZ}} \tag{7.8.1}$$

of the electric field gradient (EFG) tensor at the ^{14}N sites (see, for example Blinc et al., 1983, Seliger et al., 1977). For free rotation around the long molecular axis, the largest principal component, V_{ZZ} will point along the long molecular axis and the EFG tensor will be axially symmetric:

$$V_{XX} = V_{YY},\ \eta = 0. \tag{7.8.2}$$

Here $|V_{ZZ}| \geq |V_{YY}| \geq |V_{XX}|$ represent the principal components of the EFG tensor. Any anisotropy or biasing of the rotation will destroy the axial symmetry so that $\eta \neq 0$. This leads to the appearance of three ^{14}N NQR lines:

$$\nu_{1.2} = (3/4)(eQV_{ZZ}/h)(1 \pm \eta/3). \tag{7.8.3a}$$

$$\nu_3 = \nu_2 - \nu_1 \tag{7.8.3b}$$

instead of a single line, $\nu_1 = \nu_2, \nu_3 = 0$. Here $eQV_{zz}/h = e^2Qq/h$ is the ^{14}N quadrupole coupling constant. The biasing of the rotation can therefore be expressed in frequency units. It should be stressed that the ^{14}N quadrupole coupling constants are generally an order of magnitude larger than the deuteron quadrupole coupling constants. ^{14}N experiment is therefore more sensitive to polar or quadrupolar biasing of the rotation than quadrupole perturbed deuteron NMR.

Let us now discuss this problem in some detail. If the correlation frequency of the molecular motion is much larger than the ^{14}N NQR frequencies, which is usually the case in liquid crystals, the observed value of the EFG tensor represents a ***time average*** over the molecular motion. This value of the EFG tensor is smaller than the rigid lattice value observed in solids. The central problem therefore concerns the nature of the molecular motion producing a given time average. Some qualitative facts on the molecular motion can be immediately obtained from the measured values of e^2qQ/h and η. A comparison between the observed and rigid-lattice values of e^2qQ/h characterizes the degree of motional averaging and reflects the amplitudes of the molecular motion. The asymmetry parameter η is zero for unhindered rotation around the long molecular axis, while a non-zero value indicates a deviation from cylindrical symmetry at the nucleus and thus an anisotropy in the molecular motion.

In order to be able to quantitatively describe the molecular motion, the relations which are linking the values of the quadrupole coupling constant e^2qQ/h and η to the order parameters of liquid crystals should be properly deduced. In the rigid crystal the ^{14}N EFG tensor can be written (Seliger et al., 1978) in the molecular frame $x_\circ, y_\circ, z_\circ$ as

$$\underline{V}_{ij} = \begin{pmatrix} V_{x_\circ x_\circ} & V_{x_\circ y_\circ} & V_{x_\circ z_\circ} \\ V_{x_\circ y_\circ} & V_{y_\circ y_\circ} & V_{y_\circ z_\circ} \\ V_{x_\circ z_\circ} & V_{y_\circ z_\circ} & V_{z_\circ z_\circ} \end{pmatrix} \tag{7.8.4}$$

The molecular frame is normally chosen with the $z_\circ$ axis parallel to the long axis of the molecule. The choice of the $y_\circ$ direction depends on the particular molecular group under study. The $x_\circ$ axis is perpendicular to $y_\circ$ and $z_\circ$ so that $x_\circ, y_\circ, z_\circ$ represent an orthogonal right-handed system.

If the observed ^{14}N nucleus belongs to the nearly planar C-N=CH-C linkage group, as in the present case of TBBA and TBACA , it is convenient to choose $y_\circ$ normal to this plane. It is known from work on other Schiff 's bases that the largest principal axis of the ^{14}N EFG tensor lies then in the $x_\circ z_\circ$ plane and makes an angle of about 60° with the $z_\circ$ axis, whereas $V_{x_\circ y_\circ}$ and $V_{y_\circ z_\circ}$ should be small. The exact values of the V_{ij} tensor can be determined from a complete set of single-crystal quadrupole perturbed NMR rotation data. If the five independent EFG tensor components in the molecular frame are known and a model for the molecular motion is assumed, the components of the time averaged EFG tensor can be evaluated (Seliger et al., 1978). The principal values of the time averaged tensor, obtained by diagonalization, provide the relations between the observed values of e^2qQ/h and η on one side and the order parameters on the other.

In the smectic phases the molecular rotation around the long axis is expected to govern the time averaging of the EFG tensor at ^{14}N sites. In some phases, the fluctuations in the direction of the long axis can be significant as well. Let us assume that a molecule in the smectic phase performs a hindered rotation around the long molecular axis, i.e. it jumps between several equilibrium orientations. To take into account this motion, the EFG tensor Eq.7.8.4 is transformed from the molecular $x_\circ, y_\circ, z_\circ$ frame into a frame x, y, z, where $z \| z_\circ$ is the axis of rotation and $\psi = \psi(t)$ is the angle between x and $x_\circ$ (see Fig.7.8.1 in the following text). In the x, y, z frame the EFG tensor is

$$V_{xx} = -(1/2)V_{z_\circ z_\circ} + (1/2)(V_{x_\circ x_\circ} - V_{y_\circ y_\circ})\cos 2\psi - V_{x_\circ y_\circ} \sin 2\psi \quad (7.8.5a)$$

$$V_{xy} = (1/2)(V_{x_\circ x_\circ} - V_{y_\circ y_\circ})\sin 2\psi + V_{x_\circ y_\circ} \cos 2\psi \quad (7.8.5b)$$

$$V_{xz} = V_{x_\circ z_\circ} \cos\psi - V_{y_\circ z_\circ} \sin\psi \quad (7.8.5c)$$

$$V_{yy} = -(1/2)V_{z_\circ z_\circ} - (1/2)(V_{x_\circ x_\circ} - V_{y_\circ y_\circ})\cos 2\psi + V_{x_\circ y_\circ} \sin 2\psi \quad (7.8.5d)$$

$$V_{yz} = V_{x_\circ z_\circ} \sin\psi + V_{y_\circ z_\circ} \cos\psi \tag{7.8.5e}$$

$$V_{zz} = V_{z_\circ z_\circ}. \tag{7.8.5f}$$

When the expressions 7.8.5a-f are averaged over the molecular motion, we have to replace $\cos\psi$ by the time-average value, i.e. $\langle\cos\psi\rangle$ etc. The mean values $\langle\cos\psi\rangle$, $\langle\cos 2\psi\rangle$, etc. play the role of order parameters which are characteristic for the given type of orientational ordering.

In the case when there is ***no biasing of molecular rotation***, all equilibrium orientations which are involved in the molecular rotation around the long axis are equivalent, and the above order parameters $\langle\cos\psi\rangle$, $\langle\cos 2\psi\rangle$, etc., equal zero

$$\langle\cos\psi\rangle = \langle\cos 2\psi\rangle = \langle\sin\psi\rangle = \langle\sin 2\psi\rangle = 0\,. \tag{7.8.6}$$

The largest eigenvalue of the time averaged EFG tensor is in this case $V_{ZZ} = \langle V_{zz}\rangle = V_{z_\circ z_\circ}$. This yields a temperature independent quadrupole coupling constant e^2qQ/h and a zero value of the asymmetry parameter $\eta = 0$.

In the case of a ***molecular rotation with polar biasing***, the effective potential "seen" by each molecule has the form $V = -V'(\cos\psi)$. The resulting orientational ordering is polar with characteristic order parameters $\langle\cos\psi\rangle \neq 0$, $\langle\cos 2\psi\rangle \neq 0$, and $\langle\sin\psi\rangle = 0$, $\langle\sin 2\psi\rangle = 0$. The values of $\langle\cos\psi\rangle$ and $\langle\cos 2\psi\rangle$ can be expressed in terms of modified Bessel functions as (see Blinc et al., 1983)

$$\langle\cos\psi\rangle = I_1(a)/I_0(a),\; a = \frac{V'}{k_B T} \tag{7.8.7a}$$

$$\langle\cos 2\psi\rangle = I_2(a)/I_0(a),\; a = \frac{V'}{k_B T}\,. \tag{7.8.7b}$$

If the orientational ordering is weak, i.e. if $V' << k_B T$, the Bessel functions may be expanded and we obtain:

$$\langle \cos 2\psi \rangle \approx (1/2)\langle \cos\psi \rangle^2 << \langle \cos\psi \rangle. \tag{7.8.8}$$

The only relevant order parameter for weak polar ordering is thus $\langle \cos \Psi \rangle$, whereas $\langle \cos 2\psi \rangle$ can be neglected. The resulting values for e^2qQ/h and η are:

$$e^2qQ/h = (eQV_{z_\circ z_\circ}/h)\left[1 + (2/3)\left(\left(V_{x_\circ z_\circ}{}^2 + V_{y_\circ z_\circ}{}^2\right)/V_{z_\circ z_\circ}{}^2\right)\langle \cos\psi \rangle^2\right] = eQV_{z_\circ z_\circ}(1+\eta),$$

$$\eta = (2/3)\left[\left(V_{x_\circ z_\circ}{}^2 + V_{y_\circ z_\circ}{}^2\right)/V_{z_\circ z_\circ}{}^2\right]\langle \cos\psi \rangle^2. \tag{7.8.9}$$

We thus see that the asymmetry parameter η is different from zero for biased rotation. The temperature dependence of η reflects the temperature change of the square of the polar order parameter $\langle \cos\psi \rangle^2$.

Let us now consider the effects of uniaxial rotation with quadrupolar biasing. This quadrupolar orientational ordering takes place in a potential of the form $V = -V'(\cos 2\psi)$. In this case $\langle \cos 2\psi \rangle \neq 0$, $\langle \cos\psi \rangle = 0$, $\langle \sin\psi \rangle = 0$ and $\langle \sin 2\psi \rangle = 0$. If the quadrupolar orientational ordering is weak, $\langle \cos 2\psi \rangle << 1$, the largest principal axis of the ^{14}N EFG tensor still points along the axis of rotation $(z_\circ)$. The quadrupole coupling constant and the asymmetry parameter are:

$$eQV_{zz}/h = eQV_{z_\circ z_\circ}/h, \quad \eta = r\langle \cos 2\psi \rangle, \tag{7.8.10}$$

where

$$r = 2\sqrt{V_{x_\circ y_\circ}{}^2 + (1/4)\times\left(V_{x_\circ x_\circ} - V_{y_\circ y_\circ}\right)^2} \,/\, V_{z_\circ z_\circ}. \tag{7.8.11}$$

We see that η depends linearly on $\langle\cos 2\psi\rangle$ and hence is temperature dependent, whereas e^2qQ/h is constant. For large values of $\langle\cos 2\psi\rangle$, the largest principal axis of the ^{14}N EFG tensor no longer points along the axis of rotation. For the nitrogen atom in the C-CH=N-C group it lies in the $x_\circ z_\circ$ plane. For $r\langle\cos 2\psi\rangle > 1$ we find:

$$eQV_{zz}/h = -(1/2h)eQV_{z_\circ z_\circ}\left(1 + r\langle\cos 2\psi\rangle\right) \tag{7.8.12}$$

and

$$\eta = \frac{\left|r\langle\cos 2\psi\rangle - 3\right|}{\left|r\langle\cos 2\psi\rangle + 1\right|}, \tag{7.8.13}$$

so that both e^2qQ/h and η depend on $\langle\cos 2\psi\rangle$ and are temperature dependent.

Besides the motional averaging due to individual molecular rotation, additional averaging due to collective director fluctuations is always present in liquid crystalline phases. The fluctuations in the direction of the long molecular axis are taken into account by performing an additional transformation of the EFG tensor given by equation 7.8.5 to a fixed x', y', z' frame, where z' is normal to the smectic planes, x' is the direction of the average projection of the long molecular axis on the smectic plane, and y' is perpendicular to x' and z'. The angles $\theta = \theta(t)$ and $\Phi = \Phi(t)$ are the polar and azimuthal angles, respectively, which determine the direction of the long molecular axis in the x', y', z' frame. $\theta(t)$ can be written as $\theta(t) = \theta_\circ + \delta\theta(t)$, where θ denotes the tilt angle and is shown in Figure 7.8.1.

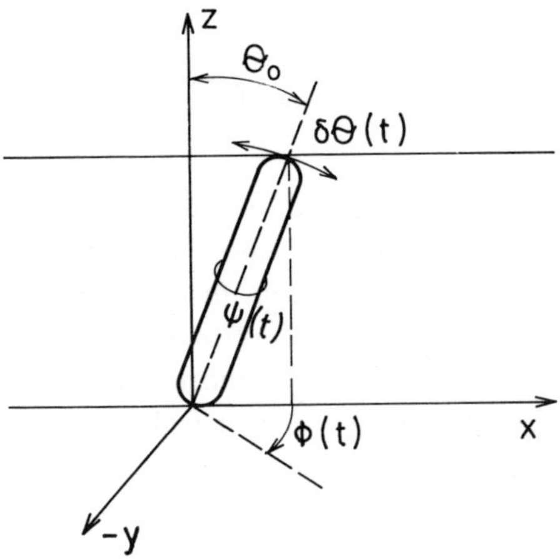

Fig.7.8.1 Schematic presentation of a tilted molecule within a smectic layer. $\theta_{\circ}$ is the average tilt angle, $\delta\theta(t)$ denotes the fluctuations in the orientation of the long molecular axis in the polar direction, and $\Phi(t)$ the fluctuations in the azimuthal direction. $\psi(t)$ denotes the orientation of the short molecular axes during the molecular rotation around its long axis.

After this second transformation has been performed the time averaged values of the ^{14}N EFG tensor are:

$$\langle V_{x'x'}\rangle = \langle (V_{zz} - V_{xx})\sin^2\theta\cos^2\Phi\rangle + \langle V_{xx}\cos^2\Phi\rangle + \langle V_{yy}\sin^2\Phi\rangle + \langle V_{xz}\sin 2\theta\cos^2\Phi\rangle \tag{7.8.14a}$$

$$\langle V_{x'y'}\rangle = \langle V_{yz}\sin\theta\cos 2\Phi\rangle + \langle V_{xy}\cos\theta\cos 2\Phi\rangle \tag{7.8.14b}$$

$$\langle V_{x'z'}\rangle = \langle (V_{zz} - V_{xx})\sin\theta\cos\theta\cos\Phi\rangle + \langle V_{xz}\cos 2\theta\cos\Phi\rangle \tag{7.8.14c}$$

$$\langle V_{y'y'}\rangle = \langle (V_{zz} - V_{xx})\sin^2\theta\sin^2\Phi\rangle + \langle V_{xx}\sin^2\Phi\rangle + \langle V_{yy}\cos^2\Phi\rangle + \langle V_{xz}\sin 2\theta\sin^2\Phi\rangle \tag{7.8.14d}$$

$$\langle V_{y'z'} \rangle = -\langle V_{xy} \sin\theta \cos\Phi \rangle + \langle V_{yz} \cos\theta \cos\Phi \rangle \tag{7.8.14e}$$

$$\langle V_{z'z'} \rangle = -\langle (V_{zz} - V_{xx}) \sin^2\theta \rangle + \langle V_{zz} \rangle - \langle V_{xz} \sin 2\theta \rangle. \tag{7.8.14f}$$

In this set of equations a large number of different order parameters appear. Those that are functions of ψ describe the order around the long molecular axis, whereas those that are functions of θ and Φ describe the ordering of the long molecular axis itself. The algebraic relations between e^2qQ/h and η and the values of the EFG tensor in the x', y', z' frame are rather complicated. If, however, the fluctuations in θ, Φ and ψ are assumed to be independent and are small enough to allow the expansion of the sine and cosine of θ and Φ in a power series, relatively simple expressions for e^2qQ/h and η can be obtained. For the case of weak bipolar ordering we find:

$$e^2qQ \approx eQV_{z_\circ z_\circ}\left[1 - (3/2)\langle \delta\theta^2 \rangle - (3/2)\theta_\circ^2\left(1 - \langle \cos\Phi \rangle^2\right)\right] \tag{7.8.15}$$

and

$$\begin{aligned} \eta \approx\ & (3/2)\theta_\circ^2\left(\langle \cos\Phi \rangle^2 - \langle \cos 2\Phi \rangle\right) - (3/2)\langle \delta\theta^2 \rangle\langle \cos 2\Phi \rangle - \\ & - \left[\left(V_{x_\circ x_\circ} - V_{y_\circ y_\circ}\right)/V_{z_\circ z_\circ}\right]\cdot\langle \cos 2\Phi \rangle\langle \cos 2\psi \rangle \end{aligned} \tag{7.8.16}$$

Equation (7.8.16) shows the combined effects of both bipolar biasing and anisotropic fluctuations in the direction of the long axis on the values of e^2qQ/h and η. In the first approximation e^2qQ/h does not depend on the orientational ordering $\langle \cos 2\psi \rangle$ but is mainly determined by the mean square fluctuations in the polar angle $\langle \delta\theta^2 \rangle$. In contrast to e^2qQ/h, η does depend on the orientational order parameter $\langle \cos 2\psi \rangle$ through the last term in equation (7.8.16). This term is however rather small in the *smectic* $-C$ phase for which the above model is appropriate. The prevailing term which produces the characteristic maximum in the temperature dependence of η is

$(3/2)\langle\delta\theta^2\rangle\langle\cos 2\Phi\rangle$. In the low temperature part of the *smectic* − *C* phase, $\langle\cos 2\Phi\rangle \approx 1$ and η increases with increasing temperature due to the increase in $\langle\delta\theta^2\rangle$. At higher temperatures, η decreases as $\langle\cos 2\Phi\rangle \to 0$, so that there has to be a maximum in between.

Anisotropic fluctuations in the direction of the long molecular axis by themselves result in a nonzero value of η though there is no biasing of the rotation around the long axis. In this case we find:

$$e^2qQ = eQV_{z^\circ z^\circ}\left[1 + (3/2)\left(\langle\sin\theta\ \cos\theta\rangle^2 - \langle\cos^2\theta\rangle\langle\sin^2\theta\ \cos^2\Phi\rangle - \langle\sin^2\theta\ \sin^2\Phi\rangle\right)\right. \tag{7.8.17}$$

and

$$\eta = (3/2)\left(\langle\sin\theta\ \cos\theta\rangle^2 - \langle\cos^2\theta\rangle\langle\sin^2\theta\ \cos^2\theta\rangle + \langle\sin^2\theta\ \sin^2\Phi\rangle\right) \tag{7.8.18}$$

The two different mechanisms which are leading to a finite value of η, i.e. orientational ordering and anisotropic fluctuations in the long molecular axis, can be easily discriminated in view of the different temperature dependencies as reflected in $\eta(T)$and $e^2qQ/$.

A different approach to the problem of relations among e^2qQ/h, η, and order parameters has been developed by Allender and Doane (1978). They obtained the expression for η, without assuming a specific model for molecular motion and evaluated η for the general case and for the limiting cases of uniform rotation or biased rotation without fluctuations.

It should be stressed that it is not possible to determine by NQR the values of all possible different order parameters as there are at each temperature only two experimental parameters available (e^2qQ/h and η). The problem is often simplified by additional information on the symmetry of the molecular motion deduced from X-ray or other experiments.

Let us now briefly review some of the experimental results, starting from the achiral tilted smectic phase of terephtal-bis-butylaniline (TBBA). This liquid crystal material has been one of the most extensively studied smectic liquid crystals by NMR. The structure and the sequence of the smectic phases in this system are shown in Fig.7.8.2. TBBA possesses two equivalent nitrogen nuclei in the central part of the molecule, so that the conclusions drawn from ^{14}N NQR data will concern the orientational ordering of the central molecular part or more precisely the ordering of the C-CH=N-C linkage group. As the ordering of the C=N electric dipoles in the linkage group is of prime importance in the microscopic theory of the smectic phases, ^{14}N is an excellent probe for this study. The ordering of the molecular tails will not be discussed at present.

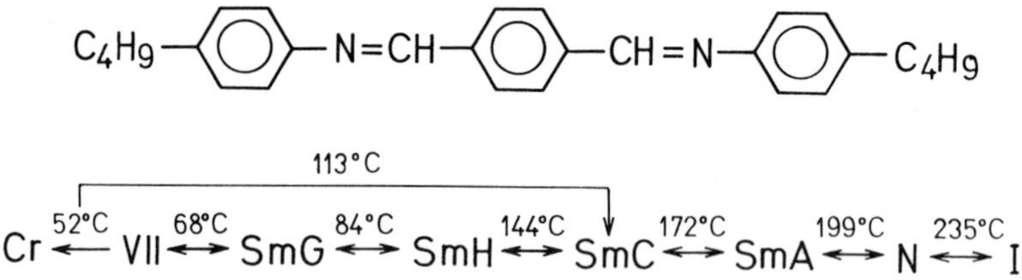

C_4H_9—⟨○⟩—N=CH—⟨○⟩—CH=N—⟨○⟩—C_4H_9

113°C (Cr → SmC)

Cr $\xleftarrow{52°C}$ VII $\xleftrightarrow{68°C}$ SmG $\xleftrightarrow{84°C}$ SmH $\xleftrightarrow{144°C}$ SmC $\xleftrightarrow{172°C}$ SmA $\xleftrightarrow{199°C}$ N $\xleftrightarrow{235°C}$ I

Fig.7.8.2. The molecular structure and liquid crystalline phases of TBBA.

The measurements of e^2qQ/h and η in TBBA (Seliger et al., 1977, 1978, Blinc et al., 1979) are presented in Figure 7.8.3. One can see that η is zero in the uniaxial $smectic-A$ phase. It is small, but non zero, in the $smectic-C$ phase and exhibits a maximum in the middle of this phase. On going to the $smectic-H$ phase, η discontinuously increases and reaches a very high value of about 0.7 at the low temperature end of this phase. Proceeding to the smectic G phase η abruptly decreases to about 0.18, and then slowly increases with decreasing temperature. In the solid phase the value of η is 0.28 and is practically temperature independent.

The quadrupole coupling constant shows a completely different temperature behavior. It increases slowly and continuously with decreasing temperature throughout the $smectic-A$, $smectic-C$ and $smectic-H$ phases. On going to the $smectic-G$ phase, e^2qQ/h discontinuously increases by nearly 300%. In spite of

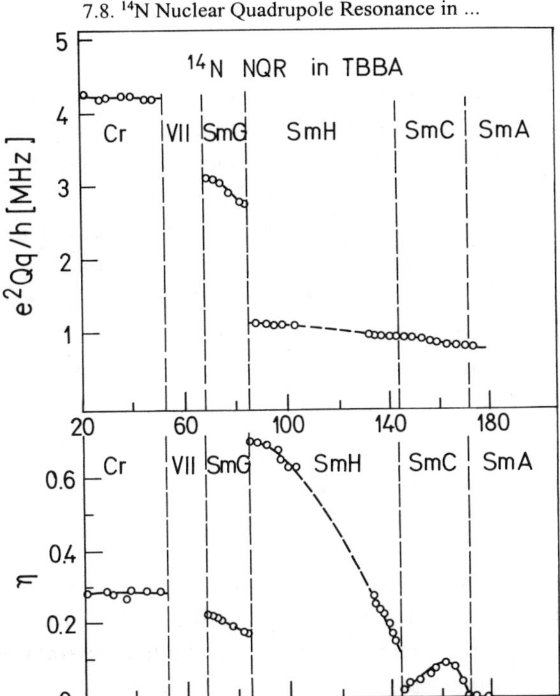

Fig.7.8.3. Temperature dependence of the ^{14}N quadrupole coupling constant e^2qQ/h and asymmetry parameter η in the smectic and crystalline phases of TBBA (Blinc et al., 1983).

this huge jump, the value of e^2qQ/h at the end of the *smectic* $-G$ phase (3.28 MHz) is still significantly smaller than in the solid (4.22 MHz). This demonstrates that motional averaging is still present. In the solid state the quadrupole coupling constant and η are temperature independent over the range +50° to -20° C, which suggests that the orientational order is saturated.

A comprehensive analysis of the experimental data in the *smectic* $-C$, *smectic* $-H$ and *smectic* $-G$ phases shows that the measured values of e^2qQ/h and η, and, in particular the large and temperature dependent value of η in the *smectic* $-H$ phase, are not compatible with either free rotation of the molecule around its long axis or polar ordering where one position would be preferred. The data are consistent with a quadrupolar ordering with $\langle \cos 2\psi \rangle \neq 0$, $\langle \cos\psi \rangle = 0$, $\langle \sin\psi \rangle = 0$ and $\langle \sin 2\psi \rangle = 0$. The physical model for such motion, which is in agreement with NQR (Seliger et al., 1977, 1978, Blinc et al., 1979) and neutron scattering data (Dianoux and Volino, 1979), is as follows. Let us assume that the rigid "body" of the TBBA molecule is reorienting in a six-well potential around the long molecular axis. Four equilibrium sites are equivalent and have an occupation probability p_2 whereas the other two, which are separated by 180° have a lower energy and an occupation probability $p_1 > p_2$. The quadrupolar order parameter is now given by

$$\langle \cos 2\psi \rangle = 2(p_1 - p_2), \quad 2p_1 + 4p_2 = 1 \tag{7.8.19}$$

The degree of orientational ordering, which is given by the values of $\langle \cos 2\psi \rangle$ for ordering of the short molecular axis and by $\langle \cos \Phi \rangle$ and S for "nematic" ordering of the long molecular axis, increases from the high temperature smectic phases towards the solid crystalline state.

The zero value of η in the $smectic-A$ phase shows that the molecules are freely rotating in this phase, or reorient between six equivalent potential wells around their long axes. The uniaxiality of the $smectic-A$ phase is thus not an "collective- averaging" phenomenon but also holds for each individual molecule. The fluctuations in the direction of the long axis are isotropic. Their magnitude $\langle \delta\theta^2 \rangle^{1/2} \approx \langle \sin^2 \delta\theta \rangle^{1/2}$ can be straightforwardly determined from the quadrupole coupling constant through the relation

$$e^2 qQ / h = (eqV_{z_\circ z_\circ} / h) \times \left[1 - (3/2)\langle \delta\theta^2 \rangle\right] \tag{7.8.20}$$

which follows from equations (7.8.16) if it is taken into account that in the $smectic-A$ phase $\eta = 0$, $\langle \cos\Phi \rangle = \langle \cos 2\Phi \rangle = \langle \cos 2\psi \rangle = \langle \cos\psi \rangle = 0$, $\theta_\circ = 0$ and $\langle \delta\theta^2 \rangle \neq 0$. In TBBA, the value of $V_{z_\circ z_\circ}$ equals 1.17 MHz in frequency units. This yields $\langle \delta\theta^2 \rangle^{1/2} \approx 23^\circ$ at 175° C and corresponds to the nematic order parameter

$S = 0.75$ for C-N=CH-C groups. This value is in excellent agreement with the results that were obtained from neutron scattering and nuclear magnetic resonance experiments (Dianoux and Volino, 1979).

In the biaxial $smectic-C$ phase the non-zero value of η demonstrates the occurrence of anisotropy in the molecular motion. A detailed analysis of e^2qQ/h *and* η shows that both partial rotational biasing of reorientations around the long axis and anisotropic fluctuations of this axis contribute to the non-zero value of η. The degree of quadrupolar biasing is very small, the upper limit of $\langle \cos 2\psi \rangle$ being 0.005. This small value of the order parameter, though indicating a non-equivalency among the six-equilibrium positions, shows that the molecules still rotate nearly uniformly around their long axis (Seliger et al., 1978).

The characteristic maximum in the temperature dependence of η shows that the orientational ordering of the long molecular axis in the $smectic-C$ phase is strongly perturbed by thermal agitation. The order parameters for the orientational ordering of the long molecular axis, $\langle \cos \Phi \rangle$ and $S' = \frac{1}{2}\langle 3\cos^2(\theta - \theta_M)\rangle$ which describes the polar fluctuations of the long molecular axis similarly as the "nematic order parameter" S have been calculated by assuming that (Dianoux and Volino, 1979):

(i) The θ and Φ motions are independent,

(ii) The Φ distribution is of the form $g(\Phi) \propto \exp(\gamma' \cos\Phi)$ (γ' describes the width of azimuthal fluctuations, i.e. the motion of long axes inside the smectic plane).

(iii) The θ distribution is of the form $f(\theta) \propto \exp[\delta' \cos(\theta - \theta_M)]$. Here θ_M is the angle where the θ -distribution is peaked, and δ' describes the width of the θ -distribution around θ_M , i.e. the motion of the long axes outside the smectic plane.

(iv) The root mean square fluctuations in the polar angle $\langle \delta\theta^2 \rangle^{1/2} \approx 22^\circ$ are constant through the whole $smectic-C$ phase.

The results of this analysis are presented in Figure 7.8.4. (Dianoux and Volino, 1979). The temperature dependence of *S*, which increases from 0.75 at the $smectic-A \rightarrow smectic-C$ transition to 0.82 at the $smectic-C \rightarrow smectic-H$ transition, is mainly due to the temperature dependence of θ_M. θ_M increases in the $smectic-C$ phase with decreasing temperature similarly to the average tilt angle $\theta_\circ$. Fluctuations in the polar angle $\langle\delta\theta^2\rangle^{1/2} \approx 22^\circ$ are smaller than those in the azimuthal angle Φ, where the root mean square fluctuation is of the order of ~60° at T = 165° C $(\langle\cos\Phi\rangle = 0.5)$ and approximately ~30° at T = 145° C $(\langle\cos\Phi\rangle \approx 0.85)$.

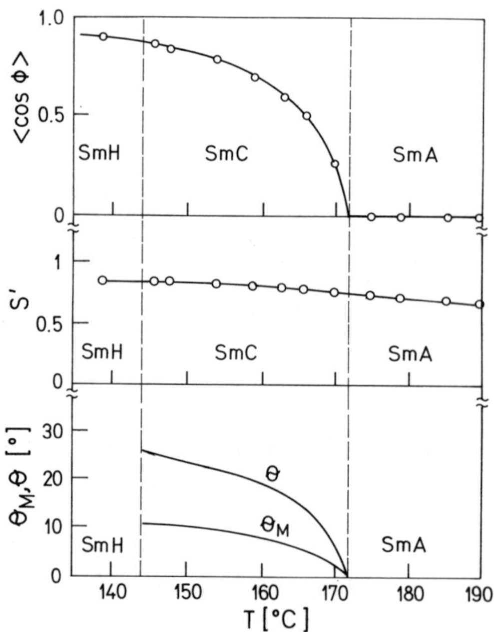

Fig.7.8.4. Temperature dependence of the azimuthal $(\langle\cos\Phi\rangle)$ and polar (S') fluctuations in the direction of the long molecular axis in the smectic phases of TBBA (after Dianoux and Volino). The values of the average tilt angle θ and the tilt angle θ_M, at which the θ distribution is peaked, are plotted as well.

When crossing the $smectic-C$ - $smectic-H$ phase transition, substantial changes are observed in the NQR spectrum of TBBA. The large temperature variation in η and the small temperature variation in e^2qQ/h in this phase demonstrate a significant quadrupolar biasing of the rotation around the long molecular axis.

The six potential wells for this reorientation are not all equivalent, which is in agreement with the X-ray structure (Levelut, 1976). Equation 7.8.11 with $r = 4.49$, which has been deduced for the case of uniaxial rotation with bipolar biasing without any fluctuations in the direction of the long molecular axis (S = 1), is already a rather good approximation to the experimental data. The calculated values of the quadrupolar order parameter $\langle\cos 2\psi\rangle$ vary from 0.03 at T =140° C to 0.16 at the low temperature end of the $smectic-H$ phase (see Figure 7.8.5). For $\langle\cos 2\psi\rangle \approx 0.1$, the occupation probability of each of the two low energy sites is $p_1 = 0.2$ whereas the occupation probability of each of the four high energy sites is $p_2 = 0.15$. The occupancy of the two preferred potential wells is thus about 30% more probable than that of the other four. It should be pointed out that neutron quasielastic scattering and X-ray data cannot discriminate between such small bipolar biasing and a uniform rotation model. Thus NQR, where the rotational bias is expressed in terms of a splitting frequency, provided the first direct evidence for biasing of the uniaxial rotation of the central part of the TBBA molecules in the biaxial smectic phases.

The slight change (~20%) of e^2qQ/h in the $smectic-H$ phase indicates that fluctuations in the direction of the long axis are still present. If these fluctuations are taken into account in the analysis of the experimental data, the calculated values of $\langle\cos 2\psi\rangle$ are somewhat, though not significantly, higher at the high temperature part of the $smectic-H$ phase. The corresponding fluctuations in the direction of the long axis, which have been assumed to be roughly isotropic, are in the range from $\approx 15^\circ$ close to the $smectic-C$ phase down to only $\approx 2^\circ$ at the boundary of the $smectic-G$ phase, where they freeze out completely.

In the $smectic-G$ phase the ordering of the long molecular axis is perfect and the averaging of the EFG tensor is entirely due to the molecular reorientations around its long axis. The huge discontinuity in both e^2qQ/h and η at the $smectic-H \rightarrow smectic-G$ transition (Fig.7.8.3) shows that the direction of the largest principal axis of the EFG tensor has changed due to the large increase of the quadrupolar biasing. It is not anymore parallel to the axis of reorientation. The value of $\langle \cos 2\psi \rangle$ is already as high as 0.85 at the high temperature end of the $smectic-G$ phase (Fig.7.8.5) and increases to 1 at the low temperature end (Blinc et al., 1979). The four high energy states in the six-fold potential well are now nearly empty as $p_2 = 0.02$ at 82.5° C and $p_2 = 0.007$ at 72° C, whereas p_1 approaches 0.5. The molecular motion in the $smectic-G$ phase can be thus indeed described as 180° orientational jumps of the central part of the TBBA molecule.

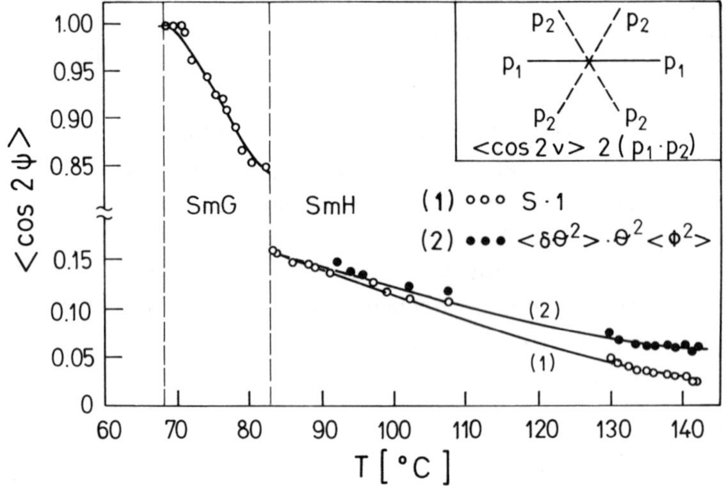

Fig.7.8.5. Temperature dependence of the quadrupolar orientational order parameter ($\langle \cos 2\psi \rangle$) in the $smectic-G$ and $smectic-H$ phases of TBBA. The values (1) have been calculated under the assumption of rigid long molecular axes, and (2) if the fluctuations in the direction of the long molecular axis are taken into account as well (Blinc et al., 1983).

On going from the $smectic-G$ to the crystalline phase, the 180° jumps freeze out so that $\langle \cos \psi \rangle = \langle \cos 2\psi \rangle = 1$ and $\langle \sin \psi \rangle = \langle \sin 2\psi \rangle = 0$. The freeze out of the 180° jumps probably happens in the $smectic-VII$ phase where no ^{14}N NQR double resonance measurements could be made because the proton T_1 is too short. In the crystalline solid phase the C-N=CH-C groups are already completely rigid, as demonstrated by the fact that both the ^{14}N quadrupole coupling constant and the ^{14}N asymmetry parameter are temperature independent.

The central part of the molecular structure of terephtal-bis-amino-methyl-butyl cinnamate (TBACA) is the same as that of TBBA. The tails contain the asymmetric carbon which is responsible for the chirality and optical activity of this substance in its right- and left-handed optically active forms. In the vicinity of the chiral group is the strongly polar -COO group. The structure and the sequence of phase transitions in TBACA, first reported by Urbach and Billard (1972), is shown in Fig.7.8.6. The temperature dependence of e^2qQ/h and η of both equivalent nitrogen atoms in the central part of TBACA (Seliger et al., 1978) is presented in Figure 7.8.7 for the $smectic-A$, $smectic-C^*$ and $smectic-H^*$ phases. The ^{14}N quadrupole coupling constant slowly increases with decreasing temperature without showing any discontinuities at the $smectic-A \rightarrow smectic-C^*$ or $smectic-C^*$ $\rightarrow$ $smectic-H^*$ transitions. The values of η in the $smectic-H^*$ phase of TBACA are less than one tenth of those in TBBA.

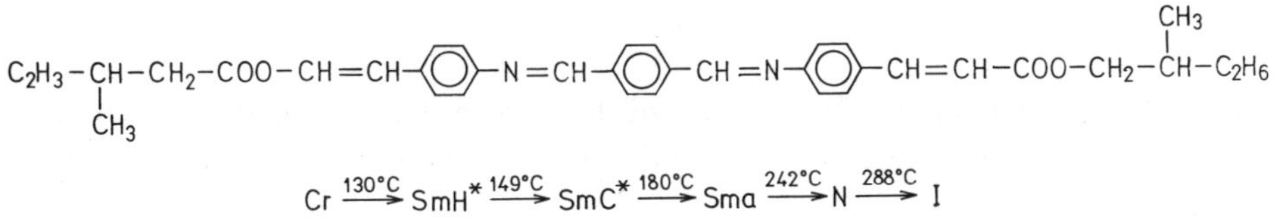

$$C_2H_3-\underset{\displaystyle CH_3}{\underset{|}{CH}}-CH_2-COO-CH{=}CH-\langle\bigcirc\rangle-N{=}CH-\langle\bigcirc\rangle-CH{=}N-\langle\bigcirc\rangle-CH{=}CH-COO-CH_2-\overset{\displaystyle CH_3}{\overset{|}{CH}}-C_2H_6$$

$$Cr \xrightarrow{130°C} SmH^* \xrightarrow{149°C} SmC^* \xrightarrow{180°C} Sma \xrightarrow{242°C} N \xrightarrow{288°C} I$$

Fig.7.8.6. The structure of chiral TBCA and its phase transitions.

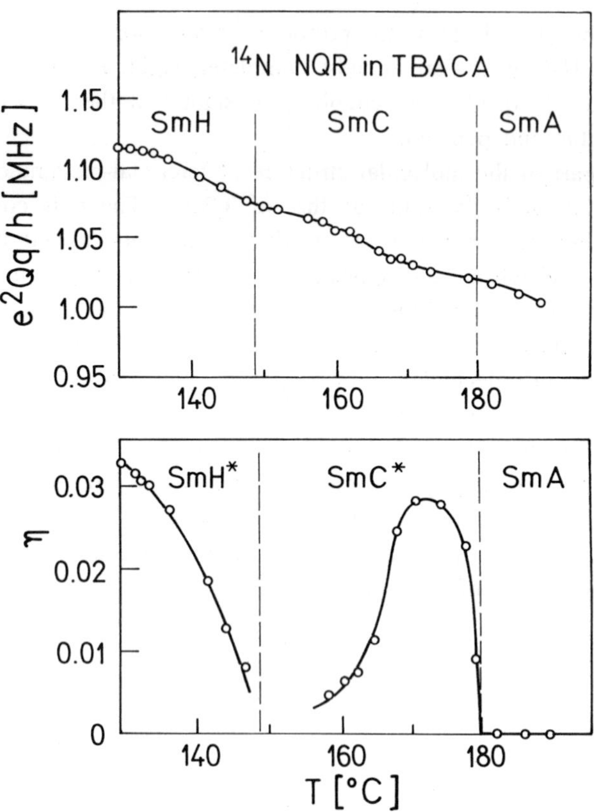

Fig.7.8.7. Temperature dependence of the ^{14}N quadrupole coupling constant e^2qQ/h and asymmetry parameter η in the smectic phase of TBACA(Seliger et al., 1978).

For the analysis of the ^{14}N data in TBACA we again assume that the θ and Φ motions are independent and that the magnitude of fluctuations in the long molecular axis is small enough to allow the expansion of sine and cosine of $\delta\theta$ and Φ in a power series with two or three terms. Thus one is able to determine the values of the order parameters $\langle \cos\psi \rangle$ or $\langle \cos 2\psi \rangle$, as well as the magnitude of the mean square fluctuations in the long molecular axis $\langle \delta\theta^2 \rangle$ and $\langle \Phi^2 \rangle$.

In the *smectic* $- A$ phase of TBACA the value of η is zero as in TBBA. The uniaxial rotation around the long molecular axis and the "nematic" fluctuations in the direction of this axis are isotropic.

In the *smectic* $- C^*$ phase of TBACA the value of η is smaller than in TBBA but non-zero. The occurrence of a maximum in the temperature dependence of η indicates that the anisotropy in molecular motion is produced mainly by anisotropic fluctuations in the direction of long molecular axis. If the orientational ordering of the short molecular axes is neglected, $\langle \delta\theta^2 \rangle^{1/2}$ and $\langle \Phi^2 \rangle^{1/2}$ can be calculated at each temperature. In Figure 7.8.8. these are plotted for $V_{zozo} = 1.15 MHz$.

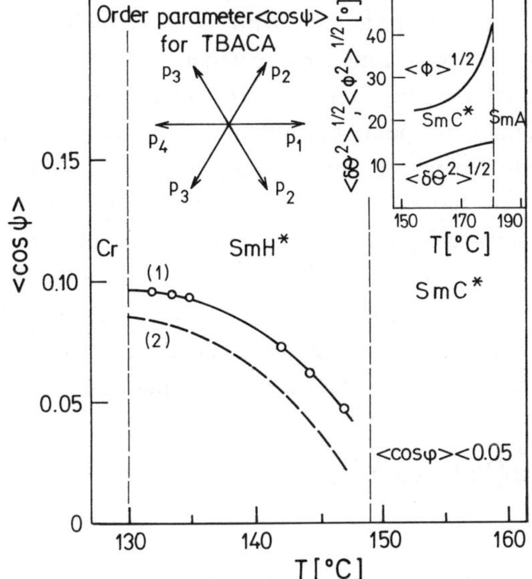

Fig.7.8.8. The temperature dependence of the polar order parameter $\langle \cos\psi \rangle$ and of the fluctuations in the direction of the long molecular axis ($\langle \delta\theta^2 \rangle^{1/2}$ and $\langle \Phi^2 \rangle^{1/2}$ in the *smectic* $- C^*$ and *smectic* $- H^*$ phases of TBACA. Full line represents the calculated values of $\langle \cos\psi \rangle$ for $V_{zozo} = 1.08 MHz$ and the broken line for $V_{zozo} = 1.15 MHz$. (Seliger et al., 1978).

The fluctuations in the azimuthal angle are larger than in the polar angle throughout the $smectic-C^*$ phase. The root mean square value of the polar fluctuations, $\langle\delta\theta^2\rangle^{1/2}$, increases slowly from $\sim 10°$ close to the boundary between the $smectic-H^*$ and the $smectic-C^*$ phase to $\sim 15°$ at the $smectic-A \rightarrow smectic-C^*$ transition. This value is in good agreement with that calculated from e^2qQ/h at the boundary between the $smectic-C^*$ and $smectic-A$ phase (where $\langle\cos\Phi\rangle = \langle\cos\psi\rangle = \langle\cos 2\psi\rangle = 0$), which turns out to be 15.5°. The fluctuations in the azimuthal angle, $\langle\Phi^2\rangle^{1/2}$, increase rapidly in the $smectic-C^*$ phase and diverge as the $smectic-A$ phase boundary is approached.

On the basis of the NQR data in the $smectic-C^*$ phase of TBACA it is not possible to determine to what extent the rotation around the long axis in the $smectic-C^*$ phase of TBACA is biased. Nevertheless the biasing should be polar and very small, in accordance with the conclusions drawn for the $smectic-H^*$ phase. Its upper limit is estimated as $\langle\cos\psi\rangle < 0.05$.

The NQR data in the $smectic-H^*$ phase of TBACA (M.Vilfan et al., 1980) can be explained by polar biasing of the uniaxial rotation around the long molecular axis with $\langle\cos\psi\rangle \neq 0$ and $\langle\cos 2\psi\rangle \approx 0$. If, on the other hand, bipolar ordering with only $\langle\cos 2\psi\rangle \neq 0$ or a uniform rotation around the long molecular axis is assumed, no physically reasonable solution which would agree with the experimental data, can be obtained. The physical model for polar ordering is that the molecule reorients around the long axis in a six-fold potential which is biased due to the coupling of the electric dipole with the local electric field. Due to coupling of the dipole to the local electric field $(-\vec{P}\vec{E})$ the six orientations are nonequivalent, with

$$p_i \propto \left(\tfrac{1}{6}\right)\exp\left[PE(\cos\psi_i)/k_BT\right] \qquad (7.8.21)$$

For such a set of probabilities $\langle\cos\psi\rangle \neq 0$ while $\langle\cos 2\psi\rangle << \langle\cos\psi\rangle$ if the coupling term is smaller than k_BT.

The temperature dependence of the polar order parameter $\langle\cos\psi\rangle$ in the $smectic-H^*$ phase is shown in Figure 7.8.8. The $\langle\cos\psi\rangle$ values have been calculated for two different values of $V_{z^o z^o}$: 1.15 MHz and 1.08 MHz. The higher value is close to that determined for TBBA (see above), and the lower one is the lowest which still gives reasonable solutions. In the whole $smectic-H^*$ phase the values of $\langle\cos\psi\rangle$ are small, ranging from ~0.1 at the low temperature end to ~0.02 at the high temperature end of this phase. For $\langle\cos\psi\rangle = 0.07$ the difference in the probabilities between the most favorable position ($p_1 = 0.19$) and the least favorable position ($p_2 = 0.14$) is thus only ~30%. In the chiral $smectic-H^*$ phase of TBACA the polar biasing of the molecular rotation around its long axis is thus rather weak. This is compatible with the small value of the spontaneous polarization observed in ferroelectric liquid crystals. The accompanying fluctuations in the direction of the long molecular axis are smaller than 10°. They can be ignored and the experimental data explained by polar ordering only.

The above results have shown that in the biaxial smectic phases of TBBA and TBACA the tilting of the molecules is associated with a biasing of the molecular rotation around the long axis. This is a rather general behavior and was also observed in the untilted two dimensional translationally ordered $smectic-B$ phase of isobutyl-4-(4'-phenylbenzyl-ideneamino)-cinnamate (IBPBAC). The structure and the sequence of the smectic phases in this liquid crystal are shown in Fig.7.8.9 (Leadbetter et al., 1979).

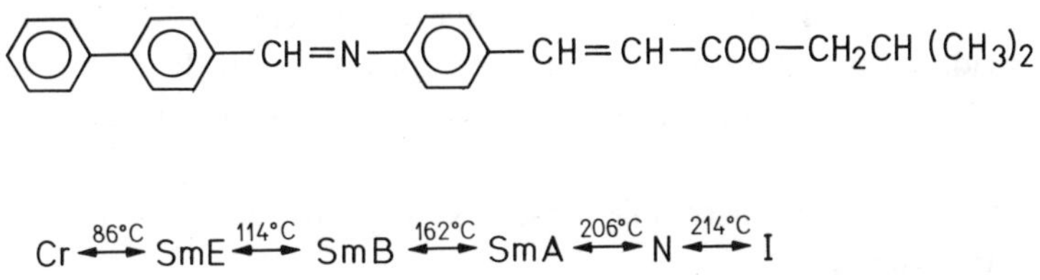

$C_6H_5-C_6H_4-CH=N-C_6H_4-CH=CH-COO-CH_2CH(CH_3)_2$

$$Cr \overset{86°C}{\longleftrightarrow} SmE \overset{114°C}{\longleftrightarrow} SmB \overset{162°C}{\longleftrightarrow} SmA \overset{206°C}{\longleftrightarrow} N \overset{214°C}{\longleftrightarrow} I$$

Fig.7.8.9. The molecular structure and the phases of IBPBAC.

The ^{14}N quadrupole coupling constant in the $smectic-B$ phase of IBPBAC (1.11 MHz at 137° C and 1.09 MHz at 146° C) is comparable (Vilfan et al., 1980b) to the values found in TBBA, whereas the value of the asymmetry parameter is much smaller (≤ 0.015 at 137° C and ~ 0 at 146° C). This demonstrates that the rotation around the long molecular axis is practically "free" in the $smectic-B$ phase, i.e. the molecules seem to reorient between six equivalent sites in a hexagonally packed structure. The tilt fluctuations are estimated to ~ 12° at 140° C. This shows that there is indeed a close relation between the molecular tilt and the biasing of molecular rotation around the long axes in smectic systems. Biasing is mainly polar in tilted chiral systems, quadrupolar in tilted achiral systems, and absent in untilted systems.

7.9. ^{13}C NMR in Ferroelectric Liquid Crystals

A major advantage of ^{13}C NMR over ^{14}N NQR or deuteron NMR, is the possibility to determine the order parameters without selective deuteration at each carbon site in the molecule and not only at one site as for ^{14}N NQR or deuteron NMR (see, for example Blinc et al., 1983 Luzar et al., 1984). If the ^{13}C chemical shift tensor in the molecular frame is known for a given site, all the components of the time averaged ^{13}C tensor can be calculated for a given model of molecular motion, and the order parameters can be determined by comparing the experimental chemical shifts with the theoretical values. Unfortunately the ^{13}C chemical shift tensors are at present known only for a few molecules and no angular dependent ^{13}C NMR measurements have been made on liquid crystals so far in contrast to ^{2}H NMR. In this Section we shall discuss the chemical shift tensor in ferroelectric liquid crystals DOBAMBC and HOBACPC.

In the isotropic phase one observes only the isotropic part of the ^{13}C chemical shift tensor $\underline{\sigma}$

$$\sigma_i = \frac{1}{3} Tr\sigma \tag{7.9.1}$$

whereas the anisotropic part is averaged out to zero due to the fast and random rotational and translational molecular motion.

When the sample is slowly cooled from the isotropic into the $smectic-A$ phase, the molecules order in such a way that the normals to the smectic layers are parallel to the direction of the external magnetic field $\vec{H}$. The chemical shift tensor is not anymore isotropic. The relevant molecular motions, which are averaging the ^{13}C chemical shift tensor, are fast molecular rotations around the long molecular axis, and fluctuations of the long molecular axis around the molecular director.

The component of the chemical shift tensor in the direction of the external magnetic field $\sigma = \langle \sigma_{z'z'} \rangle \equiv \langle \sigma_{H} \rangle$ is

$$\sigma = \sigma_i + \frac{2}{3}\left[\sigma_{zz} - \frac{1}{2}(\sigma_{xx} + \sigma_{yy})\right] S_{zz} + \frac{1}{3}(\sigma_{xx} - \sigma_{yy})(S_{xx} - S_{yy}) \tag{7.9.2.}$$

where

$$S_{zz} = S = \left\langle \frac{3}{2}\cos^2 \vartheta - \frac{1}{2} \right\rangle \tag{7.9.3a}$$

measures the amount of "nematic" ordering of the long molecular axis around the molecular director $\vec{n}$, and

$$S_{xx} - S_{yy} = \frac{3}{2}\langle \sin^2 \vartheta \cos 2\psi \rangle \tag{7.9.3b}$$

measures the asymmetry in the fluctuations of the long molecular axis around the x and y axis. The time independent coordinate frame x^*, y^*, z^* and the molecular fixed frame x, y, z are illustrated in Figure 7.9.1. The z^* axis is parallel to the normal to the smectic layers whereas the x axis points along the direction of the projection of the long molecular axis on the smectic layer. The angle $\vartheta(t)$ measures the fluctuations of the long molecular axis in the polar and the angle $\Phi(t)$ in the azimuthal direction. The angle $\psi(t)$ measures the orientation of the short molecular axis as the molecule rotates around its long axis. $\sigma_{xx} - \sigma_{yy}$ is of the order of 100 ppm for the aromatic carbons. Fast conformational changes, like 90 degrees flips of benzene rings around the para-axis effectively average out this difference so that

$$\sigma_{xx} \approx \sigma_{yy}. \tag{7.9.3c}$$

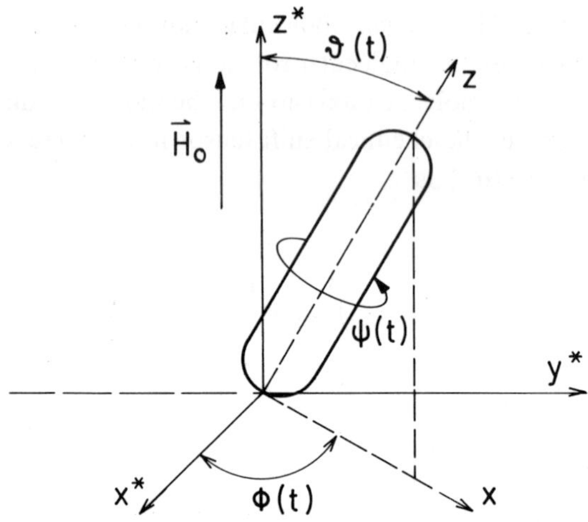

Fig.7.9.1. Relation between the molecular frame *x, y, z* and the time independent frame $x^*, y^*, z^* \| H_\circ$ in the *smectic – A* phase.

Expression (7.9.2) thus simplifies to

$$\sigma = \sigma_i + \frac{2}{3} S(\sigma - \sigma_\perp) \tag{7.9.4.}$$

where $\sigma_{//} = \sigma_{zz}$ is the component along the direction of the long molecular axis and $\sigma_\perp = \frac{1}{2}(\sigma_{xx} + \sigma_{yy})$ is the average component in the *x-y* plane.

For the aromatic ^{13}C nuclei the heaviest shielding occurs perpendicular to the aromatic plane, whereas for the aliphatic groups the most shielded element of σ lies parallel to the chain axis. We expect that at the isotropic - *smectic – A* transition a downfield shift of the aromatic $(\sigma_{//} - \sigma_\perp > 0)$ and an upfield shift $(\sigma_{//} - \sigma_\perp < 0)$ of the aliphatic lines would occur. This is exactly what has been indeed observed in the experiments of Luzar et al. (1984). Their ^{13}C NMR spectra and the molecular structure of DOBAMC and HOBACPC with assignation of the 22 resolved chemically non-equivalent ^{13}C lines are shown in Fig. 7.9.2. and 7.9.3.

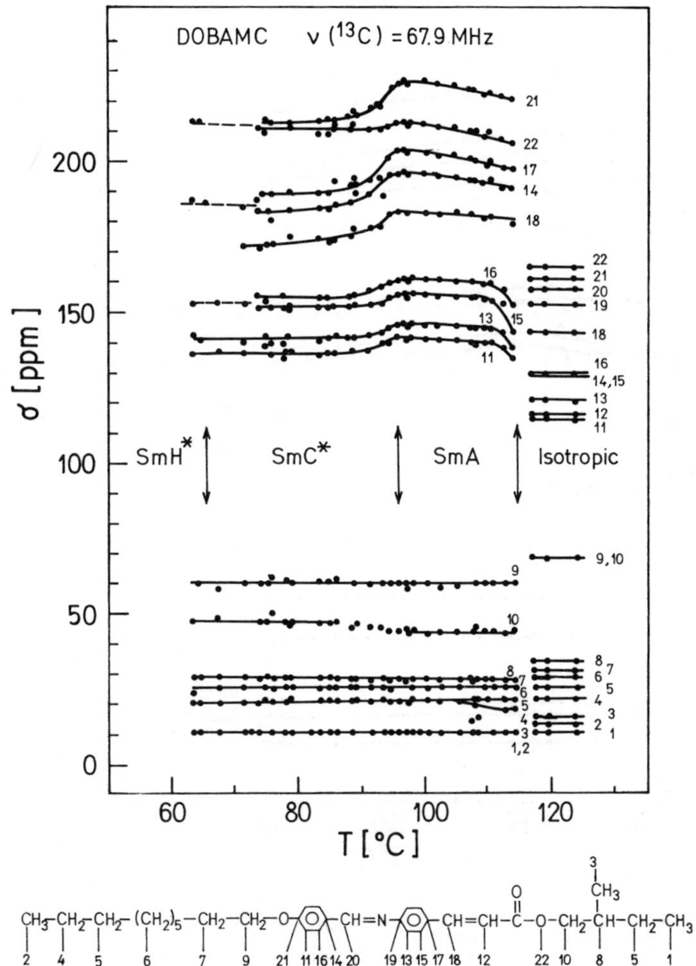

Fig.7.9.2. Temperature dependence of the ^{13}C chemical shifts of DOBAMBC with respect to tetra-methyl-silane (TMS) in the paraelectric and ferroelectric phase. The numbering of DOBAMBC carbon atoms is shown as well (after Luzar et al., 1984).

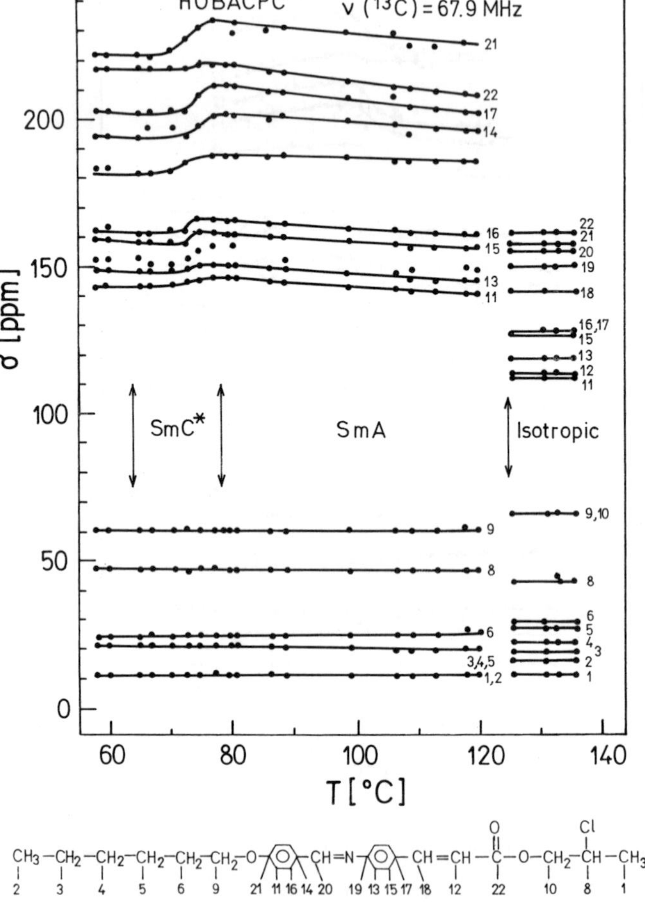

Fig.7.9.3. Temperature dependence of the ^{13}C chemical shifts of HOBACPC with respect to tetra-methyl-silane (TMS) in the paraelectric and ferroelectric phase. The numbering of HOBACPC carbon atoms is shown as well (after Luzar et al., 1984).

On going from the isotropic to the paraelectric $smectic-A$ phase, one observes a large and abrupt change in the ^{13}C chemical shifts. Clearly, as one can see form Fig.7.9.2.-3., there is a ***downfield shift of the aromatic lines*** and an ***upward shift of the aliphatic lines***, as expected. From the measured values of σ and σ_i, one can determine the nematic order parameter S if $\sigma_{\parallel}-\sigma_{\perp}$ is known.

A major obstacle for the use of ^{13}C spectroscopy in liquid crystal studies has been the traditional lack of precise knowledge of ^{13}C chemical shift tensors and their orientation. This situation has been somewhat changed in recent years. The widespread use of ^{13}C -proton double resonance has provided ^{13}C chemical shift tensors for different compounds and has shown that ^{13}C tensors for a specific molecular group can be to a rather good approximation transferred from one compound to another. This is the approach which we shall use here. Since the ^{13}C chemical shift tensors for ferroelectric liquid crystals DOBAMBC or HOBACPC are not known, we shall use the data from related solid compounds containing the relevant ^{13}C groups. It should be stressed that the thus obtained results must be therefore treated with some caution. For the aromatic carbons one finds from the published data the following eigenvalues of σ : $\sigma_{11}=275\,ppm$, $\sigma_{22}=190\,ppm$ and $\sigma_{33}=69\,ppm$, all measured with respect to tetra-methyl-silane (TMS) (see Pines et al., 1972, Pausak et al., 1973, Pines and Chang, 1974). The corresponding directions of the principal axes of σ are illustrated in Figure 7.9.4a. For an accurate determination of $\sigma_{//}-\sigma_{\perp}$ one has to take into account also the angle $\gamma\approx 10^{\circ}$ between the para-axis and the rotation axis of the molecule (see Figure 7.9.4b).

For the aromatic carbons at the para-positions one thus finds the components of the chemical shift tensor:

$$\sigma_{//}^{p}=\sigma_{11}\cos^2\gamma+\sigma_{22}\sin^2\gamma \qquad (7.9.5a)$$

$$\sigma_{\perp}^{p}=\frac{1}{2}\left(\sigma_{11}\sin^2\gamma+\sigma_{22}\cos^2\gamma+\sigma_{33}\right) \qquad (7.9.5b)$$

whereas

$$\sigma_{//}^{\circ}=\frac{1}{4}\sigma_{11}\left(1+2\sin^2\gamma\right)+\frac{1}{4}\sigma_{22}\left(1+2\cos^2\gamma\right) \qquad (7.9.6a)$$

$$\sigma_{\perp}^{\circ} = \frac{1}{2}\left[\frac{1}{4}\sigma_{11}\left(1+2\cos^2\gamma\right)+\frac{1}{4}\sigma_{22}\left(1+2\sin^2\gamma\right)+\sigma_{33}\right] \quad (7.9.6b)$$

are the corresponding values for the aromatic carbons at the ortho-positions.

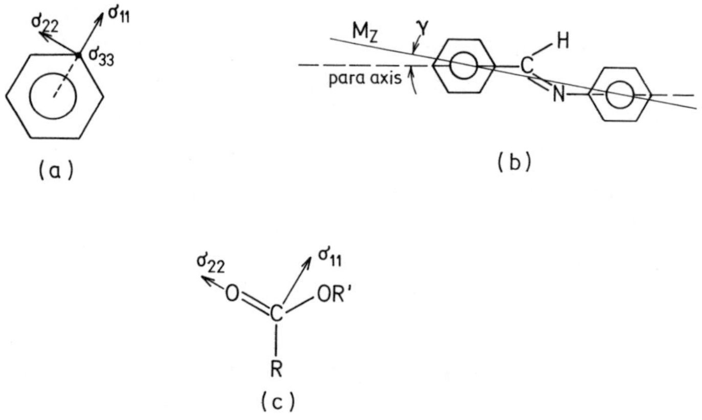

Fig.7.9.4. (a) Directions of the principal axis of the chemical shift tensor of aromatic carbons (Pines et al., 1972, Pausak et al., 1973). (b) The directions of the para-axis of aromatic rings and the molecular rotation axis. (c) The directions of the principal axis of the chemical shift tensor of the carbonyl carbons (Pines et al., 1974b).

Using Eqs.7.9.5a-7.9.6b) and the published data (Pines et al., 1972, Pausak et al., 1973, Pines and Chang, 1974) one obtains $\Delta\sigma^{\circ} = \sigma_{//}^{\circ} - \sigma_{\perp}^{\circ} = 52\,ppm$ and $\Delta\sigma^{p} = \sigma_{//}^{p} - \sigma_{\perp}^{p} = 142\,ppm$. From Eq.7.9.4. we now find that the observed S values at the aromatic sites vary from $S \approx 0.7$ near the $smectic-A$ - isotropic transition up to $S \approx 0.8$ at the $smectic-A$ - $smectic-C^*$ transition. The values of S on going from the end of the aliphatic chains towards the aromatic rings for the aliphatic carbons are significantly smaller and range from $S \approx 0.3$ to $S \approx 0.45$. In view of the small anisotropy, $|\Delta\sigma| < 30\,ppm$ (van der Hart, 1976), the accuracy is here not very high.

The value of *S* for the C=O group which is the main ferroelectric dipole in the structure can also be obtained from the ^{13}C data. The directions of the principal axes of the σ tensor for this group are illustrated in Figure 7.9.4c (Pines et al., 1974). $\sigma_{33} = 167\,ppm$ is here perpendicular to the COO^- plane, whereas $\sigma_{22} = 207\,ppm$ is nearly parallel to the C=O bond direction. σ_{11} equals 317 *ppm* (Pines et al., 1972, Pausak et al., 1973) and is approximately parallel to the long molecular axis. One thus finds that $\Delta\sigma = 130\,ppm$ and the values of *S* in both DOBAMBC and HOBACPC vary from $S \approx 0.55$ at the $smectic-A$-isotropic transition to $S \approx 0.6$ at the $smectic-A$ - $smectic-C^*$ transition. The observation that for the C=O group the values of *S* are significantly smaller than for the aromatic carbons agrees with the fact that the ordering of the aliphatic tail is always smaller than the ordering of the central aromatic part.

On going from the $smectic-A$ to the ferroelectric $smectic-C^*$ phase a steady increase in the nematic order parameter *S* and a corresponding increase in the aromatic ^{13}C chemical shifts is expected. This effect is indeed seen in achiral HOAB (Luzar et al., 1980) but not in chiral DOBAMBC and HOBACPC where a significant decrease of the ring and C=O ^{13}C chemical shifts is observed, as shown in Fig.7.9.2.-3. This effect can be understood if one assumes that in chiral systems the helicoidal axis and the normals to the smectic layers remain parallel to the direction of the external magnetic field, whereas the long molecular axis are tilted with respect to the field direction. This is opposite to the behavior of achiral $smectic-C$ systems where the long molecular axes are oriented parallel to the field whereas the layer normals are tilted.

If the onset of the molecular tilt with respect to the magnetic field direction would be the only change which takes place at the $smectic-A$ - $smectic-C^*$ transition, the observed chemical shifts in the $smectic-C^*$ phase would be:

$$\sigma = \sigma_i + \frac{2}{3} S \cdot (\sigma_{//} - \sigma_{\perp}) \cdot \left(\frac{3}{2} \cos^2 \theta - \frac{1}{2} \right) \tag{7.9.7}$$

Here θ is the molecular tilt angle and the fluctuations in the direction of the long molecular axis are assumed to be isotropic.

In the $smectic-C^*$ phase Eq.7.9.7. therefore predicts that the temperature dependence of

$$\frac{\sigma-\sigma_i}{\frac{3}{2}\cos^2\theta-\frac{1}{2}}=\frac{2}{3}\left(\sigma_{\parallel}-\sigma_{\perp}\right)\cdot S \tag{7.9.8.}$$

should reflect the temperature dependence of S which is known to increase monotonically with decreasing temperature. Since the temperature dependence of the molecular tilt angle is known from independent optical measurements for both DOBAMBC (Martinot-Lagarde, 1976, Ostrovski et al., 1978) and HOBACPC (Martinot-Lagarde, 1976) the above prediction can be easily checked.

Figures 7.9.5a, b, c show the temperature dependence of $\sigma-\sigma_i$ near the $smectic-A$ - $smectic-C^*$ transition for the aromatic carbon in the para-position $^{13}C(17)$, the aromatic carbon in the ortho-position $^{13}C(11)$, and the carbonyl carbon $^{13}C(22)$ in DOBAMBC. The corresponding data for the aliphatic carbons $^{13}C(18)$ and $^{13}C(10)$ are shown in Figures 7.9.5d and e. Figures 7.9.6a, b show the data for the $^{13}C(21)$ carbon and the C=O group $^{13}C(22)$ in HOBACPC. In all above figures the experimental data $(\sigma-\sigma_i)$ are designated by circles whereas the crosses show the calculated temperature dependence of $(\sigma-\sigma_i)/(\frac{3}{2}\cos^2\theta-\frac{1}{2})$. The dotted line shows the temperature dependence of $\frac{3}{2}S(\sigma_{\parallel}-\sigma_{\perp})$ obtained by linear extrapolation of S from the $smectic-A$ into the $smectic-C^*$ phase.

The comparison with theory demonstrates that Eq.7.9.7 describes adequately the experimental situation for most aliphatic carbons, e.g. C(18). It can be therefore used to determine the temperature dependence of the tilt angle. For HOBACPC we find $\theta=\theta_o(T_c-T)^{0.35\pm0.05}$ with $\theta_o=0.18$, which is in relatively good agreement with the optical tilt angle measurements (Martinot-Lagarde, 1976). This also agrees with the deuteron NMR results in ferroelectric CE8 and antiferroelectric MHPOBC discussed in previous Sections. A similar relation $\theta=\theta_o(T_c-T)^{0.30\pm0.05}$ with $\theta_o=0.21[K^{0.3}]^{-1}$ holds for DOBAMBC and is in excellent agreement with the optical measurements (Ostrovski et al., 1978), as shown in Fig.7.9.7.

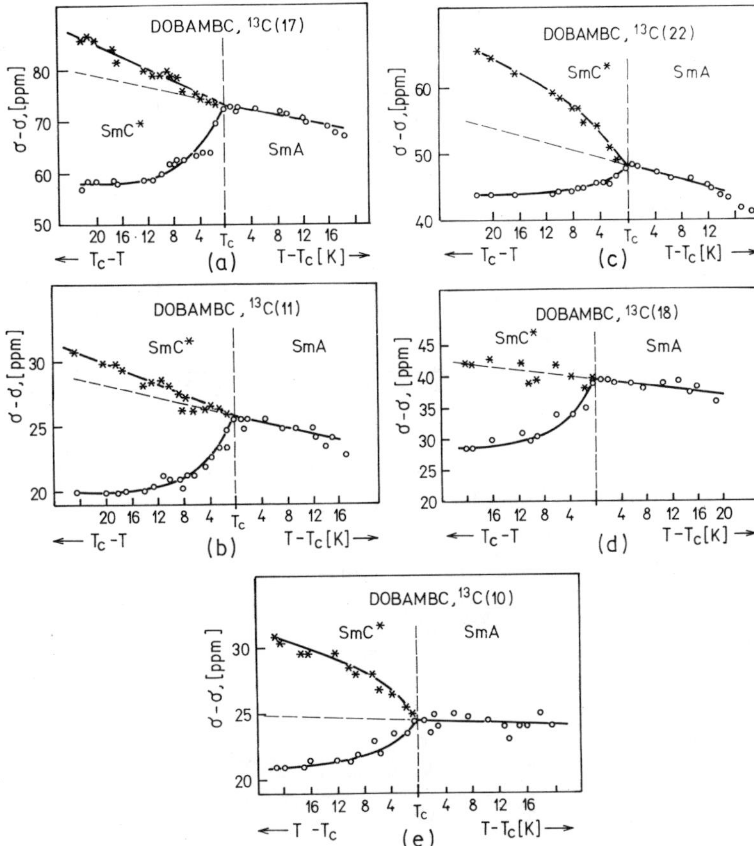

Fig.7.9.5. Temperature dependence of $\sigma - \sigma_i$ (circles) near the *smectic – A - smectic –* C^* transition in DOBAMBC for: (a) The aromatic carbon in the para-position, C(17). (b) The aromatic carbon in the ortho-position, C(11). (c) The carbonyl carbon, C(22). (d) The aliphatic carbon, C(18). (e) The aliphatic carbon C(10). The experimental data are designated by circles. The crosses show the calculated temperature dependence of $(\sigma - \sigma_i)/(\frac{3}{2}\cos^2\theta - \frac{1}{2})$, whereas the dotted line shows the temperature dependence of $\frac{2}{3}S(\sigma_{\parallel} - \sigma_{\perp})$, obtained by a linear extrapolation of S from the *smectic – A* into the *smectic –* C^* phase (Luzar et al., 1984).

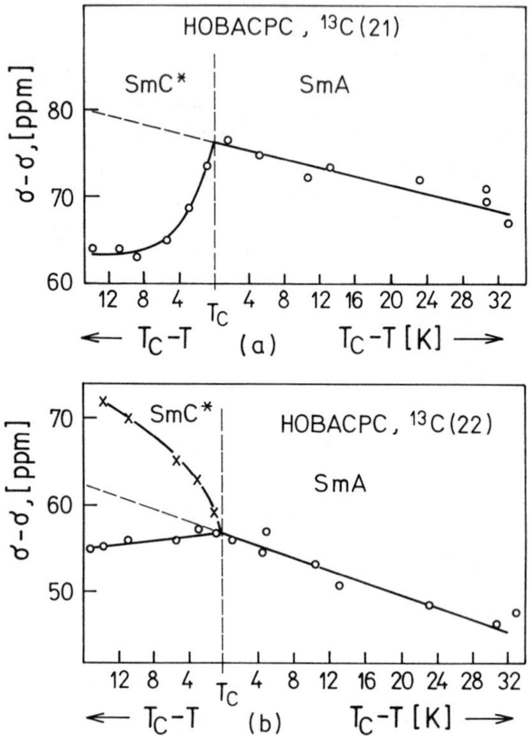

Fig.7.9.6. Temperature dependence of $\sigma - \sigma_i$ (circles), near the *smectic* − *A* - *smectic* − C^* transition in HOBACPC. (a) The C(21) carbon. (b) The carbonyl carbon C(22). The crosses (lower figure) show the calculated temperature dependence of $(\sigma - \sigma_i)/(\frac{3}{2}\cos^2\theta - \frac{1}{2})$ whereas the dotted line shows the temperature dependence of $\frac{2}{3}S(\sigma_{\parallel} - \sigma_{\perp})$ obtained by a linear extrapolation of S from the *smectic* − *A* into the *smectic* − C^* phase (Luzar et al., 1984).

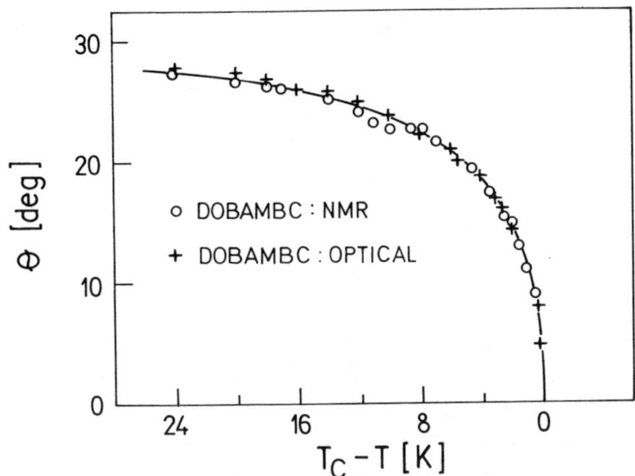

Fig.7.9.7. Temperature dependence of the molecular tilt angle $\theta(T)$ in DOBAMBC as derived from ^{13}C NMR of aliphatic carbons. The optical data of Ostrovski et al. (1978) are shown for comparison (after Luzar et al., 1984).

There are however deviations from the relation 7.9.8. for the aromatic carbons and for the C=O group in both DOBAMBC and HOBACPC. This discrepancy can be the result of:

(i) An anisotropy in the fluctuations of the long molecular axis around the director

(ii) A polar biasing of the molecular rotation around the long molecular axis

(iii) A quadrupolar biasing of the molecular rotation around the long molecular axis.

All of these effects can of course also occur simultaneously. In such case Eq.7.9.8. has to be extended to

$$\frac{\sigma-\sigma_i}{\frac{3}{2}\cos^2\theta-\frac{1}{2}}=\frac{2}{3}\left(\sigma_{//}-\sigma_{\perp}\right)\cdot S+\Delta(T) \tag{7.9.9a}$$

where

$$\Delta=\Delta_a+\Delta_p+\Delta_q \tag{7.9.9b}$$

Here Δ_a,Δ_p and Δ_q are standing for the contributions due to anisotropic fluctuations, polar and quadrupolar biasing, respectively. For DOBAMBC at $T=T_c-20K$, $\Delta\approx 10\,ppm$ for the C=O group, 3 *ppm* for the aromatic carbons at the ortho- and $\Delta=1.5\,ppm$ for the aromatic carbons at the para-sites. It is negligible for aliphatic carbons. For HOBACPC, $\Delta=9\,ppm$ for the C=O group at $T=T_c-10K$.

If the deviations from Eq.7.9.8. are due to the anisotropy in the fluctuations around the molecular director, $\Delta(T)$ is given by:

$$\Delta_a(T)=\left(\sigma_{//}-\sigma_{\perp}\right)\cdot\frac{\sin^2\theta}{3\cos^2\theta-1}\left\langle\sin^2\vartheta\cos 2\Phi\right\rangle \tag{7.9.10a}$$

whereas

$$\Delta_p(T)=4\sigma_{xz}\frac{\sin^2\theta}{3\cos^2\theta-1}\left\langle\cos\psi\right\rangle \tag{7.9.10b}$$

if the observed deviation is the result of a polar biasing of the molecular rotation around the long molecular axis. Here ϑ denotes the angular deviation of the long molecular axis with respect to the director and θ is the tilt angle.

If the observed deviation is due to a quadrupolar biasing of the molecular rotation we get

$$\Delta_q(T) = 2(\sigma_{xx} - \sigma_{yy}) \frac{\sin^2\theta}{3\cos^2\theta - 1} \langle \cos 2\psi \rangle . \tag{7.9.10c}$$

Let us now analyze each of these contributions separately. We shall first treat the case when the rotation around the long molecular axis is free, i.e. $\langle \cos\psi \rangle = 0$, $\langle \cos 2\psi \rangle = 0$ and $\Delta \cong \Delta_a$ is due to anisotropic fluctuations of the long molecular axis around the director $\vec{n}$. For DOBAMBC at $T = T_c - 20K$ one finds within this model from the experimental data for the C=O group:

$$\frac{\langle \sin^2 \vartheta \cos 2\Phi \rangle}{S} = 0.73 \tag{7.9.11a}$$

For strongly anisotropic fluctuations, where $\cos 2\Phi = 1$, we have

$$\langle \sin^2\theta \cos 2\Phi \rangle_{max} = \langle \sin^2\theta \rangle = \frac{2}{3}(1 - S). \tag{7.9.11b}$$

This yields $\langle \sin^2\theta \cos 2\Phi \rangle_{max} = 0.35$ and $S = 0.48$. The value of S derived from this model is much too small as S for the C=O group equals $\sim$0.6 even in the $smectic-A$ phase. Anisotropic fluctuations are thus not capable of explaining the observed value of Δ for the C=O group though they yield a large enough effect to be capable of explaining the observed Δ-values for the aromatic carbons. The same conclusion is true for HOBACPC.

The observed values of $\Delta(T)$ can be however satisfactorily explained by polar biasing of the molecular rotation around the long molecular axis. For the C=O group one can get from the literature (Pines et al., 1972) the off-diagonal component $\sigma_{xz} \approx 80 ppm$. Evaluating the numerical values of the prefactors in the expressions (7.9.10b) and (7.9.10c) one finds for the polar term $\Delta_p \approx 320 \cdot \langle \cos\psi \rangle \cdot \theta \; ppm$ whereas $\Delta_q \approx 40 \cdot \langle \cos 2\psi \rangle \cdot \theta \; ppm$, so that $\Delta_q << \Delta_p$, though $\langle \cos\psi \rangle$ and $\langle \cos 2\psi \rangle$ would be comparable. Expression (7.9.10b) now yields for $\Delta \approx \Delta_p$ for DOBAMBC at $T = T_c - 20K$ an upper limit for the polar ordering $\langle \cos\psi \rangle \approx 5 \times 10^{-2}$. On the other hand, we obtain for HOBACPC at the low temperature edge of the $smectic-C^*$ phase, i.e. at $T = T_c - 10K$ an estimation $\langle\cos\psi\rangle \approx 7\times10^{-2}$

The above "microscopic" values of $\langle\cos\psi\rangle$ should be compared to "macroscopic" values of $\langle\cos\psi\rangle$, which are obtained, for example, from the measurements of the spontaneous polarization. It turns out that microscopic values of $\langle\cos\psi\rangle$ are somewhat higher than those obtained from macroscopic measurements of the spontaneous polarization where the pitch is unwound. For DOBAMBC at $T = T_c - 20K$ one finds from macroscopic measurements $\langle\cos\psi\rangle = 4\times10^{-2}$ whereas one gets for HOBACPC $\langle\cos\psi\rangle = 1.3\times10^{-2}$ .The orders of magnitude of $\langle\cos\psi\rangle$ are therefore the same in both cases. The above values also agree with $\langle\cos\psi\rangle$, estimated from ^{14}N NQR in the $smectic-C^*$ phase of chiral TBACA.

The C(8)-C(1) carbon NMR line could not be observed in HOBACPC. This is probably due to line-broadening in view of the magnetic dipolar coupling with the chlorine nucleus. Though the electric dipole moment of the C-Cl bond is larger than that of the C=O group it is probable that the contribution of this group to the in-plane spontaneous polarization is smaller than that of the C=O group. Since this group is closer to the molecular tail where S is small and the mobility is relatively large, $\langle\cos\psi\rangle$ is probably much less than 7×10^{-2} thus explaining the fact that the macroscopic polarizations of HOBACPC and DOBAMBC are practically equal.

For the aromatic carbons in the para-positions $\sigma_{xz} \approx 0$. For the ortho-carbons $\sigma_{xz} \approx 100$ ppm. From the measured $\Delta = 2ppm$ for the ortho-carbons one thus finds for the aromatic rings in DOBAMBC an upper limit $\langle\cos\psi\rangle \approx 8\times10^{-3}$. This demonstrates that the rotation of the central part of the DOBAMBC molecule in the $smectic-C^*$ phase is nearly free. The above conclusion agrees with the ^{14}N quadrupole coupling measurements of $\langle\cos\psi\rangle$ in the $smectic-C^*$ phase of chiral TBACA. The molecular rotation is however biased in the neighborhood of the polar C=O group which is the main ferroelectric dipole in the structure. This is clearly demonstrated in the case of C(10) carbon in DOBAMBC (Figure 7.9.5e) where, in contrast to other aliphatic carbons, Δ is rather large and clearly results from polar ordering, $\langle\cos\Psi\rangle \neq 0$.

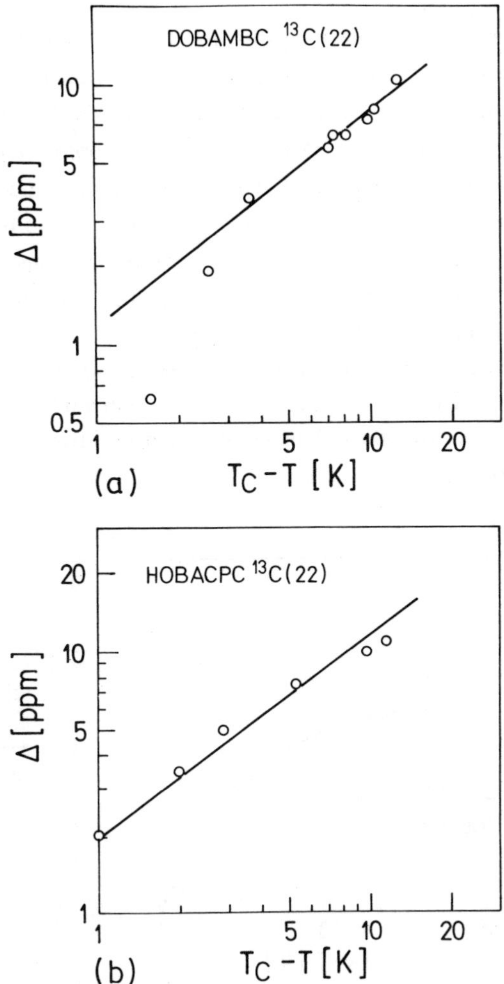

Fig.7.9.8. Temperature dependence of the "biasing" shift Δ for C(22) near the *smectic – A - smectic – C** transition in DOBAMBC (a) and HOBACPC (b). The critical exponent $\Delta \propto (T_c - T)^{0.7}$ is the one expected for "polar" biasing.

The temperature dependence of the "biasing" shift for the C=O group in DOBAMBC and HOBACPC close to the *smectic – A - smectic – C** transition provides an additional argument for the validity of our approximation $\Delta \approx \Delta_p$ (Figures 7.9.8). Over most of this temperature range $\Delta \propto (T_c - T)^{0.7}$ for both DOBAMBC and HOBACPC. This agrees with the expected temperature dependence of the polar "biasing" shift $\Delta_p \propto \theta\langle\cos\psi\rangle$ as both θ and $\langle\cos\psi\rangle$ is expected to be proportional to $(T_c - T)^{\beta}$ with $\beta \approx 0.35$. The exponent for Δ_p should be thus 0.7 as indeed observed. However, it clearly does not agree with the expected temperature dependence of the quadrupolar shift $\Delta_q \propto \theta^2\langle\cos 2\psi\rangle$.

The temperature dependence of $\langle\cos\psi\rangle$ for the C = 0 group determined by ^{13}C NMR can be compared with the values of $\langle\cos\psi\rangle$, determined by macroscopic spontaneous polarization measurements. This comparison is shown in Fig.7.9.9. and the agreement is satisfactory.

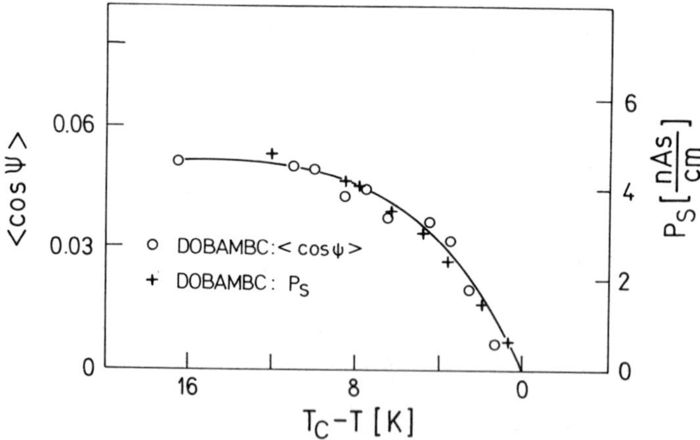

Fig.7.9.9. Temperature dependence of the polar order parameter $\langle\cos\psi\rangle$ of the C=O group in DOBAMBC. The temperature dependence of the macroscopic spontaneous polarization (Ostrovski et al., 1978) is shown for comparison.

The analysis of the ^{13}C NMR (Luzar et al., 1984) and the ^{14}N NQR spectra of ferroelectric liquid crystals and their non-chiral analogues in the tilted smectic phases thus show that:

(i) The molecular tilt induces in the $smectic-C$ and $smectic-C^*$ phases an anisotropy in the fluctuations of the long molecular axis which is practically the same for chiral and non-chiral molecules. It is responsible for the temperature variation of the ^{14}N asymmetry parameter η of TBBA and TBACA in the $smectic-C$ and $smectic-C^*$ phases, i.e. for the apparent quadrupolar ordering of the transverse molecular axes. Intrinsic polar or quadrupolar orientational ordering is here rather small for the central part of the molecule.

(ii) In the low temperature tilted smectic phases like the $smectic-H$ phases with a two dimensional translational ordering of the molecules, non-chiral molecules show a pronounced quadrupolar, and chiral molecules a pronounced polar orientational ordering. Fluctuations of the long molecular axes are not important at least for the central part of the molecules where the ^{14}N nuclei in TBBA or TBACA are located.

(iii) Polar orientational ordering of the transverse outboard molecular dipoles close to the chiral group, like the C=O group in DOBAMBC and HOBACPC, can be clearly observed by ^{13}C NMR. The polar orientational biasing of the rotation around the long molecular axis is rather small thus accounting for the small value of the spontaneous polarization in the $smectic-C^*$ phase of ferroelectric liquid crystals. The chiral polar ordering coexists with the non-chiral quadrupolar order which is produced by the biaxiality of the $smectic-C$ phase, i.e. tilt-induced anisotropy in the fluctuations of the long molecular axis. Polar ordering is not important for the aliphatic part of the molecule.

(iv) The anomalously large value of the biquadratic coupling between the polarization and the tilt is to a large extent the result of the non-chiral quadrupolar ordering induced by the biaxiality of the $smectic-C^*$ phase and not just a higher power of the chiral bilinear coupling between the polarization and the tilt. The ^{13}C NMR and ^{14}N NQR results allow for a microscopic estimate of the relative importance of these terms and thus provide a microscopic basis for the extended Landau model of ferroelectric liquid crystals.

7.10. Synclinic versus Anticlinic Ordering in Tilted Smectics

The majority of tilted smectic systems exhibit synclinic ordering where molecules in adjacent smectic layers tilt in the same direction. Anticlinic ordering where the tilt direction alternates from layer to layer occurs in antiferroelectric liquid crystalline phases such as in the $smectic-C_A^*$ phase. The microscopic mechanism leading to anticlinic respectively synclinic ordering are still not completely understood.

One possibility is that synclinic inter-layer ordering is entropically favored over anticlinic one by out of layer molecular fluctuations and molecular exchange. These fluctuations and exchange may be quenched in MHPOBC and other antiferroelectric liquid crystals due to a bent chiral tail configuration. The suppression of the cores position entropic term may allow weaker dipole-dipole and other interactions that favor anticlinic ordering to take over.

The fact that molecular exchange and self diffusion are indeed strongly reduced in anticlinic $smectic-C_A^*$ phase of MHPOBC as compared to the $smectic-C^*$ ferroelectric synclinic phase of CE-8 is shown by deuteron NMR (see Chapter 7.6) and direct self-diffusion studies. The out-of-plane component of the self-diffusion term is $D_\perp \approx 10^{-9} cm^2 s^{-1}$ in the anticlinic $smectic-C_A^*$ and $D_\perp \approx 10^{-7} m^2 s^{-1}$ in synclinic $smectic-C^*$ phase. As a direct result of reduced self-diffusion in the anticlinic phases, the smectic layers are much more ordered, which is also shown in the harmonic analysis of X-ray spectra (Fukuda et al., 1994). Higher harmonics in the X-ray spectra are much more pronounced and indicate significantly increased and "sharp" ordering of smectic layers. It should be noted, that this can lead to decoupling of orientational fluctuations in neighboring smectic layers, thus resulting in enhanced tilt fluctuations in the *smectic* – *A* phase of antiferroelectric liquid crystals (Škarabot et al., 1999b).

Recent deuteron NMR studies (Yoshida et al., 1999) have revealed very small quadrupolar splitting for the chiral chain methylene deuteron, which can be explained by the fact that the chiral chain is bent away from the long molecular axis. The "bent angle" is estimated as 42°. This may be also responsible for the reduced self-diffusion. The achiral chain, on the other hand, extends parallel to the long molecular axis. The deuteron results agree very well with a recent high resolution ^{13}C NMR study of MHPOBC (Nakai et al., 1999a,b). The ^{13}C NMR results show that the average direction of the chiral chain deviates from the long molecular axis by $\approx 43^\circ$, while the achiral chain is parallel to the long molecular axis.

Still another important result obtained from the above deuteron NMR study is that two different quadrupolar splittings are seen for each methylene unit close to the chiral center. This means that the molecular rotation is asymmetrically hindered close to the chiral center in contrast to the methylenes of the achiral group. This could contribute to the relatively large in-plane polarizations leading to anticlinic and antiferroelectric ordering in the absence of fast self-diffusion.

Chapter 8

Antiferroelectricity in Liquid Crystals

8.1. Introduction

Although the first antiferroelectric liquid crystalline substance was synthesized by Levelut et al. already in 1983, it was Chandani and co-workers, who in 1989 first realized the existence of an antiferroelectric ordering in 4-(1-methylheptyloxycarbonyl) 4'-octylbiphenyl-4-carboxylate (MHPOBC). Almost simultaneously with the work of Chandani et al. (1989) and Takezoe et al. (1989), Galerne and Liebert (1989) have proposed the "herring-bone" structure for the so-called $smectic\text{-}O$ phase, which was latter recognized (Heppke et al., 1993; Takanishi et al., 1993; Cladis and Brand, 1993) as equivalent to the antiferroelectric $smectic\text{-}C_A^*$ phase.

According to Chandani et al. (1989), the antiferroelectric liquid crystalline $smectic-C_A^*$ phase is characterized by an alternation of the tilt direction of the average molecular orientation. As a consequence of this, also the direction of the in-plane spontaneous polarization $P(\vec{r})$ alternates by nearly 180 degrees on going from one smectic layer to another (Fig.8.1.1a). Two neighboring layers thus form an antiferroelectric unit cell with two nearly antiparallel electric dipoles and a very small value of the equilibrium electric polarization $P_\circ(\vec{r}) = P_i + P_{i+1} \approx 0$. Because of chirality, the directions of the spontaneous tilt and the in-plane polarization slowly precess around the layer normal as one moves along the direction perpendicular to the smectic plane. This causes a small deviation from the 180° alternation in the tilt between two consecutive layers and the formation of a modulated, helicoidal structure. Since the basic structural unit of the $smectic\text{-}C_A^*$ phase are two neighboring layers, we have here in fact a double-twisted helicoidal structure, formed by two identical ferroelectric $smectic\text{-}C^*$ helices gearing into each other as shown in Fig.8.1.1b. The periodicity of this helical modulation is of the order of the wavelength of the visible light and is in general incommensurate to the basic antiferroelectric unit cell of the *smectic* - C_A^* phase.

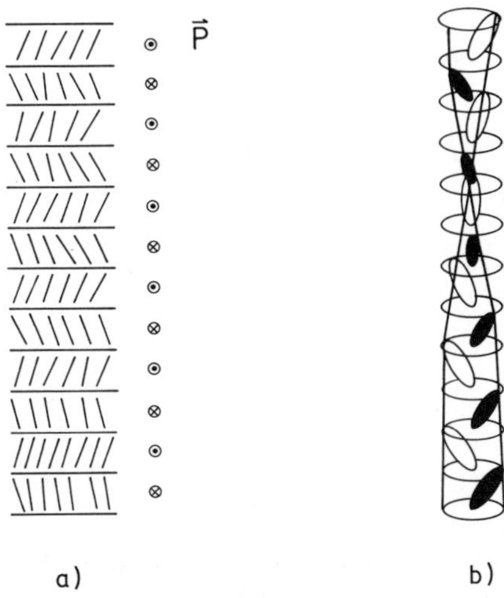

Fig.8.1.1. (a) The alternating-tilt model of the homogeneous antiferroelectric *smectic* - C_A^* phase of liquid crystals and (b) the antiferroelectric double helix.

What is the evidence for this extraordinary alternating-tilt structure of the antiferroelectric phase? Most of the experiments, which have been performed in the bulk antiferroelectric phase give only an indirect evidence for this alternating-tilt structure. However, there are several experiments on freely-suspended or floating antiferroelectric films of the thickness of only several smectic layers thickness, which directly confirm the alternate-tilt structure. Let us now overview briefly the most significant physical properties of the antiferroelectric *smectic* - C_A^* phase and the relevant experiments:

(i) The dielectric constant of the antiferroelectric phase is very small and nearly equal to the dielectric constant of the non-polar *smectic* - *A* phase of the same material, as shown in Fig.8.1.2. This suggests that the orientational susceptibility of the permanent electric polarization is highly suppressed in this phase, as expected if the period of the precession of the

polarization structure is very small, presumably of the order of the thickness of the smectic layers.

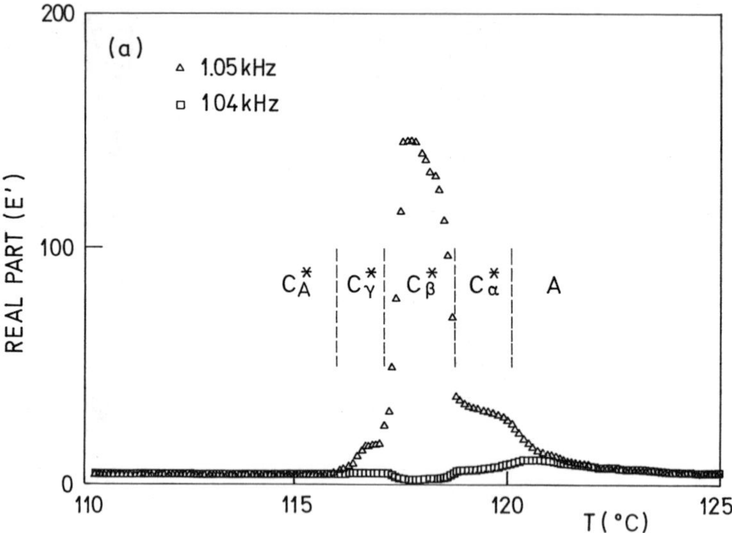

Fig.8.1.2. Dielectric constant of 4-(1-methylheptyloxycarbonyl) 4'-octylbiphenyl-4-carboxylate (MHPOBC), after Hiraoka et al. (1990). Note the very low value of the dielectric constant in the antiferroelectric $smectic\text{-}C_A^*$ phase.

(ii) In very thin, surface stabilized cells, the so-called tristable switching is observed, as shown in Fig.8.1.3.(a). Three stable, optically distinct states are observed when one applies an external DC electric field. Whereas two states are identified as unwound, electric-field induced ferroelectric states, the third state (i.e. antiferroelectric, alternating) has the same optical properties as the $smectic-A$ phase. This can be explained if the antiferroelectric phase has an intrinsic, very short helical modulation. The period of this modulation must be much smaller than the wavelength of visible light. As a conjecture, the shortest possible modulation equals two smectic layers and results in a herringbone-like arrangement of the neighboring layers, as shown in Fig.8.1.3.(b).

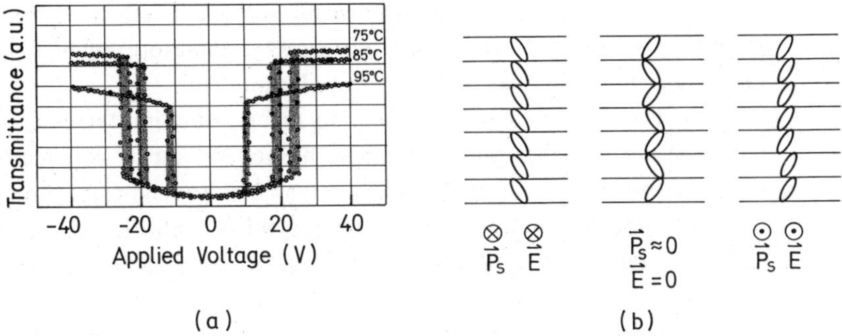

Fig.8.1.3. (a) The tristable switching in a thin layer of an antiferroelectric $smectic\text{-}C_A^*$ liquid crystal R-TFMNPOBC. At small voltages, the intensity of light, transmitted through crossed polarizers increases quadratically with the applied field up to a threshold electric field, where the intrinsic antiferroelectric structure is "unwound" and we have a ferroelectric structure. Similar switching can be obtained by spontaneous polarization measurements. (b) The unwinding of an antiferroelectric smectic - C_A^* structure in an external electric field. After Johno et al. (1990).

(iii) For oblique incidence, the reflection spectrum of the antiferroelectric phase for the visible light reveals the existence of a single reflection peak, which is located at wavelength of the Bragg reflection (Chandani et al., 1989). A polar helicoidal smectic structure should exhibit two reflection peaks (see Chapter5.): The first peak, which originates from the "polar" part of the dielectric tensor corresponds to a Bragg reflection with a wave-vector $q_c = 2\pi/p$, where p is the period of the helical superstructure. This is the so-called full-pitch band. The second reflection peak corresponds to a Bragg scattering with a wave-vector $2q_c$, which is the so-called half-pitch band. Formally, it originates from the so-called "quadrupolar" part of the dielectric tensor. In the antiferroelectric $smectic\text{-}C_A^*$ phase only this latter, half-pitch band is observed. This

clearly indicates that in the antiferroelectric phase we have a helicoidal superstructure (because there is a Bragg scattering), which is locally non-polar, as evidenced by the absence of the full-pitch reflection band. The symmetry of the antiferroelectric structure is therefore equal to that of cholesteric liquid crystals, with a non-polar unit cell.

The difference of the reflection spectra of chiral ferroelectric and chiral antiferroelectric phases can also be understood by considering the dielectric properties of both phases when an electromagnetic light wave is incident at some angle, as shown in Fig.8.1.4. Fig.8.1.4.(a) shows the dielectric tensor ellipsoid for each smectic layer of the ferroelectric $smectic\text{-}C^*$ phase. Fig.8.1.4(a) also shows molecules that are half a helical period away and their directions are therefore rotated by 180 degrees around the smectic normal. Let us now consider what dielectric constant is experienced by the electric field of an electromagnetic wave, which is incident at some oblique angle. Obviously, the dielectric constant for this oblique incidence strongly varies as we move along the normal to the smectic layers. As shown, it is rather large for the molecules at the top and rather small for the molecules at the bottom. This means, that in the wave equation for the propagation of a light wave at oblique incidence, there is a component of a dielectric constant that has a period equal to the full period of the helix.

On the other hand, the situation is quite different for oblique propagation of light in the chiral antiferroelectric $smectic\text{-}C_A^*$ phase, as shown in Fig.8.1.4.(b). Here, we have a sequence of two smectic layers with nearly antiparallel orientation of the molecules. The dielectric constant, experienced by an electromagnetic field with a wavelength much larger than the thickness of the smectic layer, is an average over two neighboring smectic layers. The dielectric tensor has a period of one half of the helical period. This means, that in the wave equation for the oblique propagation of light in the antiferroelectric $smectic\text{-}C_A^*$ phase, there is no component of the dielectric constant with a period equal to the period of the helix. Hence, there is no Bragg reflection at the wave vector $q_c = 2\pi/p$.

This reasoning motivated Chandani et al. (1989) to suggest a locally alternating-tilt structure. The unit cell of such a structure consists of two smectic layers with antiparallel orientation of the molecular dipole moments, respectively.

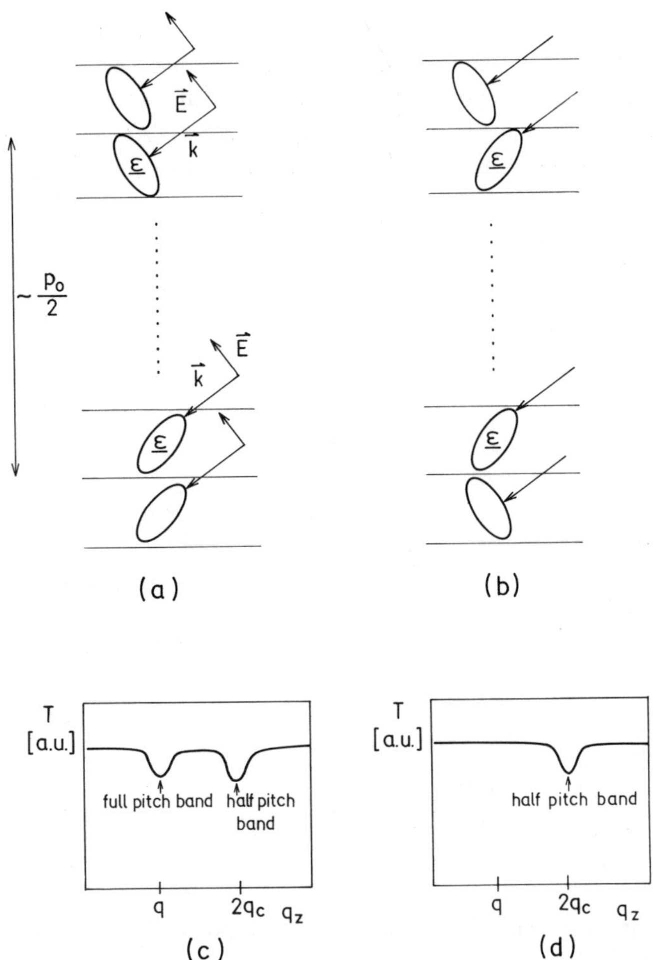

Fig.8.1.4. (a) The illustration of a light wave, propagating at an oblique angle in a ferroelectric $smectic-C^*$ phase. (b) The illustration of light propagation in an antiferroelectric $smectic-C_A^*$ liquid crystal at oblique incidence. $\vec{k}$ is the wave-vector and $\vec{E}$ is the electric field of light. (c) The transmission spectrum of a ferroelectric $smectic-C^*$ phase for oblique incidence. (d) The transmission spectrum of an antiferroelectric $smectic-C_A^*$ liquid crystal for oblique incidence. Here, $q_c = 2\pi/p$ is the wave-vector of the helical structure with a period p.

(iv) A more direct proof of the alternating-tilt structure of the $smectic\text{-}C_A^*$ antiferroelectric phase has been given by linear-electrooptic-response experiments of Galerne and Liebert (1990) on freely floating films of antiferroelectric MHTAC and by the experiments of Bahr and Fliegner (1993a) on freely suspended films with varying numbers of layers, as shown in Fig.8.1.5.

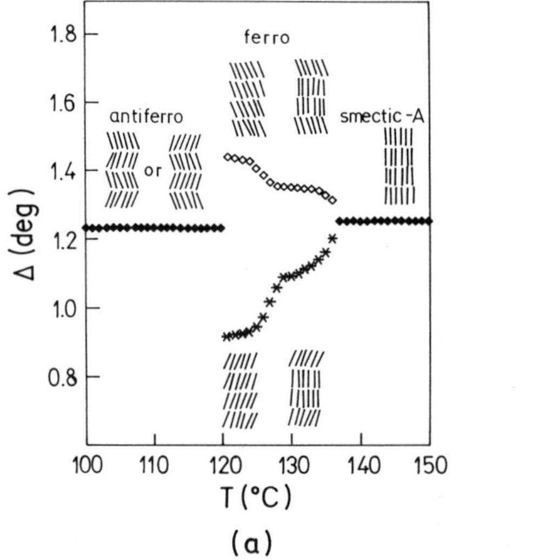

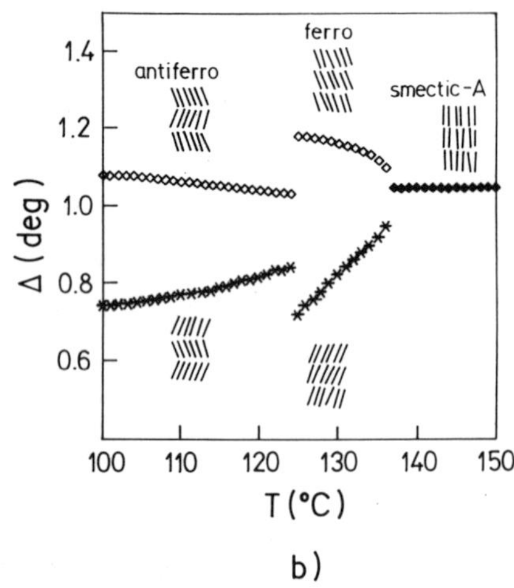

Fig.8.1.5. The linear electrooptic response for four (a) and three (b) smectic layers of freely suspended 4-(1-methylheptyloxycarbonyl) 4'-octylbiphenyl-4-carboxylate (MHPOBC), after Bahr and Fliegner, (1993a). Note the absence of the linear response for an even number of smectic layers in the antiferroelectric $smectic - C_A^*$ phase.

Both groups have observed the so-called parity effects, i.e. a strong changes of the linear coupling of the $smectic\text{-}C_A^*$ phase to an external field when the number of layers was gradually increased in steps of a single molecular layer. They have observed that for an odd number of smectic layers there is a strong linear electrooptic response, whereas for an even number of layers, there is practically no linear response. This directly confirms the conjecture that the spontaneous polarization is compensated in the two neighboring layers for an even number of layers, as predicted by the alternating-tilt model. For an odd number of layers, the film is polar due to polarization of unpaired layer and shows ferroelectric-like response.

(v) The most direct proof of the alternating-tilt structure of the $smectic\text{-}C_A^*$ antiferroelectric phase was given by the resonant X-ray scattering experiment by Mach et al. in 1998. Whereas the normal X-ray scattering is insensitive to the direction of the long molecular axis of a given smectic layer, in the resonant X-ray scattering experiment the cross-section depends on the orientation of the director with respect to the X-ray beam. In such an experiment, the energy of the X-rays is tuned close to the absorption spectrum of some atom in the liquid crystal molecule. In this case, the cross-section for the X-ray scattering becomes orientation-sensitive and enables one to monitor the orientational distribution on the molecular level. Using this technique, Mach et al. (1998) have measured the dependence of the resonant-scattering intensity on the magnitude of the scattering wave-vector q_z along the smectic normal, as shown in Fig.8.1.6. For example, in the scan that was measured in the antiferroelectric $smectic-C_A^*$ phase (Fig.8.1.6.(d)), one can see satellite peaks at $q_z/q_\circ = 1.5$ and $q_z/q_\circ = 2$, respectively. This means that there is an extra modulation in a system and the wave-vector of this modulation is $q_l = 2\pi/l = 0.5 \cdot q_\circ$. This corresponds to modulation length $l = 2d$, which is exactly an antiferroelectric unit cell, composed of two antiparallel ferroelectric and tilted layers. One can see a small splitting of $q_z/q_\circ = 1.5$ satellite, which is due to helical modulation. This is absent in racemic liquid crystal, shown in Fig.8.1.6e.

(vi) It should be also mentioned that the deuteron NMR spectra of antiferroelectric liquid crystals as well provide a direct microscopic proof of the alternating tilt model of the $smectic-C_A^*$ phase as discussed in Section 7.6.

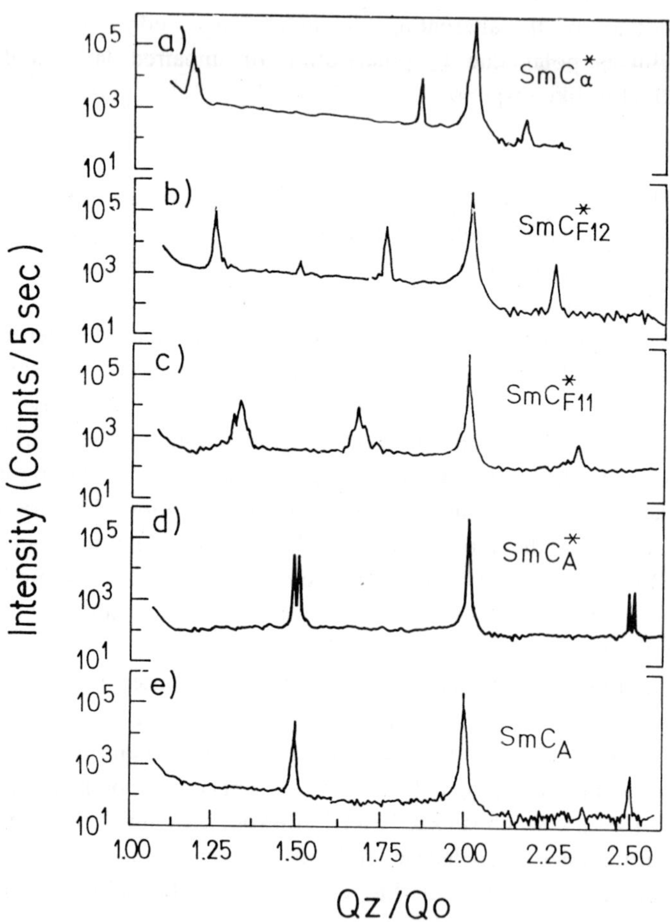

Fig.8.1.6. Resonant X-ray intensity scans in various phases of pure (R) (a-d) and racemic 10OTBBB1M7 (e). Here, q_z is the component of the wave-vector of the X-ray along the normal to the smectic layers and $q_\circ = 2\pi/d$, where d is the thickness of a smectic layer. After Mach et al., 1998.

Note 8.1.1. Polarization of antiferroelectric liquid crystals:
Typically, the antiferroelectric phase of liquid crystals is observed in materials with rather large polarization. For example, the series of MHPnCBC liquid crystals exhibits the following value of spontaneous polarization (Rauch, 1996):

n	8	9	10	11	12	14	16
P_s /nCcm^{-2}	61	71	85	·91	103	102	86

8.2. Landau Continuum Theory of the Antiferroelectric *smectic-A-smectic-C_A^** Phase Transition

The first theoretical description of the antiferroelectric ordering in smectic liquid crystals was given by Orihara and Ishibashi in 1990. Their work was followed by a variety of different formulations of either continuous (Orihara and Ishibashi, 1990; Žekš and Čepič, 1993; Lorman et al. 1994) or discrete models (Yamashita, 1991; Yamashita and Miyazima,1993; Yamashita, 1996a-c; Čepič and Žekš 1995, 1996; Žekš et al., 1991, 1993) in order to explain phase sequences, which were observed in antiferroelectric liquid crystalline substances.

Orihara and Ishibashi have defined the ferroelectric order parameter ξ_f and the antiferroelectric order parameter ξ_a with respect to a pair of two adjacent smectic layers labeled with indices j and $j+1$, respectively. The ferroelectric order parameter is a sum of the two order parameters in the individual neighboring layers,

$$\vec{\xi}_f = \tfrac{1}{2}\left(\vec{\xi}_i + \vec{\xi}_{i+1}\right) \tag{8.2.1}$$

whereas the antiferroelectric order parameter is a difference of the two order parameters

$$\vec{\xi}_a = \tfrac{1}{2}\left(\vec{\xi}_i - \vec{\xi}_{i+1}\right) \tag{8.2.2}$$

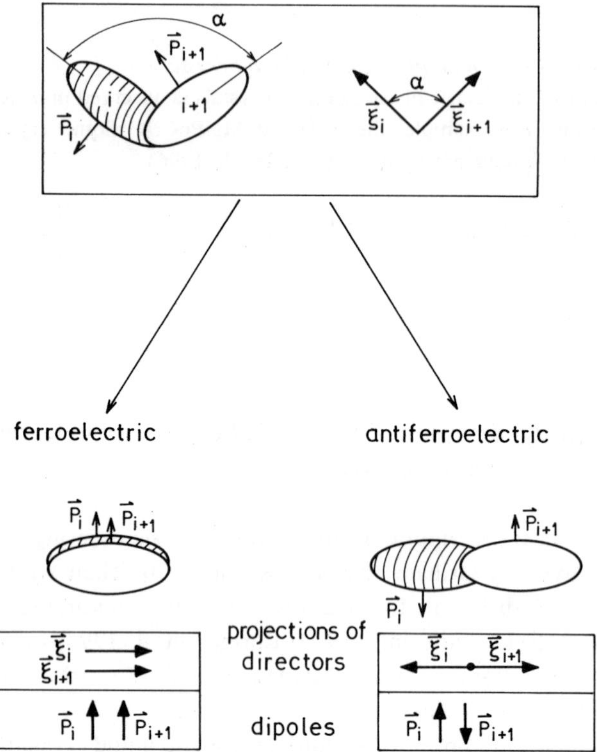

Fig.8.2.1. The ferroelectric structure is obtained in the limit of a very small angle α between the projections of the molecular directors in the two neighboring layers. The antiferroelectric structure is obtained in the limit of an antiparallel ordering of projections of molecular directors in the two neighboring layers, resulting in antiparallel dipoles. The tilt is constant throughout the sample.

Here, ξ_i is the projection of the director $\vec{n}$ in the i-th smectic layer onto the i-th smectic layer. They have also defined two additional secondary order parameters, i.e. the ferroelectric polarization

$$\vec{P}_f = \tfrac{1}{2}\left(\vec{P}_i + \vec{P}_{i+1}\right) \tag{8.2.3}$$

and the "antiferroelectric" polarization

$$\vec{P}_a = \tfrac{1}{2}\left(\vec{P}_i - \vec{P}_{i+1}\right) \tag{8.2.4}$$

Similarly to the theory of the $smectic-C^*$ phase, $\xi_i = (n_{ix}n_{iz}, n_{iy}n_{iz})$ is the projection of the director field of the i-th layer onto the smectic plane, and $\vec{P}_i = (P_{ix}, P_{iy}, 0)$ is the in-plane polarization of the i-th smectic layer.

What is the physical meaning of these order parameters? Let us consider that the magnitude of the molecular tilt is equal throughout the sample and the projections $\vec{\xi}_i$ and $\vec{\xi}_{i+1}$ are at an angle α, as shown on the top of Fig.8.2.1.

If the angle α is very small, we have the ferroelectric structure, shown on the bottom left-side of Fig.8.2.1. Here, the two ferroelectric order parameters ξ_f and P_f are large, whereas the two antiferroelectric order parameters are negligibly small, $\vec{\xi}_a \approx 0$ and $\vec{P}_a \approx 0$. On the other hand, if the angle α is close to π, we have an antiferroelectric ordering, where the two antiferroelectric order parameters $\vec{\xi}_a$ and $\vec{P}_a$ are large, whereas the two ferroelectric order parameters $\vec{\xi}_f$ and $\vec{P}_f$ are very small. In principle, we may also have stable phases, where both order parameters are finite, i.e. $\xi_a \neq 0$ and $\vec{\xi}_f \neq 0$. This is the so-called ***ferrielectric phase***

It should be stressed that within the above definition of the order parameters, the unit cell ***always consists of only two smectic layers or a single smectic layer***. This model therefore cannot reproduce structures, where a unit cell is formed of more than two smectic layers. For this purpose, a discrete lattice model has to be used.

The nonequilibrium free-energy density is expanded in terms of combinations of order parameter components which represent scalar invariants of the D_∞ point group of the $smectic-A$ phase

$$g = g_A + \tfrac{1}{2}a_a\vec{\xi}_a^{\,2} + \tfrac{1}{4}b_a\vec{\xi}_a^{\,4} + \Lambda_a\left(\xi_{ax}\frac{\partial\xi_{ay}}{\partial z} - \xi_{ay}\frac{\partial\xi_{ax}}{\partial z}\right) + \tfrac{1}{2}K_{3a}\left(\frac{\partial\vec{\xi}_a}{\partial z}\right)^2 +$$

$$+ \tfrac{1}{2}a_f\vec{\xi}_f^{\,2} + \tfrac{1}{4}b_f\vec{\xi}_f^{\,4} + \Lambda_f\left(\xi_{fx}\frac{\partial\xi_{fy}}{\partial z} - \xi_{fy}\frac{\partial\xi_{fx}}{\partial z}\right) + \tfrac{1}{2}K_{3f}\left(\frac{\partial\vec{\xi}_f}{\partial z}\right)^2 +$$

$$+ \tfrac{1}{2}\gamma_1\vec{\xi}_a^{\,2}\vec{\xi}_f^{\,2} + \tfrac{1}{2}\gamma_2\left(\vec{\xi}_a\vec{\xi}_f\right)^2 +$$

$$+\tfrac{1}{2\varepsilon_a}\vec{P}_a^2 + C_a\left(\vec{P}_a \times \vec{\xi}_a\right)_z + \tfrac{1}{2\varepsilon_f}\vec{P}_f^2 + C_f\left(\vec{P}_f \times \vec{\xi}_f\right)_z \qquad (8.2.5)$$

The free-energy density is composed of two equivalent expansions, one for the ferroelectric and the other for the antiferroelectric order parameter. Each of them is similar to the simple free-energy density expansion for the ferroelectric $smectic-C*$ phase (see Eq.2.4.1). We have therefore two harmonic coefficients $a_a(T)$ and $a_f(T)$ which are both temperature dependent, whereas the other coefficients are temperature independent. The b_a and b_f terms are added for stability, the two Lifshitz coefficients Λ_a and Λ_f induce the helicoidal precession of either the ferro- or the antiferroelectric order parameter and the torsional elastic constants K_a and K_f describe the increase of the free-energy density due to elastic torsional deformations. Similarly to the ferroelectric liquid crystals, the two polarizations are here secondary order parameters, which are coupled to the primary order parameters ξ_a and ξ_f via the two piezoelectric coefficients C_a and C_f. The flexoelectric term has been here omitted for simplicity.

The two order parameters ξ_a and ξ_f are coupled via the lowest order invariants with the coefficients γ_1 and γ_2, which represent competition between the ferroelectric and the antiferroelectric helical ordering. For positive $\gamma_1 > 0$, the system tries to reduce the magnitudes of the two order parameters, whereas for negative $\gamma_1 < 0$ the two order parameters are enhanced. The γ_2 coefficient dictates the mutual orientation of the two order parameters, as will be discussed further on.

In the absence of any coupling coefficients γ_1 and γ_2, the free-energy density (Eq.8.2.5) induces either the $smectic-A \rightarrow$ $smectic-C_A^*$ or the $smectic-A-smectic-C^*$ transition, depending on the temperature dependence of the two harmonic coefficients. We are therefore free to choose

$$a_f = \alpha_f\left(T - T_f^\circ\right) \text{ and } a_a = \alpha_a\left(T - T_a^\circ\right) \qquad (8.2.6)$$

where T_f° and T_a° are the transition temperatures into the spatially homogeneous ferroelectric and antiferroelectric phases, respectively. If $T_f^\circ > T_a^\circ$, the ferroelectric phase is below the $smectic-A$ phase and vice versa. In the following we shall assume that the two coefficients α_f and α_a are equal, which expresses the fact that the entropy contribution to a, (i.e. the part proportional to the temperature), is the same both for the ferroelectric and the antiferroelectric ordering. On the contrary, the two transition temperatures are different, i.e. $T_f^\circ \neq T_a^\circ$, expressing the fact that the energies of the two states are different.

In the absence of the coupling terms, we would have a phase transition $smectic-A-smectic-C^*$ for $T_f^\circ > T_a^\circ$ and the $smectic-A-smectic-C_A^*$ transition for $T_f^\circ < T_a^\circ$. The coupling term locks-in the two helices and may induce incommensurably modulated structures (Čepič, 1993). In the following we shall discuss the main features of the $smectic-A-smectic-C_A^*$ transition. Other possible phase sequences are discussed elsewhere (Žekš and Čepič, 1993).

In analogy to the phase transition into the ferroelectric $smectic-C^*$ phase we use here the ***adiabatic approximation*** for the motion of the two polarization fields P_a and P_f. This means that the polarization motion is so fast, that it is ***always in thermal equilibrium*** with the director field. The difference between the two time scales is several orders of magnitude. This implies the minimization of the free-energy density (Eq.8.2.5) with respect to P_a and P_f. This leads to a set of equations, linking the instantaneous value of the polarizations P_a and P_f with the two order parameters ξ_a and ξ_f:

$$P_{f,x} = -\varepsilon_f C_f \cdot \xi_{f,y} \tag{8.2.7a}$$

$$P_{f,y} = \varepsilon_f C_f \cdot \xi_{f,x} \tag{8.2.7b}$$

and

$$P_{a,x} = -\varepsilon_a C_a \cdot \xi_{a,y} \tag{8.2.8a}$$

$$P_{a,y} = \varepsilon_a C_a \cdot \xi_{a,x} \tag{8.2.8b}$$

After inserting the equilibrium values of the two polarization fields into the free-energy expansion, we obtain the so-called polarization-renormalized free-energy density:

$$g = g_A + \tfrac{1}{2}a_a\bar{\xi}_a^2 + \tfrac{1}{4}b_a\bar{\xi}_a^4 + \Lambda_a\left(\bar{\xi}_a \times \partial\bar{\xi}_a/\partial z\right)_z + \tfrac{1}{2}K_{3a}\left(\frac{\partial\bar{\xi}_a}{\partial z}\right)^2 +$$

$$+ \tfrac{1}{2}a_f\bar{\xi}_f^2 + \tfrac{1}{4}b_f\bar{\xi}_f^4 + \Lambda_f\left(\bar{\xi}_f \times \partial\bar{\xi}_f/\partial z\right)_z + \tfrac{1}{2}K_{3f}\left(\frac{\partial\bar{\xi}_f}{\partial z}\right)^2 +$$

$$+ \tfrac{1}{2}\gamma_1\bar{\xi}_a^2\bar{\xi}_f^2 + \tfrac{1}{2}\gamma_2\left(\bar{\xi}_a\bar{\xi}_f\right)^2 \tag{8.2.9}$$

Here only the coefficients a_a and a_f are renormalized by the polarizations. This can be interpreted as a small increase of the two transition temperatures T_a and T_f (see also Chapter 2)

Similarly to the stability analysis of the ferroelectric $smectic-C^*$ phase (see Section 2.1), we look at the stability of the $smectic-A$ phase, when small helicoidal fluctuations of the two order parameters $\delta\xi_a$ and $\delta\xi_f$ are created due to the thermal fluctuations in the system:

$$\delta\xi_{f,x} = \sum_q \left(x_q \cos qz - y_q \sin qz\right) \tag{8.2.10a}$$

$$\delta\xi_{f,y} = \sum_q \left(x_q \sin qz + y_q \cos qz\right) \tag{8.2.10b}$$

$$\delta\xi_{a,x} = \sum_q \left(u_q \cos qz - v_q \sin qz\right) \tag{8.2.10c}$$

$$\delta\xi_{a,y} = \sum_q \left(u_q \sin qz + v_q \cos qz\right) \tag{8.2.10d}$$

The corresponding increase of the total free-energy of the $smectic-A$ phase is

$$G = \tfrac{1}{L}\int_0^L \Delta g \cdot dz = \sum_q \left[\left(\tfrac{1}{2}a_f(T) + \Lambda_f q + \tfrac{1}{2}K_{3f}q^2\right)\left(x_q^2 + y_q^2\right)\right] +$$

$$+ \sum_q \left[\left(\tfrac{1}{2}a_a(T) + \Lambda_a q + \tfrac{1}{2}K_{3a}q^2\right)\left(u_q^2 + v_q^2\right)\right] \tag{8.2.11}$$

For each wave-vector q we have two uncoupled fluctuations in the *smectic* $-A$ phase of an antiferroelectric liquid crystal. The first contribution favors the ferroelectric ordering, whereas the second one favors the antiferroelectric ordering. If all harmonic coefficients of the free-energy (8.2.11) are positive,

$$\forall q: \quad A_{f,q} = \left(\tfrac{1}{2} a_f(T) + \Lambda_f q + \tfrac{1}{2} K_{3f} q^2\right) > 0 \tag{8.2.12a}$$

$$\forall q \quad : A_{a,q} = \left(\tfrac{1}{2} a_a(T) + \Lambda_a q + \tfrac{1}{2} K_{3a} q^2\right) > 0 \tag{8.2.12b}$$

the equilibrium values of the order parameters would always be zero and there would be no transition.

The phase transition into the ferroelectric or antiferroelectric phase is induced when the smallest of the coefficients $A_{f,q}$ or $A_{a,q}$ becomes negative. Therefore, for $T_a > T_f$, the smallest of the coefficients $A_{a,q}$ has to be considered and we find from Eq.8.2.12. that it corresponds to the wave-vector

$$q_a = -\frac{\Lambda_a}{K_{3a}} \tag{8.2.13}$$

and becomes negative at

$$T_a^* = T_a + \frac{\Lambda_a^2}{\alpha_a K_{3a}} \tag{8.2.14}$$

The phase transition temperature for the helicoidally modulated antiferroelectric *smectic* $-C_A^*$ phase is therefore slightly higher than for the homogeneous phase. This is a general phenomenon, which is observed in all twisted, helical phases.

For $T \geq T_a^*$, we have therefore a stable phase with

$$x_q^\circ = y_q^\circ = u_q^\circ = v_q^\circ = 0 \tag{8.2.15}$$

which represents the $smectic - A$ phase. For $T < T_a^*$ the helicoidally modulated $smectic - C_A^*$ phase is stable with

$$x_q = y_q = 0 \qquad \text{and} \qquad u_q \neq 0 \ , \ v_q = 0 \tag{8.2.16}$$

or

$$x_q = y_q = 0 \qquad \text{and} \qquad u_q = 0 \ , \ v_q \neq 0 \tag{8.2.17}$$

The Eq.8.2.16. and Eq.8.2.17. describe two antiferroelectric helicoidally modulated structures, which differ only by the value of the phase angle at $z = 0$. We have therefore for $T \leq T_a^*$ a helicoidal $smectic - C_A^*$ phase

$$\xi_{a,x} = \theta_a \cdot \cos q_a z \tag{8.2.18a}$$

$$\xi_{ay} = \theta_a \cdot \sin q_a z \tag{8.2.18b}$$

with the equilibrium tilt angle given by

$$\theta_a = \sqrt{(\alpha_a / b_a)(T_a^* - T)} \tag{8.2.19}$$

The $smectic - A \rightarrow smectic - C_A^*$ phase transition is here completely analogous to the Indenbom-Pikin-Loginov model of the ferroelectric phase transition. Both models predict a temperature independent period of the modulation of the low-temperature phase and mean-field critical exponents. The only difference is here the structure of a unit cell. Whereas in the modulated ferroelectric $smectic - C$ phase the unit cell is formed by a single tilted smectic layer, this unit cell is doubled in the modulated antiferroelectric $smectic - C_A^*$ phase.

8.3. Order Parameter Dynamics and the Doubling of the Smectic Unit Cell

The onset of the antiferroelectric ordering at the *smectic* $-$ *A* - *smectic* $- C_A^*$ phase transition is characterized by a condensation of a non-polar antiferroelectric soft mode, that breaks the continuous symmetry of the *smectic* $-$ *A* phase. This can be most easily visualized by showing the effects of the soft antiferroelectric fluctuations in the *smectic* $-$ *A* phase in Figures 8.3.1(a) and (b). The soft mode can be here considered as a herring-bone-like collective orientational fluctuation. As we approach the *smectic* $-$ *A* - *smectic* $- C_A^*$ phase transition from above, these fluctuations become slower and slower and finally freeze-out at the phase transition.

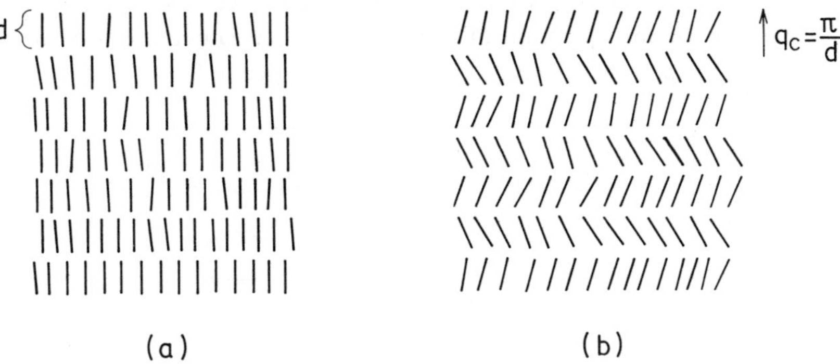

Fig.8.3.1. (a) The equilibrium *smectic* $-$ *A* phase. (b) The eigenvector of the soft antiferroelectric fluctuation, superposed to the equilibrium director field of the *smectic* $-$ *A* phase. The critical wave-vector is close to $q = \pi/d$, where d is the layer spacing.

The wave-vector of the antiferroelectric soft mode is $q = \pi/d$, where d is a distance between the two neighboring smectic layers. If we consider the *smectic* $-$ *A* a one-dimensional solid, the Brillouin zone (BZ) of this linear chain is $(-\pi/d, \pi/d)$ and the antiferroelectric soft mode is therefore the zone-boundary mode that condenses at the edge of the smectic BZ.

The condensation of a zone-boundary soft mode has the effect of doubling of the smectic unit cell in the $smectic-C_A^*$ phase and the original BZ is reduced to:

$$(-\pi/d,\pi/d) \rightarrow (-\pi/2d,\pi/2d) \quad (8.3.1)$$

The doubling of a smectic unit cell at the $smectic-A$-$smectic-C_A^*$ phase transition has implications for the spectrum of collective excitations of the antiferroelectric phase (Sun et al., 1993a,b). These can be most easily understood by considering analogous phenomena in the solid-state theory, i.e. the excitation phonon spectrum of a 3-D crystalline lattice.

Let us consider lattice vibrations of a solid crystal with a unit cell that consists of N atoms. There are $3N$ degrees of freedom per each unit cell of a crystal. It is well known from the solid state theory (see, for example Ziman, 1962) that the phonon spectrum of such a crystalline lattice consists of three acoustic branches and $3n-3$ optical branches, as shown in Fig.8.3.2.

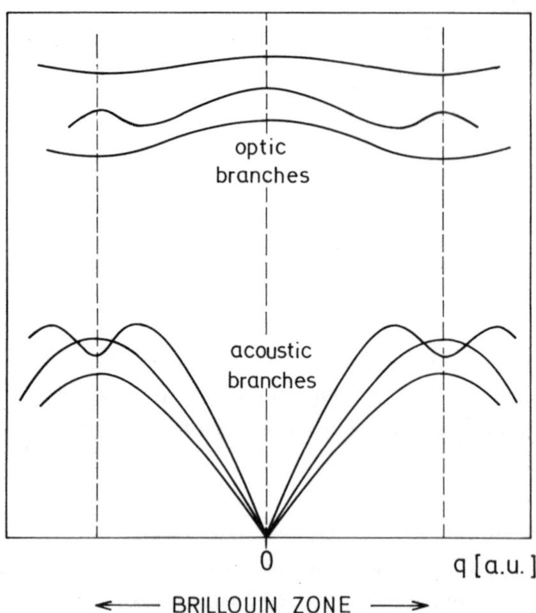

Fig.8.3.2. The phonon spectrum of a 3-D crystal with n=2 atoms in the unit cell. The drawing is schematic.

The three acoustic branches can be roughly classified into two transversal and one longitudinal branch of excitations. The frequency of each of these three branches vanishes in the limit of small wave-vectors, or infinite wavelengths. The phonon spectrum is periodic in the reciprocal space and the maximum phonon frequency is limited by the material properties. We therefore use the repeated Brillouin zone concept to represent the phonon spectrum.

The phonon spectrum of a linear chain of identical atoms is shown in Fig.8.3.3. The spectrum shows a single excitation branch, which is also known as the ***acoustic phonon branch*** and corresponds to an "along-the-chain" collective movement of the atoms. The spectrum changes significantly for diatomic linear chain. Here, an additional, the so-called ***optic phonon branch*** appears in the spectrum, reflecting the presence of two different atoms per unit cell.

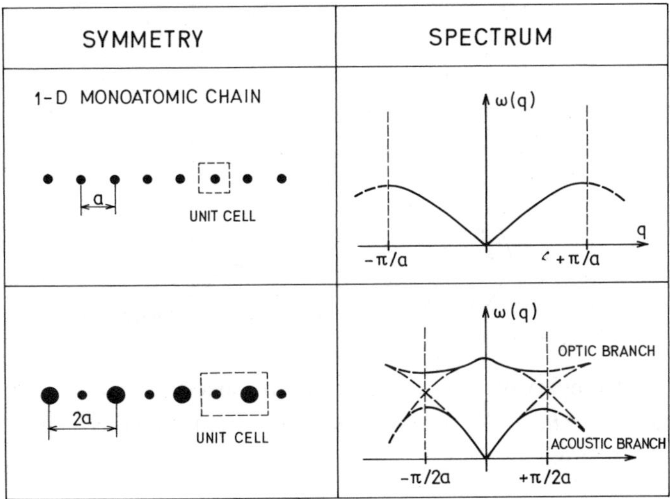

Fig.8.3.3. Phonon spectrum of a linear chain of identical atoms and the phonon spectrum of a diatomic chain.

Similar to a monatomic linear chain, the unit cell of the $smectic-C^*$ phase is a single smectic layer, as shown in Fig.8.3.4. The spectrum consists of two dispersion branches because there are two degrees of freedom for the orientational motion: a gapless phason branch, which is analogous to the acoustic phonon branch in solids and the amplitudon branch, as shown in Fig.8.3.4. Similar to the diatomic linear chain, the unit cell of the homogeneous antiferroelectric $smectic-C_A^*$ phase consists of two smectic layers, as shown in Fig.8.3.4. Because of this doubling of the unit cell, the number of dispersion branches is doubled in the antiferroelectric phase and we expect two dispersion branches for phase fluctuations and two dispersion branches for amplitude fluctuations.

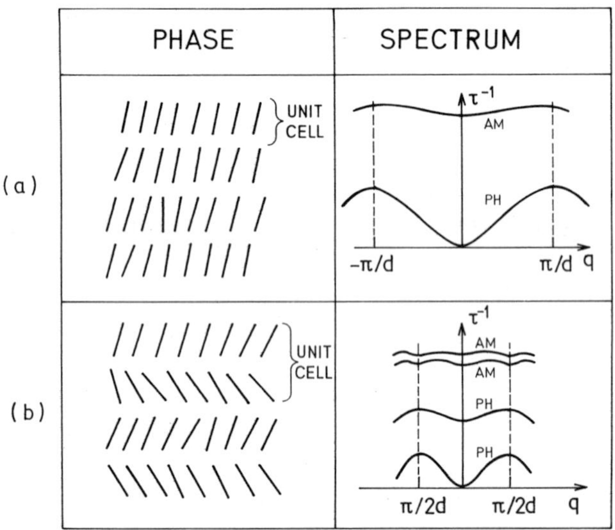

Fig.8.3.4. Comparison of the excitation spectrum of a homogeneous ferroelectric $smectic-C^*$ phase and the homogeneous antiferroelectric $smectic-C_A^*$ phase.

The onset of four dispersion branches in the antiferroelectric $smectic-C_A^*$ phase can be understood by looking at the soft mode condensation at the $smectic-A$ - $smectic-C_A^*$ phase transition. In the $smectic-A$ phase of an antiferroelectric liquid crystal there is a doubly degenerate branch of antiferroelectric collective excitations that critically slows-down at the zone-boundary, as shown in Fig.8.3.5(a). At the phase transition, the zone-boundary modes appear as the zone-center modes of the low-temperature phase due to the doubling of a unit cell, as shown in Fig.8.3.5(b). As a consequence, there are therefore four dispersion branches in the $smectic-C_A^*$ phase. The lowest branch is the *gapless phason branch*, with a zero-frequency, symmetry restoring Goldstone mode, that is analogous to the acoustic phonon branch of a diatomic lattice (Figure.8.3.4).

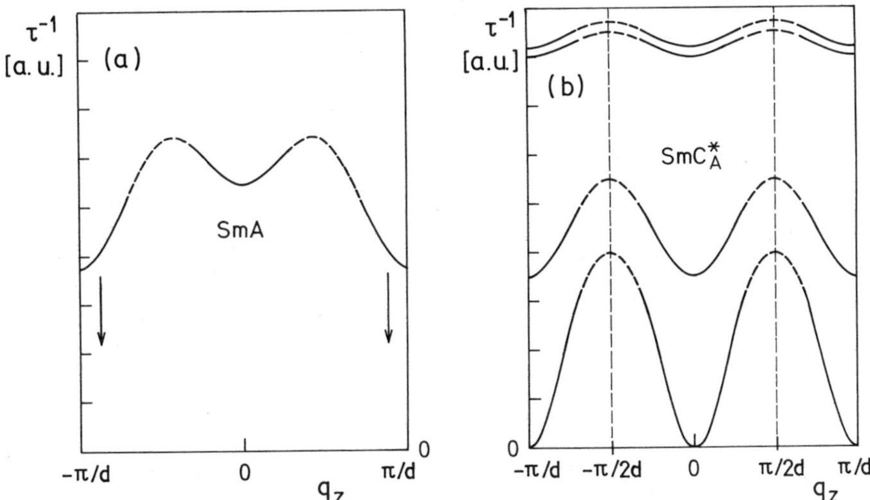

Fig.8.3.5. (a) The excitation spectrum of the $smectic-A$ phase near the phase transition into the antiferroelectric $smectic-C_A^*$ phase. The soft mode branch is doubly-degenerate and critical at the zone-boundary. (b) The excitation spectrum of the $smectic-C_A^*$ phase in the reduced zone representation. The two upper branches represent the amplitudon excitations and both show a strong temperature dependence. The two lower branches are the phase excitations branches. The lowest is *gapless (acoustic)* whereas the upper-one always has a finite frequency gap and is called the *optical branch*. The doubling of the number of the phason and amplitude modes is due to the doubling of the unit cell and the occurrence of "in-phase" and "out-of-phase" director motions.

In addition to the two amplitude branches that are strongly temperature dependent, an *optic-like branch* of phase excitations is expected to appear in the antiferroelectric phase. Similarly to the optical phonon branch in a diatomic solid crystalline lattice (see Figure 8.3.3.), the appearance of this optical phason branch is here a direct consequence of doubling of smectic unit cell.

The excitation spectrum in the vicinity of the $smectic-A$-$smectic-C_A^*$ phase transition can be calculated either within a continuous (Žekš et al., 1991; Čepič, 1993) or discrete lattice model (Sun et al., 1993a,b). Following the continuous model approach of Čepič (1993), we rewrite the polarization-renormalized free energy density (Eq.8.2.9) as

$$g = g_A + \tfrac{1}{2}a_a\vec{\xi}_a^{\,2} + \tfrac{1}{4}b_a\vec{\xi}_a^{\,4} + \Lambda_a\left(\vec{\xi}_a \times \partial\vec{\xi}_a/\partial z\right)_z + \tfrac{1}{2}K_{3a}\left(\frac{\partial\vec{\xi}_a}{\partial z}\right)^2 +$$

$$+\tfrac{1}{2}a_f\vec{\xi}_f^{\,2} + \tfrac{1}{4}b_f\vec{\xi}_f^{\,4} + \Lambda_f\left(\vec{\xi}_f \times \partial\vec{\xi}_f/\partial z\right)_z + \tfrac{1}{2}K_{3f}\left(\frac{\partial\vec{\xi}_f}{\partial z}\right)^2 +$$

$$+\tfrac{1}{2}\gamma_1\vec{\xi}_a^{\,2}\vec{\xi}_f^{\,2} + \tfrac{1}{2}\gamma_2\left(\vec{\xi}_a\vec{\xi}_f\right)^2 \qquad (8.3.2.)$$

Here only the harmonic coefficients a_a and a_f are temperature dependent,

$$a_a = \alpha\left(T - T_a\right) \text{ and } a_f = \alpha\left(T - T_f\right) \qquad (8.3.3.)$$

whereas the rest of the coefficients is constant. Furthermore we have $T_a > T_f$, which leads to the $smectic-A$-$smectic-C_A^*$ transition at $T_a^* = T_a + \Lambda_a^2/(\alpha K_{3a})$.

The dynamics of chiral systems can be most conveniently analyzed by introducing a twisting reference frame (see Section 2.2), that follows the helix, as we move from one layer to another. The two order parameters are written as a sum of a static and a time-fluctuating part

$$\vec{\xi}_a(z,t) = \vec{\xi}_a^{\,\circ}(z) + \delta\vec{\xi}_a(z,t) \qquad (8.3.4a)$$

$$\vec{\xi}_f(z,t) = \delta\vec{\xi}_f(z,t) \qquad (8.3.4b)$$

Here, we have taken into account that the static part of the ferroelectric order parameter equals zero both in the $smectic-A$ and $smectic-C_A^*$ phases. In the rotating frame, the fluctuating part of the order parameters is further divided into components which are parallel and perpendicular to the local direction of the order parameter, as illustrated in Fig.8.3.6.

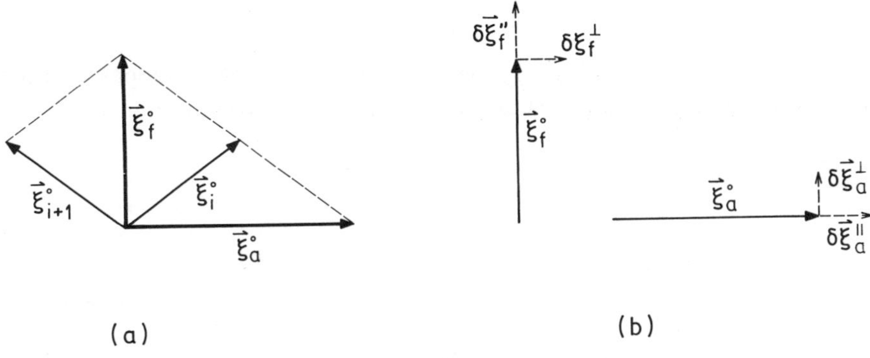

Fig.8.3.6. (a) The definition of the static order parameters ξ_a° and ξ_f°. (b) the two orthogonal components $\delta\xi_{a,f}^{//}$ and $\delta\xi_{a,f}^{\perp}$ of the fluctuating part of the two order parameters ξ_a and ξ_f. We consider the case $\xi_f=0$ and $\xi_a\neq 0$.

The two orthogonal components of the fluctuating part of the order parameters are expanded in terms of plane waves

$$\delta\xi_a^{//}=\sum_q \delta\xi_{a,q}^{//}\cdot e^{iqz} \tag{8.3.5a.}$$

$$\delta\xi_a^{\perp}=\sum_q \delta\xi_{a,q}^{\perp}\cdot e^{iqz} \tag{8.3.5b.}$$

$$\delta\xi_f^{//}=\sum_q \delta\xi_{f,q}^{//}\cdot e^{iqz} \tag{8.3.5c.}$$

$$\delta\xi_f^{\perp}=\sum_q \delta\xi_{f,q}^{\perp}\cdot e^{iqz} \tag{8.3.5d.}$$

where $\delta\xi^j_{i,-q} = \left(\delta\xi^j_{i,q}\right)^*$ to keep the physical variables real and q is given in the rotating frame.

The nonequilibrium value of the free energy $G = \frac{1}{L}\int_0^L g(z)dz$ per unit lengths is

$$\Delta G = G - G_\circ = \sum_q \vec{\Psi}^+_q \cdot \underline{D}(q) \cdot \vec{\Psi}_q \tag{8.3.6}$$

where $\Psi^+_q = \left(\delta\xi^{//}_{a,q}, \delta\xi^{\perp}_{a,q}, \delta\xi^{//}_{f,q}, \delta\xi^{\perp}_{f,q}\right)$ is a 4-dimensional vector, composed of the fluctuating parts of both order parameters and $\underline{D}(q)$ is a "dynamical matrix"

$$\underline{D}(q) = \begin{bmatrix} A_q & iE_q & 0 & 0 \\ -iE_q & B_q & 0 & 0 \\ 0 & 0 & C_q & iF_q \\ 0 & 0 & -iF_q & D_q \end{bmatrix} \tag{8.3.7}$$

with elements

$$A_q = \tfrac{1}{2}\alpha(T - T_a) + \Lambda_a q_a + \tfrac{1}{2}K_{3a}q_a^2 + \tfrac{1}{2}K_{3a}q^2 + \tfrac{3}{2}b_a\theta_a^2(T) \tag{8.3.8a}$$

$$B_q = \tfrac{1}{2}\alpha(T - T_a) + \Lambda_a q_a + \tfrac{1}{2}K_{3a}q_a^2 + \tfrac{1}{2}K_{3a}q^2 + \tfrac{1}{2}b_a\theta_a^2(T) \tag{8.3.8b}$$

$$C_q = \tfrac{1}{2}\alpha(T - T_f) + \Lambda_f q_a + \tfrac{1}{2}K_{3f}q_a^2 + \tfrac{1}{2}K_{3f}q^2 + \tfrac{1}{2}(\gamma_1 + \gamma_2)\theta_a^2(T) \tag{8.3.8c}$$

$$D_q = \tfrac{1}{2}\alpha(T - T_f) + \Lambda_f q_a + \tfrac{1}{2}K_{3f}q_a^2 + \tfrac{1}{2}K_{3f}q^2 + \tfrac{1}{2}\gamma_1\theta_a^2(T) \tag{8.3.8d}$$

$$E_q = q(\Lambda_a + q_a K_{3a}) \tag{8.3.8e}$$

$$F_q = q(\Lambda_f + q_a K_{3f}) \tag{8.3.8f}$$

We can see from the structure of the nonequilibrium free-energy density, that plane waves with different wave-vectors $\vec{q}$ are uncoupled. This means that the eigenwaves, which describe the thermal fluctuations of the order parameter, have the form of monochromatic plane waves. Note that the dynamical matrix is Hermitian,

$$\underline{D}^{+}(q) = \underline{D}(q) \tag{8.3.9}$$

and the corresponding eigenvalues are always real, whereas the eigenvectors are orthogonal (Killingbeck and Cole, 1971).

The dynamical properties are governed by the Landau-Khalatnikov equations of motion

$$\frac{d\delta\vec{\xi}}{dt} = -\Gamma\frac{\partial\Delta G}{\partial\delta\vec{\xi}} \tag{8.3.10a}$$

where Γ is the corresponding kinetic coefficient. Using the matrix formalism, the Landau-Khalatnikov equations of motion of Ψ_q can be expressed as

$$\frac{d\vec{\Psi}_q}{dt} = -2\underline{\Gamma}\cdot\underline{D}(q)\cdot\vec{\Psi}_q \tag{8.3.10b}$$

where $\underline{\Gamma}$ is a matrix, composed of the reciprocal generalized viscosity,

$$\underline{\Gamma} = \begin{bmatrix} \Gamma_a & 0 & 0 & 0 \\ 0 & \Gamma_a & 0 & 0 \\ 0 & 0 & \Gamma_f & 0 \\ 0 & 0 & 0 & \Gamma_f \end{bmatrix}, \quad \Gamma_a = 1/\eta_a \text{ and } \Gamma_f = 1/\eta_f \tag{8.3.11}$$

Here η_a and η_f are the viscosities for the motion of the antiferroelectric and the ferroelectric order parameter, respectively. Note that we have here neglected all the cross-coupling viscosity terms and we have also set equal the viscosity for the amplitude and phase fluctuations.

The solutions of the Landau-Khalatnikov equations of motion are overdamped waves

$$\vec{\Psi}_q(t) = \vec{\Psi}_q^\circ \cdot e^{-t/\tau} \tag{8.3.12.}$$

where $\tau^{-1}(q)$ is the corresponding relaxation rate, which is q-dependent. The Landau-Khalatnikov equations then reduce to an eigenvalue problem

$$\left(2 \cdot \underline{\Gamma} \cdot \underline{D}(q) - \lambda \cdot \underline{I}\right)\vec{\Psi}_q^\circ = 0 \tag{8.3.13.}$$

which has nontrivial solutions if the corresponding determinant is zero,

$$\left|2 \cdot \underline{\Gamma} \cdot \underline{D}(q) - \lambda \cdot \underline{I}\right| = 0 \tag{8.3.14}$$

For a given wave-vector of the excitation there are ***four real eigenvalues***, which correspond to four dispersion branches. This number of excitation branches may be further reduced due to the symmetry of the phase, as we shall see in what follows.

In the *smectic-A phase*, the equilibrium tilt angle is zero, $\theta_a\left(T \geq T_a^*\right) = 0$ and the dynamical matrix $\underline{D}(q)$ is more symmetric. This has a consequence, that there are two doubly degenerate soft branches of excitations in the $smectic - A$ phase. One of them, i.e. the ***antiferroelectric soft mode*** branch of excitations critically slows down at the phase transition temperature T_a^* :

$$\tau_{s,a}^{-1} = \frac{\alpha}{\eta_a}\left(T - T_a^*\right) + \frac{K_{3a}}{\eta_a} \cdot q^2 \tag{8.3.15.}$$

whereas the so-called ***ferroelectric soft mode*** has a finite frequency at the phase transition temperature T_a^* :

$$\tau_{s,f}^{-1} = \frac{\alpha}{\eta_f}\left(T - T_f^*\right) + \frac{K_{3f}}{\eta_f} \cdot q^2 \tag{8.3.16}$$

Here, the phase transition temperature $T_f^* = T_f + \Lambda_f^2 / (\alpha K_{3f})$ represents a hypothetical phase transition into the ferroelectric phase. Both temperature dependencies are shown in Fig.8.3.7.

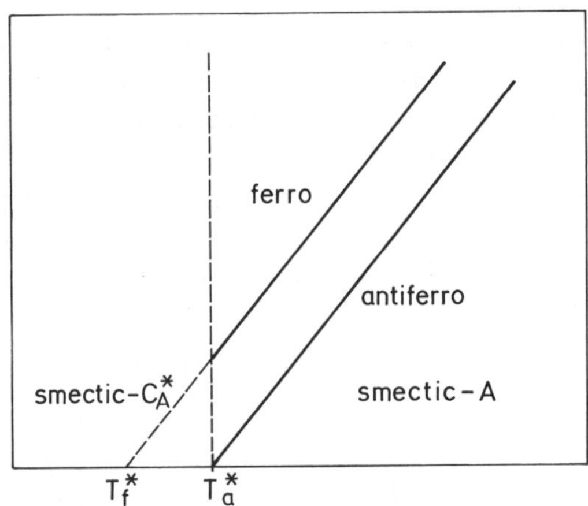

Fig.8.3.7. Critical slowing down of the doubly degenerate antiferroelectric and the ferroelectric soft modes at the phase transition into the antiferroelectric $smectic-C_A^*$ phase. Note that the antiferroelectric soft mode condenses at $q = \pi / d$, whereas the ferroelectric soft mode condenses at $q = 0$.

This is the competition between the ferroelectric and the antiferroelectric order: the mode that condenses first induces the ordering of the low-temperature phase. It is important to note that whereas the antiferroelectric soft mode has a zero frequency at the phase transition point, the ferroelectric soft mode is nearly critical and reaches a finite relaxation rate

$$\tau_{s,f}^{-1}\left(T_a^*\right)=\frac{\alpha}{\eta_f}\left(T_a^*-T_f^*\right) \qquad (8.3.17)$$

It is instructive to look in more detail at the wave-vectors of these soft elementary excitations. For the homogeneous $smectic-C_A^*$ phase, the wave-vector of the antiferroelectric soft mode equals to

$$q_c=\frac{\pi}{d} \qquad (8.3.18.)$$

It is a zone boundary mode. In order to avoid confusion, it is important to note that the definition of the antiferroelectric order parameter (Eq.8.2.2) corresponds to the zone-boundary wave-vector, as shown in Fig.8.3.8.

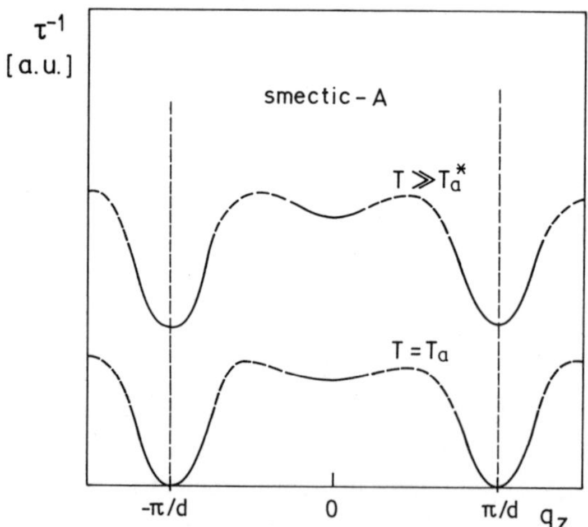

Fig.8.3.8. Dispersion relation for elementary excitations in the first Brillouin zone of a $smectic-A$ liquid crystal near the phase transition into the homogeneous antiferroelectric $smectic-C_A$ phase. The Orihara-Ishibashi continuous model is valid only very close to the edge or to the center of the Brillouin zone, as indicated by the full line. Note that by cooling down, the dispersion shifts down as a whole. The plot is illustrative and not to scale.

It is clear from Figure 8.3.8. that we have in reality only a single, doubly degenerate branch of soft excitations in the $smectic-A$ phase of an antiferroelectric liquid crystal. Within the Orihara-Ishibashi continuous, Landau-type model, the dispersion has two asymmetric minima in the reciprocal space. The first minimum is at the center of the BZ and therefore represents the ***ferroelectric soft excitations***. The second minimum is close to the zone boundaries, i.e. at $q = \pm\pi/d$ and corresponds to the ***antiferroelectric soft excitations***. If for $T_a > T_f$, the zone-boundary minimum is slightly lower than the zone-center minimum, the antiferroelectric mode condenses first and results in a herring-bone structure of the antiferroelectric $smectic-C_A^*$ phase.

In the antiferroelectric $smectic-C_A^*$ phase, the onset of the non-zero equilibrium tilt angle substantially alters the spectrum of elementary excitations. The spectrum is also more complicated because of doubling of the smectic unit cell at the phase transition, which leads to a splitting of the Brillouin zone into two halves, as has been shown in Fig.8.3.4. We have therefore in the $smectic-C_A^*$ phase

(i) A doubling of the number of the excitation branches due to the doubling of a unit cell.

(ii) A splitting of the originally doubly degenerate soft branch of excitation into the amplitudon and the phason branches due to the spontaneous breaking of the continuous D_∞ symmetry of the $smectic-A$ phase.

The calculation of the four dispersion branches in the $smectic-C_A^*$ phase is rather straightforward within the dynamical matrix approach in the helicoidal reference frame. The antiferroelectric soft mode splits at the phase transition temperature in the antiferroelectric amplitude branch of excitations

$$\tau_{am,a}^{-1} = \frac{2\alpha}{\eta_a}\left(T_a^* - T\right) + \frac{K_{3a}}{\eta_a}\cdot q^2 \qquad (8.3.19)$$

and the gapless branch of antiferroelectric phase excitations

$$\tau^{-1}_{ph,a}(q) = \frac{K_{3a}}{\eta_a} \cdot q^2 \tag{8.3.20}$$

Here, q is the wave-vector of the excitation in a helicoidal reference frame. Similarly, the ferroelectric soft branch splits into a ferroelectric amplitudon

$$\tau^{-1}_{a,f} = \Delta(T) + \frac{K_{3f}}{\eta_f} \cdot q^2 + \frac{\sqrt{\frac{1}{4}\gamma_2^2\theta_a^4(T) + 4K_{3f}^2(q_a - q_f)^2 \cdot q^2}}{\eta_f} \tag{8.3.21}$$

and a ferroelectric phason branch

$$\tau^{-1}_{ph,f} = \Delta(T) + \frac{K_{3f}}{\eta_f} \cdot q^2 - \frac{\sqrt{\frac{1}{4}\gamma_2^2\theta_a^4(T) + 4K_{3f}^2(q_a - q_f)^2 \cdot q^2}}{\eta_f} \tag{8.3.22}$$

Here, $\Delta(T) = \left(a_f(T) + \left(\gamma_1 + \frac{1}{2}\gamma_2\right)\cdot\theta_a^2(T) + \left(2\Lambda_f + K_{3f}q_a\right)q_a\right)/\eta_f$ is a temperature dependent frequency gap and $q_f = -\Lambda_f / K_{3f}$. Note that we have here considered only the fluctuations with a wave-vector $\vec{q} = (0,0,q)$, which are therefore spreading along the helical axis.

Similarly to the elementary excitations in the ferroelectric $smectic - C^*$ phase, the two antiferroelectric dispersion branches show a parabolic dispersion. In contrast, the two ferroelectric-like branches are expected to show a double-minima dispersion, as shown schematically in Figure.8.3.9. The separation between the two minima is of the order of $(q_a - q_f) = \Lambda_f / K_{3f} - \Lambda_a / K_{3a}$, whereas the minima are quite distinctive close to T_a^* and fade away due to the onset of a tilt angle in the $smectic - C_A^*$ phase (see Eqs.8.3.21. and 8.3.22.). It should be stressed that the only temperature independent branch of excitations within Orihara-Ishibashi model is the gapless phason branch, whereas all other branches have an intrinsic temperature dependence.

A further insight is given by analyzing the structure of the corresponding eigenvectors for a given wave-vector q.

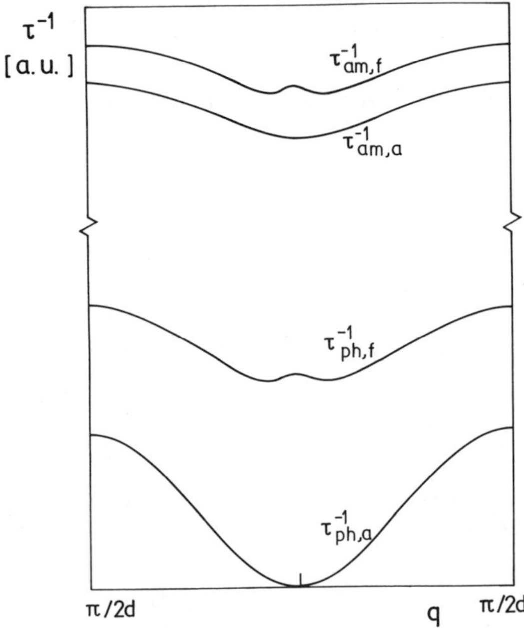

Fig.8.3.9. The spectrum of elementary excitations in the *smectic* $-C_A^*$ phase of an antiferroelectric liquid crystal. Note that there are four excitations for each wave-vector. This means that for $q \to 0$, we can observe in principle four different relaxations. We have used a repeating Brillouin zone scheme and the drawing is illustrative only.

In the *smectic* $-A$ phase we obtain a zone-center ferroelectric, nearly critical soft mode, which is shown in Fig.8.3.10(a) and a truly soft antiferroelectric mode, which is a zone-boundary mode, shown in Fig.8.3.10(b). Whereas the ferroelectric soft mode represents coherent and collective tilting of the molecules in a given direction, the antiferroelectric soft mode represents herring-bone-like fluctuations.

In the homogeneous antiferroelectric *smectic* $-C_A^*$ phase, the eigenvectors of fluctuations are shown in Figs.8.3.11. and 8.3.12. for $\vec{q} = 0$. Figures 8.3.11(a) and (b) show the two amplitude excitations: the ferroelectric amplitudon (a) and the antiferroelectric amplitudon (b). Whereas the first mode represents in-phase fluctuations of the magnitude of the tilt angle of the two neighboring molecules in a unit cell, the antiferroelectric amplitudon represents an out-of-phase fluctuation of the tilt. As a result, the ferroelectric amplitudon is here a polar mode and the antiferroelectric amplitudon is a non-polar mode for $\vec{q} = 0$.

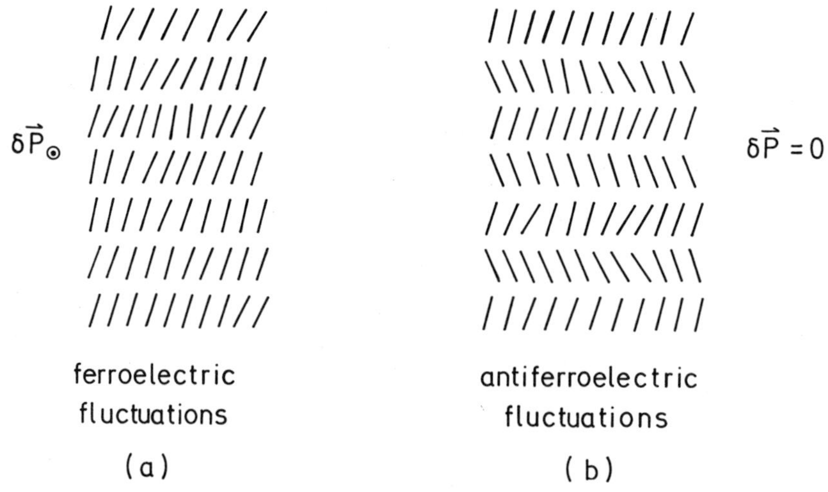

Fig.8.3.10. (a) The polar ferroelectric fluctuations in the *smectic* – *A* phase of an antiferroelectric liquid crystal. This mode represents collective fluctuations of a space-homogeneous dipole moment and can therefore couple to an external field, for example in a dielectric experiment. (b) The non-polar antiferroelectric excitations in the *smectic* – *A* phase critically slow down at the phase transition. This mode is non-polar and has a small scattering cross-section for a light scattering experiment.

Figure 8.3.12(a) shows a ferroelectric phase excitation in the $smectic - C_A^*$ phase. It represents a local rotation (therefore a change in phase) of molecules at constant tilt angle, where the molecules in the neighboring layers rotate in opposite directions. As a result, there will be a small, time-fluctuating dipole moment, because molecular directions are slightly different than 180° in successive layers and the mode is polar. Note that from purely symmetry reasons, this mode cannot be gapless and it will always have a finite relaxation rate.

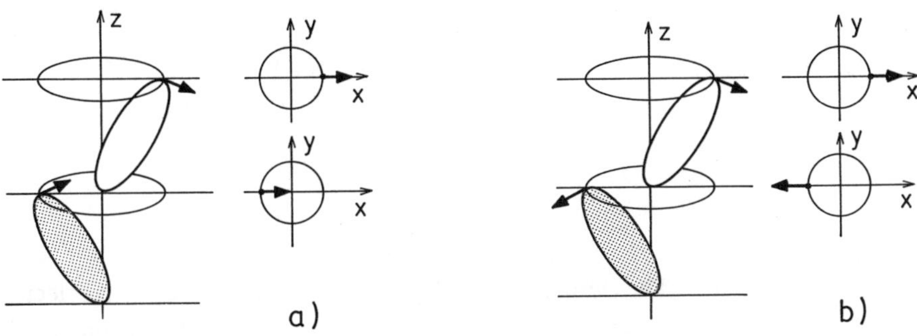

Fig.8.3.11. (a) The eigenvector of a polar ferroelectric amplitudon shown for a single unit cell of antiferroelectric $smectic-C_A^*$ phase. (b) The eigenvector of the non-polar antiferroelectric amplitudon. Note that both modes represent a homogeneous fluctuation with a wave-vector $\vec{q}=0$, as the displacements of the molecules, shown above for a given unit cell, repeat uniformly throughout the liquid crystal.

Figure 8.3.12(b) shows the antiferroelectric phase excitation. It represents a local rotation of neighboring molecules *in the same direction.* It therefore results in a rotation of a unit cell and also a sample as a whole. We realize that this kind of motion costs no energy. As a consequence, there is no restoring force and this is therefore the Goldstone mode, recovering the spontaneously broken continuous symmetry of the $smectic-A$ phase. The Goldstone mode is here of non-polar character.

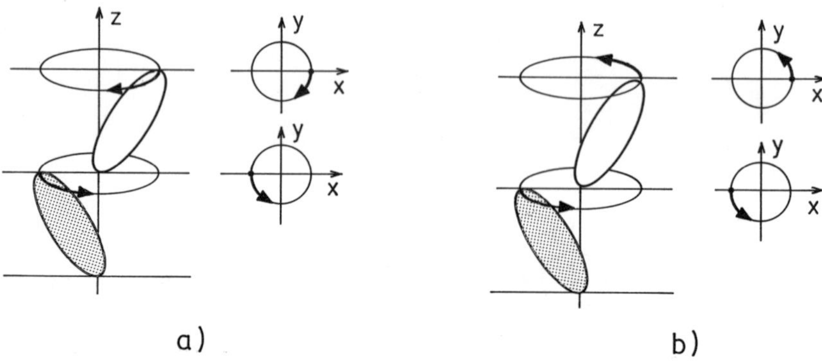

Figure 8.3.12. (a) The ferroelectric phase mode. (b) The antiferroelectric phase mode for $\vec{q}=0$. The antiferroelectric phase mode is the Goldstone mode of the *smectic – A - smectic –* C_A^* phase transition.

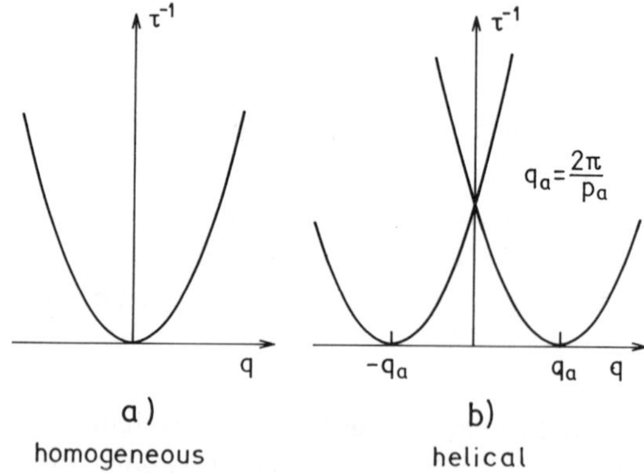

Figure 8.3.13. The effect of a helical modulation on the dispersion relation for order parameter excitations in modulated smectics. (a) The dispersion relation for a homogeneous phase and, (b) the corresponding shift of the dispersion relation for the helicoidally modulated phase. Note the change in the relaxation frequencies for $\vec{q}=0$.

So far we have considered excitations, as they appear in the "rotating" reference frame, which follows the local direction of the projection of the molecular tilt as we move along the layer normal. Whereas for the case of a spatially homogeneous (non-helical) $smectic - C_A^*$ phase this will cause no change of the dispersion relations after transformation into the laboratory frame, this will have the well known effect (see Chapter 2) of rescaling the wave-vector of excitation ,

$$q \to q \pm q_a \tag{8.3.23}$$

where q_a is the wave-vector of the helical modulation of the antiferroelectric $smectic - C_A^*$ phase. The transformation into the laboratory reference frame therefore implies a change of the wave-vector in the dispersion relations. As a consequence, the dispersion relations are shifted by the wave-vector q_a of the modulation in the reciprocal space, as shown in Fig.8.3.13. Note that the minima of the dispersion relations then appear at the position of the Bragg scattering in the reciprocal space.

In conclusion, the analysis of the dynamics in the vicinity of the antiferroelectric $smectic - A$ - $smectic - C_A^*$ phase transition within the Orihara-Ishibashi continuous model can be summarized as follows:

(i) The transition is characterized by a doubling of a smectic unit cell and a breaking of a continuous D_∞ symmetry of the $smectic - A$ phase.

(ii) The elementary excitations are always of plane-wave, monochromatic character.

(iii) In the $smectic - A$ phase there is a single, doubly degenerate branch of soft excitations. The dispersion branch has two minima, i.e. the zone-center and the zone-boundary minima, which are slightly deeper.

(iv) The zone-boundary antiferroelectric soft modes have a parabolic dispersion in the $smectic - A$ phase and critically slow-down as the phase transition into the antiferroelectric phase is approached

$$\tau_{s,a}^{-1} = \frac{\alpha}{\eta_a}\left(T - T_a^*\right) + \frac{K_{3a}}{\eta_a}\cdot\left(q \pm q_a\right)^2 \tag{8.3.24.}$$

The zone-center ferroelectric soft modes also show a parabolic dispersion in the *smectic* $- A$ phase but have a finite frequency gap at the phase transition temperature T_a^*

$$\tau_{s,f}^{-1} = \frac{\alpha}{\eta_f}\left(T - T_f^*\right) + \frac{K_{3f}}{\eta_f}\cdot\left(q \pm q_a\right)^2 \tag{8.3.25}$$

(v) There are four dispersion branches in the antiferroelectric *smectic* $- C_A^*$ phase. The antiferroelectric amplitudon branch

$$\tau_{am,a}^{-1} = \frac{2\alpha}{\eta_a}\left(T_a^* - T\right) + \frac{K_{3a}}{\eta_a}\cdot\left(q \pm q_a\right)^2 \tag{8.3.26}$$

and the gapless phason branch

$$\tau_{ph,a}^{-1}(q) = \frac{K_{3a}}{\eta_a}\cdot\left(q \pm q_a\right)^2 \tag{8.3.27}$$

both show parabolic dispersion, whereas the ferroelectric amplitudon

$$\tau_{a,f}^{-1} = \Delta(T) + \frac{K_{3f}}{\eta_f}\cdot\left(q \pm q_a\right)^2 + $$

$$+\frac{\sqrt{\frac{1}{4}\gamma_2^2\theta_a^4(T) + 4K_{3f}^2\left(q_a - q_f\right)^2\cdot\left(q \pm q_a\right)^2}}{\eta_f} \tag{8.3.28}$$

and the ferroelectric phason branch

$$\tau_{ph,f}^{-1} = \Delta(T) + \frac{K_{3f}}{\eta_f} \cdot (q \pm q_a)^2 - \\ - \frac{\sqrt{\frac{1}{4}\gamma_2^2\theta_a^4(T) + 4K_{3f}^2(q_a - q_f)^2 \cdot (q \pm q_a)^2}}{\eta_f} \qquad (8.3.29)$$

both show a characteristic double-minima dispersion relation. Here, $\Delta(T) = \left(a_f(T) + \left(\gamma_1 + \frac{1}{2}\gamma_2\right)\cdot\theta_a^2(T) + \left(2\Lambda_f + K_{3f}q_a\right)q_a\right)/\eta_f$.

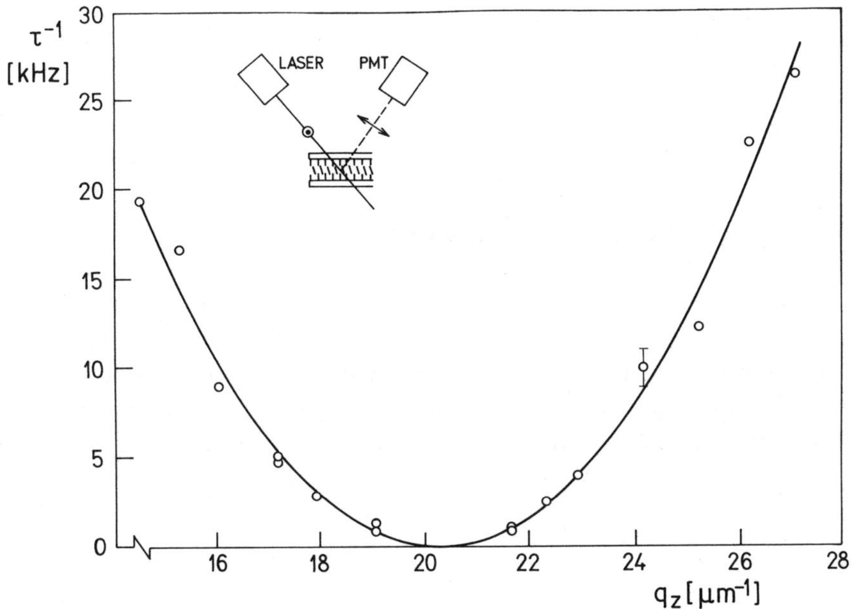

Fig.8.3.14. The gapless phason dispersion in the *smectic* $-C_A^*$ phase of MHPOBC (Muševič et al., 1993c).

The experimental observations of the order parameter dynamics in the *smectic* $-C_A^*$ phase are almost exclusively limited to a large number of dielectric experiments, which will be discussed in Section 8.6. Here we shall briefly report the observations of the quasielastic light scattering experiments, performed in the antiferroelectric $smectic-C_A^*$ phases of MHPOBC and EHPOCBC.

The first observation of the gapless antiferroelectric phason branch was reported in 1993 by Muševič et al. They have performed a quasielastic light scattering experiment in the $smectic-C_A^*$ phase of 4-(1-methylheptyl-oxycarbonyl)phenyl 4'-octyloxybiphenyl-4-carboxylate (MHPOBC) and their results are shown in Fig.8.3.14. The results are in clear agreement with the predictions of the Orihara-Ishibashi model and give the value of the phason diffusion coefficient $K_{3a}/\eta = 5.8\cdot 10^{-10}\, m^2 s^{-1}$. Similar values were observed in the quasielastic light scattering experiment of Sun et al.(1993a).

The ferroelectric and the antiferroelectric phase mode dispersion as well as the ferroelectric soft mode were observed in the antiferroelectric $smectic-C_A^*$ phase of EHPOCBC (Muševič et al., 1996b), see Fig.8.5.4 in Section 8.5.

8.4. Optical Properties of the Antiferroelectric Phase

The unit cell of the antiferroelectric $smectic-C_A^*$ phase is many orders of magnitude smaller than the wavelength of visible light. This means that the optical properties of the antiferroelectric $smectic-C_A^*$ phase can be very well described with a dielectric tensor, which is an average of a dielectric tensor over two neighboring smectic layers,

$$\underline{\varepsilon}_{CA*} = \tfrac{1}{2}\left(\underline{\varepsilon}_i + \underline{\varepsilon}_{i+1}\right) \tag{8.4.1}$$

Here, $\underline{\varepsilon}_i$ is a dielectric tensor of a single, tilted ferroelectric smectic layer, which is explicitly given by the Eq.5.1.2 of Chapter 5. The change of the local phase for the two neighboring layers are

$$\Phi_{i+1} = \Phi_i + \pi + \delta \tag{8.4.2}$$

Here Φ_i is the phase of the *i*-th layer. The molecular direction reverses by nearly 180° as we move to the neighboring layer, so that the residual phase difference δ is very small, $\delta \ll 1$. The dielectric tensor of the antiferroelectric $smectic-C_A^*$ phase is

$$\underline{\varepsilon}_{CA^*} = \begin{bmatrix} \frac{1}{2}\left[(\varepsilon_1+\varepsilon_2)+(\varepsilon_3-\varepsilon_2)\sin^2\theta\right] & 0 & 0 \\ 0 & \frac{1}{2}\left[(\varepsilon_1+\varepsilon_2)+(\varepsilon_3-\varepsilon_2)\sin^2\theta\right] & 0 \\ 0 & 0 & \varepsilon_3-(\varepsilon_3-\varepsilon_2)\sin^2\theta \end{bmatrix} +$$

$$+\frac{1}{2}\left[(\varepsilon_2-\varepsilon_1)+(\varepsilon_3-\varepsilon_2)\sin^2\theta\right] \begin{bmatrix} \cos 2\Phi(z) & \sin 2\Phi(z) & 0 \\ \sin 2\Phi(z) & -\cos 2\Phi(z) & 0 \\ 0 & 0 & 0 \end{bmatrix} +$$

$$+\frac{1}{2}\left[(\varepsilon_2-\varepsilon_1)+(\varepsilon_3-\varepsilon_2)\sin^2\theta\right]\cdot\delta\cdot \begin{bmatrix} -\sin 2\Phi(z) & \cos 2\Phi(z) & 0 \\ \cos 2\Phi(z) & -\sin 2\Phi(z) & 0 \\ 0 & 0 & 0 \end{bmatrix} +$$

$$+(\varepsilon_3-\varepsilon_2)\sin\theta\cos\theta\cdot\delta\cdot \begin{bmatrix} 0 & 0 & \sin\Phi(z) \\ 0 & 0 & -\cos\Phi(z) \\ \sin\Phi(z) & -\cos\Phi(z) & 0 \end{bmatrix} \quad (8.4.3.)$$

For a perfect antiparallel ordering of the two neighboring layers, i.e. $\delta = 0$ there are only two terms in the above expression. In this case the dielectric tensor of the $smectic-C_A^*$ phase is ***identical to the dielectric tensor of cholesteric liquid crystals*** (deGennes and Prost, 1993). In this approximation, the two phases should therefore have identical optical properties. In particular,

(i) The antiferroelectric $smectic-C_A^*$ phase should exhibit a single Bragg reflection band, which is characteristic for non-polar,

quadrupolarly ordered helical structures. This is the so-called half-pitch band, which arises in helical non-polar structures because the dielectric tensor repeats every half of the helical period of the structure. The absence of the full-pitch band in reflection experiments of Chandani et al.(1989) was in fact a clue for the proposed alternate-tilt structure.

(ii) The antiferroelectric $smectic-C_A^*$ phase should exhibit a rotatory power (see Section 5.4. for ferroelectric liquid crystals), which is identical to the rotatory power Ψ/d of a ferroelectric $smectic-C^*$ or a cholesteric phase:

$$\frac{\Psi}{d}=-\frac{2\pi}{p}\cdot\frac{\alpha^2}{8\lambda^2\left(1-\lambda^2\right)} \tag{8.4.4}$$

Here Ψ is the angle of rotation of polarization of light, propagating along the helical axis of a $smectic-C_A^*$ phase with thickness d and other quantities being defined in Section 5.4. The equivalency follows from the fact, that the cholesteric, the ferrolectric $smectic-C^*$ and the antiferroelectric $smectic-C_A^*$ phases cannot be distinguished by light, which is propagating along the helix.

We have seen in Section 5.1. that a first order approximation to the wave equation gives a good description of the optical properties of chiral smectics. In this approximation, the optical properties of the antiferroelectric $smectic-C_A^*$ phase are given by the space-averaged dielectric tensor $\underline{\varepsilon}_{AF}$

$$\left\langle\underline{\varepsilon}_{AF}\right\rangle=\begin{bmatrix}\langle\varepsilon_{xx}\rangle & 0 & 0\\ 0 & \langle\varepsilon_{yy}\rangle & 0\\ 0 & 0 & \langle\varepsilon_{zz}\rangle\end{bmatrix} \tag{8.4.5.}$$

where the space-averaged components $\langle\varepsilon_{ii}\rangle$ are

$$\langle \varepsilon_{xx} \rangle = \tfrac{1}{2}\left[(\varepsilon_1 + \varepsilon_2) + (\varepsilon_3 - \varepsilon_2)\sin^2\theta\right] \tag{8.4.6a}$$

$$\langle \varepsilon_{yy} \rangle = \tfrac{1}{2}\left[(\varepsilon_1 + \varepsilon_2) + (\varepsilon_3 - \varepsilon_2)\sin^2\theta\right] \tag{8.4.6b}$$

$$\langle \varepsilon_{zz} \rangle = \varepsilon_3 - (\varepsilon_3 - \varepsilon_2)\sin^2\theta \tag{8.4.6c}$$

The off-diagonal terms average to zero because of the axial symmetry of the system,

$$\langle \cos 2\Phi \rangle = \langle \cos \Phi \rangle = \langle \sin 2\Phi \rangle = \langle \sin \Phi \rangle = 0 \tag{8.4.7}$$

If this axial symmetry is broken, for example, by an external field or by the finite thickness of a sample (free standing smectic films), the off-diagonal terms will appear, representing an induced biaxiality and tilting of the effective optical axis (see Section 5.5. and 5.6.).

It can be seen from the dielectric tensor (Eqs.8.4.5-7) that a bulk antiferroelectric $smectic-C_A^*$ phase is an ***optically uniaxial phase*** with a temperature dependent birefringence, reflecting the temperature dependence of the tilt angle $\theta(T)$. The measurements of the temperature dependence of birefringence have been performed in the bulk antiferroelectric $smectic-C_A^*$ phase of MHPOBC by Philip et al.(1994,1995) and are in qualitative agreement with the first order approximation. High resolution measurements of the birefringence in the antiferroelectric $smectic-C_A^*$ phase of many antiferroelectric materials has been performed by Škarabot et al. (1997). Their conclusions are:

(i) The first order approximation to the wave equation gives a quantitatively good description of the optics of antiferroelectric phases.

(ii) A large pretransitional decrease of the birefringence is systematically observed in the $smectic-A$ phase of antiferroelectric materials, as shown in Fig.8.4.1. This pretransitional decrease of birefringence is much larger then in ferroelectric materials, as for example CE-8, and has been analyzed elsewhere (Škarabot et al., 1999b). It can be attributed to large orientational fluctuations of the director field in the $smectic-A$ phase and is experimentally consistent with similar phenomena that were

observed in heat capacity studies of antiferroelectric materials by K.Ema et al. (1995, 1996, 1997).

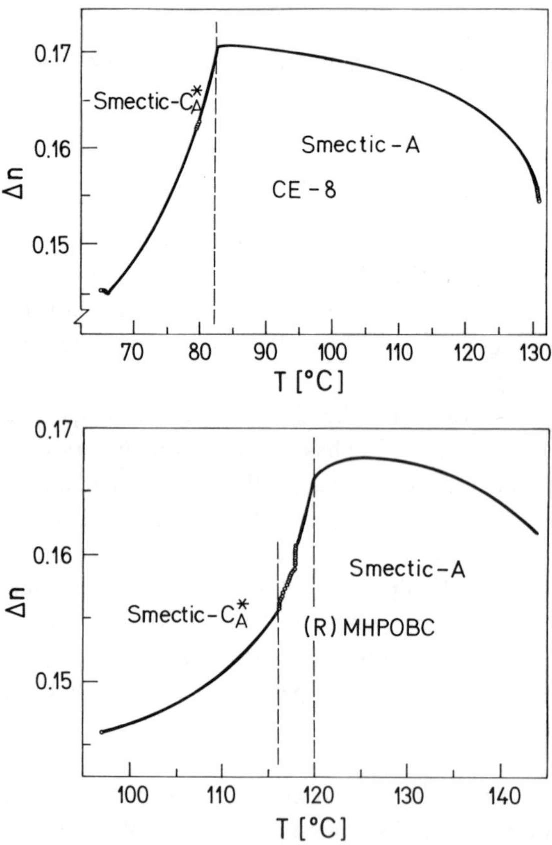

Fig.8.4.1. The pretransitional behavior of the birefringence in the *smectic – A* phase of a ferroelectric (upper curve) and an antiferroelectric liquid crystal (lower curve).

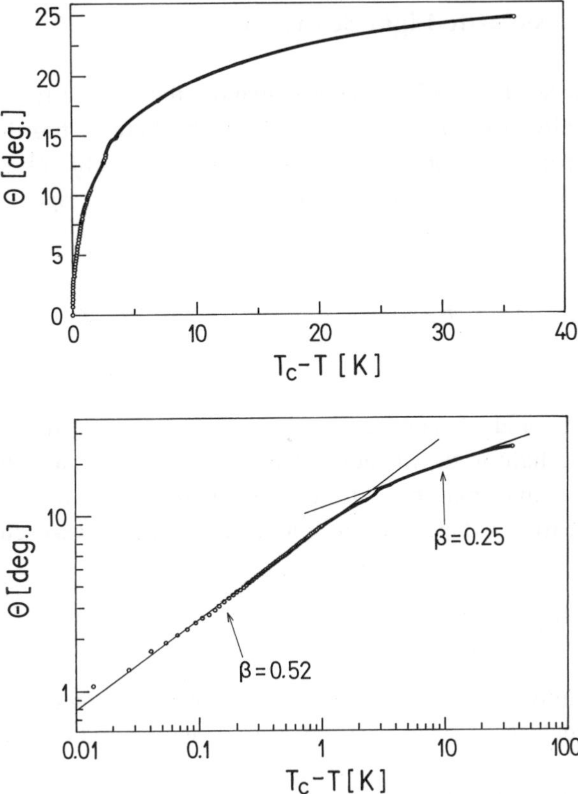

Fig.8.4.2. Temperature dependence of the tilt angle in (R)-MHPOBC, as deduced from the birefringence data. Note the crossover in the exponent β at $T_c - T \approx 1K$. After Škarabot et al., 1998.

(iii) The temperature dependence of the tilt angle, as deduced from the birefringence data, clearly shows a crossover of the exponent β from the value of $\beta \approx 0.5$ close to the *smectic* $-A$ phase, to the value of $\beta \approx 0.25$ far away, as shown in Fig.8.4.2. This is similar to the temperature behavior of the tilt angle in the ferroelectric phases and indicates the importance of the sixth order terms in the Landau free-energy expansion.

8.5. Quasielastic Light Scattering

In the single scattering (Born) approximation, the electric field amplitude of the quasielastically scattered light is given by the Fourier transform of the time-fluctuating part of the dielectric tensor of a medium $\delta\underline{\varepsilon}(\vec{q}_s,t)$, projected onto the incoming and scattered polarizations of light $\vec{i}$ and $\vec{f}$, respectively:

$$E_s(\vec{r},t)=\vec{f}\cdot\vec{E}_s=-\frac{|\vec{k}_i|E_\circ\cdot e^{i\vec{k}_f\vec{R}-i\omega t}}{4\pi\varepsilon_\circ R}\cdot\vec{f}\cdot\delta\underline{\varepsilon}(\vec{q}_s,t)\cdot\vec{i} \qquad (8.5.1)$$

Here, $|k_i|$, ω_i and $E_\circ$ are the wave-vector, frequency and a field amplitude of the incoming light in the medium, respectively, and R is a distance between the scattering volume and the detector. The Fourier transform $\delta\underline{\varepsilon}(\vec{q}_s,t)$ of the fluctuating part of a dielectric tensor at the scattering wave-vector $\vec{q}_s$ is

$$\delta\underline{\varepsilon}(\vec{q}_s,t)=\int_V\delta\underline{\varepsilon}(\vec{r},t)\cdot e^{i\vec{q}_s\vec{r}}\cdot d^3\vec{r} \qquad (8.5.2)$$

The contributions of collective elementary excitations $\delta\vec{n}(\vec{q},t)$ to the fluctuating part of the dielectric tensor $\delta\underline{\varepsilon}(\vec{q}_s,t)$ can be easily found by remembering that

(i) The dielectric tensor, which is responsible for the optical properties of the antiferroelectric $smectic-C_A^*$ is an average over the antiferroelectric unit cell:

$$\underline{\varepsilon}_{CA*}=\tfrac{1}{2}\left(\underline{\varepsilon}_i+\underline{\varepsilon}_{i+1}\right) \qquad (8.5.3)$$

(ii) The dielectric tensor $\underline{\varepsilon}_i$ of the *i*-th smectic layer depends linearly on the tensor $\vec{n}_i\otimes\vec{n}_i$, where $\vec{n}_i$ is the director of the *i*-th layer (de Gennes, 1974)

$$\varepsilon_{\alpha\beta}=\varepsilon_1\delta_{\alpha\beta}+(\varepsilon_3-\varepsilon_1)\cdot n_\alpha n_\beta \qquad (8.5.4.)$$

Any fluctuation of the director therefore also represents a fluctuation of the dielectric tensor and gives rise to a strong scattering of light. Here we have used simple uniaxial approximation $\varepsilon_1 = \varepsilon_2$.

Conveniently, the dielectric tensor of the antiferroelectric $smectic-C_A^*$, averaged over a unit cell, is expressed in terms of the ferroelectric and antiferroelectric order parameters as

$$\underline{\varepsilon}_{CA*} = \varepsilon_1 \cdot \underline{I} +$$

$$+\left(\varepsilon_3-\varepsilon_1\right)\begin{bmatrix} \xi_{ax}^2+\xi_{fx}^2 & \xi_{ax}\xi_{ay}+\xi_{fx}\xi_{fy} & \xi_{fx} \\ \xi_{ax}\xi_{ay}+\xi_{fx}\xi_{fy} & \xi_{ay}^2+\xi_{fx}^2 & \xi_{fy} \\ \xi_{fx} & \xi_{fy} & 1-\left(\xi_{ax}^2+\xi_{fx}^2+\xi_{ay}^2+\xi_{fx}^2\right) \end{bmatrix} \quad (8.5.5.)$$

where $\xi_f = \left(\xi_{fx}, \xi_{fy}\right)$ and $\xi_a = \left(\xi_{ax}, \xi_{ay}\right)$. The fluctuations $\delta\varepsilon_{\alpha\beta}$ of the dielectric tensor are derived by differentiating the Equation 8.5.4.

$$\delta\varepsilon_{\alpha\beta} = \left(\varepsilon_3-\varepsilon_1\right)\cdot\left(\delta n_\alpha \cdot n_\beta^\circ + n_\alpha^\circ \cdot \delta n_\beta\right) \quad (8.5.6.)$$

Here $\delta n_\alpha\left(\vec{r},t\right)$ is the fluctuation of the projection of the director and $n_\alpha^\circ\left(\vec{r}\right)$ is the corresponding equilibrium value. In the $smectic-C_A^*$ phase, the above expression for the fluctuating part of the dielectric tensor can be conveniently separated into two parts, i.e. antiferro- and ferro- fluctuations:

$$\delta\underline{\varepsilon}_{CA*} = \delta\underline{\varepsilon}_a + \delta\underline{\varepsilon}_f =$$

$$= \left(\varepsilon_3-\varepsilon_1\right)\cdot\begin{bmatrix} 2\xi_{ax}^\circ\cdot\delta\xi_{ax} & \xi_{ax}^\circ\cdot\delta\xi_{ay}+\xi_{ay}^\circ\cdot\delta\xi_{ax} & 0 \\ \xi_{ax}^\circ\cdot\delta\xi_{ay}+\xi_{ay}^\circ\cdot\delta\xi_{ax} & 2\xi_{ay}^\circ\cdot\delta\xi_{ay} & 0 \\ 0 & 0 & 1-2\xi_{ax}^\circ\cdot\delta\xi_{ax}-2\xi_{ay}^\circ\cdot\delta\xi_{ay} \end{bmatrix} +$$

$$+\left(\varepsilon_3-\varepsilon_1\right)\begin{bmatrix} 0 & 0 & \delta\xi_{fx} \\ 0 & 0 & \delta\xi_{fy} \\ \delta\xi_{fx} & \delta\xi_{fy} & 0 \end{bmatrix} \tag{8.5.7}$$

The above expression is of formal character and gives the structure of dielectric tensor fluctuations in terms of both ferroelectric and antiferroelectric order parameters. A more concise form can be found by expressing ferro- and antiferroelectric order parameters in terms of fluctuations of $\xi_i(t)=(\sin\theta_i(t)\cdot\cos\Phi_i(t),\sin\theta_i(t)\cdot\sin\Phi_i(t))$, where both the local tilt magnitude $\theta_i(t)$ and the phase $\Phi_i(t)$ are allowed to fluctuate in time and space. This leads to four dispersion branches in the antiferroelectric $smectic-C_A^*$ phase that are characterized by their eigenvectors, which have already been considered in Section 8.3. (see Figs.8.3.11.-12).

The fluctuations of the dielectric tensor which correspond to these four types of excitations can be expressed as

$$\delta\underline{\varepsilon}_{CA^*}=\delta\underline{\varepsilon}_a+\delta\underline{\varepsilon}_f=$$

$$=\left(\varepsilon_3-\varepsilon_1\right)\cdot\begin{bmatrix} \sin\theta_\circ\cdot\left(1+\cos 2q_c z\right) & \sin\theta_\circ\cdot\cos 2q_c z & 0 \\ \sin\theta_\circ\cdot\cos 2q_c z & \sin\theta_\circ\cdot\left(1-\cos 2q_c z\right) & 0 \\ 0 & 0 & 1-2\sin\theta_\circ \end{bmatrix}\cdot\delta\theta_a+$$

$$+\left(\varepsilon_3-\varepsilon_1\right)\cdot\sin^2\theta_\circ\cdot\begin{bmatrix} -\sin 2q_c z & \cos 2q_c z & 0 \\ \cos 2q_c z & \sin 2q_c z & 0 \\ 0 & 0 & 0 \end{bmatrix}\cdot\delta\Phi_a+$$

$$+\left(\varepsilon_3-\varepsilon_1\right)\cdot\begin{bmatrix} 0 & 0 & \cos q_c z \\ 0 & 0 & \sin q_c z \\ \cos q_c z & \sin q_c z & 0 \end{bmatrix}\cdot\delta\theta_f+$$

$$+\left(\varepsilon_3-\varepsilon_1\right)\cdot\sin\theta_\circ\cdot\begin{bmatrix}0 & 0 & -\sin q_c z\\ 0 & 0 & \cos q_c z\\ -\sin q_c z & \cos q_c z & 0\end{bmatrix}\cdot\delta\Phi_f \qquad (8.5.8)$$

Here, $\theta_\circ(T)$ is the tilt angle, which is temperature dependent and we have assumed a simple, plane wave modulation of the equilibrium phase, $\Phi_\circ = q_c z$. The fluctuations $\delta\theta_a, \delta\Phi_a, \delta\theta_f$, and $\delta\Phi_f$ denote here the antiferroelectric amplitudon, the antiferroelectric phason, the ferroelectric amplitudon and the ferroelectric phason, respectively. These fluctuations have the form of a monochromatic plane wave, $\delta\theta \approx e^{i\vec{q}\vec{r}}$, when the helical structure is "smooth", i.e. when the equilibrium phase is a linear function $\Phi_\circ = q_c z$ of the spatial coordinate *z*.

The above expression for the fluctuations of the dielectric tensor shows the following interesting features:

(i) The antiferroelectric modes have the center of the dispersion at $\vec{q} = 2\vec{q}_c$ in the reciprocal space of the laboratory system. This is in contrast to the ferroelectric modes that have the center of the dispersion at $\vec{q} = \vec{q}_c$, as shown in Fig.8.5.1.

(ii) The scattering cross section for the elementary excitations in the $smectic - C_A^*$ phase is proportional to the square of the equilibrium tilt angle, except for the antiferroelectric phason. This branch of excitations has a scattering cross-section which is proportional to the fourth power of the tilt angle and is therefore more difficult to observe.

The predicted dispersion of the order parameter excitations has been observed in a light scattering experiment, performed in the antiferroelectric $smectic - C_A^*$ phase of 4-(1-ethylheptyloxycarbonyl)phenyl-4'-alkylcarbonyloxybiphenyl-4-carboxylate (EHPOCBC) (Muševič et al., 1996b). The results are shown in Fig.8.5.2. and 8.5.3.

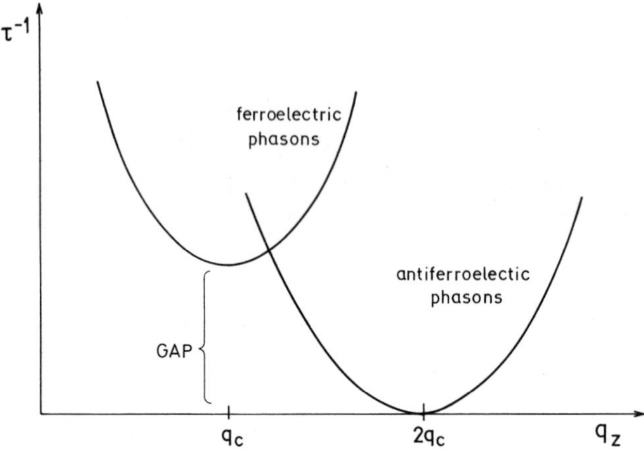

Fig.8.5.1. The dispersion of the phase excitations in the antiferroelectric $smectic - C_A^*$ phase. Note that only the antiferroelectric branch of phase excitations is gapless.

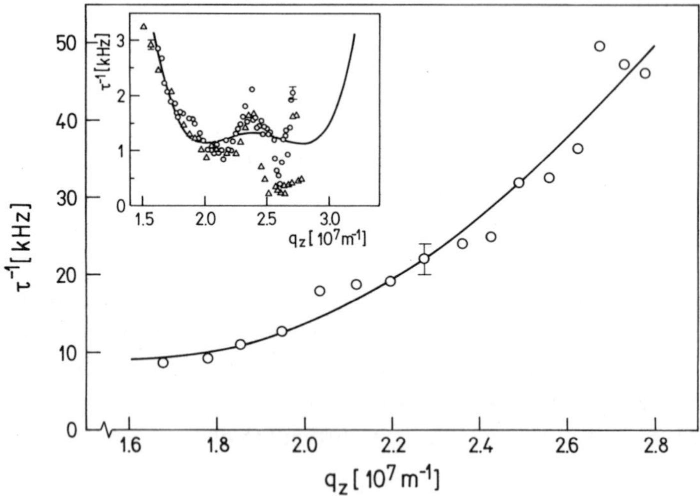

Fig.8.5.2. The dispersion relation of the optic-like phase mode (ferroelectric phason) in the $smectic - C_A^*$ phase of EHPOCBC at $T_a^* - T \approx 10K$. The solid line is the best parabolic fit with $K_{3f}/\eta = 2.6 \cdot 10^{-10} m^2 s^{-1}$ and the inset shows the dispersion of the same mode very near the $smectic - A$ phase, $T_a^* - T \approx 80mK$.

One can clearly see from the dispersion of the optic-like, ferroelectric phason modes that they always have a finite relaxation rate. The frequency gap of these optic-like ferroelectric phase excitations is of the order of 10kHz. This is in contrast to the acoustic-like, antiferroelectric phasons, which show a nearly gapless character (300Hz) within the experimental accuracy. The temperature dependencies of the two phason modes, as determined at a fixed scattering wave-vector, are shown in Fig.8.5.4.(a) and (b). One can observe a rather strong temperature dependence of both modes. Close to the $smectic-A$-$smectic-C_A^*$ phase transition, a soft mode is observed (far-right side of Fig.8.5.4.(a)), which slows-down at the phase transition. In the $smectic-C_A^*$ phase this mode is masked by a very intense optic-like, ferroelectric phason, which shows a strong temperature dependence. Than, at $T_a^* - T \approx 7K$, another mode is observed in a narrow temperature interval, and this mode disappears at very low temperatures. The second mode was identified as the acoustic-like, antiferroelectric phason.

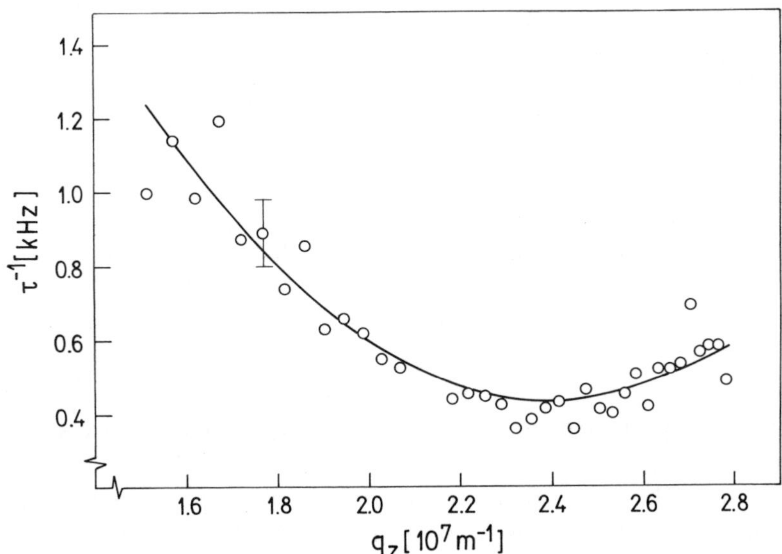

Fig. 8.5.3. The dispersion relation of the slow, acoustic-like phase mode in the $smectic-C_A^*$ phase of EHPOCBC at $T_a^* - T \approx 14K$. The dispersion is practically gapless within the experimental error. The solid line is the best fit to the Eq.8.3.27 with $K_{3a}/\eta = 1\cdot10^{-11} m^2 s^{-1}$.

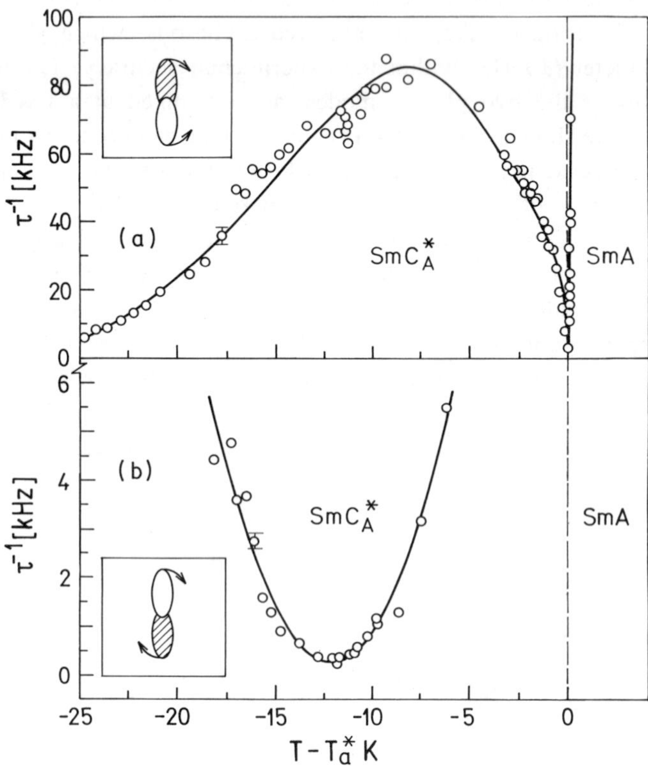

Fig. 8.5.4. Temperature dependence of (a) the optic-like phase mode and (b) the acoustic-like phason in the $smectic-C_A^*$ phase of EHPOCBC at $q_z = 2.8 \cdot 10^7\,m^{-1}$. The soft mode in the $smectic-A$ phase is shown at the far-right side of (a). The insets show the corresponding molecular motion in a unit cell as seen along the layer normal. The solid lines represent for $T < T_a^*$ the best fit to the Eq.8.2.28 and Eq.8.2.27, with $K_{3a}/\eta = 2.2 \cdot 10^{-11}\,m^2 s^{-1}$, $K_{3f}/\eta = 2 \cdot 10^{-10}\,m^2 s^{-1}$, $\alpha/\eta = 1.1 \cdot 10^6\,s^{-1} K^{-1}$, $\gamma_1/b_a = 0.9$, $\gamma_2/b_a = 0.1$ and $q_f = 1 \cdot 10^7\,m^{-1}$. The temperature dependence of the helical period $q_q(T)$ was measured independently and included in the fit (Muševič et al., 1996b).

The temperature dependencies of both modes can be understood by remembering that in a quasielastic like scattering experiment in helicoidally modulated phases, such is the $smectic-C_A^*$ phase of EHPOCBC, one observes the minimum of the dispersion of order parameter excitations at the position of the Bragg angle, i.e. it is shifted off center of the reciprocal space in the laboratory system. This has a consequence that for a given scattering wave-vector $\vec{q}_s$ in the laboratory system, we observe the phase modes with the wave-vectors $\vec{q} \pm n \cdot \vec{q}_a(T)$, where $\vec{q}$ denotes the wave-vector in the "twisted" reference frame. This introduces an apparent temperature dependence of the relaxation rates due to the temperature dependence of the helical period $q_a(T)$. The measurements of the helical period indeed show a very strong temperature dependence. The period increases by lowering the temperature below T_a^*, reaches a maximum at $T_a^* - T \approx 8K$ and then gradually decreases. This behavior explains the extraordinary strong temperature dependencies of both the optic-like and the acoustic-like phase modes.

The measurements were fitted to the predictions of a simple Orihara-Ishibashi model and a very good agreement of the fitting parameters was found. The values of the coefficients, averaged over different samples and experimental runs are $K_{3a}/\eta = 1.6(1 \pm 0.3) \cdot 10^{-11} m^2 s^{-1}$ which is an order of magnitude smaller than in MHPOBC and $K_{3f}/\eta = 2(1 \pm 0.5) \cdot 10^{-10} m^2 s^{-1}$. The rest of the coefficients are $\alpha/\eta = 1.1(1 \pm 0.1) \cdot 10^6 s^{-1} K^{-1}$, $\gamma_1/b_a = 0.9(1 \pm 0.2)$, $\gamma_2/b_a = 0.1(1 \pm 0.2)$ and $q_f = 1(1 \pm 0.5) \cdot 10^7 m^{-1}$. The results clearly indicate that only one coupling term $(\gamma_1 >> \gamma_2)$ is dominant in the expansion 8.2.5.

The experiment pointed out yet another very important observation, which is relevant for the antiferroelectric order in liquid crystals. As one may note, the frequency gap of the optic-like phason branch is exceptionally small in the $smectic-C_A^*$ phase, i.e. it is of the order of $10kHz$. This indicates that the ferroelectric fluctuations are quite soft and well developed in the antiferroelectric phase, or, in another words, the zone-center ferroelectric soft mode has nearly condensed at the $smectic-A$ - $smectic-C_A^*$ phase transition point. This is consistent with the nature of the long-range orientational order in tilted smectics, as the phase transition into the tilted smectic is driven by the in-plane, short-range van der Waals forces. In this sense, there should not be much difference between the dynamics of the ferroelectric and antiferroelectric fluctuations in the $smectic-A$ phase of an antiferroelectric liquid crystal, because they both imply spontaneous and collective tilting of the molecules.

Note 8.5.1. Diffusivity and other thermodynamic model parameters of the antiferroelectric liquid crystals:
Using light scattering technique, diffusivity and the model parameters of antiferroelectric liquid crystals have been determined, which are listed in the table below.

Substance	K_{3a}/γ $[cm^2s^{-1}]$	K_{+}/γ $[cm^2s^{-1}]$	K_{3f}/γ $[cm^2s^{-1}]$	α/γ $[s^{-1}K^{-1}]$	γ_1/b_a	γ_1/b_a	q_f $[m^{-1}]$
MHPOBC (Muševič,1993c)	5.8×10^{-6}	1×10^{-5}	---	---	---	---	---
MHPOBC (Sun et al., 1993a)	6.0×10^{-6}	5×10^{-5}	---	---	---	---	---
EHPOCBC (Rastegar, 1996)	5×10^{-7}	---	5×10^{-7}	1.1×10^{6}	0.9	0.1	1×10^{7}

8.6. Dielectric Properties of the Antiferroelectric Phase: Linear and Non-Linear Response

When a small electric field is applied in a direction perpendicular to the smectic layer normal of an homogeneous antiferroelectric $smectic - C_A^*$ phase, the linear electric polarization can be induced via two mechanisms, as illustrated in Figs.8.6.1. and 8.6.2.

> *(i)* The spontaneous polarization of a unit cell can be increased by changing the magnitude of the spontaneous polarization in a smectic layer, as shown in Fig.8.6.1.Because of the piezoelectric coupling, the overall increase of electric polarization is accompanied by the increase of the tilt angle in the *i*-th layer and decreases of the tilt angle in the neighboring two layers. This gives rise to a small additional electric polarization in the antiferroelectric unit cell. Note that this mechanism can be interpreted as a coupling of a homogeneous external field to an ferroelectric amplitudon mode at $\vec{q}=0$.

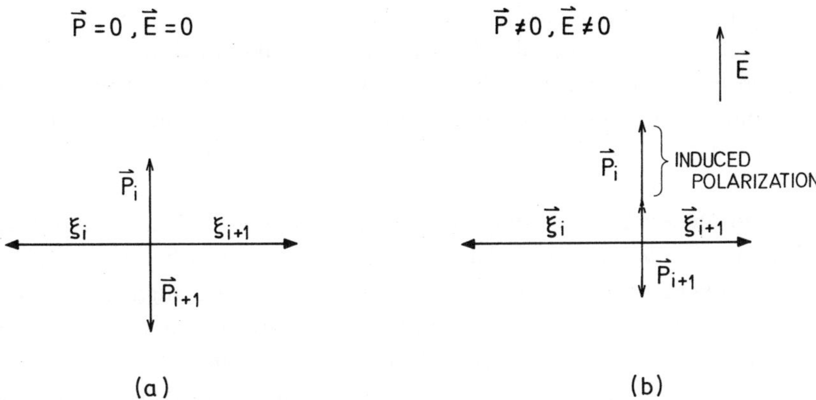

Fig.8.6.1. The increase of the spontaneous polarization of an antiferroelectric $smectic-C_A^*$ phase due to the change of the magnitude of the tilt angle.

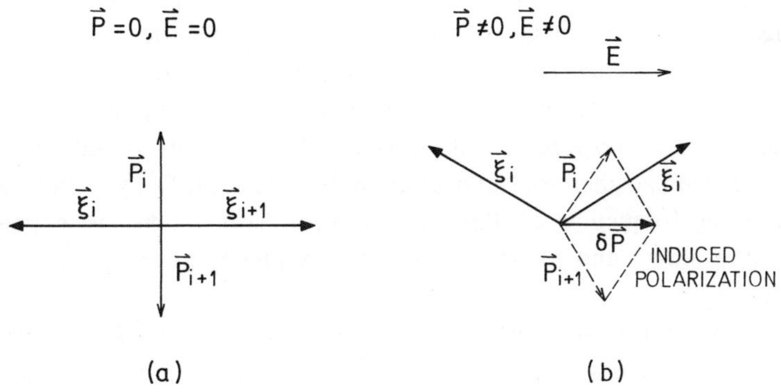

Fig.8.6.2. The increase of the spontaneous polarization of an antiferroelectric $smectic-C_A^*$ phase due to the change of the phase of the tilt angle in the two neighboring layers.

(ii) The polarization can be increased by slightly changing the directions of the two dipole moments, as shown in Fig.8.6.2. The electric field therefore forces the dipoles to align into the field direction via the reorientation along the tilt cone. Note that this can be interpreted as a coupling of a homogeneous electric field with the ferroelectric phason at $\vec{q}=0$.

These considerations are valid only in the limit of a truly homogeneous antiferroelectric $smectic-C_A^*$ phase. As soon as there is a helical modulation, the angle between the two neighboring molecules deviates slightly from 180°. This results in the appearance of small but finite polarization in the unit cell, and can give rise to an additional mode in the dielectric response.

The temperature dependence of the linear response in the vicinity of the $smectic-A$ - $smectic-C_A^*$ can be qualitatively analyzed by considering the coupling of the measuring electric field to the polar modes in the system.

In the $smectic-A$ phase the spectrum of elementary excitations consists of a single, doubly degenerate soft mode branch, which has a characteristic two minima in the dispersion relation. The minimum at the edge of the Brillouin zone is slightly deeper than the minimum in the center of the Brillouin zone leading to the condensation of a non-polar, antiferroelectric soft mode and consequently to the onset of the antiferroelectric ordering. On the other side, the zone-center, $q=0$ mode is here a polar, non-critical ferroelectric soft mode that couples to an external homogeneous electric field and gives rise to a linear dielectric response. In the $smectic-A$ phase of an antiferroelectric liquid crystal we can therefore expect a single-relaxation Debye-like spectrum. The relaxation frequency of this "ferroelectric soft mode" is expected to critically slow down and saturate at a finite frequency at the phase transition temperature T_a^*.

The excitation spectrum in the antiferroelectric $smectic-C_A^*$ phase on the other hand, consists of four branches and therefore four modes at $q=0$. However, two of them are non-polar, antiferroelectric modes and can be observed only in a light-scattering experiment. The other two modes, the so-called ferroelectric phason and the ferroelectric amplitudon are both polar modes and can couple linearly to an external measuring field. We therefore expect a splitting of a single Debye spectrum of the $smectic-A$ phase into two modes in the antiferroelectric $smectic-C_A^*$ phase.

The linear response in the vicinity of the $smectic-A$-$smectic-C_A^*$ phase transition can be conveniently analyzed within the Landau-Khalatnikov dynamical response in the presence of a small, time varying electric field, applied perpendicular to the smectic layer normal

$$\vec{E} = E_{\circ}(1,0,0)\cdot e^{i\omega t} \tag{8.6.1}$$

This electric field couples linearly to the electric polarization and generates a small time-varying contribution to the free-energy density

$$g_E = -\vec{P}\cdot\vec{E} \qquad (8.6.2.)$$

In the "adiabatic approximation", the free-energy density of an antiferroelectric liquid crystal (Eq.8.2.5) is minimized with respect to the electric polarization, which leads to a non-equilibrium free-energy density

$$g = g_A + g(\xi_a,\xi_f) + \varepsilon_f C_f \xi_{fy} E_{\circ}\cdot e^{i\omega t} \qquad (8.6.3.)$$

Here, g_A is the equilibrium density of the $smectic-A$ phase, $g(\xi_a,\xi_f)$ is given by Eq.8.2.9. and represents the free-energy density contribution due to the two order parameters and the last term represents the linear coupling to the external electric field. Here, the ferroelectric polarization is related to the ferroelectric order parameter and the field via expressions

$$P_x = -\varepsilon_f C_f\cdot\xi_{f,y} + \varepsilon_f E_{\circ} \qquad (8.6.4a)$$

$$P_x = \varepsilon_f C_f\cdot\xi_{f,y} \qquad (8.6.4b)$$

A linear response of the system to an external field is analyzed within the "twisted" reference frame, or, what is equivalent, by introducing the so-called "parallel" and "perpendicular" components of the two order parameters (see Section 8.2.). The presence of an external electric field, applied in the x-direction will, in general, induce changes of both order parameters, $\vec{\xi}_a$ and $\vec{\xi}_f$ and therefore give rise to a linear response.

The magnitude of the induced order parameters $\delta\vec{\xi}_a$ and $\delta\vec{\xi}_f$ is easily calculated within the four-component vector formalism using the dynamical matrix. By expanding $\delta\vec{\xi}_a$ and $\delta\vec{\xi}_f$ in terms of plane, monochromatic waves (Eqs.8.3.5a-d) the increase of the total free-energy G per unit length of an antiferroelectric liquid crystal is

$$\Delta G = G - G_\circ =$$

$$= \sum_q \vec{\Psi}_q^+ \cdot \underline{D}(q) \cdot \vec{\Psi}_q + \tfrac{1}{2}\varepsilon_f C_f E_\circ \cdot e^{i\omega t} \left[\vec{\Psi}_{q_\circ}^+ \begin{pmatrix} 0 \\ 0 \\ -i \\ 1 \end{pmatrix} + \vec{\Psi}_{q_\circ} \begin{pmatrix} 0 \\ 0 \\ i \\ 1 \end{pmatrix} \right] \tag{8.6.5.}$$

Here, Ψ_q is a four-vector, composed of the induced order parameters

$$\vec{\Psi}_q = \begin{pmatrix} \delta\xi_{a,q}^{//} \\ \delta\xi_{a,q}^{\perp} \\ \delta\xi_{f,q}^{//} \\ \delta\xi_{f,q}^{\perp} \end{pmatrix} \tag{8.6.6.}$$

Ψ_q^+ is its conjugate and $\underline{D}(q)$ is the dynamical matrix of the antiferroelectric liquid crystal (see Eqs.8.3.7. and 8.3.8a-f). The last term represents the free-energy change due to the coupling to an external electric field. The field induces changes of the order parameter components with the wave vector $q_\circ$, which in our case corresponds to the wave-vector of the helical modulation of the antiferroelectric phase, $q_\circ = q_a$.

The dynamics of the linear response of an antiferroelectric liquid crystal to an electric field is obtained after applying Landau-Khalatnikov equations of motion to the free-energy (Eq.8.6.5.)

$$\frac{d\vec{\Psi}_q}{dt} = -\underline{\Gamma} \frac{\partial \Delta G}{\partial \vec{\Psi}_q^*} \tag{8.6.7.}$$

Here $\underline{\Gamma}$ is a matrix of reciprocal viscosities (see Eq.8.3.11). We are looking for the response of the form

$$\vec{\Psi}_q(t) = \vec{\Psi}_q \cdot e^{i\omega t} \tag{8.6.8.}$$

which leads to the following set of equations for the induced order parameter:

$$\left(2\underline{\Gamma} \cdot \underline{D}(q_a) + i\omega \cdot \underline{I}\right)\vec{\Psi}_{q_a} = \vec{f} \tag{8.6.9.}$$

Here, $\underline{I}$ is the identity matrix and the vector $\vec{f}$ on the right-side represents the linear coupling to the external field,

$$\vec{f} = -\tfrac{1}{2}\Gamma\varepsilon_f C_f E_\circ \begin{pmatrix} 0 \\ 0 \\ -i \\ 1 \end{pmatrix} \tag{8.6.10}$$

Here we have set for simplicity all the viscosity coefficients equal and the matrix $\underline{\Gamma}$ is diagonal.

The dynamic equation (Eq.8.6.9.) which describes the motion of the induced order parameter is

$$\left(\hat{A} + \lambda_\circ\right)|\,\Psi\rangle = |\,f\rangle \tag{8.6.11}$$

Because $\hat{A}$ is Hermitian, the solution $|\Psi\rangle$ can be expanded in terms of the orthogonal eigenfunctions $|u_j\rangle$ of this operator,

$$|\,\Psi\rangle = \sum_j b_j |u_j\rangle \tag{8.6.12}$$

The coefficients b_j are given by

$$b_j = \frac{1}{\lambda_j - \lambda_\circ} \cdot \frac{\langle u_j | f \rangle}{\langle u_j | u_j \rangle} \tag{8.6.13}$$

In our case, the linear response Ψ_{q_a} to the external field is therefore given by

$$\vec{\Psi}_{q_a} = \sum_{j=1}^{4} \frac{1}{\tau_j^{-1}(q_a) + i\omega} \cdot \frac{\vec{\Psi}^{+}_{j,q_a} \cdot \vec{f}}{\left|\vec{\Psi}_{j,q_a}\right|^2} \tag{8.6.14}$$

Here, Ψ_{j,q_a} is the j-th eigenvector at the wave-vector q_a and $\tau_j^{-1}(q_a)$ is the corresponding relaxation rate.

The interpretation of the linear response vector is straightforward. For a given electric field we have a vector $\vec{f}$ and the coupling to the system is described by the scalar product

$$\langle \Psi | f \rangle = \vec{\Psi}^{+}_{j,q_a} \cdot \vec{f} \tag{8.6.15}$$

The electric field therefore couples to those eigenmodes of the system that have a finite scalar product with the vector f. These are the excitations of the ferroelectric order parameter ξ_f, whereas the modes representing the excitations of the antiferroelectric order parameter ξ_a remain unexcited. In the $smectic-A$ phase of the antiferroelectric liquid crystal we have a response of a single, doubly degenerate ferroelectric soft mode, whereas in the $smectic-C_A^*$ phase the linear response consists of the ferroelectric phase mode and splitting of a single Debye-peak in the dielectric spectrum into two distinct relaxations below T_a^*.

In order to calculate the magnitude of the linear response Ψ_{q_a} and the corresponding linear dielectric susceptibility, one has to know the structure of the eigenvectors, which have been calculated in Section 8.3. First, the vector of the induced order parameter

$$\vec{\Psi}_{q_a} = \begin{pmatrix} 0 \\ 0 \\ \delta\xi_f^{//}(\omega, E_\circ) \\ \delta\xi_f^{\perp}(\omega, E_\circ) \end{pmatrix} \qquad (8.6.16)$$

is calculated from the Eq.8.6.14. This allows us to determine the space-average values $\langle\delta P_x\rangle$ and $\langle\delta P_y\rangle$ of the induced polarization via Equations 8.1.7. and 8.1.8 that link the polarization and the tilt

For $T > T_a^*$, the frequency dispersion of the linear susceptibility of the $smectic - A$ phase is

$$\chi_{xx}(\omega) = \varepsilon_f + \frac{\varepsilon_f^2 C_f^2}{\alpha_f(T - T_f) + K_{3f} q_a^2} \cdot \frac{1}{1 + i\omega\tau_{s,f}(q_a)} \qquad (8.6.17)$$

where $\tau_{s,f}^{-1}(q_a)$ is the relaxation rate of the ferroelectric soft mode at the wave-vector $q = q_a$, (see Eq.8.3.16)

$$\tau_{s,f}^{-1}(q_a) = \frac{\alpha_f}{\eta_f}(T - T_f^*) + \frac{K_{3f}}{\eta_f} \cdot q_a^2 \qquad (8.6.18)$$

The temperature dependence of a static $(\omega = 0)$ dielectric constant of the $smectic - A$ phase of an antiferroelectric liquid crystal is shown in Fig.8.6.3. Due to the softening of the ferroelectric soft mode, the static dielectric constant increases as we approach the phase transition into the antiferroelectric phase. Because the softening is incomplete and as we are measuring the response at a noncritical wave-vector $q = 0 \neq q_a$, the static linear response of the antiferroelectric $smectic - A$ phase remains finite at T_a^*

$$T = T_a^* \quad : \qquad \chi_{xx}\left(T_a^*, \omega = 0\right) = \varepsilon_f + \frac{\varepsilon_f^2 C_f^2}{\alpha_f\left(T_a^* - T_f\right) + K_{3f} q_a^2} \qquad (8.6.19)$$

The static dielectric constant therefore exhibits a "quenched", -or "incomplete"- Curie-Weiss law at the phase transition into the chiral antiferroelectric phase. The corresponding relaxation frequency of the ferroelectric soft mode is finite at the phase transition,

$$\tau_{s,f}^{-1}\left(T_a^*\right) = \frac{\alpha_f}{\eta_f}\left(T_a^* - T_f^*\right) + \frac{K_{3f}}{\eta_f} \cdot q_a^2 \qquad (8.6.20)$$

For $T < T_a^*$ there is a splitting of the ferroelectric fluctuations into the ferroelectric amplitude fluctuations and ferroelectric phase fluctuations. The splitting between these two modes is given by the Eqs.8.3.21 and 8.3.22. There are therefore two distinct relaxations in the antiferroelectric $smectic - C_A^*$ phase. They both give rise to the linear response and the corresponding linear susceptibility is

$$\chi_{xx}(\omega, T) = \chi_{f,am} + \chi_{f,ph} \qquad (8.6.21)$$

Here $\chi_{f,am}$ denotes the contribution of the ferroelectric amplitudon and $\chi_{f,ph}$ is the contribution of the ferroelectric phason.

The calculation of the scalar product (Eq.8.6.15), which determines the magnitude of the linear response in the antiferroelectric $smectic - C_A^*$ phase is quite cumbersome. Here we shall discuss briefly the magnitude of the response for the limiting values of the coupling coefficient γ_2, which couples the directions of mutual orientation of the ferroelectric and the antiferroelectric sublattices via the coupling term $\frac{1}{2}\gamma_2\left(\xi_a \xi_f\right)^2$.

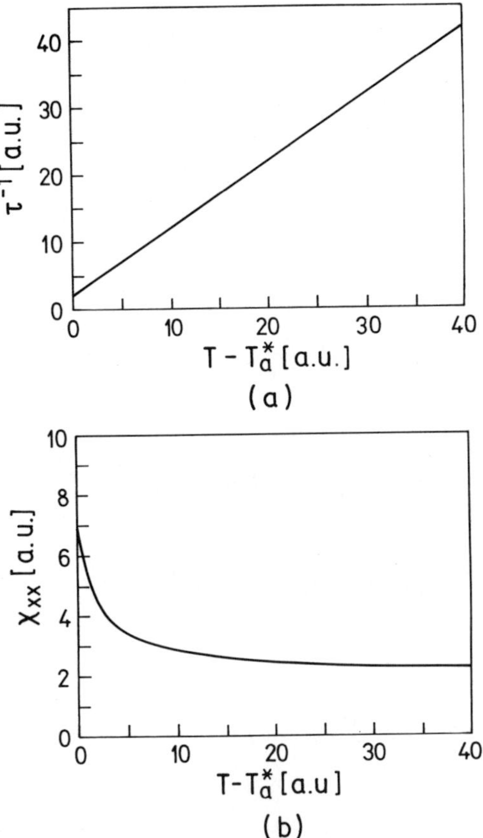

Fig.8.6.3. (a) The "nearly" critical slowing down of the ferroelectric soft mode in the *smectic* – *A* phase of an antiferroelectric liquid crystal. (b) The dielectric constant in the vicinity of the *smectic* – *A* - *smectic* – C_A^* transition of an antiferroelectric liquid crystal.

For $T < T_a^*$ and in the limit of very small $\gamma_2 \to 0$,

$$\tfrac{1}{2}\gamma_2^2\theta_a^2(T) << 4K_{3f}^2 q_a^2 (q_a - q_f)^2 \tag{8.6.22}$$

the contributions of the ferroelectric amplitudon and phason are

$$\chi_{f,am} = \varepsilon_f + \frac{\varepsilon_f^2 C_f^2}{\alpha_f\left(T - T_f\right) + \gamma_1 \dfrac{\alpha_a}{b_a}\left(T_a^* - T\right) + 4K_{3f} q_a\left(q_a - q_f\right)} \cdot \frac{1}{1 + i\omega\tau_{f,am}\left(q_a\right)}$$

(8.6.23a)

and

$$\chi_{f,ph} = \varepsilon_f + \frac{\varepsilon_f^2 C_f^2}{\alpha_f\left(T - T_f\right) + \gamma_1 \dfrac{\alpha_a}{b_a}\left(T_a^* - T\right)} \cdot \frac{1}{1 + i\omega\tau_{f,ph}\left(q_a\right)} \qquad (8.6.23b)$$

In this limit, the corresponding relaxation rates are

$$\tau_{f,am}^{-1}\left(q_a\right) = \frac{\alpha_f}{\eta_f}\left(T - T_f\right) + \frac{\gamma_1}{\eta_f}\frac{\alpha_a}{b_a}\left(T_a^* - T\right) + 4\frac{K_{3f}}{\eta_f} q_a\left(q_a - q_f\right) \qquad (8.6.24a)$$

$$\tau_{f,ph}^{-1}\left(q_a\right) = \frac{\alpha_f}{\eta_f}\left(T - T_f\right) + \frac{\gamma_1}{\eta_f}\frac{\alpha_a}{b_a}\left(T_a^* - T\right) \qquad (8.6.24b)$$

The splitting between the two ferroelectric modes depends on the period of helical modulation q_a

$$\Delta\tau^{-1} = \tau_{f,am}^{-1} - \tau_{f,ph}^{-1} = 4\frac{K_{3f}}{\eta_f} \cdot q_a^2(T)\left(1 - \frac{q_f}{q_a}\right) \qquad (8.6.25)$$

Because the helical period is usually temperature dependent in chiral smectics, this splitting should be temperature dependent, too. The temperature dependence of the dielectric constant is shown schematically in Fig.8.6.4.

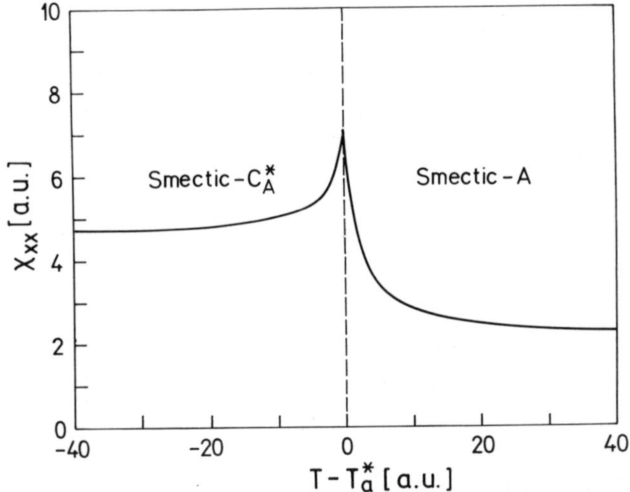

Fig.8.6.4. The temperature dependence of the dielectric constant in the vicinity of the *smectic – A - smectic* – C_A^* phase transition. The drawing is schematic and not to scale.

For $T < T_a^*$ and in the limit of large γ_2,

$$\tfrac{1}{2}\gamma_2^2\theta_a^2(T) >> 4K_{3f}^2 q_a^2 (q_a - q_f)^2 \tag{8.6.26}$$

the contributions of the ferroelectric amplitudon and the phason are

$$\chi_{f,am} = \varepsilon_f + \frac{\varepsilon_f^2 C_f^2}{2\alpha_f(T - T_f) + 2(\gamma_1 + \gamma_2)\frac{\alpha_a}{b_a}(T_a^* - T) + 4K_{3f}(q_a^2 - q_a q_f)} \cdot \frac{1}{1 + i\omega\tau_{f,am}(q_a)}$$

(8.6.27a)

and

$$\chi_{f,ph} = \varepsilon_f + \frac{\varepsilon_f^2 C_f^2}{2\alpha_f(T - T_f) + 2\gamma_1 \frac{\alpha_a}{b_a}(T_a^* - T) + 4K_{3f}(q_a^2 - q_a q_f)} \cdot \frac{1}{1 + i\omega\tau_{f,ph}(q_a)}$$

(8.6.27b)

In this limit, the corresponding relaxation rates are

$$\tau_{f,am}^{-1}(q_a) = \frac{\alpha_f}{\eta_f}(T - T_f) + \frac{\gamma_1 + \gamma_2}{\eta_f} \cdot \frac{\alpha_a}{b_a}(T_a^* - T) + 2\frac{K_{3f}}{\eta_f} q_a(q_a - q_f)$$

(8.6.28a)

$$\tau_{f,ph}^{-1}(q_a) = \frac{\alpha_f}{\eta_f}(T - T_f) + \frac{\gamma_1}{\eta_f} \cdot \frac{\alpha_a}{b_a}(T_a^* - T) + 2\frac{K_{3f}}{\eta_f}(q_a^2 - q_a q_f) \quad (8.6.28b)$$

whereas the frequency splitting linearly increases with decreasing temperature

$$\Delta\tau^{-1} = \tau_{f,am}^{-1} - \tau_{f,ph}^{-1} = \frac{\gamma_2}{\eta_f}\frac{\alpha_a}{b_a}(T_a^* - T) \quad (8.6.29)$$

So far, we have considered the linear dielectric response of the antiferroelectric $smectic-C_A^*$ phase to a small external electric field. However, when the strength of the measuring electric field is increased, the dielectric response of the system becomes non-linear and one can observe higher harmonics in the response. These non-linear properties are described by the dielectric constant, which becomes a function of the external electric field and the electric displacement is a non-linear function of the measuring field:

$$D = \varepsilon_\circ \varepsilon(E)E \quad (8.6.30)$$

Formally, the nonlinear dielectric response of a system is described by expanding the induced electric polarization in a power series in a field (see Bloembergen, 1965). For pure electric dipoles, one has, for example

$$\vec{P} = \chi^{(1)} \cdot \vec{E} + \chi^{(2)} \cdot \vec{E}\vec{E} + \chi^{(3)} \cdot \vec{E}\vec{E}\vec{E} + \ldots \tag{8.6.31}$$

The first term describes linear susceptibility, the second term represents the lowest order nonlinear susceptibility etc. $\chi^{(1)}$ is a second-rank tensor of the linear susceptibility, $\chi^{(2)}$ is a third-rank tensor of the second-order susceptibility, $\chi^{(3)}$ is a fourth-rank tensor describing the third-order nonlinearity and the dots indicate corresponding tensor products. One may note that the nonlinearity of the system generates electric dipoles, which oscillate at many different frequencies, including higher-harmonics and mixed frequencies.

The nonlinear dielectric response of ferroelectric and antiferroelectric liquid crystals has been first treated in a series of papers by Orihara et al. (1993a-b, 1995a-c). As an example, a detailed calculation of the SHG electrooptic response of the antiferroelectric phase based on their approach is given in Section 8.8. Here, we shall briefly explain the underlying physical mechanisms that lead to the nonlinear dielectric response. Let us consider a free energy density in a simple form:

$$g(\xi, P) = \frac{1}{2} a\xi^2 + c\xi P^2 + \frac{1}{2\varepsilon} P^2 - PE \tag{8.6.32}$$

Here, ξ is a scalar order parameter, P is the electric polarization and E is the external electric field. The polarization is coupled to the order parameter via the coupling term $c\xi P^2$, which is essential for the non-linear response. For simplicity, we calculate the static dielectric constant by minimizing $g(\xi, P)$ with respect to ξ and P:

$$\left.\frac{\partial g}{\partial \xi}\right|_{\xi_\circ, P_\circ} = 0 \quad \text{and} \quad \left.\frac{\partial g}{\partial P}\right|_{\xi_\circ, P_\circ} = 0 \tag{8.6.33}$$

This leads to a set of coupled equations

$$a\xi_\circ + cP_\circ^2 = 0 \tag{8.6.34a}$$

$$2c\xi_\circ P_\circ + \frac{1}{\varepsilon}P_\circ - E = 0 \tag{8.6.34b}$$

In the approximation of small coupling, $c \to 0$, the second equation yields linear response to external field,

$$P_\circ = \varepsilon E \tag{8.6.35}$$

whereas the order parameter increases as a square of the applied field,

$$\xi = -\frac{c\varepsilon^2}{a}\cdot E^2 \tag{8.6.36}$$

By substituting Eq.8.6.35 into Eq.8.6.34b, we obtain a third order correction to the polarization, which is now nonlinear in the field:

$$P_\circ = \varepsilon \cdot E + \frac{2c^2\varepsilon^4}{a}\cdot E^3 \tag{8.6.37}$$

This simple example shows that the origin of the nonlinear response of a system can be a coupling of two order parameters, or better say, two modes. In an antiferroelectric liquid crystals, it has been shown by Orihara and Ishibashi (1995b,c) that the nonlinear response is due to nonlinear coupling of the ferroelectric and antiferroelectric order parameters ξ_f and ξ_a. In particular, a coupling term

$$g_c \propto \delta\xi_a(\delta\xi_f)^2 \tag{8.6.38}$$

between the antiferroelectric fluctuations $\delta\xi_a$ and the square of the ferroelectric fluctuations $(\delta\xi_f)^2$ is completely analogous to the coupling term $c\xi P^2$ in Eq.8.6.33, because the ferroelectric order parameter is proportional to the electric polarization. This term is therefore responsible for the third order non-linear dielectric susceptibility of the $smectic-C_A^*$ phase of antiferroelectric liquid crystals. The nonlinear dielectric response of the antiferroelectric $smectic-C_A^*$ phase can be calculated within matrix formalism developed in Section 8.8. The basic steps are the same as those used in our simple calculation, apart from the fact that calculations are far more complicated.

The physical explanation of the coupling is as follows: Since the antiferroelectric modes modulate the dielectric constant in proportion to the square of the applied field, $\varepsilon(E) \propto E^2$, the polarization $P = \varepsilon(E)E$ is proportional to the third power of the applied field, $P \propto E^3$. This gives rise to the third-order dielectric nonlinearity, which is most easily observed by detecting third-harmonic response to an oscillating electric field.

Third order nonlinear dielectric susceptibility in antiferroelectric liquid crystals is important as it gives us access to the dynamics of non-polar modes, such as the antiferroelectric Goldstone (phase) mode. In a linear regime, this mode can be observed only in light scattering experiments, because it is non-polar. However, due to the coupling to the ferroelectric modes via Eq.8.6.38, this non-polar mode becomes observable in an experiment, where the third order non-linear dielectric susceptibility is measured. These experiments have been performed by Obayashi et al. (1995) in MHPOBC. They have observed Debye-like dispersion of the third-order dielectric response, which led to the conclusion that the third order nonlinear dielectric susceptibility of the antiferroelectric $smectic-C_A^*$ phase is proportional to the linear susceptibility of the non-polar antiferroelectric phase mode.

8.7. Linear Electrooptic Response of the Antiferroelectric Phase

The linear electrooptic response of the antiferroelectric $smectic-C_A^*$ phase to an external field is calculated from the theory of the linear response (see Section 8.6.) and considering the corresponding optical changes.

In the paraelectric $smectic-A$ phase, there is a single polar ferroelectric soft mode that couples to the external electric field (see Eqs.8.6.17. and 8.6.18.) and the corresponding molecular motion is shown in Fig.8.7.1(a). On the other hand, this single ferroelectric soft mode splits into an ferroelectric amplitudon and a ferroelectric phason in the antiferroelectric $smectic-C_A^*$ phase. The relaxation rates of these two polar modes are given by Eqs.8.6.24a. and 8.6.24b., respectively, whereas the corresponding molecular motions are shown in Figs.8.7.1(b) and (c).

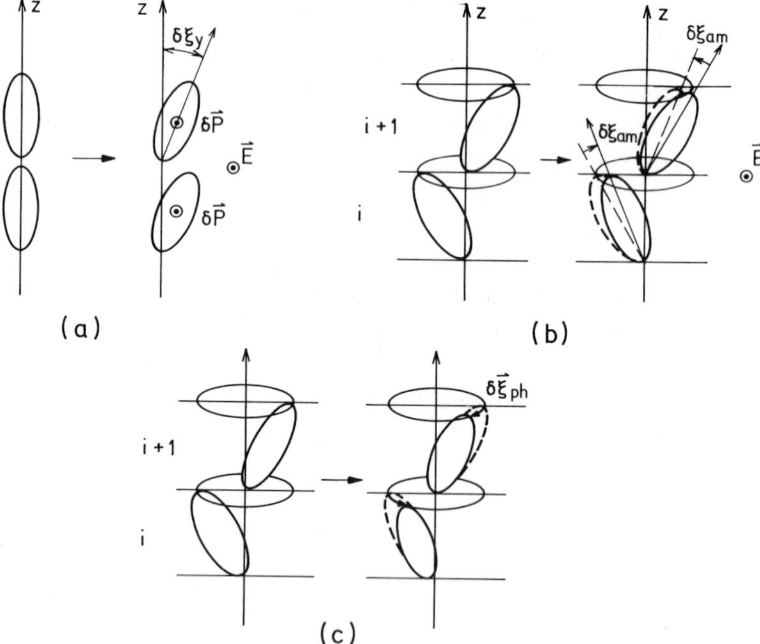

Fig.8.7.1. (a) The ferroelectric soft mode in the $smectic-A$ phase of an antiferroelectric liquid crystal. (b) The ferroelectric amplitudon in the $smectic-C_A^*$ phase of an antiferroelectric liquid crystal. (c) The ferroelectric phason in the $smectic-C_A^*$ phase of an antiferroelectric liquid crystal.

The linear electrooptic response of an antiferroelectric liquid crystal to a small electric field, applied in the direction of the smectic layers is derived in the following way

i) Similar to the dielectric response of the $smectic-C_A^*$, we calculate the vector of the electric-field induced order parameter

$$\Psi_{q_a} = \begin{pmatrix} 0 \\ 0 \\ \delta\xi_f^{//}(\omega, E) \\ \delta\xi_f^{\perp}(\omega, E) \end{pmatrix} \tag{8.7.1}$$

Here, $\delta\xi_f^{//}(\omega, E)$ and $\delta\xi_f^{\perp}(\omega, E)$ denote the change of the tilt amplitude and the phase, when a small electric field is applied.

ii) We calculate the change in the dielectric tensor of the antiferroelectric $smectic-C_A^*$ phase due to the change in the amplitude and the phase of the ferroelectric order parameter

$$\delta\underline{\varepsilon}_{FE} = (\varepsilon_3 - \varepsilon_1)\cdot\begin{pmatrix} 0 & 0 & \delta\xi_{fx} \\ 0 & 0 & \delta\xi_{fy} \\ \delta\xi_{fx} & \delta\xi_{fy} & 0 \end{pmatrix} \tag{8.7.2}$$

Note that the appearance of the off-diagonal terms represents the tilting of the optical axis, as discussed in Section 5.6. For example, the x-z component of the tensor, $(\varepsilon_3 - \varepsilon_1)\cdot\delta\xi_{fx}$, represents a small rotation of the tensor around the y-axis. This small rotation can be detected using electrooptic response techniques.

For $T > T_a^*$, both ferroelectric modes at $q = 0$ are degenerate in the paraelectric $smectic-A$ phase. The response of the soft mode to an electric field $E = E_\circ(1 \quad 0 \quad 0)\cdot e^{i\omega t}$ applied in the x-direction is

$$\delta\xi_{fx} = 0 \tag{8.7.3a}$$

$$\delta\xi_{fy} = -\frac{\varepsilon_f C_f E_\circ}{\alpha_f\left(T - T_f\right) + K_{3f} q_a^2} \cdot \frac{e^{i\omega t}}{1 + i\omega\tau_{s,f}\left(q_a\right)} \tag{8.7.3b}$$

Here, $\tau_{s,f}^{-1}\left(q_a\right)$ is the relaxation rate of the ferroelectric soft mode, given by Eq.8.6.19. The electric field in the x-direction induces a collective tilt of the molecules in the y-direction, which is similar to the electroclinic response of the ferroelectric liquid crystals.

For $T < T_a^*$, the ferroelectric amplitudon and the ferroelectric phason at $q = 0$ both contribute to the linear response. Due to the symmetry, the electric field applied in the x-direction again induces a space-averaged macroscopic tilt of the molecules in the y-direction. This tilt can be expressed as a sum of two contributions

$$\left\langle \delta\xi_{fx} \right\rangle = 0 \tag{8.7.4a}$$

$$\left\langle \delta\xi_{fy} \right\rangle = \left\langle \delta\xi_{fy}^{am} \right\rangle + \left\langle \delta\xi_{fy}^{ph} \right\rangle \tag{8.7.4b}$$

Here, $\langle \delta\xi_{fy}^{am} \rangle$ represents the contribution of the ferroelectric amplitude mode, $\langle \delta\xi_{fy}^{ph} \rangle$ is the contribution of the ferroelectric phase mode to the molecular tilt and $\langle \ \rangle$ represents the space-average.

Following expression 8.6.23, the electric-field induced molecular tilt in the antiferroelectric $smectic - C_A^*$ phase due to the coupling to the ferroelectric amplitudon is in the limit of small $\gamma_2 \to 0$

$\gamma_2 \to 0:$

$$\left\langle \delta\xi_{fy}^{am} \right\rangle = -\frac{\varepsilon_f C_f E_\circ}{\alpha_f\left(T - T_f\right) + \gamma_1 \dfrac{\alpha_a}{b_a}\left(T_a^* - T\right) + 4K_{3f}\left(q_a^2 - q_a q_f\right)} \cdot \frac{e^{i\omega t}}{1 + i\omega\tau_{f,am}} \tag{8.7.5}$$

whereas the contribution of the ferroelectric phason is

$$\left\langle \delta\xi_{fy}^{ph} \right\rangle = -\frac{\varepsilon_f C_f E_\circ}{\alpha_f\left(T - T_f\right) + \gamma_1 \frac{\alpha_a}{b_a}\left(T_a^* - T\right)} \cdot \frac{e^{i\omega t}}{1 + i\omega\tau_{f,ph}} \tag{8.7.6}$$

Here, $\tau_{f,am}^{-1}(q_a)$ and $\tau_{f,ph}^{-1}(q_a)$ are the relaxation rates for the ferroelectric amplitudon and phason, which are given by the Eqs.8.6.24a and b, respectively.

Similarly, we obtain in the limit of large γ_2 the contribution of the ferroelectric amplitudon (see Eq.8.6.27):

$$\left\langle \delta\xi_{fy}^{am} \right\rangle = -\frac{\varepsilon_f C_f E_\circ}{2\alpha_f\left(T - T_f\right) + 2\left(\gamma_1 + \gamma_2\right)\frac{\alpha_a}{b_a}\left(T_a^* - T\right) + 4K_{3f}\left(q_a^2 - q_a q_f\right)} \cdot \frac{e^{i\omega t}}{1 + i\omega\tau_{f,am}} \tag{8.7.7}$$

and of the ferroelectric phason:

$$\left\langle \delta\xi_{fy}^{ph} \right\rangle = -\frac{\varepsilon_f C_f E_\circ}{2\alpha_f\left(T - T_f\right) + 2\gamma_1 \frac{\alpha_a}{b_a}\left(T_a^* - T\right) + 4K_{3f}\left(q_a^2 - q_a q_f\right)} \cdot \frac{e^{i\omega t}}{1 + i\omega\tau_{f,ph}} \tag{8.8.8}$$

In these expressions, $\tau_{f,am}^{-1}(q_a)$and$\tau_{f,ph}^{-1}(q_a)$ are the relaxation rates for the ferroelectric amplitudon and phason, which are in this case given by the Eqs.8.6.28a and b, respectively.

Similarly to the linear electrooptic response of the antiferrolectric liquid crystal in the $smectic - A$ phase, the electric-field induced molecular tilt in the antiferroelectric $smectic - C_A^*$ phase is reflected in the appearance of non-zero off-diagonal terms in the average dielectric tensor of this phase:

$$\langle \delta \underline{\varepsilon}_{FE} \rangle \approx (\varepsilon_3 - \varepsilon_1) \cdot \begin{pmatrix} 0 & 0 & \langle \delta \xi_{fx}^{am} \rangle \\ 0 & 0 & \langle \delta \xi_{fy}^{am} \rangle \\ \langle \delta \xi_{fx}^{am} \rangle & \langle \delta \xi_{fy}^{am} \rangle & 0 \end{pmatrix} + (\varepsilon_3 - \varepsilon_1) \cdot \begin{pmatrix} 0 & 0 & \langle \delta \xi_{fx}^{ph} \rangle \\ 0 & 0 & \langle \delta \xi_{fy}^{ph} \rangle \\ \langle \delta \xi_{fx}^{ph} \rangle & \langle \delta \xi_{fy}^{ph} \rangle & 0 \end{pmatrix} \tag{8.7.9}$$

The off-diagonal terms are oscillating at the frequency of the applied electric field. These small oscillations of the dielectric tensor are visible in an electrooptic experiments as small oscillations of the index of refraction and can be detected very accurately with the linear electrooptic technique, discussed in Section 5.6. A review of experimental data for various antiferroelectric liquid crystal materials can be found in the work of Hiraoka et al.(1993) and Glogarova et al.(1993).

8.8. Non-Linear Electrooptic Response of the Antiferroelectric Phase

The electrooptic response experiments in the antiferroelectric $smectic - C_A^*$ (Hiraoka et al., 1993) have proved the existence of a second harmonic signal in the intensity of the transmitted light, when a strong DC bias field was applied to a homogeneously aligned sample. This second harmonic signal is due to the quadratic coupling of the electric field to the nonpolar modes of the antiferroelectric phase, which can be observed in the experiments on tristable switching in a thin layer of the antiferroelectric liquid crystal TFMNOPBC (Johno et al., 1990), shown in Fig.8.8.1. The transmitted light intensity is shown as a function of a strong, low frequency electric field, applied along the smectic layers. At small fields the parabolic shape of the response curve indicates the quadratic coupling of the structure to an external DC electric field. This nonlinear response of the antiferroelectric $smectic - C_A^*$ phase was theoretically first analyzed in a series of papers by Orihara et al.(1993(a),1993(b), 1995(a), 1995(b), 1995(c)).

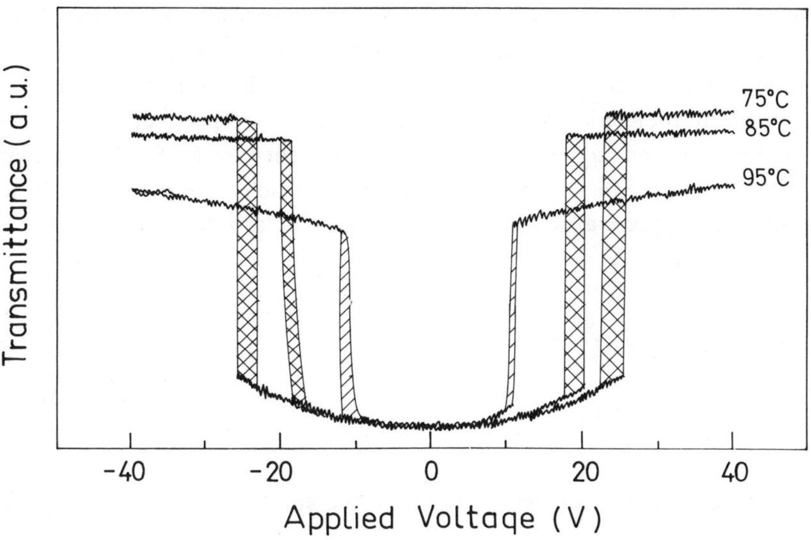

Fig.8.8.1. The intensity of light, transmitted through the sample of an antiferroelectric liquid crystal between the crossed polarizers, as a function of the applied electric field (after Johno et al., 1990).

The polarization-renormalized free energy density of the antiferroelectric liquid crystal is (Eq.8.3.2.)

$$g = g_A + \tfrac{1}{2} a_a \vec{\xi}_a^{\,2} + \tfrac{1}{4} b_a \vec{\xi}_a^{\,4} + \Lambda_a \left(\vec{\xi}_a \times \partial \vec{\xi}_a / \partial z\right)_z + \tfrac{1}{2} K_{3a} \left(\frac{\partial \vec{\xi}_a}{\partial z} \right)^2 +$$

$$+ \tfrac{1}{2} a_f \vec{\xi}_f^{\,2} + \tfrac{1}{4} b_f \vec{\xi}_f^{\,4} + \Lambda_f \left(\vec{\xi}_f \times \partial \vec{\xi}_f / \partial z\right)_z + \tfrac{1}{2} K_{3f} \left(\frac{\partial \vec{\xi}_f}{\partial z} \right)^2 + \tfrac{1}{2} \gamma_1 \vec{\xi}_a^{\,2} \vec{\xi}_f^{\,2} + \tfrac{1}{2} \gamma_2 \left(\vec{\xi}_a \vec{\xi}_f\right)^2 +$$

$$- \vec{P}_f \vec{E} - \tfrac{1}{2} \varepsilon_\circ \Delta\varepsilon \left(\left(\vec{n}_i \vec{E}\right)^2 + \left(\vec{n}_{i+1} \vec{E}\right)^2 \right) \qquad (8.8.1)$$

Here the last two terms represent the ferroelectric (linear) and the dielectric (quadratic) coupling of the order parameter to an external field, respectively. The director of the *i*-th layer is $\vec{n}_i$ and can be expressed in terms of the ferroelectric and the antiferroelectric order parameters ξ_f and ξ_a, respectively. The harmonic coefficients a_a and a_f are temperature dependent (see Eq.8.3.3), whereas the rest of the coefficients is constant. For an electric field $E = E_\circ(1\,0\,0)\cdot e^{i\omega t}$ which is applied along the x-axis, the induced electrical polarization is

$$P_x = -\varepsilon_f C_f \xi_{f,y} + \varepsilon_f E_\circ \tag{8.8.2a}$$

$$P_y = \varepsilon_f C_f \xi_{f,x} \tag{8.8.2b}$$

The nonequilibrium free-energy density of the antiferroelectric phase (Eq.8.8.1) is then

$$g = g_A + g(\xi_a, \xi_f) + \varepsilon_f C_f E_\circ e^{i\omega t}\cdot \xi_{f,y} - \tfrac{1}{2}\varepsilon_\circ \Delta\varepsilon\left(\xi_{a,x}^2 + \xi_{f,x}^2\right)E_\circ^2 \cdot e^{i2\omega t} +$$

$$+ \tfrac{1}{2}\gamma_1 \xi_a^2 \xi_f^2 + \tfrac{1}{2}\gamma_2 (\xi_a \xi_f)^2 \tag{8.8.3}$$

Here, g_A is the equilibrium free-energy density of the $smectic-A$ phase and $g(\xi_a, \xi_f)$ is the non-equilibrium free-energy density of the antiferroelectric $smectic-C_A^*$ phase. It is given by Eq.8.2.9., but the coupling terms between the two order parameters are not included, as they are treated separately. The last two terms in the first line represent the linear and the quadratic coupling of the structure to an external electric field and are oscillating at the fundamental and the second harmonic frequency of the applied field, respectively. The last term of this first line reveals the origin of the second harmonic signal: It is due to the quadratic coupling of the field to the antiferroelectric and the ferroelectric order parameter via the dielectric anisotropy. There is, however, another contribution to the non-linear response, which is not so evident from the above equation. The two terms in the second line of the Equation 8.8.3 are both proportional to the square of the ferroelectric order parameter and thus oscillate at a second harmonic frequency.

Let us now discuss the second harmonic response of the antiferroelectric $smectic-C_A^*$ phase. The equilibrium value of the ferroelectric order parameter is here zero, whereas the equilibrium value of the antiferroelectric order parameter is non-zero:

$$\xi_f^\circ = 0 \tag{8.8.4a}$$

$$\xi_a^\circ = \vartheta_\circ (\cos q_a z, \sin q_a z) \tag{8.8.4b}$$

$\vartheta_\circ$ is the tilt angle of the antiferroelectric phase, given by the Eq.8.2.19, $q_a = -\Lambda_a / K_{3a}$ is the wave-vector of the helix of the antiferroelectric phase.

The dynamics of the non-linear response is most easily calculated in the twisted reference frame, which follows the antiferroelectric helix with the wave-vector q_a. Here, the two non-equilibrium order parameters are written as a sum of a static and time-fluctuating part (see Fig.8.8.2.):

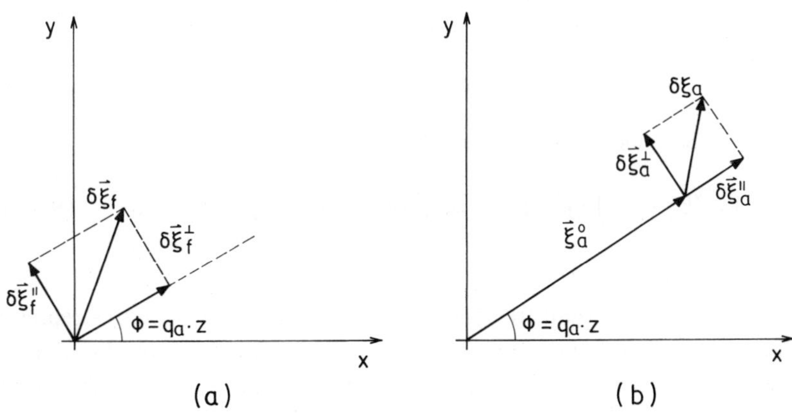

Fig.8.8.2. (a) The definition of the fluctuating part of the ferroelectric order parameter in the $smectic-C_A^*$ phase. (b) The definition of the static and fluctuating part of the antiferroelectric order parameter in the $smectic-C_A^*$ phase.

$$\vec{\xi}_f(z,t) = \delta\vec{\xi}_f(z,t) \tag{8.8.5a}$$

$$\vec{\xi}_a(z,t) = \vec{\xi}_a^{\circ} + \delta\vec{\xi}_a(z,t) \tag{8.8.5b}$$

The non-equilibrium free-energy density (Eq.8.8.3) is then

$$\begin{aligned}
g(z,t) = g_\circ &+ \tfrac{1}{2}\left(a_a + 3b_a\vartheta_\circ^2 - K_{3a}q_a^2\right)\cdot\left(\delta\xi_a^{/\!/}\right)^2 + \tfrac{1}{2}\left(a_a + b_a\vartheta_\circ^2 - K_{3a}q_a^2\right)\cdot\left(\delta\xi_a^{\perp}\right)^2 + \\
&+ \tfrac{1}{2}K_{3a}\left(\left(\partial\delta\xi_a^{/\!/}/\partial z\right)^2 + \left(\partial\delta\xi_a^{\perp}/\partial z\right)^2\right) + \\
&+ \tfrac{1}{2}\left(a_f + \gamma_1\vartheta_\circ^2 + K_{3f}q_a^2 - 2K_{3f}q_f q_a\right)\cdot\left(\delta\xi_f^{/\!/}\right)^2 + \\
&+ \tfrac{1}{2}\left(a_f + (\gamma_1 + \gamma_2)\vartheta_\circ^2 + K_{3f}q_a^2 - 2K_{3f}q_f q_a\right)\cdot\left(\delta\xi_f^{\perp}\right)^2 + \\
&+ \tfrac{1}{2}K_{3f}\left(\left(\partial\delta\xi_f^{/\!/}/\partial z\right)^2 + \left(\partial\delta\xi_f^{\perp}/\partial z\right)^2\right) - \\
&- K_{3f}\left(q_a - q_f\right)\left(\delta\xi_f^{/\!/}\frac{\partial\delta\xi_f^{\perp}}{\partial z} - \delta\xi_f^{\perp}\frac{\partial\delta\xi_f^{/\!/}}{\partial z}\right) + \\
&+ \varepsilon_f C_f\left(\delta\xi_f^{\perp}\cdot\sin q_a z + \delta\xi_f^{/\!/}\cdot\cos q_a z\right)E_\circ\cdot e^{i\omega t} - \\
&- \varepsilon_\circ\Delta\varepsilon\vartheta_\circ\left(\delta\xi_a^{/\!/}\cdot\cos^2 q_a z - \tfrac{1}{2}\delta\xi_a^{\perp}\cdot\sin 2q_a z\right)\cdot E_\circ^2\cdot e^{i2\omega t} + \\
&+ \gamma_1\left(\vec{\xi}_a^{\circ}\cdot\delta\vec{\xi}_a\right)\cdot\left(\delta\vec{\xi}_f\right)^2 + \gamma_2\left(\vec{\xi}_a^{\circ}\cdot\delta\vec{\xi}_f\right)\cdot\left(\delta\vec{\xi}_a\cdot\delta\vec{\xi}_f\right)
\end{aligned} \tag{8.8.6}$$

Here, $g_\circ$ is the equilibrium free energy density and the rest of terms represents the fluctuation contributions to the free energy. The term in the first line represents the increase of the free-energy density due to fluctuations of the amplitude and phase of the antiferroelectric order parameter. The term in the second line represents a contribution due to spatial variations (elastic deformation) of the antiferroelectric amplitude and phase fluctuations. The two terms in the third and fourth line represent the contributions of the fluctuations of the amplitude and the phase of the ferroelectric order parameter, respectively. The term in fifth line is again the elastic deformation term, whereas the term in the sixth line is the remnant of the Lifshitz term for the ferroelectric order parameter. Here, $q_a = -\Lambda_a / K_{3a}$ and $q_f = -\Lambda_f / K_{3f}$ are the wave-vectors of the antiferroelectric and ferroelectric helices, respectively, which are in general of different magnitude. There is no Lifshitz term for the antiferroelectric order parameter.

The terms in the last three lines in the Eq.8.7.6. describe the coupling of the ferroelectric and antiferroelectric order parameters to the external field. The term

$$g_E = \varepsilon_f C_f \left(\sin q_a z \cdot \delta\xi_f^{\perp} + \cos q_a z \cdot \delta\xi_f^{//} \right) E_\circ \cdot e^{i\omega t} \tag{8.8.7}$$

describes the linear coupling of the ferroelectric order parameter to an external field. A similar term has been found in the discussion of the dielectric properties of the ferroelectric $smectic-C^*$ phase, see Eq.6.1.6 in Section 6.1. This term is oscillating at the frequency of the applied field and induces a polarization and the tilt of the molecules that oscillate at the frequency of the external field. The coupling term

$$g_{E^2} = -\varepsilon_\circ \Delta\varepsilon \vartheta_\circ \left(\cos^2 q_a z \cdot \delta\xi_a^{//} - \tfrac{1}{2} \sin 2q_a z \cdot \delta\xi_a^{\perp} \right) \cdot E_\circ^2 \cdot e^{i2\omega t} +$$

$$+ \gamma_1 \left(\vec{\xi}_a^{\circ} \cdot \delta\vec{\xi}_a \right) \cdot \left(\delta\vec{\xi}_f \right)^2 + \gamma_2 \left(\vec{\xi}_a^{\circ} \cdot \delta\vec{\xi}_f \right) \cdot \left(\delta\vec{\xi}_a \cdot \delta\vec{\xi}_f \right) \tag{8.8.8}$$

describes the onset of a second harmonic signal.

The way of obtaining the second harmonic response from the above expression was first described by Orihara and Ishibashi (1995b) and we shall here follow their analysis. First, one has to calculate the linear response, because the induced ferroelectric order parameter contributes to the second harmonic response as well. This step is formally analogous to our calculation of the linear response of the antiferroelectric liquid crystal, see Section 8.5. In the second step, the expression for the linearly induced order parameter is put into the free-energy expansion 8.8.6. Furthermore, the two orthogonal components of the fluctuating part of the antiferroelectric order parameter are expanded in terms of plane waves

$$\delta\xi_a^{//} = \sum_q \delta\xi_{a,q}^{//} \cdot e^{iqz} \tag{8.8.9a}$$

$$\delta\xi_a^{\perp} = \sum_q \delta\xi_{a,q}^{\perp} \cdot e^{iqz} \tag{8.8.9b}$$

where $\delta\xi_{i,-q}^{j} = \left(\delta\xi_{i,q}^{j}\right)^*$ to keep physical variables real.

The increase of the total free-energy density G per unit length of an antiferroelectric liquid crystal is

$$\Delta G = G - G_\circ = \sum_q \bar{\Psi}_q^+ \cdot \underline{D}(q) \cdot \bar{\Psi}_q -$$

$$-\tfrac{1}{4}\varepsilon_\circ\vartheta_\circ\left(\Delta\varepsilon + \gamma_2\left(\chi_{xy}^{\xi}(\omega)\right)^2\right)E_\circ^2 \cdot e^{i2\omega t} \cdot \left[\begin{pmatrix}1 & 0 & 0 & 0\end{pmatrix}\bar{\Psi}_{2q_a}^+ + \bar{\Psi}_{2q_a}\begin{pmatrix}1\\0\\0\\0\end{pmatrix}\right] -$$

$$-\tfrac{1}{4}\varepsilon_\circ\vartheta_\circ\left(\Delta\varepsilon - \left(\chi_{xy}^{\xi}(\omega)\right)^2\right)E_\circ^2 \cdot e^{i2\omega t} \cdot \left[\begin{pmatrix}0 & i & 0 & 0\end{pmatrix}\bar{\Psi}_{2q_a}^+ + \bar{\Psi}_{2q_a}\begin{pmatrix}0\\-i\\0\\0\end{pmatrix}\right] \tag{8.8.10}$$

Here the first term

$$G_1 = \sum_q \bar{\Psi}_q^+ \cdot \underline{D}(q) \cdot \bar{\Psi}_q \tag{8.8.11}$$

represents the non-equilibrium free-energy of the antiferroelectric $smectic-C_A^*$ phase due to spontaneous fluctuations of the antiferroelectric and the ferroelectric order parameters. The second term

$$G_{2q_a}^{am} = -\varepsilon_\circ\vartheta_\circ\left(\Delta\varepsilon + \gamma_2\left(\chi_{xy}^{\xi}(\omega)\right)^2\right)E_\circ^2 \cdot e^{i2\omega t} \cdot \left[\begin{pmatrix}1 & 0 & 0 & 0\end{pmatrix}\bar{\Psi}_{2q_a}^+ + \bar{\Psi}_{2q_a}\begin{pmatrix}1\\0\\0\\0\end{pmatrix}\right] \tag{8.8.12}$$

represents the quadratic coupling of the field to the antiferroelectric amplitudon, which is evident from the structure of the vectors on the right side of this expression. Note that the wave-vector of this coupling is $2q_a$, which is characteristic for a quadrupolar coupling. Here, we have introduced for simplicity the "tilt susceptibility" $\chi_{xy}^{\xi}(\omega)$, which describes the magnitude of the space-averaged tilt, induced by the electric field. It is related to the dielectric susceptibility χ_{xx} by

$$\chi_{xy}^{\xi}(\omega)=\frac{\langle\delta\xi_{f,y}\rangle}{E_x}=-\frac{\chi_{xx}(\omega)-\varepsilon_f}{\varepsilon_f C_f} \tag{8.8.13}$$

The third term

$$G_{2q_a}^{ph}=-\tfrac{1}{4}\varepsilon_\circ\vartheta_\circ\left(\Delta\varepsilon-\left(\chi_{xy}^{\xi}(\omega)\right)^2\right)E_\circ^2\cdot e^{i2\omega t}\cdot\left[\begin{pmatrix}0 & i & 0 & 0\end{pmatrix}\vec{\Psi}_{2q_a}^{+}+\vec{\Psi}_{2q_a}\begin{pmatrix}0\\-i\\0\\0\end{pmatrix}\right] \tag{8.8.14}$$

represents the quadratic coupling of the field to the antiferroelectric phason and is of similar structure as the coupling of the field to the antiferroelectric amplitudon.

The non-linear response of the antiferroelectric $smectic-C_A^*$ phase is now calculated from the Landau-Khalatnikov equations for the non-equilibrium free-energy (Eq.8.8.10)

$$\frac{d\vec{\Psi}_q}{dt}=-\underline{\Gamma}\cdot\frac{\partial\Delta G}{\partial\vec{\Psi}_q^*} \tag{8.8.15}$$

We are looking for the time-oscillating response of the system at a second harmonic frequency of the applied field

$$\vec{\Psi}_q(t)=\vec{\Psi}_q^\circ\cdot e^{i2\omega t} \tag{8.8.16}$$

From the structure of the free-energy (Eq.8.8.10) it is clear, that the response of the system would be at the wave-vector $2q_a$. This means that the corresponding changes in the chiral structure have a periodicity, which is two times smaller than the pitch of the antiferroelectric structure $p_a = 2\pi/q_a$. The Landau-Khalatnikov equation (Eq.8.8.15) for the excitations with the wave-vector $2q_a$ can be expressed as

$$\left(2\underline{\Gamma}\underline{D}(2q_a)+i2\omega\,\underline{I}\right)\cdot\vec{\Psi}_{2q_a}=\vec{g} \tag{8.8.17}$$

Here, $\underline{I}$ is the identity matrix and the vector $\vec{g}$ on the right side represents a quadratic coupling to the external electric field

$$\vec{g}=\varepsilon_\circ\Delta\varepsilon\vartheta_\circ\cdot\begin{pmatrix}1 & i & 0 & 0\end{pmatrix}\frac{E_\circ^2}{4}\cdot e^{i2\omega t}+$$
$$+\left\{\gamma_2\vartheta_\circ\left(\chi_{xy}^{\xi}(\omega)\right)^2\cdot\begin{pmatrix}1 & 0 & 0 & 0\end{pmatrix}-\vartheta_\circ\left(\chi_{xy}^{\xi}(\omega)\right)^2\cdot\begin{pmatrix}0 & i & 0 & 0\end{pmatrix}\right\}\cdot\frac{E_\circ^2}{4}\cdot e^{i2\omega t} \tag{8.8.18}$$

The Landau-Khalatnikov Equation 8.8.17 is in fact the equation for forced oscillations of the system imposed by the nonlinear coupling of the system to the external electric field. Here, the vector $\vec{g}$ represents the external driving force of the overdamped system. We have seen that this problem can be solved using operator algebra. The Equation (8.8.17) is an inhomogeneous operator equation of the form

$$\left(\hat{A}+\lambda_\circ\right)\cdot|\Psi\rangle=|g\rangle \tag{8.8.19}$$

The operator $\hat{A}\equiv 2\underline{\Gamma}\cdot\underline{D}(2q_a)$ is Hermitian and the solution $|\Psi\rangle$ can be expanded in terms of the orthogonal eigenfunctions $|u_j\rangle$ of this operator. Following similar procedure as in the case of the linear response (see Eqs.8.6.11 to 8.6.15), we obtain the vector Ψ_{2q_a} of the non-linear response that satisfies the Landau-Khalatnikov Eq.8.8.18

$$\bar{\Psi}_{2q_a} = \frac{\varepsilon_\circ \Delta\varepsilon \vartheta_\circ + \gamma_2 \vartheta_\circ \left(\chi_{xy}^{\xi}(\omega)\right)^2}{\frac{2\alpha}{\eta}\left(T_a^* - T\right) + 4\frac{K_{3a}}{\eta} q_a^2} \cdot \frac{E_\circ^2}{4} \cdot \frac{e^{i2\omega t}}{1 + i2\omega\tau_{am,a}} \cdot \bar{\Psi}_{2q_a}^{am} +$$

$$+ i \cdot \frac{\varepsilon_\circ \Delta\varepsilon \vartheta_\circ - \vartheta_\circ \left(\chi_{xy}^{\xi}(\omega)\right)^2}{4\frac{K_{3a}}{\eta} q_a^2} \cdot \frac{E_\circ^2}{4} \cdot \frac{e^{i2\omega t}}{1 + i2\omega\tau_{ph,a}} \cdot \bar{\Psi}_{2q_a}^{ph} \tag{8.8.20}$$

Here, η is the rotational viscosity for the motion of the antiferroelectric amplitude and phase excitations. The first contribution is the non-linear response of the antiferroelectric amplitudon, with the normalized vector $\Psi_{2q_a}^{am} = (1\ 0\ 0\ 0)$ and the relaxation rate

$$\tau_{am,a}^{-1} = \frac{2\alpha}{\eta}\left(T_a^* - T\right) + 4\frac{K_{3a}}{\eta} q_a^2 \tag{8.8.21}$$

The second contribution is the nonlinear response of the antiferroelectric phason with the normalized vector $\Psi_{2q_a}^{ph} = (0\ 1\ 0\ 0)$ and the relaxation rate

$$\tau_{aph,a}^{-1} = 4\frac{K_{3a}}{\eta} q_a^2 \tag{8.8.22}$$

The changes in the antiferroelectric amplitude and the phase due to the nonlinear coupling to the external electric field are:

$$\delta\xi_a^{//} = \frac{\varepsilon_\circ \Delta\varepsilon \vartheta_\circ + \gamma_2 \vartheta_\circ \left(\chi_{xy}^{\xi}(\omega)\right)^2}{\frac{2\alpha}{\eta}\left(T_a^* - T\right) + 4\frac{K_{3a}}{\eta} q_a^2} \cdot \frac{E_\circ^2}{2} \cdot \frac{e^{i2\omega t}}{1 + i2\omega\tau_{am,a}} \cdot \cos 2q_a \tag{8.8.23a}$$

$$\delta\xi_a^{\perp} = -\frac{\varepsilon_\circ \Delta\varepsilon \vartheta_\circ - \vartheta_\circ \left(\chi_{xy}^{\xi}(\omega)\right)^2}{4\frac{K_{3a}}{\eta} q_a^2} \cdot \frac{E_\circ^2}{2} \cdot \frac{e^{i2\omega t}}{1 + i2\omega\tau_{ph,a}} \cdot \sin 2q_a \tag{8.8.23b}$$

As expected, both perturbations are periodic in space and time and have a double helical period along the antiferroelectric helical structure.

The observability of this nonlinear response by optical methods can be calculated by considering the corresponding changes in the space-averaged index of refraction. The change of the dielectric tensor of the antiferroelectric $smectic-C_A^*$ phase due to a small change of the antiferroelectric order parameter is (Eq.8.5.7)

$$\delta \underline{\varepsilon}_{AFE} \approx (\varepsilon_3 - \varepsilon_1) \begin{pmatrix} 2\xi^{\circ}_{ax}\delta\xi_{ax} & \xi^{\circ}_{ay}\delta\xi_{ax} + \xi_{ax}\delta\xi_{ay} & 0 \\ \xi^{\circ}_{ay}\delta\xi_{ax} + \xi_{ax}\delta\xi_{ay} & 2\xi^{\circ}_{ay}\delta\xi_{ay} & 0 \\ 0 & 0 & 1 - 2\xi^{\circ}_{ax}\delta\xi_{ax} - 2\xi^{\circ}_{ay}\delta\xi_{ay} \end{pmatrix} \tag{8.8.24}$$

Here, $\vec{\xi}^{\circ}_a = \vartheta_{\circ}(\cos q_a z, \sin q_a z)$ is the equilibrium value of the antiferroelectric order parameter and $\delta\vec{\xi}_a$ is a small perturbation, given by Eq.8.8.23. The space-averaged contribution to the dielectric tensor due to nonlinear coupling of the antiferroelectric $smectic-C_A^*$ phase to an external electric field is

$$\langle \delta \underline{\varepsilon}_{AFE}(2\omega) \rangle \approx \frac{(\varepsilon_3 - \varepsilon_1)\varepsilon_{\circ}\Delta\varepsilon E_{\circ}^2 e^{i2\omega t}}{\tau_{am}^{-1} + i2\omega} \cdot \begin{pmatrix} \Delta_{am} & 0 & 0 \\ 0 & -\Delta_{am} & 0 \\ 0 & 0 & 0 \end{pmatrix} +$$

$$+ \frac{(\varepsilon_3 - \varepsilon_1)\varepsilon_{\circ}\Delta\varepsilon E_{\circ}^2 e^{i2\omega t}}{\tau_{ph}^{-1} + i2\omega} \cdot \begin{pmatrix} \Delta_{ph} & 0 & 0 \\ 0 & -\Delta_{ph} & 0 \\ 0 & 0 & 0 \end{pmatrix} \tag{8.8.25}$$

Here, Δ_{am} and Δ_{ph} are given by

$$\Delta_{am} = \frac{\vartheta_{\circ}^2}{4} \left(1 + \gamma_2 \frac{(\chi^{\xi}_{xy}(\omega))^2}{\Delta\varepsilon} \right) \tag{8.8.26a}$$

$$\Delta_{ph} = \frac{\vartheta_{\circ}^2}{4} \left(1 - \frac{(\chi^{\xi}_{xy}(\omega))^2}{\Delta\varepsilon} \right) \tag{8.8.26b}$$

The nonlinear coupling of the external electric field to the antiferroelectric $smectic-C_A^*$ phase is therefore reflected in ***the onset of biaxiality*** of the originally uniaxial space-averaged dielectric tensor for the optical frequencies. The effect is to some extent similar to the magnetic-field induced biaxiality, which has been discussed in Section 5.7. This is not surprising, because the physical mechanism and the formal expression of the free-energy coupling between the external field and the structure are in both cases the same. Let us remember that in the case of an external magnetic field, the coupling was

$$g_H = -\tfrac{1}{2}\mu_\circ\Delta\chi\left(\vec{n}\cdot\vec{H}\right)^2 \tag{8.8.27}$$

which is in the case of the electric field replaced by

$$g_E = -\tfrac{1}{2}\varepsilon_\circ\Delta\varepsilon\left(\vec{n}\cdot\vec{E}\right)^2 \tag{8.8.28}$$

The distortion is in both cases the same if the diamagnetic anisotropy and the dielectric anisotropy are of equal sign. For positive anisotropy, $\Delta\varepsilon > 0$, the effect of the electric field on the helical structure of the $smectic-C_A^*$ phase is shown in Fig.8.8.3. As a result of a global distortion, the dielectric constant increases along the direction of the field and decreases in the transverse direction, which gives rise to globally biaxial optical structure.

The second-harmonic electrooptical response in the antiferroelectric $smectic-C_A^*$ phase was observed by Y.Kimura et al., 1996, and K.Hiraoka et al., 1996. The second-harmonic technique can give valuable information on the dynamics of the antiferroelectric order parameter. The dynamics of the antiferroelectric order parameter is not observable with a linear dielectric response technique, because the modes in question are not polar. It should be stressed, that these nonpolar antiferroelectric modes are however visible in the non-linear dielectric spectroscopy experiments, which are related to the nonlinear electrooptic experiments (see Section 8.6). For details, the reader should see the work of K.Obayashi et al., 1995, Orihara and Ishibashi, 1995(b) and 1993(a) and the references therein.

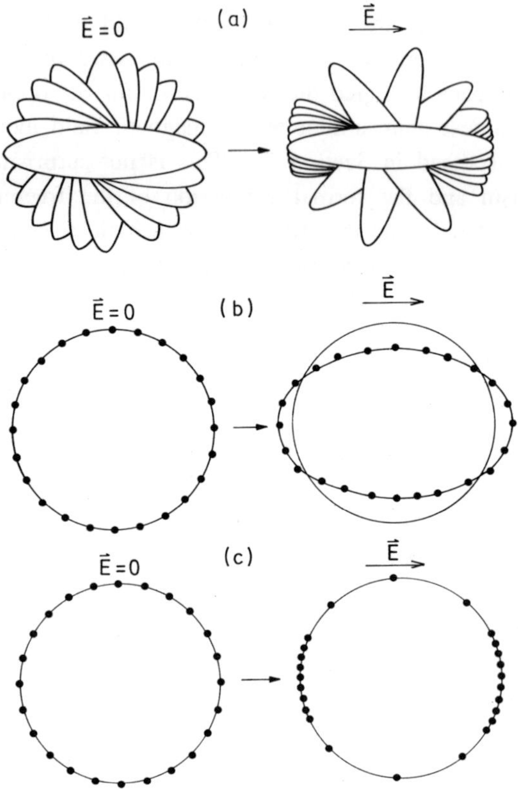

Fig.8.8.3. (a) As a global effect of an external electric field to the antiferroelectric $smectic-C_A^*$ structure, the helical structure is deformed in such a way, that most of the molecules are aligned either into $+\vec{E}$ or $-\vec{E}$ directions. These are energetically equivalent because the quadratic $(\vec{n}\cdot E)$ coupling does not distinguish between the "+" and "-" directions. This elastic deformation can be considered as a superposition of the distortions due to antiferroelectric amplitudon and the antiferroelectric phason. (b) The antiferroelectric amplitudon at a wave-vector $2q_a$ represents a distortion of the magnitude of the tilt angle. Here, the molecules preferentially tilt into the field direction. (c) The antiferroelectric phason at a wave-vector $2q_a$ represents a reorientation around the cone of a constant tilt into the field direction.

8.9. Ferroelectricity, Antiferroelectricity and Intermediate Phases

The first antiferroelectric liquid crystal 4-(1-methyl-heptyloxycarbonyl-phenyl) 4'-octylbiphenyl-4-carboxylate (MHPOBC) shows the following phase sequence, as we cool it down from the isotropic phase:

$$I \xrightarrow{156^\circ C} SmA \xrightarrow{122^\circ C} SmC^*_\alpha \xrightarrow{120.7^\circ C} SmC^* \xrightarrow{119^\circ C} SmC^*_\gamma \xrightarrow{118.3^\circ C}$$

$$\xrightarrow{118.3^\circ C} SmC^*_A \xrightarrow{66^\circ C} SmI^*_A \xrightarrow{30^\circ C} Cryst.$$

Here, $smectic-C^*$ is the ferroelectric phase and $smectic-C^*_A$ is the antiferroelectric phase. At the time of discovery of antiferroelectricity in liquid crystals in 1989, only the structure of the antiferroelectric $smectic-C^*_A$ could be determined without doubt, whereas the structure of the $smectic-C^*_\alpha$ and $smectic-C^*_\gamma$ phases remained unexplained until 1998. The reason for this uncertainty was the fact that only optical conoscopy, dielectric and heat capacity methods were used to investigate these new phases. These methods give only an indirect indication of the microscopic structure. On the basis of conoscopic observations it was conjectured (Gorecka et al. 1990) that the $smectic-C^*_\gamma$ was a ferrielectric phase, with a structure that is intermediate between pure ferroelectric and the antiferroelectric order.

Later on, several new intermediate phases were discovered either in mixtures of materials that exhibit antiferroelectric order (Isozaki et al. 1992, 1993a,b) or in single compound materials (Okabe et al., 1992; Nguyen et al., 1994; Gisse et al., 1996).

For example, Isozaki et al. (1993) report the following sequence of phases in binary mixtures of MHPOCBC and MHPOOCBC, which is also shown in the (E,T) phase diagram in Fig.8.9.1.:

$$SmA \to SmC^*_\alpha \to SmC^* \to SmC^*_{AF} \to SmC^*_{FIH} \to SmC^*_\gamma \to SmC^*_{FIL} \to SmC^*_A$$

Here, the $smectic-C^*_{FIH}$ and $smectic-C^*_{FIL}$ were tentatively designated as ferrielectric phases that appear above and below the ferrielectric $smectic-C^*_\gamma$ phase, respectively. Based on the conoscopic observations in an external field, the $smectic-C^*_{AF}$ phase was tentatively designated as the new antiferroelectric phase with critical fields that were much lower than those in the traditional antiferroelectric $smectic-C^*_A$ phase. Figure 8.9.1. represents the phase diagram (Isozaki et al., 1993) for a binary mixture 70% of MHPOCBC and 30% of MHPOOCBC, where this rather general phase sequence is observed.

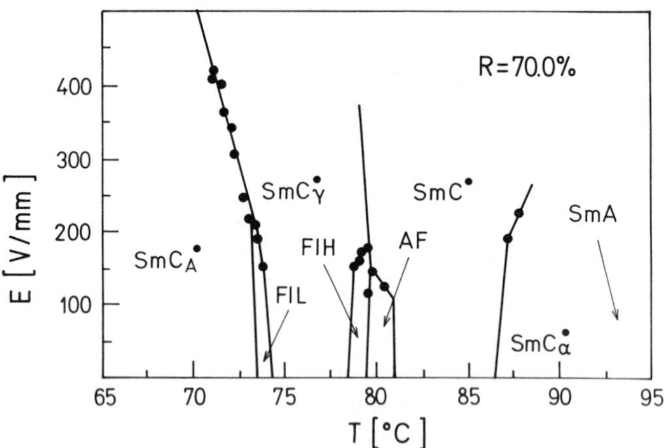

Fig.8.9.1. The electric-field-temperature phase diagram of a binary mixture of 70% MHPOCBC and 30% of MHPOOCBC (Isozaki et al., 1993). One can observe a rather general sequence of phases that were identified according to the response of a conoscopic image to an external DC field. These phases need to be confirmed with a microscopic method that is capable of determining the unit cell.

Nguyen et al. (1994) reported two tentatively designated ferrielectric phases $smectic-C^*_{FI1}$ and $smectic-C^*_{FI2}$ in thiobenzoate material 10OTBBB1M7. They have proved that the $smectic-C^*_{FI2}$ phase of this material is equivalent to $smectic-C^*_\gamma$ phase of MHPOBC.

So far, there are problems in characterizing the phase boundaries between these intermediate phases. Most of them were characterized with macroscopic methods, (optical conoscopy, dielectric response, switching experiments) which do not give direct insight into the structure on a microscopic level. Only recently, the microscopic structure of two of these intermediate phases ($smectic-C^*_{FI1}$ and $smectic-C^*_{FI2}$) were determined using resonant X-ray scattering (Mach et al., 1998). In the following, we shall present the results of different experimental methods that were used in characterizing the phase transitions in intermediate phases:

(i) ***Conoscopic observations of helical smectic phases in external electric field.***

In a conoscopic experiment, one observes an interference figure of an optically anisotropic crystal in a convergent polarized light (see, for example, Hartshorne, N.H., and Stuart, A, 1964). A typical conoscopic figure of the $smectic-A$ phase is shown in Figure 8.9.2.

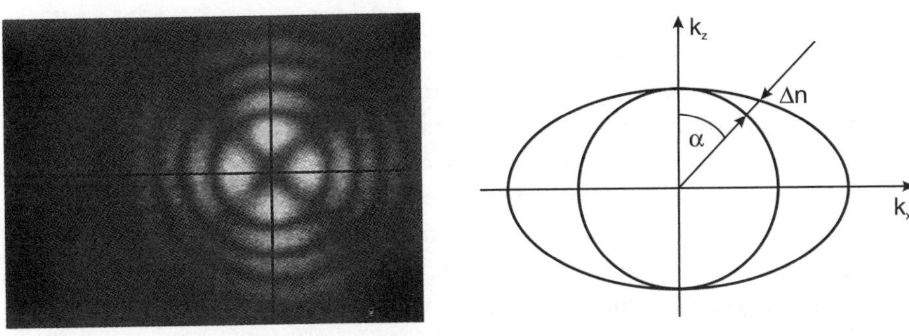

Figure 8.9.2. (a) Conoscopic figure of the $smectic-A$ phase of liquid crystals. There is a single optical axis in a direction perpendicular to the plane of the figure. The interference rings appear due to optical anisotropy. (b) The optical indicatrix of the $smectic-A$ phase is typical for uniaxial crystals with positive anisotropy, $n_e > n_o$.

When the chiral smectic liquid crystal is cooled into one of the tilted and helicoidal smectic phases (such as ferroelectric, ferrielectric or antiferroelectric), the uniaxial figure does not change significantly. This can be understood in view of the ***space-averaging*** of the dielectric tensor due to helical order and is a direct confirmation of the validity of the first order approximation to the wave equation (see Chapter 5. Optical properties). However, conoscopic figures of various tilted smectic phases appear differently, when a DC external field is applied in a direction of the smectic layers. This method was used for the classification of most antiferroelectric and intermediate phases.

Figure 8.9.3. shows a typical response of the conoscopic figure of the ferroelectric $smectic-C^*$ phase to an external DC electric field, applied perpendicular to the helix. For small fields, one can observe that the conoscopic figure tilts as a whole in a direction which is perpendicular to the direction of the field. Then, close to the critical electric field for the unwinding of the helix (Fig.8.9.3(c)), the conoscopic figure gets blurred and distorted in a plane perpendicular to the field. This indicates the onset of process of unwinding of the helical structure. Finally, one can clearly observe a biaxial conoscopic figure of the unwound $smectic-C^*$ phase above the critical field (Fig.8.9.3(d)). The sequence of these images can be easily understood: for small fields, the helix is soliton-like distorted with a series of 2π domain walls. The corresponding dielectric tensor of sample as a whole can be considered as a nearly uniaxial tensor with the optical axis which is tilted due to the action of the field. The direction of the tilt of the optical axis is in a plane perpendicular to the field, because electric dipoles are for a given smectic layer oriented perpendicularly to the direction of the tilt. Above the critical field, the structure is spatially homogeneous and the biaxiality originates from the intrinsic biaxiality of liquid crystalline molecules, which are aligned with their electric dipoles in a field direction. In this regime, the two optical axes are for many materials (CE-8, MHPOBC) aligned in a plane (i.e. optical plane) perpendicular to the field direction. The response is qualitatively different in $smectic-C_\gamma^*$ and $smectic-C_A^*$ phases, as shown in a series of images in Fig.8.9.4. (Gorecka et al., 1990). In the $smectic-C_\gamma^*$ phase of MHPOBC, the two "eyes" of the conoscopic image appear first in a plane parallel to the electric field and then rotate and shift as a whole at larger fields in a plane perpendicular to the field, when the structure is unwound (Fig.8.9.4(b)). This behavior has motivated Gorecka et al., to speculate on the ferrielectric nature of the $smectic-C_\gamma^*$ phase. The detailed structure has however remained unknown until the X-ray experiment of Mach et al. (1998) who have shown that this is in fact a four-layer smectic structure. The experiment of Mach et al. has shown that it is very difficult to conjecture on the microscopic nature of more complex phases from macroscopic experiments. In the case of experiments in electric fields, there is an additional complication due to the complicated process of unwinding of helical structures, which appears both on the mesoscopic (unwinding of a unit cell) and microscopic level (unwinding of a helix).

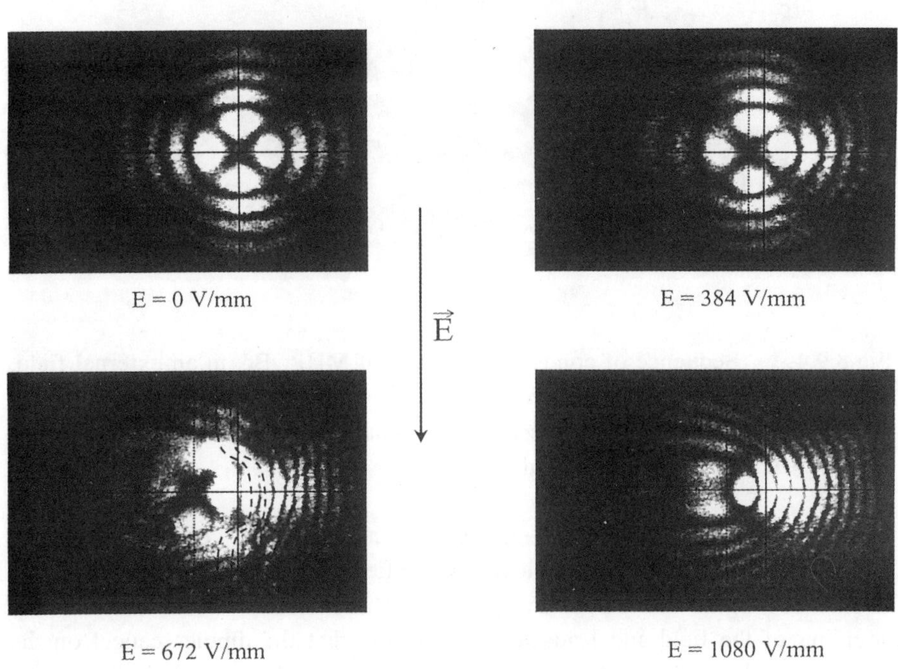

Fig.8.9.3. Conoscopic figure in the *smectic* $-C^*$ phase of a mixture of 35% ferroelectric CE-8 and 65% of racemic CE-8 in an external electric field, perpendicular to the helix. (a) $0V/mm$, (b) $384V/mm$, (c) $672V/mm$ and (d) $1080V/mm$ (Conradi et al., 1999).

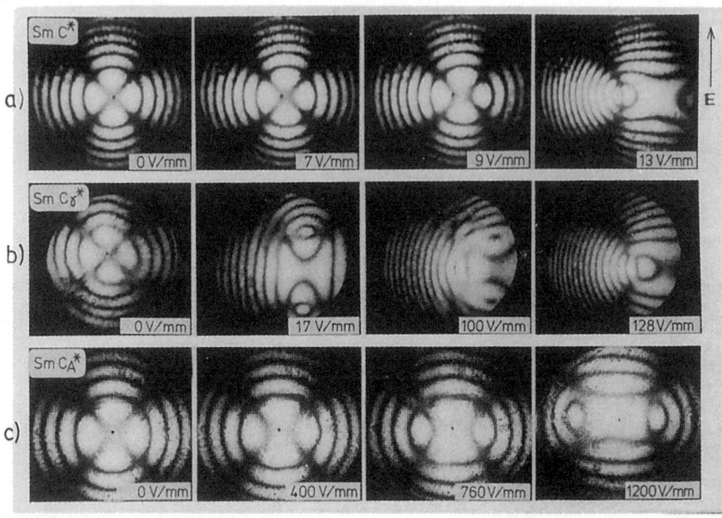

Fig.8.9.4. Sequence of conoscopic images of MHPOBC in an external field, applied perpendicular to the smectic normal. (a) Ferroelectric $smectic-C^*$ phase, (b) ferrielectric $smectic-C_\gamma^*$ phase and (c) antiferroelectric $smectic-C_A^*$. After Gorecka et al., 1990.

In the antiferroelectric $smectic-C_A^*$ phase, the two "eyes" of the conoscopic image appear in a plane perpendicular to the field, but there is no linear rotation or shift of a figure as a whole (Fig.8.9.4(c)). This indicates the absence of linear coupling of the field and leads to a conclusion that the dipoles cancel on the microscopic level. From this experiment, we can therefore conclude that in the antiferroelectric phase, the dipoles in the two neighboring smectic layers are aligned nearly perpendicularly and therefore cancel. The electric field distorts the structure due to the quadratic dielectric coupling and induces a π-soliton structure, similar to the action of a magnetic field in ferroelectric liquid crystals. At very high fields, one expects that the last image in Fig.8.9.4(c) would shift as a whole to the right, when the structure would be completely unwound. This was indeed observed in another antiferroelectric $smectic-C_{AF}^*$ phase of MHPBC, which is much easier distorted by the external field than, for example, the

$smectic-C_A^*$ phase of MHPOBC. The sequence of conoscopic images for the $smectic-C_A^*$ and $smectic-C_{AF}^*$ phases of MHPBC are shown in Fig.8.9.5.

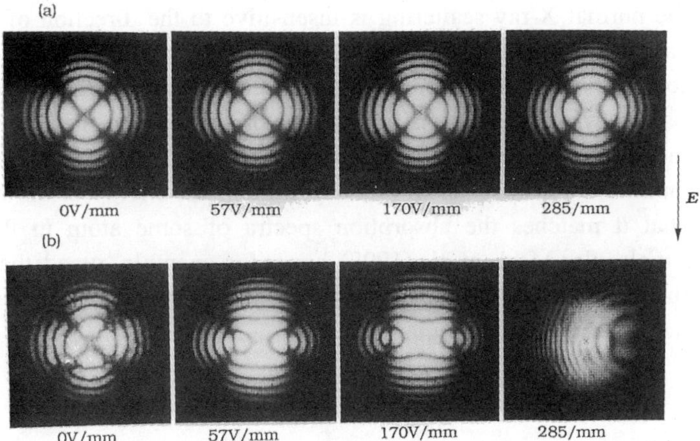

Fig.8.9.5. Conoscopic images in (a) the $smectic-C_A^*$ and (b) the $smectic-C_{AF}^*$ phases of 4-(1-methylheptyloxycarbonyl-phenyl) 4'-octylbiphenyl-4-carboxylate (MHPBC). After Okabe et al., 1992.

Finally, the conoscopic images of the so-called $smectic-C_{FIH}^*$ and $smectic-C_{FIL}^*$ phases in an external field (Isozaki et al., 1993) are very similar to those of $smectic-C_\gamma^*$ phase of MHPOBC. This raises the question whether these two "ferrielectric" phases are indeed new phases. The question should be answered by resonant X-ray scattering.

(ii) Resonant X-ray scattering

The resonant X-ray scattering technique was first applied to the study of orientational order in tilted smectic structures by Mach et al. in 1998. In a normal X-ray scattering experiment, the X-rays are elastically scattered by the electron distribution in the sample. This enables us to measure, for example, the onset of a smectic modulation in smectic phases, as the X-rays are at a certain angle Bragg-reflected from the periodic distribution of smectic layers. There are several experiments that were performed in antiferroelectric materials (Suzuki et al., 1990, Takanishi et al., 1995). In an experiment, Takanishi et al. (1995) investigate even higher-order Bragg reflections from the smectic layering, which enable us to see the "strength" of the smectic order.

The normal X-ray scattering is insensitive to the direction of the long molecular axis of a given smectic layer, because the corresponding scattering cross section is a scalar quantity. This has a serious drawback, as it does not allow to observe by normal X-ray scattering the orientational distribution (if any) along the normal to the smectic layers. This problem can be circumvented by using resonant X-ray scattering. In this experiment, the energy of the X-rays is tuned so that it matches the absorption spectra of some atom in the liquid crystalline molecule. Mach et al. (1998) have, for example, tuned their X-ray energy to the K-edge absorption spectra of the sulfur atom, which is a part of the liquid crystal molecule, as shown in Fig.8.9.6.

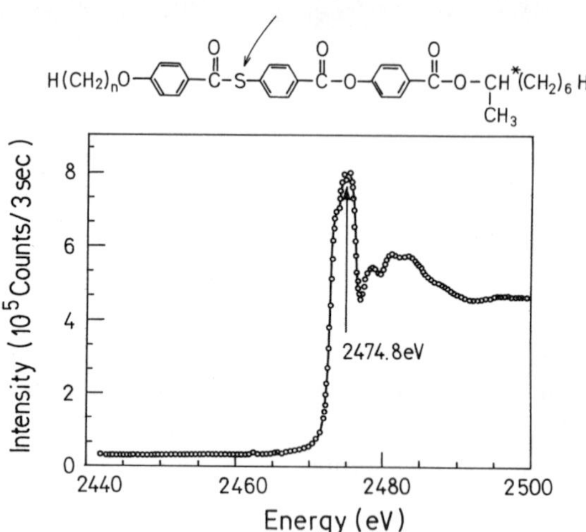

Fig.8.9.6. The liquid crystal molecule 10OTBBB1M7, first synthesized by Nguyen et al. (1994), together with its fluorescence intensity spectrum. The X-ray beam energy was tuned to 2474,8eV, thus matching the K-edge absorption peaks of sulfur atom. After Mach et al., 1998.

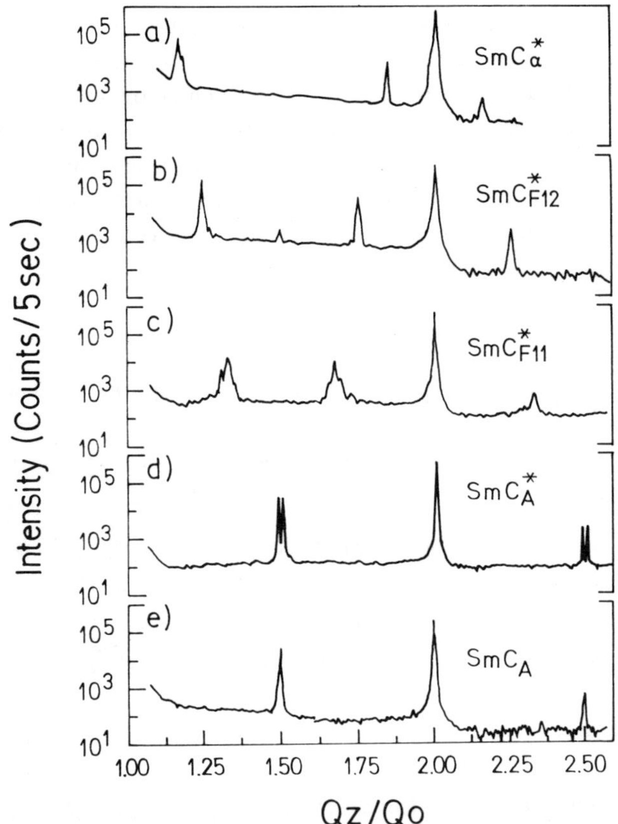

Fig.8.9.7. Resonant X-ray intensity scans in various phases of pure (R) (a-d) and racemic 10OTBBB1M7 (e). Here, q_z is the component of the wave-vector of the X-ray along the normal to the smectic layers and $q_{\circ} = 2\pi/d$, where d is the thickness of a smectic layer. After Mach et al., 1998.

In this case, the cross-section for the X-ray scattering becomes a tensorial property which is orientation-sensitive and enables us to monitor the orientational distribution on the molecular level. Using this technique, Mach et al. have measured the dependence of the resonant-scattering intensity on the magnitude of the scattering wave-vector q_z along the normal to the smectic layers, as shown in Fig.8.9.7. This allowed them to determine the molecular orientation in subsequent smectic layers in various smectic phases.

For example, in the scan that was measured in the antiferroelectric $smectic-C_A^*$ phase (Fig.8.9.7(d)), the satellite peaks are at $q_z/q_\circ = 1.5$ and $q_z/q_\circ = 2$, respectively. This means that there is an extra modulation in system with the wave-vector of $q_l = 2\pi/l = 0.5 \cdot q_\circ$. This corresponds to modulation length $l = 2d$, which is exactly an antiferroelectric unit cell, composed of two antiparallel ferroelectric tilted layers. Similarly, one can deduce a three-layer unit cell for the ferrielectric $smectic-C_{FI1}^*$ and a four-layer unit cell for the ferrielectric $smectic-C_{FI2}^*$ phase.

Symbol	Type	Unit cell	Top view of a unit cell
smectic - C*	ferroelectric	one smectic layer	1
smectic - C_A^*	antiferroeletric	two smectic layers	2 1
smectic - C_{FI1}^* (≡ smectic - C_Y^*)	ferrielectric 1	three smectic layers	~120 2 1 3
smectic - C_{FI2}^*	ferrielectric 2	four smectic layers	~90° 2 3 1 4
smectic - C_α^*	presumably incommensurate	none	3 2 4 1 7 5 6

Fig.8.9.8. List of the tilted smectic structured, deduced from the resonant X-ray scattering.

Finally, this experiment indicates that the $smectic-C_\alpha^*$ phase of 10OTBBB1M7 could be an incommensurate phase. In the orientationally incommensurate phase, the period of the helical modulation is on the molecular length-scale and is not a multiple of the interlayer distance. The experiment shows that the modulation, which is close to a unit cell, is close to six smectic layers and changes with temperature. Due to finite resolution, it is however impossible to claim true incommensurability of the orientational order in the $smectic-C_{\alpha}^{*}$ phase of 10OTBBB1M7.

The structures of intermediate phases, as deduced from the resonant X-ray scattering experiment of Mach et al. are listed in Figure 8.9.8. and can be described as follows.

(i) In the chiral ferroelectric $smectic-C^{*}$, the length of the unit cell of a crystal a is equal to a single smectic layer of thickness, $a=d$. The molecules are tilted at a tilt angle θ with respect to the layer normal and the spontaneous polarization appears perpendicular to the plane of the tilt. The tilt direction precesses, as we move along the smectic normal. The period of this helical arrangement is ≈ 1000 times larger than the length of a unit cell and therefore represents only a small chiral perturbation to the system.

(ii) In the chiral antiferroelectric liquid crystalline $smectic-C_{A}^{*}$ phase, the molecules are tilted in opposite directions in neighboring smectic layers. ***The unit cell*** of this phase therefore consists of ***two smectic layers*** with a length of $a=2d$. The alternation of the tilt direction of the molecular orientation in neighboring layers is accompanied by the alternation of the direction of spontaneous polarization. Two neighboring layers thus form an antiferroelectric unit cell with two nearly antiparallel electric dipoles and a very small value of the equilibrium electric polarization $P_{\circ}(\bar{r})=P_{i}+P_{i+1}\approx 0$. Because of chirality, the directions of the spontaneous tilt and the in-plane polarization slowly precess around the layer normal as one moves along the direction perpendicular to the smectic plane. This causes a small deviation from the 180° alternation in the tilt between two consecutive layers and the formation of a modulated, helicoidal structure.

(iii) In the chiral ferrielectric I. $smectic-C_{FI1}^{*}$ phase, the angle between the tilt directions in neighboring smectic layers is close to 120°. The unit cell of this phase therefore consists of three smectic layers. The length of a unit cell is $a=3d$ and the tilt angle is uniform throughout the crystal.

Note that this phase is equivalent to the $smectic-C_{\gamma}^{*}$ phase of MHPOBC (Nguyen et al., 1994). The spontaneous polarization in the three neighboring layers nearly cancels and results in a very small value, $P_{\circ}(\bar{r}) = P_i + P_{i+1} + P_{i+2} \approx 0$. It is not clear, whether due to chirality, this unit cell precesses, as we move along the smectic layer normal. In this case, the period of this modulation would be much larger than the length of a unit cell and could be considered as a combination of three helical structures, gearing into each other at an angle of 120°. Optically, this structure should be equivalent to the $smectic-A$ phase with a very small optical rotation. This is in fact in contradiction with optical experiments, which show significant optical rotation, similar to the ferroelectric phase.

(iv) In the chiral ferrielectric II. $smectic-C_{FI2}^{*}$ phase, the angle between the tilt directions in neighboring smectic layers is close to 90°. The unit cell of this phase therefore consists of four smectic layers. The length of a unit cell is $a = 4d$ and the tilt angle is uniform throughout the crystal. The spontaneous polarization in the four neighboring layers nearly cancels and results in a very small value of total polarization, $P_{\circ}(\bar{r}) = P_i + P_{i+1} + P_{i+2} + P_{i+3} \approx 0$. Again, resonant X-ray experiment does not confirm that due to chirality, this unit cell precesses as a whole, as we move along the smectic layer normal. Optically, this structure is equivalent to the $smectic-A$ phase with very small optical rotation, which is again in contradiction with optical experiments.

(v) In the chiral alpha phase, $smectic-C_{\alpha}^{*}$, the angle between the directions of the tilt in the neighboring layers is smaller than 90°. For example, the resonant X-ray scattering in the $smectic-C_{\alpha}^{*}$ phase of 10OTBBB1M7, indicates that this angle is close to 60°, which indicates that this is an incommensurate phase. The optical rotation of this phase is very small, as expected for such a short-period helix.

(iii) **Heat capacity experiments**

Several heat capacity techniques have been used in order to characterize the phase sequences in antiferroelectric liquid crystal materials, as presented in

Fig.8.9.9. Among them, the most advanced are the high resolution ac calorimetry experiments, performed by Ema et al. (1995-1998), see Fig.8.9.9(a).

From these experiments, the following two important conclusions were made: *(i)* The heat capacity peak is relatively easily observed at the $smectic-A$-$smectic-C_{\alpha}^{*}$ transition. The height of this peak is of the order of $\Delta c_p \approx 0.3JK^{-1}g^{-1}$. The anomalies are also observable at the reconstructing transitions $smectic-C_{\alpha}^{*}$-$smectic-C^{*}$ and $smectic-C^{*}$-$smectic-C_{\gamma}^{*}$. The corresponding anomalies are nearly an order of magnitude smaller, i.e. $0.01JK^{-1}g^{-1}$ and $0.003JK^{-1}g^{-1}$, respectively. *(ii)* A clearly observable pretransitional heat capacity was observed in the $smectic-A$ phase. The analysis showed different classes of critical behavior, ranging from the XY universality class to Gaussian tricritical behavior.

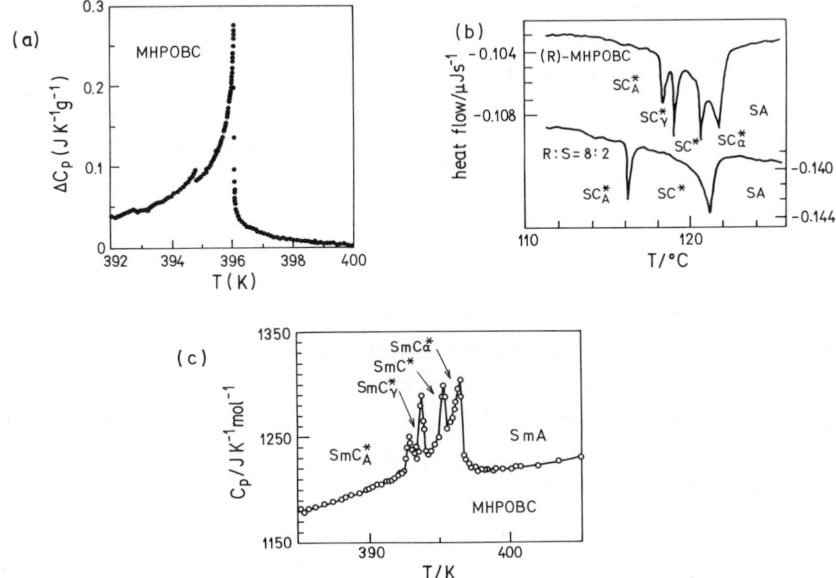

Fig.8.9.9. Heat capacity in the intermediate phases of antiferroelectric liquid crystals. (a) AC calorimetry by Ema et al. (1995), (b) DSC calorimetry by Chandani et al. (1989) and (c) adiabatic calorimetry by Asahina et al. (1993).

(iv) Birefringence and optical rotation

The birefringence of the helical phases is very sensitive to any change of the tilt angle and can in fact be used to determine the optical tilt angle with an extreme accuracy (Muševič et al., 1996a). On the other hand, the optical rotation of helical phases is highly sensitive both to the magnitude of the tilt angle and the length of the helical period. The experiments that were performed in antiferroelectric liquid crystals (Muševič et al., 1997; Škarabot et al., 1998; Philip et al., 1995) clearly indicate phase transitions between intermediate phases, as shown in Fig.8.9.10. These experiments led to the following important conclusions:

(i) The phase transition $smectic-A\text{-}smectic-C_{\alpha}^{*}$ is in most substances of second order. However, this transition is close to a tricritical point. $Smectic-A\text{-}smectic-C_{A}^{*}$ phase transitions are likely to be of first order (for example EHPOBC).

(ii) There are very strong pretransitional fluctuations of the tilt angle close to the $smectic-A\text{-}smectic-C_{\alpha}^{*}$ transition. These show a critical behavior, which is in some substances close to the expected XY -3D universality class (see Škarabot et al., 1999).

(iii) The $smectic-C_{\alpha}^{*}$ phase is definitely a tilted phase, as evidenced from the decrease of the birefringence due to the molecular tilt. It has, however, a very small and nearly undetectable rotatory power. This allows us to estimate the upper limit of the length of the helical modulation of the $smectic-C_{\alpha}^{*}$ phase (Škarabot et al., 1998), which is $p_{\circ} < 150nm$ for MHPOBC.

(iv) All the reconstructive phase transitions $smectic-C_{\alpha}^{*}\text{-}smectic-C^{*}$ and $smectic-C^{*}\text{-}smectic-C_{\gamma}^{*}$ are of first order and are accompanied by a discontinuity of the tilt angle.

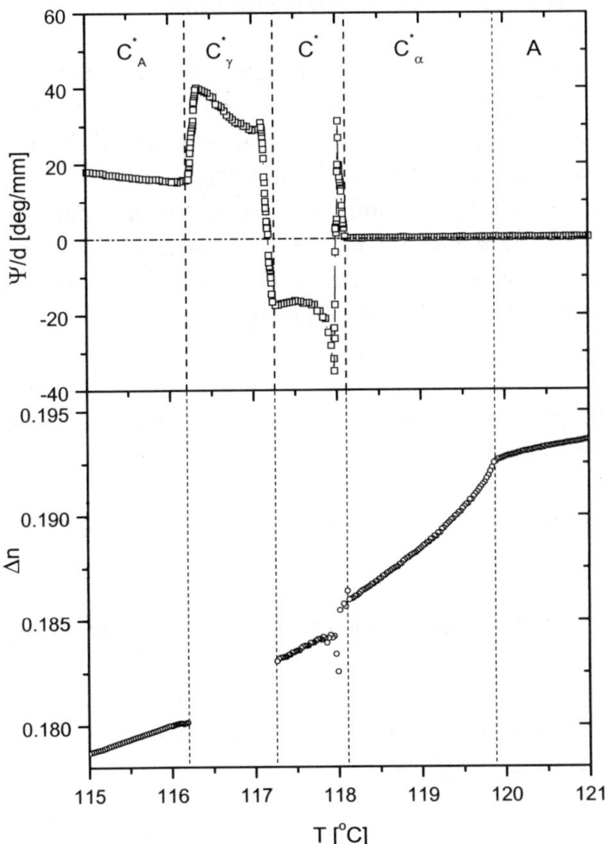

Fig.8.9.10. (a) The temperature dependence of the birefringence in the intermediate phases of (R)-MHPOBC. (b) The temperature dependence of the optical rotation per unit length in (R)-MHPOBC. After Škarabot et al., 1998a.

8.10. Discrete Models of Intermediate Phases

The thermodynamic properties and phase transitions between the antiferroelectric $smectic-C_A^*$ phase and the related ferri-, ferro-, and paraelectric phases have been first theoretically analyzed by Orihara and Ishibashi (1990) and later by Žekš and Čepič (1993). Their theoretical description was of continuous nature and implied the introduction of two order parameters, i.e. the ferroelectric and the antiferroelectric order parameters ξ_f and $\vec{\xi}_a$, respectively. In this sense, the ***continuum theory*** are of bilayer character and can reproduce only four different structures:

(i) The paraelectric $smectic-A$ phase, where both equilibrium order parameters equal zero, $\xi_f = 0, \xi_a = 0$.

(ii) The ferroelectric phase where the ferroelectric order parameter is non-zero, $\xi_f \neq 0$

(iii) The antiferroelectric $smectic-C_A^*$ phase where the antiferroelectric order parameter is non-zero, $\xi_a \neq 0$.

(iv) The ferrielectric $smectic-C_\gamma^*$ phase , where both order parameters are non zero, $\xi_f \neq 0, \xi_a \neq 0$.

The corresponding structures are schematically shown in Fig.8.10.1. It soon became clear that this simple continuous phenomenological theory of Orihara and Ishibashi cannot explain real structures, because: *(i)* it does not explain the structure of the $smectic-C_\alpha$ phase, and *(ii)* there is only one ferrielectric phase within this model.

The problem of the structure of the $smectic-C_\alpha$ phase and the existence of several ferrielectric phases have initiated the development of discrete phenomenological models for the description of intermediate phases (i.e. $smectic-C_\alpha$ and $smectic-C_\gamma$). Within these models, one considers Landau or a microscopic model theory for a single smectic layer and includes the interactions of a given smectic layer with its neighbors.

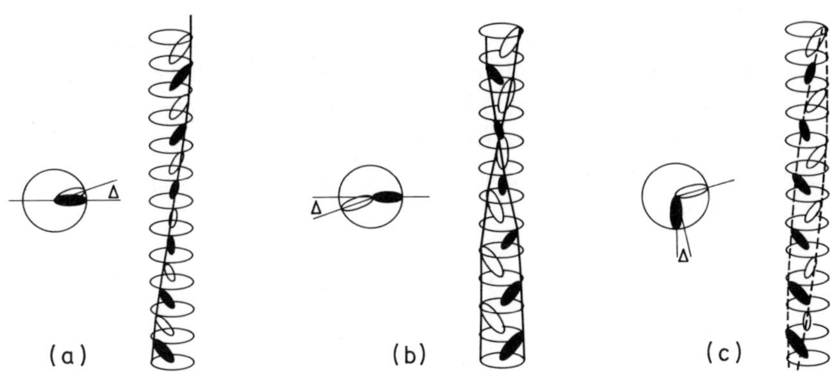

Fig.8.10.1. The ferroelectric, antiferroelectric and ferrielectric structures with a constant magnitude of the tilt, as predicted by the continuous phenomenological Landau theory.

There are several discrete model, which can be classified into two categories: *(i)* Ising models (Nakagawa, 1993a,b; Koda and Kimura, 1996; Yamashita, 1996a,b, 1997) and *(ii)* discrete XY models (Sun et al., 1993b; Čepič and Žekš, 1995, 1996; Pikin, 1995; Roy and Madhusudana, 1996). Let us briefly discuss the main features of both models:

(i) In an Ising model of tilted and polar smectics, one uses a single scalar quantity σ to describe the orientation of the molecules in a single smectic layer. For example, as shown in Fig.8.10.2. $\sigma = +1$ denotes molecules which are tilted in the plane of the paper in the right direction for a given angle and $\sigma = -1$ denotes the smectic layer where the molecules tilt in the left direction. It is important to note that this corresponds to two antiparellel orientations of the electric dipoles, or, two different states. One of the states is denoted as the ferroelectric state *F*, where the dipole is pointing into the direction of the external field and the other state is the antiferroelectric *A* state, where the dipole is pointing in the reverse direction. The magnitude of the tilt is usually temperature independent (Koda and Kimura 1996; Yamashita, 1996a,b, 1997; Nakagawa, 1993)

and the authors include different types of interactions between neighboring layers. As a result, Nakagawa (1993) reproduces ferro-, antiferro- and ferrielectric phases including exponential interlayer interaction. Yamashita (1996a,b,1997) reproduces the complete phase sequence within the axial, next nearest neighbor Ising model (ANNNI) with third-neighbor interactions. Koda and Kimura (1996) consider the head-tail asymmetry of molecules as the microscopic origin of the long range interaction.

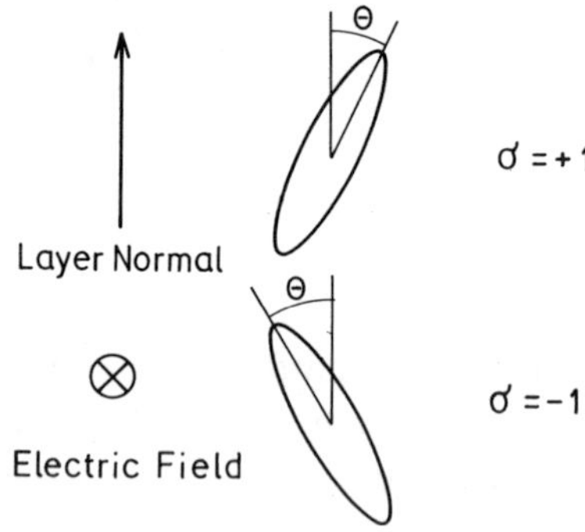

Fig.8.10.2. Within the Ising model of the intermediate phases, the molecular orientation is constrained to a single plane. In this plane, the molecules can take on two different orientations. The scalar variable $\sigma = +1$ denotes molecules which are tilted into the right direction and $\sigma = -1$ denotes those molecules, tilted in the opposite direction (Nakagawa, 1993).

As an example, Eq.8.10.1. shows the Hamiltonian of the ANNNI model as introduced by Yamashita and Miyazima (1993)

$$H = -J_1 \sum_{(i,j)} \sigma_i \sigma_j - J_2 \sum_{(i,j')} \sigma_i \sigma_{j'} - J_3 \sum_{(i,j'')} \sigma_i \sigma_{j''} \tag{8.10.1}$$

Here, the J_1 term measures the interaction between nearest neighboring layers i and j, whereas J_2 and J_3 measure the interaction between the second and third neighboring pairs of smectic layers, respectively. The in-plane interactions are not shown in this expression. As a result of minimization, one obtains a ground state of the model, which is shown in Fig.8.10.3. Here, the regions of stability of different phases are shown as a function of the ratio of the coupling constants J_i.

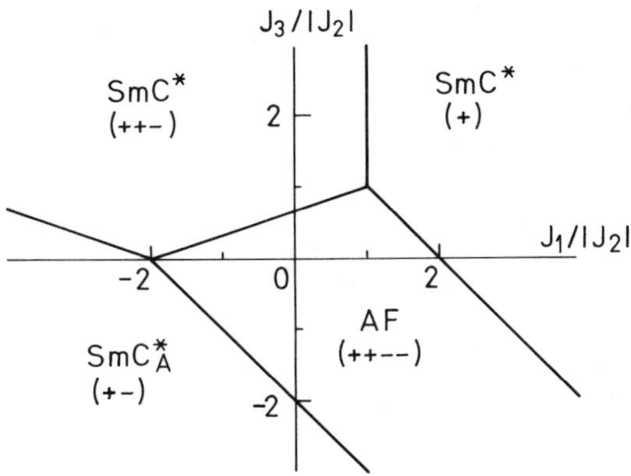

Fig.8.10.3. The phase diagram of the ground state of the ANNNI model (Yamashita and Miyazima, 1993).

The Ising models of the intermediate phase can only qualitatively describe the properties of intermediate phases. For example, Fig.8.10.4. shows the antiferroelectric $smectic-C_A^*$ and the ferrielectric $smectic-C_\gamma^*$ phase. They are denoted with a symbol $q=1/2$ and $q=1/3$, which denotes the fraction of a ferroelectric order in a unit cell, which is in this case composed of 2 and 3 smectic layers, respectively. Further details about the sequence of structures, denoted as the Devil's staircase can be found in the work of Fukuda et al. (1994).

Based on the experimental facts, it soon became clear, that the Ising models can explain the phase transitions between the intermediate phases only qualitatively. This model was in fact ruled out by the resonant X-ray experiments of Mach et al., (1998), which clearly showed that the XY, or "clock" model should be used for intermediate smectic phases.

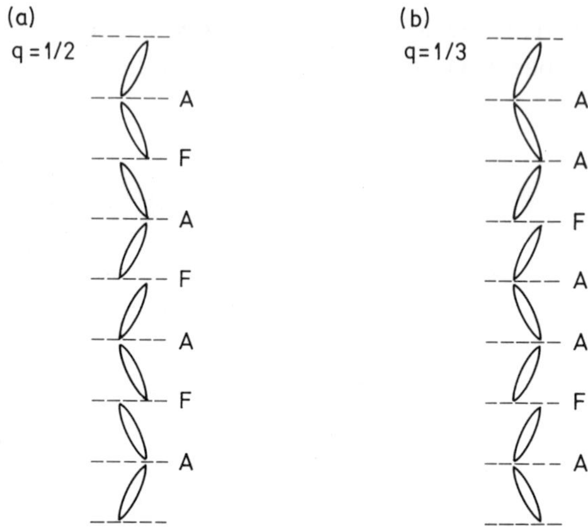

Fig.8.10.4. An example of the ground state of Ising model of intermediate smectic phases representing (a) the antiferroelectric $smectic - C_A^*$ phase and (b) the ferrielectric $smectic - C_\gamma^*$ phase (Takanishi et al., 1991). The unit cell of the antiferroelectric phase is composed of two smectic layers and the unit cell of a ferrielectric phase is composed of three layers. The spontaneous polarization is zero in the antiferroelectric phase and finite (but smaller than the ferroelectric polarization) in the ferrielectric phase. Here the two dipoles cancel and only one out of three layers contributes to the spontaneous polarization.

(ii) The constraint of keeping the directions of the molecular tilt in a single plane has been released in the XY mean-field discrete theories, first introduced by Sun et al., (1993b) and later developed by different authors (Čepič and Žekš, 1995; Pikin, 1995; Roy and Madhusudana, 1996). In these models, the tilt directions of the molecules in the neighboring layers are allowed to make arbitrary angles as one moves from one smectic layer to another. Whereas the original discrete model of Sun et al. (1993a) only reproduces structures that are obtained within the continuous model, the analysis of the influence of competition between the interlayer interactions has for the first time predicted the existence of an

incommensurate *smectic* $-C_{\alpha}^{*}$ phase (the "clock" model of Čepič and Žekš, 1995). We shall briefly describe the properties of their model, which has been quantitatively tested with high resolution tilt angle measurements, performed in the antiferroelectric liquid crystal MHPOBC (Škarabot, 1998a).

The free energy of a tilted smectic liquid crystal is within the discrete phenomenological model (Čepič and Žekš, 1996) given by

$$G=\sum_{j}\tfrac{1}{2}a_{\circ}\xi_{j}^{2}+\tfrac{1}{4}b_{\circ}\xi_{j}^{4}+\tfrac{1}{6}c_{\circ}\xi_{j}^{6}+\tfrac{1}{2}A_{1}\left(\vec{\xi}_{j}\cdot\vec{\xi}_{j+1}\right)+\tfrac{1}{8}A_{2}\left(\vec{\xi}_{j}\cdot\vec{\xi}_{j+2}\right)+$$

$$+\tfrac{1}{4}B_{1}\left(\vec{\xi}_{j}\cdot\vec{\xi}_{j+1}\right)^{2}+\tfrac{1}{2}f\left(\vec{\xi}_{j}\times\vec{\xi}_{j+1}\right)_{z} \quad (8.10.2)$$

Here ξ_j is a two-dimensional tilt order parameter that describes the magnitude and the direction of the molecular tilt of the *j*-th smectic layer. The normal to the layers is along the z-direction. The coefficient of the harmonic term is linearly temperature dependent, $a_{\circ}=a(T-T_{\circ})$ whereas other coefficients are constant. $T_{\circ}$ is here the phase transition temperature into a tilted phase for a system of smectic layers without interlayer interactions. Because the transition to the tilted phase can be either continuous or of the first order, both fourth- and sixth-order terms have to be considered. The coefficient A_1 determines the tilt orientation of the order parameters in neighboring layers and, depending on its sign, favors either ferroelectric or antiferroelectric order. Similarly, the coefficient A_2 determines the tilt orientation in next-nearest-neighboring layers. The coefficient B_1 corresponds to the interactions between quadrupolarly ordered transverse molecular dipoles in two neighboring layers and is always positive. This means that it prefers perpendicular tilt directions in neighboring layers. The coefficient *f* is of chiral origin and is expected to be small with respect to the rest of the coefficients. The summation is taken over N smectic layers in a system. Because the X-ray experiments have shown that the magnitude of the tilt is spatially homogeneous in the antiferroelectric and intermediate phases (Cluzeau et al., 1995), the order parameter can be written as

$$\vec{\xi}_{j}=\theta\left(\cos\varphi_{j},\sin\varphi_{j}\right) \quad (8.10.3)$$

Here, θ is the magnitude of the tilt and φ_j is the corresponding phase angle of the tilt in the j-th layer.

The coefficients A_1, A_2 and B_1 give the magnitude of interlayer interactions, that are composed of steric, van der Waals and electrostatic contributions. These coefficients are therefore complicated functions of θ. For small tilt angles, these functions can always be expanded as

$$A_1 = a_1 + a_1'\theta^2 + a_1''\theta^4 \tag{8.10.4a}$$

$$A_2 = a_2 + a_2'\theta^2 + a_2''\theta^4 \tag{8.10.4b}$$

$$B_1 = b_1 + b_1'\theta^2 \tag{8.10.4c}$$

and the corresponding free energy is

$$G = \sum_j \tfrac{1}{2} a_\circ \theta^2 + \tfrac{1}{4} b_\circ \theta^4 + \tfrac{1}{6} c_\circ \theta^6 + \tfrac{1}{2}\left(a_1 + a_1'\theta^2 + a_1''\theta^4\right)\theta^2 \cos\left(\varphi_{j+1} - \varphi_j\right) +$$
$$+ \tfrac{1}{8}\left(a_2 + a_2'\theta^2 + a_2''\theta^4\right)\theta^2 \cos\left(\varphi_{j+2} - \varphi_j\right) + \tfrac{1}{4}\left(b_1 + b_1'\theta^2\right)\theta^2 \cos^2\left(\varphi_{j+1} - \varphi_j\right) +$$
$$+ \tfrac{1}{2} f\theta^2 \sin\left(\varphi_{j+1} - \varphi_j\right) \tag{8.10.5}$$

One should note that this expansion introduces a temperature dependence of the interlayer interactions, because θ is temperature dependent. Stable solutions for θ and φ_j are obtained by minimizing the free energy with respect to θ and all phase angles φ_j, simultaneously. It turns out that only two classes of solutions are stable, depending on the sign of the coefficient A_2.

For a positive A_2, $A_2 > 0$, there are three characteristic stable solutions, that correspond to three different structures, as shown in Figs.8.10.5a-c. In these three solutions, the difference between the phase angles α in neighboring layers is constant. The phase angle therefore increases monotonically, as we move along the smectic layer normal and the solutions differ only in the magnitude of this phase difference. The first solution, shown in Fig.8.10.5(a), obviously corresponds to the ferroelectric $smectic-C^*$ phase. Here, the phase angle between the molecules in neighboring layers is very small and originates from the chirality of the molecules. The second solution is shown in Fig.8.10.5(b) and corresponds to the antiferroelectric $smectic-C_A^*$ phase.

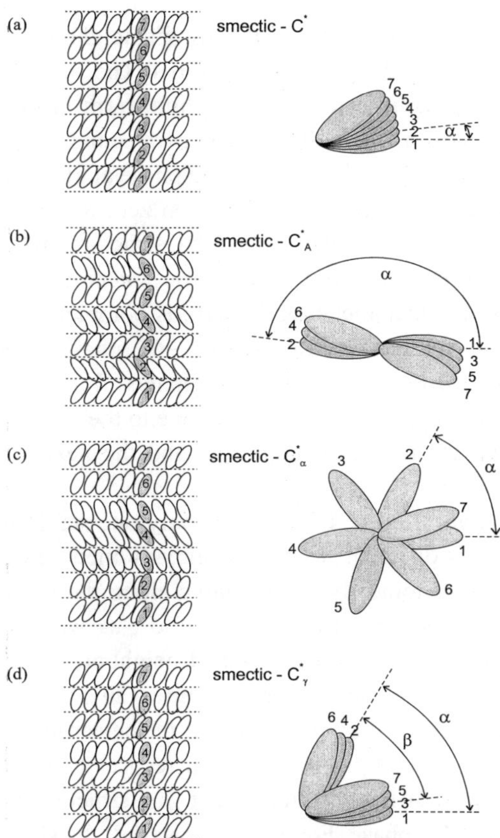

Fig.8.10.5. The equilibrium structures, as obtained within the discrete "clock" model with interlayer interactions up to the next nearest neighbor: (a) The ferroelectric *smectic* $- C^*$ phase is characterized by a small value of α, giving rise to a monotonously growing phase angle, as we move along the layer normal. (b) The antiferroelectric *smectic* $- C_A^*$ phase is characterized by a doubled unit cell, composed, say of molecule 1 and 2, as indicated. In this phase α is close to π. (c) The *smectic* $- C_\alpha^*$ phase is characterized by the "rapid winding" of the phase angle, as we move along the layer normal of these materials. This phase is in fact a short-pitch ferroelectric phase with $0 < \alpha < \pi$. (d) The ferrielectric *smectic* $- C_\gamma^*$ phase is characterized by an alternate behavior of the phase angle. The phase first increases for α, as we move to the next layer and then "flips back" for an angle β as we move to the third layer.

The phase between the directions of the tilt in neighboring layers increases nearly by π , as we move along the layer normal. The small deviation from the antiparallel ordering is again caused by the chiral term in the free energy. The third solution is shown in Fig.8.10.5(c) and corresponds to a structure that is identified as the $smectic-C_\alpha$ phase. Here, the phase angle between the directors in neighboring layers has a finite value between zero and π. The $smectic-C_\alpha$ phase is here in fact a short pitch helicoidal structure, which is structurally equivalent to the well-known chiral ferroelectric $smectic-C^*$ phase. However, the origin of this short, nanometer-sized helix is completely different from the origin of the helix in the ferroelectric $smectic-C^*$ phase. Whereas the helix in chiral ferroelectric $smectic-C^*$ phase originates from the chirality of the molecules, it arises in the $smectic-C_\alpha$ phase due to the competition between the interactions between the nearest neighboring and the next nearest neighboring layers.

For a negative A_2 , $A_2 < 0$, there is only one stable solution, shown in Fig.8.10.5(d), which is different from those already mentioned. Here, the phase angle behaves non-monotonically, as we move along the normal to the smectic layers. First, it increases for a finite value of α , as we move to the nearest neighbor, but when we reach the next-nearest neighbor, the phase angle nearly "flips-back" to the original value. The structure can also be considered as a double-twist structure, formed by two identical ferroelectric helices, gearing into each other. Here, the helices are rotated with respect to each other at a finite phase angle $0 < \alpha < \pi$. It is conjectured that this structure corresponds to the ferrielectric $smectic-C_\gamma^*$ phase. For a given material, the signs and the ratios of the coefficients in the free-energy expansion Eq.8.10.5. are determined by the sequence of phases, the phase transition temperatures and the values of the tilt angle at the phase transition temperatures between different intermediate phases. This allows for a qualitative test of the theory with an experiment that is able to measure the temperature dependence of the tilt angle with a great accuracy.

In an optical experiment, Škarabot et al. (1998b) have measured the temperature dependence of the birefringence in the various phases of MHPOBC, as shown in Fig.8.10.6.. From these data, they calculated the corresponding root-mean square tilt angle, shown for a narrow temperature interval in Fig.8.10.7. The following important features can be observed in their results, which are also characteristic for other materials, showing antiferroelectric and intermediate phases: *(i)* In the $smectic-A$ phase, the birefringence increases monotonically with decreasing temperature, which is due to a gradual increase of the nematic

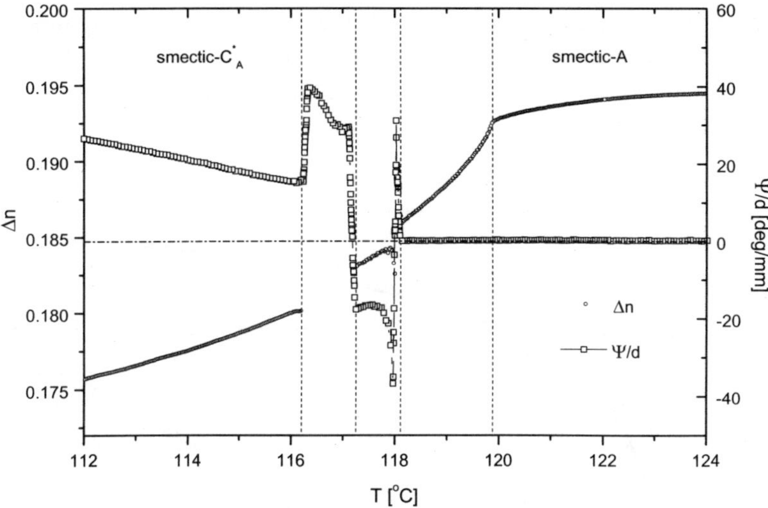

Fig.8.10.6. The temperature dependence of the birefringence and optical rotation per unit length in (R)-MHPOBC.

order parameter with decreasing temperature. *(ii)* Several degrees above the phase transition into the tilted phase, a significant deviation from this monotonous behavior is observed. The birefringence even starts to decrease gradually, as we approach the phase transition into the tilted phase. This is due to large fluctuations of the tilt angle in the pretransitional regime, which tend to decrease the birefringence.

Similar fluctuation effects were observed in heat capacity studies of these materials (Ema et al., 1995, 1996a-c, 1997, 1998), and in the first studies of the birefringence in the vicinity of the $smectic-A$ - $smectic-C$ transition (Lim and Ho, 1978). *(iii)* A strong and monotonical decrease of birefringence is observed in the tilted phase, as expected from the theory. *(iv)* Discontinuous jumps of the birefringence are clearly observed at the $smectic-C_\alpha^*$ to $smectic-C^*$ phase and between the $smectic-C^*$ and $smectic-C_A^*$ phase.

The temperature dependence of the tilt angle in different phases of (R)-MHPOBC, as determined from the birefringence data, is shown in Fig.8.10.7, together with the theoretical fit. The inset to this figure shows a log-log plot of the data and reveals an important fact. There are obviously two regimes in the temperature dependence of the tilt: Close to the *smectic – A* phase, the exponent for the tilt angle is $\beta \approx 0.5$, as expected from the Landau free-energy expansion up to the fourth order term.

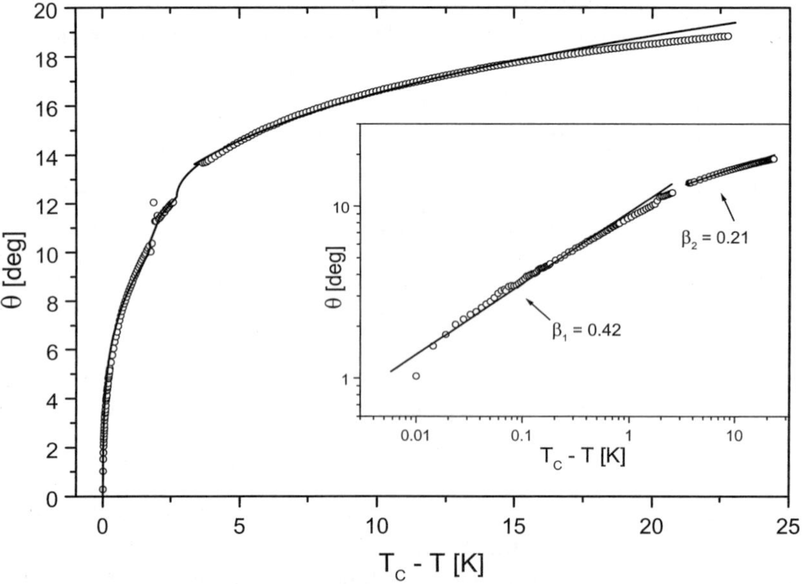

Fig.8.10.7. The temperature dependence of the tilt angle in (R)-MHPOBC. The solid line is the best fit to the discrete model (Škarabot et al., 1998). The inset shows the same data in a log-log scale to present the crossover of the power-law behavior of the tilt angle. Solid lines in the inset are the best power-law fits for the temperature interval $0.5K$ below T_c and for the temperature interval $4K < T_c - T < 15K$, respectively.

However, several degrees below the *smectic – A* phase, the tilt exponent changes to $\beta \approx 0.25$, which is characteristic of a Landau tricritical behavior and shows the importance of the sixth-order terms (Huang and Viner,1982). The crossover temperature is several degrees below the $smectic-A$ phase, i.e. in the region where the transitions to the $smectic-C^*$, $smectic-C^*_\gamma$ and $smectic-C^*_A$ phases are observed. The experimental data were fitted to the theory and the authors find a good quantitative agreement. The parameters of the fit for MHPOBC together with their estimated accuracy are: $a_1/a_2=-1.1$ $a_1'/a=30.6$, $a_1''/a=-192$, $a_2/a=2.5$ $a_2'/a=-77.8$, $a_2''/a=400$, $b_1/a=2.4$, $b_1'/a=-20.5$ and $c_\circ/a=1400$. It is important to note that the experimentally observed signs of the coefficients are in agreement with those required to obtain the experimentally observed phase sequence. This can be considered as a clear indication of the internal consistency of the Čepič and Žekš "clock" model. It should be stressed that the resonant X-ray scattering experiments of Mach et al.(1998) have clearly shown that the discrete XY model is realistic and describes the structures of the intermediate phases correctly. It seems, however, that the discrete model of Čepič and Žekš needs further refinements to describe the experiments of Mach et al. quantitatively, as it is difficult to reproduce discrete transitions between three and four layer structures.

Chapter 9

STM and AFM in Complex Liquids

9.1. Introduction

The operation of the ***Scanning Tunneling Microscope (STM)*** was first demonstrated by G.Binnig et al. in 1981. The principle of operation of an STM is based on a very sharp conductive tip, which is positioned close to a flat conducting crystal surface and measures a current as a function of lateral position of the tip. Due to the quantum nature of the motion of electrons in solids, there would be a ***tunneling current between the tip and the surface*** even at a finite separation between the tip and the surface. Moreover, in the close proximity, the tunneling current grows exponentially when approaching the surface. This is the physical reason for extremely high sensitivity of the STM: It can detect small changes of distances of the order of 0.001nm! This extreme sensitivity of the STM has opened new ways to the study of surfaces of conductive materials and has triggered a development of a series of a new experimental surface methods such as the Atomic Force Microscope etc.

One of the drawbacks of ***STM*** is the fact that it can be used only for ***conductive surfaces***. However, in 1986, G.Binnig et al. have demonstrated a new microscope that could observe fine details of surfaces of nonconductive materials with a nanometer resolution. The microscope was named ***Atomic Force Microscope (AFM)*** due to the fact that it uses ***interatomic forces*** to observe the surface morphology of nonconductive surfaces. In an AFM, the forces between the tip and the substrate are measured with a precision better than $10^{-11} N$. By keeping the force constant, the AFM tip is scanned in the close proximity of the substrate. This allows one to determine the surface topography of the sample via the forces between the tip and the substrate. These forces depend on the nature of both materials and are usually van der Waals attractive forces. However, if both materials are charged or magnetic, the forces are predominately ***long range electrostatic or magnetic forces***. This opens new possibilities of studying the electrostatic or magnetic properties of materials on a submicrometer scale.

Both the STM and AFM, as well as all other ***Surface Probe Microscopes (SPM)*** have basically the same structure, which is shown in Fig.9.1.1. The sample is mounted on a piezoelectric translator, which generates the movement of the sample along the three spatial coordinates with sub-nanometer precision. A sharp tip is positioned above the surface of the sample, so that it senses local properties of the surface. These local properties of the sample, as for example the tunneling current or the force, exerted on a tip, are sensed by a very sensitive detector. The output of this detector is led into a feed-back circuitry of an SPM, which assures that the measuring quantity remains stable in time. The surface probe is then scanned across the sample surface in a regular manner, shown in Fig.9.1.2. and the feedback electronics adjusts the height of the scanner so that the measuring quantity remains constant during the scanning.

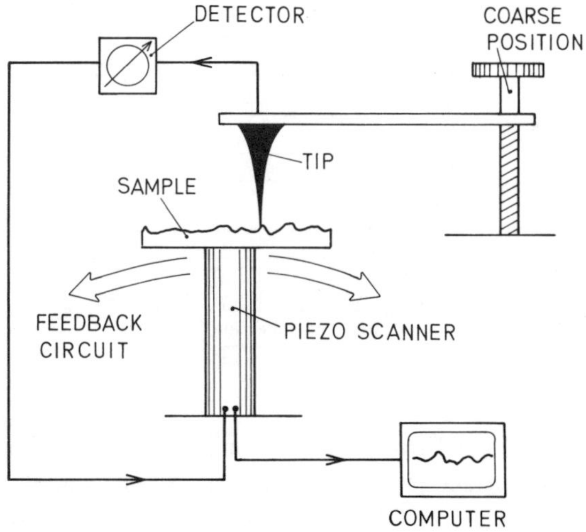

Fig.9.1.1. Schematic presentation of an Surface Probe Microscope. In STM, the surface is probed by the current between the sample and the tip whereas in AFM, the force between the tip and the surface is detected (courtesy Park Scientific).

The voltage, applied to the z-section of the scanner therefore reflects the topography of the sample surface, as seen by the detector. This voltage, together with other signals, is then sampled at pre-defined positions and are stored by a computer. This digitally-stored array or matrix therefore represents the image of the surface of the sample.

There are two crucial parts in every SPM, that determine the quality of the instrument: The piezoelectric scanner and the surface-sensing probe. The piezoelectric scanners in most modern SPM-s are made of a piezoelectric tube with electrodes on the inner and outer surfaces. For example, with a single inner electrode and four, equally spaced outer electrodes, one can generate motion of the top end of the piezoelectric tube in all three directions, as shown in Fig.9.1.3. The motion of the sample is therefore quasi-linear for small bending of the piezoelectric tube.

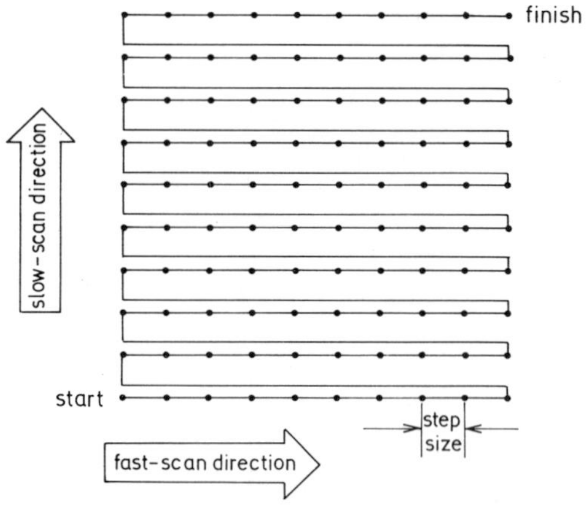

Fig.9.1.2. The signal from the detector of an SPM is sampled in regular time intervals. As the surface probe is scanned across the sample surface in a zigzag fashion, these regular time intervals correspond to well-defined points on the sample surface and represent a matrix or an array of pixels, and form a digital image of the surface.

The motion of the sample can be generated with an extreme precision of less than 0.1nm. For this reason the scanners are extremely sensitive to thermal expansion drifts as well as aging. Typically, one has to wait several hours before the scanner reaches an equilibrium to observe an image at an atomic scale. The second most important part of an SPM is the sensing probe with the detector electronics. Whereas the STM tips can be relatively easily produced by etching a thin wire, the production of AFM tips requires highly dedicated micromachining facilities.

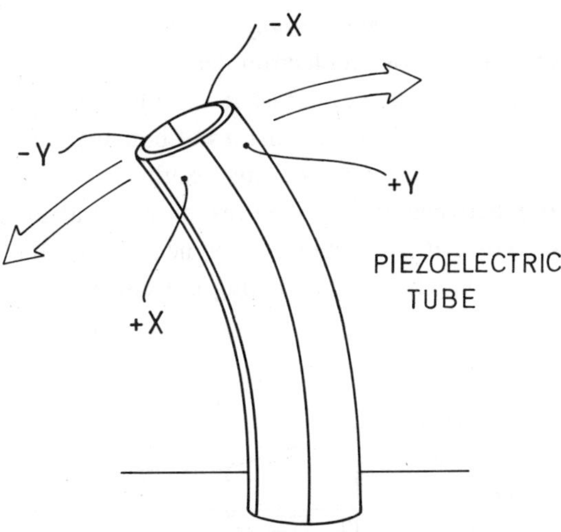

Fig.9.1.3. Motion of a sample, mounted on a piezoelectric tube. As the voltage is applied between the inner electrode and the two, oppositely faced outer electrodes, the tube bends. Because the bottom part of the tube is fixed to the base of the instrument, the upper part of the tube moves in a quasi-linear manner due to the bending of the tube.

9.2. STM of Adsorbed Liquid Crystalline Molecules

The principle of operation of the ***Scanning Tunneling Microscope*** is based on the quantum nature of the electron wave function. At a free surface of a crystal, the electron density ***decays exponentially***, as we move away from the crystal surface. This results in significant electron densities at a distance of several tens of a nanometer away from the crystal surface. When two pieces of conductive crystals are approached, the electron densities of both crystals begin to overlap at a finite separation of both surfaces. This results in a ***finite probability for electron tunneling*** between the two pieces of material and gives rise to an ***electric tunneling current***. A voltage difference of $\approx 0.1V$ between the two pieces of metal can give rise to a tunneling current of $\approx 1nA$, which is easily detected.

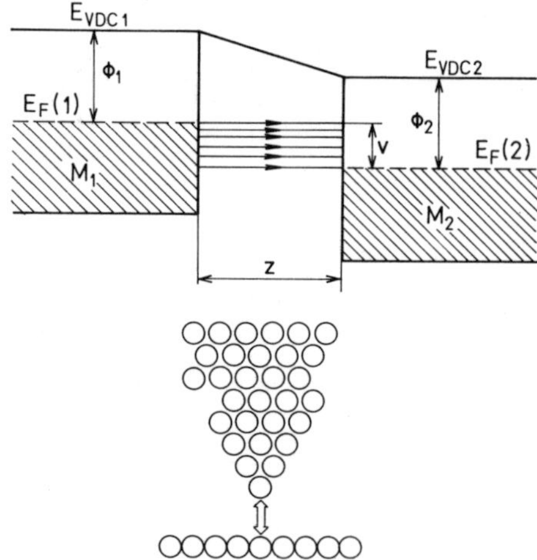

Fig.9.2.1. Electron tunneling between the two pieces of metal, separated by a gap. ε_F is the Fermi level, Φ_1 and Φ_2 are the work functions of both metals and V is the voltage difference between the two metals. The figure on the bottom shows that the electrons flow mainly via the "last" atom of the tip, which is closest to the surface.

The magnitude of the tunneling current I between the two flat pieces of metal is (J.Simmons, 1963)

$$I \approx \frac{\sqrt{\Phi}}{d} \cdot V \cdot e^{-\frac{8m^2}{h} \cdot d \cdot \sqrt{\Phi}} \qquad 9.2.1$$

Here, d is the distance between two metal surfaces, whereas $\Phi = \frac{1}{2}(\Phi_1 + \Phi_2)$ is an average height of the potential barrier between the two metals. This average potential barrier is assumed to be much larger than the applied voltage difference, $V << \Phi$. One can see from this equation, that for a change of 0.1nm of the distance d, the electric current increases nearly for an order of magnitude! This is the reason for the very high spatial resolution of the STM. Most of the tunneling current flows via that atom on the tip, which is closest to the surface, as shown in Fig.9.2.1. Theoretical considerations show (Tersoff and Hamann, 1985) that the tunneling current is also proportional to the ***local surface density of electron states*** close to the Fermi level at the place, where the tip is positioned.

There are several modes of operation of the STM and we shall consider here only the two topographic modes. In the first mode of operation, the tunneling current is set to a constant value and the tip is scanned across the surface, as shown in Fig.9.2.2a. As the detector senses variations of the tunneling current, the feedback electronics keeps it constant by changing the height of the sample. This is done by applying a voltage to the piezoscanner. A record of this voltage, as the tip is scanned across the sample surface, therefore represents a "topographic image", where the heights correspond to larger tunneling currents and the low parts of the image correspond to low tunneling currents. This image is a combination of both the "true" topographic image and the image of the density of states, which an STM cannot discriminate. In the second mode of operation, shown in Fig.9.2.2b, the sample surface is kept at a constant height and the tunneling current is measured, as the tip is scanning across the sample. This mode of operation can be used for very flat surfaces, where there is no possibility of crashing the tip into elevated parts of the sample. In both modes of operation, an STM can reach a truly atomic resolution on metallic surfaces. Since the invention of the first STM, commercial instruments have reached a very sophisticated level, where the atomically resolved image on graphite can be reached routinely, as shown in Fig.9.2.3.

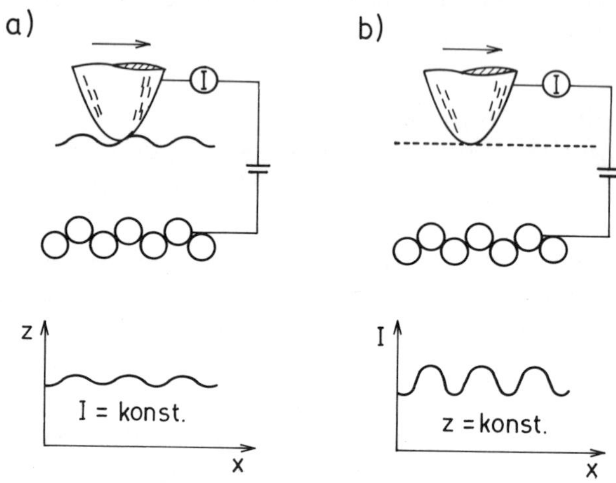

Fig.9.2.2. (a) The constant tunneling current mode of operation of a STM. The current is held at a fixed value, as the STM tip is scanning across the sample. The feedback signal that regulates the height of the position of the surface is stored as the topographic image. (b) The constant height mode of operation. Here the height of the sample is held fixed and the tunneling current is measured. This mode of operation is much faster than the constant current mode.

From the very beginning of the invention of STM, there has been a lot of speculations, whether it is possible to image individual molecules on the surface of solids with an STM. Early results on imaging CO molecules on atomically flat surfaces were rather discouraging (see S.Chiang 1992), because no truly atomic resolution could be achieved. The reason became clear after the first observation of rapid molecular movement on a crystal surface (Smith, 1987). The thermal kinetic energy of the molecules at room temperature is responsible for fast surface diffusion of the molecules. One has either to decrease this thermal energy by cooling the sample to liquid helium temperatures or immobilize the molecules to the surface by chemisorption. It has been observed that in particular the aromatic rings of the molecules strongly chemisorb on metal surfaces. This makes organic molecules with aromatic rings particularly good candidates for atomically resolved STM molecular imaging.

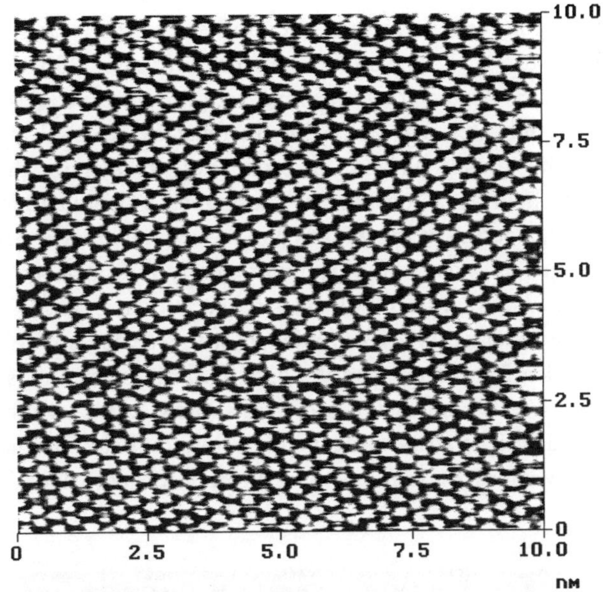

Fig.9.2.3. Atomically resolved STM image of HOPG (highly oriented pyrolithic graphite).

Imaging experiments on liquid crystalline molecules adsorbed to a crystal surface can be done in three different ways: *(i)* by placing a small drop of liquid crystal material on a crystal surface, heating it into the isotropic phase and cooling down, *(ii)* by heating a small amount of liquid crystal to $\approx 100^{\circ} C$ and evaporating it onto a crystal surface, *(iii)* by dissolving liquid crystal molecules in polar solvent such as alcohol and allowing the dilute solution to dry-out on the crystal surface. In most cases, highly oriented pyrolitic graphite (HOPG) or MoS_2 are used as substrates. These materials can be easily cleaved, so that large, atomically flat regions are routinely obtained.

The first atomically resolved image of a liquid crystal molecule on graphite has been reported by Foster and Frommer in 1988. They have placed a drop of 4-n-octyl-4'-cyanobiphenyl (8CB) onto a freshly cleaved surface of HOPG, immersed the STM tip into the drop and imaged the liquid crystal interface. As they report, they had to wait for several hours, before they could resolve individual molecules adsorbed to the HOPG surface. This indicates a slow, self-assembling process at the interface, where large regions of spontaneously ordered liquid crystal molecules are observed, as shown in

Fig.9.2.4. The imaging was performed with a direct current bias from the tip to the sample of -0.5 to $-0.7V$ and a low tunneling current of 200 to $500 pA$. This low imaging current is characteristic for imaging of large molecules, adsorbed to a solid crystal. It indicates that the STM tip must be scanned at rather large distances from the metal surface in order not to disturb the molecular arrangement at the interface. Consequently, large bias voltage and low tunneling are needed and the experiment is performed in the regime of large gap resistance.

The images of Foster and Frommer clearly indicate long range, solid-like frozen order of liquid crystal molecules at the interface. The 8CB molecules seem to be arranged in parallel rows of bilayers with a periodicity of $3.4nm$. The bilayer rows are clearly brighter than the dark space in between, as can be seen in Fig.9.2.4. The $2nm$ width of the bright rows and $1.4nm$ width of the dark rows indicated that bright rows correspond to interdigitated aromatic cores of 8CB molecules, whereas the dark regions in between correspond to aliphatic chains.

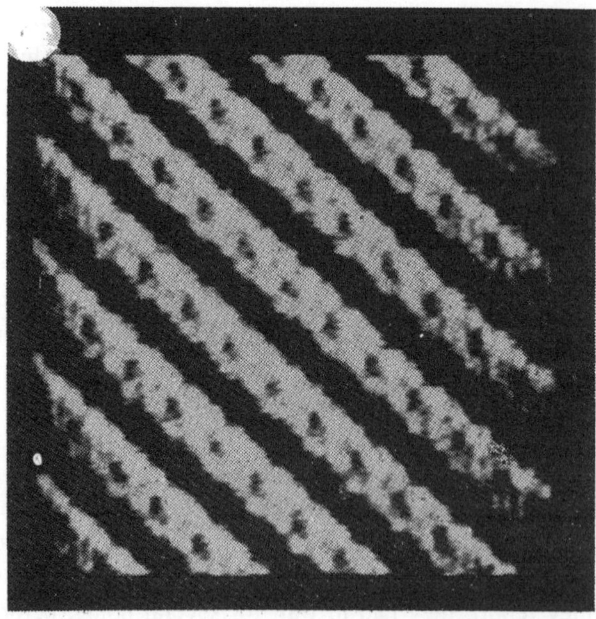

Fig.9.2.4. STM image of 8CB on a graphite substrate (Foster and Frommer, 1988). The scale is $21.5 \times 21.5nm$.

This conjecture of molecular order on HOPG was confirmed in a STM study by Smith et al. in 1989. They have studied the HOPG-liquid crystal interface for a series of n-alkylcyanobiphenyls (8CB to 12CB) which were deposited by evaporating liquid crystal material on a freshly cleaved graphite surface. These studies have confirmed the observations of Foster and Frommer and clearly showed bilayer-organized liquid crystal molecules. The long range order of liquid crystal molecules at the interface is clearly solid-like, because of translational periodicity of the molecular arrangement. This indicates strong interaction of liquid crystalline molecules with the underlying graphite lattice. However, the fact that the molecules are still organized in smectic-like bilayers with a positional order indicates the importance of interactions between the liquid crystal molecules themselves.

The molecular packing of 10CB molecules on HOPG, as observed by Smith et al., 1989, is shown in Fig.9.2.5. Again, one can clearly observe regular stacks of bright rows, separated by dark rows. Bright rows correspond to bilayers of aromatic heads of 10CB molecules. These are facing each-other with a polar cyano group and are slightly interdigitated, because cyano groups try to avoid directly facing each other because of the dipolar electrostatic energy. The authors also noted that phenyl (aromatic) rings are for 10CB oriented with their plane perpendicular to the crystal surface, which can be seen from the $0.63nm$ distance between the two neighboring 10CB molecules. Unfortunately, the STM resolution was not high enough to resolve the atomic arrangement within the phenyl rings.

The STM images of Smith et al. revealed a very important detail, which finally identified bright regions on the image as phenyl groups and dark regions as alkyl tails of the molecules. Fig.9.2.5.(b) shows a height contour, taken across one molecular pair, as indicated by a white, zigzag line in Fig.9.2.5.(a). The bottom Figure 9.2.5.(c) shows a model of two 10CB molecules, facing each other with their polar heads. The contour above this image represents a height variation, which is in striking agreement with the measured height variation, shown in Fig.9.2.5.(b). This was a clear confirmation, that bright rows on the STM images correspond to benzene rings, whereas the dark rows correspond to alkyl tails. The alkyl tails fit very well to the underlying graphite lattice, because the mismatch in the distance between two carbon atoms in the alkyl chain and HOPG differ by less than 2%.

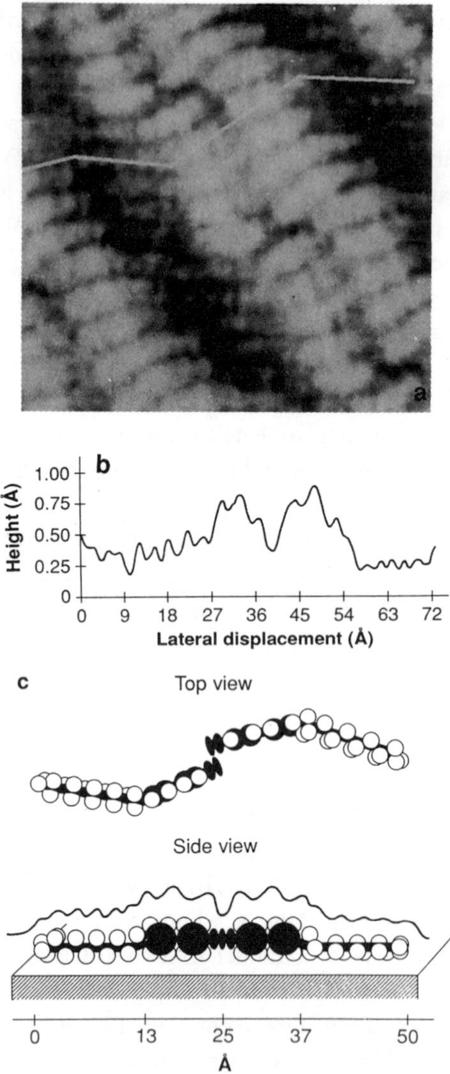

Fig.9.2.5. (a) STM image of 10CB on graphite surface. The field of view is $7.2 \times 7.2 nm^2$ (Smith et al., 1989). The white line indicates the path, where the height has been taken. (b) Contour line taken along a pair of 10CB molecules, indicated by a white line in (a). (c) Model showing the orientation of two molecules of 10CB. A trace is shown, representing a possible STM contour.

There are other STM studies of nonferroelectric materials on graphite (Shigeno et al., 1990) and MoS_2 (Hara et al., 1990; Iwakabe et al., 1990) that reveal basically the same crystalline-like order of liquid crystal molecules at a solid interface. This indicates the existence of a qualitatively different crystalline order at a solid crystal-liquid crystal interface and is very important for aligning mechanisms in liquid crystals.

The first STM images of the molecular arrangement at a solid crystal-ferroelectric liquid crystal were reported by Brandow et al. (1993) and Parks et al. (1993). These authors have observed crystalline order of the first molecular layer on graphite. The highest resolution STM images of ferroelectric liquid crystals were reported by Walba et al. (1995) and are shown in Fig.9.2.6.

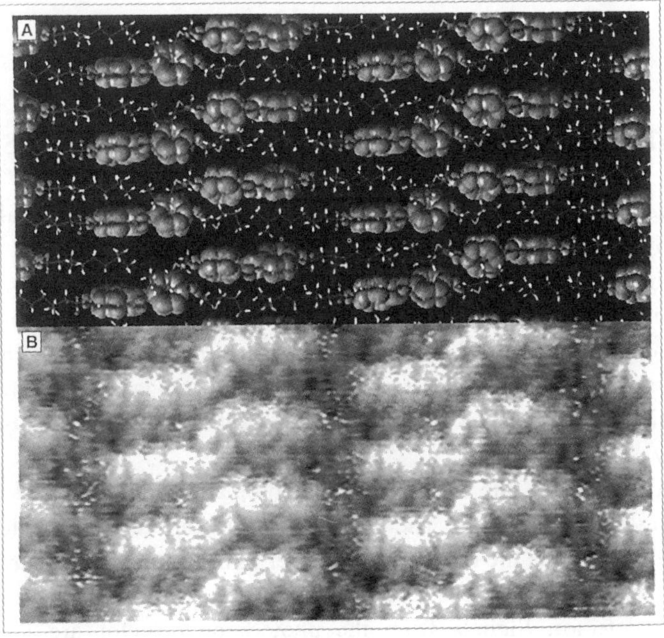

Fig.9.2.6. STM images of ferroelectric liquid crystalline molecules of MDW 74 on HOPG (Walba et al., 1995). The images are raw data, taken at constant current 1nA and positive voltage at the tip (+0.6V). (a) The molecular structure of MDW 74 on graphite. (b) STM image of MDW 74 on graphite, raw data.

The authors were actually first to report that from about 100 experimental sessions, two third of the sessions did not give any, or only poor resolution images, which is a clear indication of the difficulties of STM experiments on adsorbed molecular layers. The very high, near atomic resolution of these STM experiments showed that the enhanced tunneling efficiency of benzene rings (which therefore appear bright on a STM image) cannot be explained solely by the electron tunneling via the highest occupied molecular orbital (HOMO) or the lowest unoccupied molecular orbital (LUMO). These orbitals are each localized on one of the phenylbenzoate systems of MDW 74, whereas in the experiment, both rings appear bright, which means that both contribute to the tunneling. The authors conclude that combinations of many molecular orbitals are involved in the enhanced tunneling efficiency in an STM experiment on adsorbed molecular layers.

9.3. AFM in Complex Liquids: Forces on the Molecular Scale

The ***Atomic Force Microscope*** (AFM) was invented by G.Binnig et al. in 1986. The principle of operation of an AFM is substantially different from the operation of an STM. Unlike the STM, the AFM uses a very sharp tip that is mounted on a very thin cantilever for the detection of forces that act on a tip and causes a slight bending of the cantilever. This mechanical bending of the cantilever is detected by a photo-detector. If the tip is scanning across the sample, this would cause a bending of the cantilever, reflecting the surface topography.

In real AFM's, the sample is mounted on a piezoelectric tube, similar to the scanning tube of an STM. The deflection of a cantilever is detected in most commercial AFM's via the reflection of a laser light from the back surface of the cantilever, as shown in Fig.9.3.1. The laser light is detected by a four-quadrant photodiode detector. Signals from the photodiode detector are led into the feedback circuitry of the AFM that maintains a constant deflection of the cantilever. When the sample is scanning in the x-y direction across the surface of the sample, the forces between the surface of the sample and the atoms of the tip induce deflection of the cantilever. This small deflection is detected by the photodiode and the feedback signal is applied to the scanning tube to adjust the height of the sample to a level that causes the predetermined deflection of the

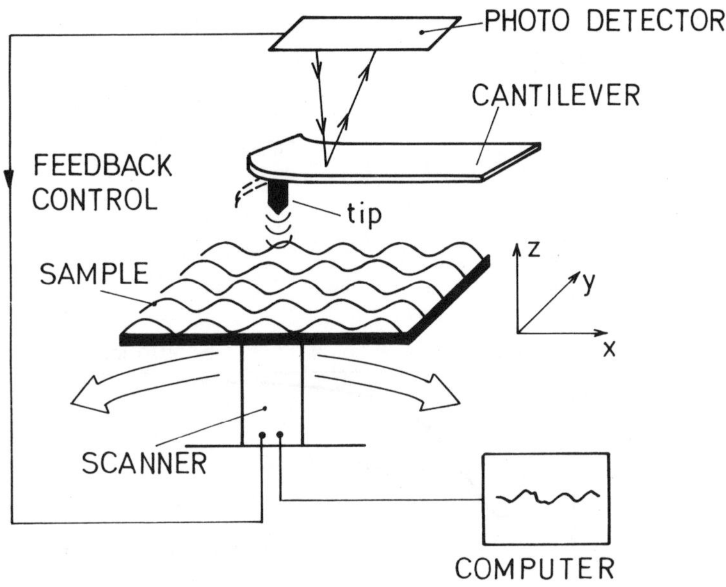

Fig.9.3.1. Principle of operation of the AFM.

cantilever. This small deflection is detected by the photodiode and the feedback signal is applied to the scanning tube to adjust the height of the sample to a level that causes the predetermined deflection of the cantilever. This feedback signal is then stored by a computer and represents a digitized topography of the sample.

There are two basic modes of operation of an AFM, depending on the force regime, as shown in Fig.9.3.2. These two modes are called the ***contact mode*** and the ***non-contact mode*** of operation, respectively. Fig.9.3.2. shows a typical dependence of the force between a tip and the surface of a non-magnetic and electrically neutral sample, as a function of the separation between them. When the tip is far away, there is practically no force acting an it. When the tip is approaching the surface, an attractive van der Waals force is observed, which tends to bend the cantilever towards the surface of the sample. Finally, when the tip touches the sample, there is a strong repulsion between the atoms in the tip and on the surface, which originates from the Pauli principle. The force therefore here increases tremendously with decreasing distance.

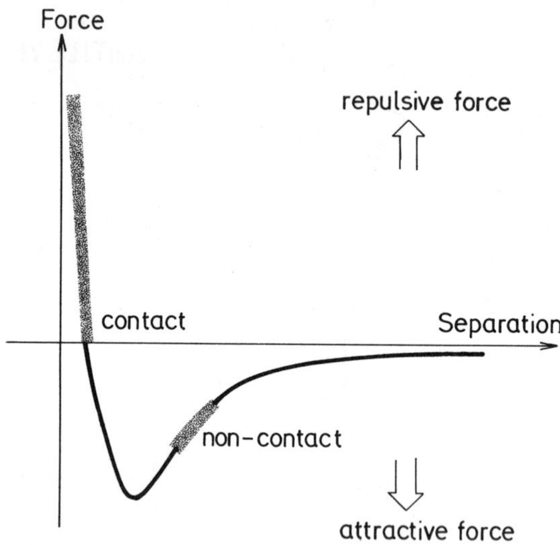

Fig.9.3.2. Schematic presentation of the separation-dependence of the force between a non-magnetic and electrically neutral AFM tip and the surface of a sample(courtesy Park Scientific).

The contact mode of operation of the AFM has the highest resolution, which can reach true atomic resolution. In this regime, the tip is in the close contact with the surface atoms, as shown in Fig.9.3.3. As a result, the forces between the tip and surface atoms are considerably larger, and are typically of the order of $10^{-9} N$. This large force can be a source of irreproducibility when imaging soft surfaces, because the tip damages the surface. To avoid surface damaging by the scanning tip, one has to move the tip away from the surfaces and use the so-called non-contact mode of operation at larger distances, typically $1nm$. This results in lower forces between the tip and the surface, which can be reduced down to 10^{-11} to $10^{-12} N$. As a consequence of larger tip-surface separation, the image resolution is worse and usually cannot reach atomic resolution.

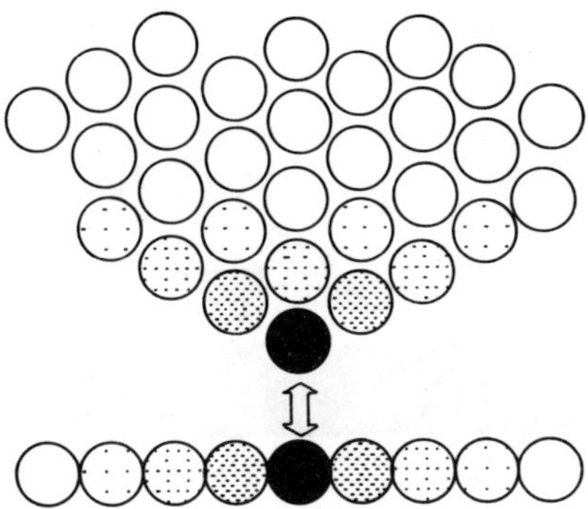

Fig.9.3.3. The contact mode of operation of an AFM. The tip is in close contact with the surface atoms. The force on the tip is due to strong interaction of the atoms at the very end of the tip and the surface atoms.

There are several methods of imaging in the non-contact regime and one of the first reported is the so-called tapping-mode operation. Here, the tip is mounted to a small piezoelectric crystal, which is vibrated close to the resonance frequency of the cantilever, as shown in Fig.9.3.4. When the tip is far away from the sample, the cantilever oscillates at its own resonance frequency. When the tip approaches the surface, there are for example gradients of Van der Waals or some other forces. This ***force gradient changes the natural resonance frequency*** of the cantilever and changes the amplitude and the phase of the cantilever vibrations, as shown in Fig.9.3.5.

In the tapping mode of operation, the AFM therefore images constant force gradient contours across the surface of the sample. There are other possibilities to induce oscillations of an AFM tip, as for example magnetic, thermal or electric forces (Florin et al., 1993; O'Shea et al., 1994; Han et al., 1996; Hiller and Bard, 1997). Usually these methods result in much better force resolution than the tapping mode itself.

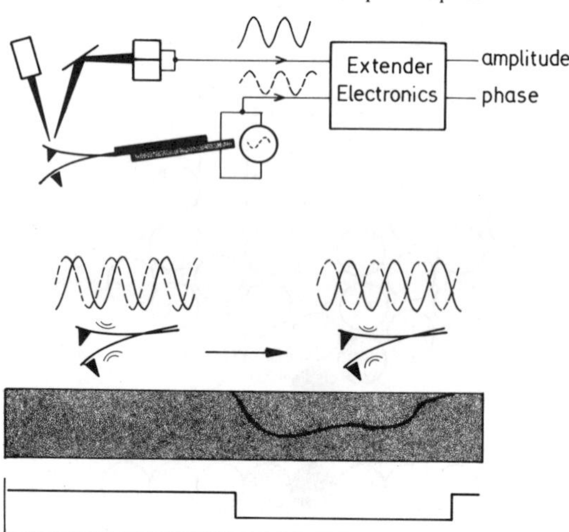

Fig.9.3.4. The tapping mode of operation of an AFM. The tip is mechanically vibrated close to its resonance frequency, which is in the range of $10kHz - 200kHz$ when operating in air. This mode of operation also works in liquids, with lower resonance frequencies and considerable damping.

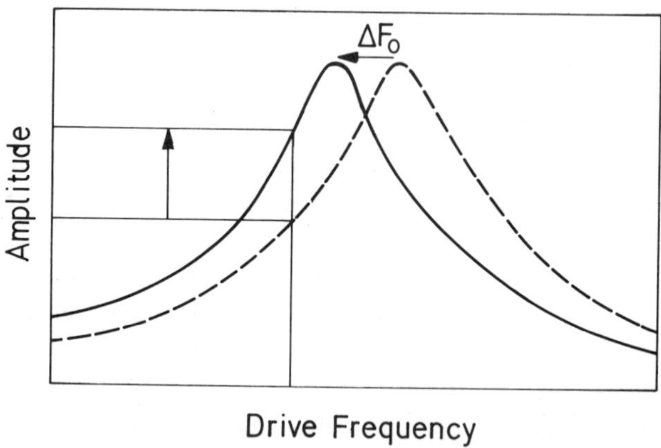

Fig.9.3.5. The principle of operation of the tapping mode of an AFM. Left figure shows the position of the cantilever natural resonance frequency with respect to the driving frequency. Force gradients change this natural resonance frequency of the cantilever and induce a change of the vibration amplitude and the phase of the cantilever.

There are only few reports of AFM studies of liquid crystalline systems so far. These have been performed either on transferred freely-suspended or Langmuir-Blodgett LC films (Overney et al., 1993) or frozen, glassy state cholesteric oligomers (Meister et al., 1996). The main problem is of course the liquid nature of the liquid crystal surface, which has to be overcome by freezing the structure into the solid phase. Nevertheless, these AFM studies can give valuable indirect indication on the structure of free surfaces of mesophases.Fig.9.3.6. shows an example of the AFM image of a frozen cholesteric structure. The image was taken across a microtome cut perpendicular to the free surface and clearly shows the interface of the cholesteric sample and the epoxy resin, which embedded the liquid crystal.

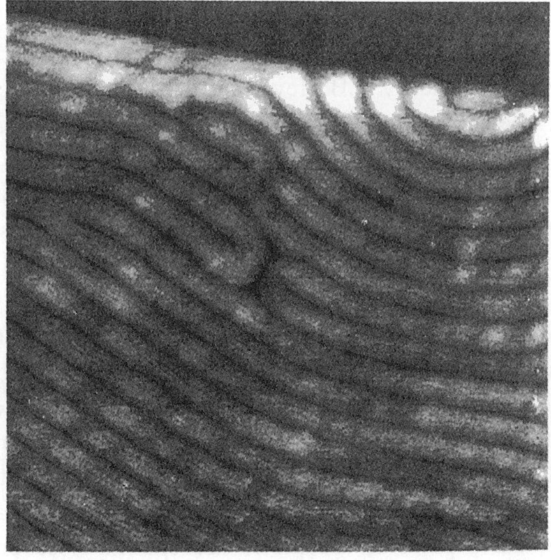

Fig.9.3.6. An AFM image of a microtome cut, perpendicular to the interface of a cholesteric and an epoxy resin. The dimensions of the image are $2407nm$ square (after R.Meister et al., 1996b).

Chapter 10

Electrooptic Effects in Ferro- and Antiferroelectric Phases and their Applications

10.1. Introduction

The first observation of a linear electrooptic effect in the bulk chiral $smectic-C^*$ was reported by R.B.Meyer and co-workers in their article on the discovery of ferroelectricity in liquid crystals in 1975 (see R.B.Meyer et al., 1975). Although the reported effect seemed not to be of practical interest, it was N.A.Clark and S.T.Lagerwall who noticed the technological importance of a large electrooptic effect, when a ferroelectric $smectic-C^*$ phase is constrained in a thin layer between two glass plates (see N.A.Clark and S.T.Lagerwall, 1980).

The principle of the operation of the so-called "Surface Stabilized Ferroelectric Liquid Crystal Display" (SSFLCD) is shown in Fig.10.1.1. When a bulk helical ferroelectric $smectic-C^*$ is constrained between the closely spaced glass plates in the so-called bookshelf geometry, the interaction forces between the liquid crystal molecules and the surface plates are transmitted through the structure by elastic stress and result in an unwinding of the $smectic-C^*$ helix. This unwinding takes place if the plates are spaced close enough, i.e. at a separation $d \le p$, where p is the helical period in the bulk. In sufficiently thin samples we thus get a homogeneously polarized ferroelectric state which can be switched by an external electric field (Fig.10.1.1.). The ferroelectric liquid crystal is therefore a dielectric in a transparent capacitor. Voltages that are applied across the capacitor plates produce linear electro-optical switching of the polarization and the tilt and thus also the switching of the principal optical axis.

Although the real structures are fairly more complicated, we shall consider here only a simple bookshelf geometry, where the smectic layers are perpendicular to the plates and the molecules are parallel to the plates. We therefore get two distinct unwound states of uniform molecular orientation with opposite signs of the spontaneous polarization pointing either ***up*** or ***down*** (Fig.10.1.1.).

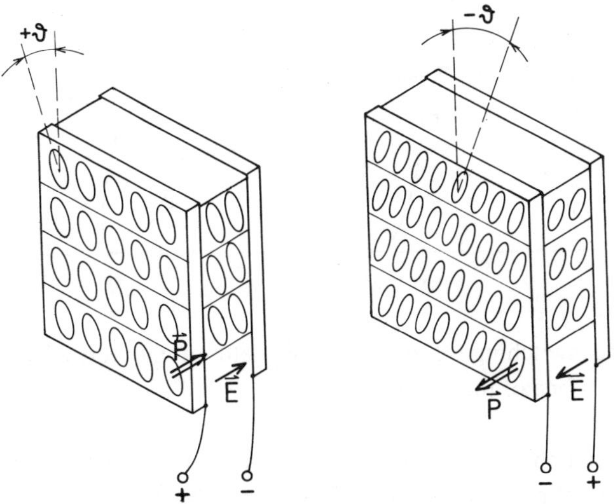

Fig.10.1.1. The principle of operation of the Surface Stabilized Ferroelectric Liquid Crystal Display (SSFLC), invented by N.A.Clark and S.T.Lagerwall. The switching is bistable if the molecules at the surface can be switched away from the surface.

For the state with the polarization pointing ***upwards***, the director field points, let say, in the "***right***" direction with respect to the smectic layer normal. On the other hand, for the ***downwards*** polarization, the director points in the "***left***" direction, making an angle of $2 \cdot \theta$ with respect to the right direction. Let us remind that the optical properties of ferroelectric liquid crystals are determined by the director field $\vec{n}(\vec{r})$ and the principal optic axis which is parallel to this director. The optical axis therefore changes its orientation by twice the tilt angle θ, when the direction of the spontaneous polarization is reversed by the external electric field. The tilt angle is in most ferroelectric liquid crystals of the order of $20\,\text{deg}$, which means that the direction of the optical axis in SSFLC switches for $\approx 40\,\text{deg}$, when an alternating voltage is applied across the SSFLC cell.

The switching proceeds via a collective motion of the director field around the surface of the cone with a tilt angle θ , as shown in Fig.10.1.2.

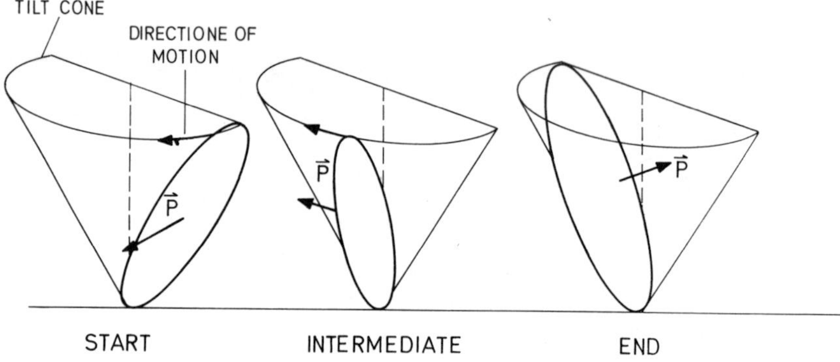

Fig.10.1.2. The switching of the director field in the ferroelectric display proceeds on the surface of the tilt cone.

The collective motion of the molecules during the switching can also be considered as a coupling of the external field to the phase (Goldstone) mode. It is physically the same mechanism. This switching can be easily observed between crossed polarizers. If these polarizers are oriented so that the input polarization is along one of the extreme directions of the director, a switching between a "dark" and a "bright" state is observed. In the "dark" state, the optical axis of the structure is aligned along the direction of the input polarizer, whereas in the "bright" state, it makes a double tilt angle with respect to the input polarizer. Due to birefringence of a liquid crystal, the ordinary and extraordinary waves travel with different speeds through the cell and interfere at the output. For a proper thickness, the phase difference will result in an elliptically polarized output light, that is transmitted through the back polarizer. The SSFLC therefore represents a birefringent plate, where the direction of the optical axis can be switched over large angles with relatively small electric fields. Typically, a $100\mu s$ switching of a half-wave retardation can be achieved with $\pm 10V$ pulses, applied across the $\approx 2\mu m$ thick cells.

Whereas the idea of fast switching in SSFLC is simple, severe technological problems are met in real applications. First, it is clear that in principle the switching that we have described allows for a bistable operation of a SSFLC. Here we have two thermodynamically stable director states, which are stabilized by the two surfaces. These two distinct director states are optically different and we can switch from one state to another by the application of an external electric field. However, note that this simplified view of ferroelectric switching in the bookshelf geometry ***implies switching of the molecules at a surface***. It turns-out that this is not a trivial technological problem and that in reality it is very difficult to control the surface anchoring. Secondly, the structure of the director field in real cells is usually far from the simple bookshelf geometry. Complications due to the thinning of the smectic layers across the $smectic-A$ - $smectic-C^*$ transition arise in the form of a chevron-shaped configuration of smectic layers. Third, the electrostatic field of ionic impurities severely influences the bistability of switching in SSFLC devices. We shall discuss these technological problems from the physical point of view in the following chapters. Advanced information on the technology and applications of ferroelectric liquid crystals can be found in the publications of N.A.Clark and S.T.Lagerwall (1984, 1991), N.A.Clark and K.M.Johnson (1992), D.Coates (1992) and J.Dijon (1992).

When the antiferroelectric $smectic-C_A^*$ phase is confined in between two closely spaced plates, the macroscopic helical modulation can be suppressed by the confining walls in a similar manner than the ferroelectric helix. However, the intrinsic alternation of the molecular tilt on going from one layer to another still remains, unless we have a very large coupling of the interfacial molecules with the surface. This means that without any electric field, the optical properties of such a thin layer of antiferroelectric $smectic-C_A^*$ phase are identical to those of the $smectic-A$ phase. This is because the optical properties of the unwound $smectic-C_A^*$ phase can be averaged over the two neighboring smectic layers, yielding in fact a biaxial $smectic-A$ phase. When the external electric field is applied to such a structure between crossed polarizers, one obtains the so-called tristable switching, shown in Fig.10.1.3. Here, the input polarizer is oriented so that it matches the direction of the normal to the smectic layers and the analyzer is crossed. For zero electric field, $E=0$, we have the alternating tilt structure with zero polarization and the optical axis is directed along the normal to the smectic layers, due to the spatial averaging of the optical properties. The cell is optically equivalent to the $smectic-A$ phase and does not transmit light, because of the orientation of the polarizers.

When an electric field is applied, it couples to the electric polarization of the individual smectic layers and at a sufficiently high field this coupling induces a phase transition into the uniformly polarized ferroelectric phase. This means that the molecules in smectic layers that have had their polarization in the opposite direction with respect to $\vec{E}$ have rotated collectively around the tilt half-cone, until their dipole moments have reached a thermodynamically stable state where $P \parallel E$. The rotation of half of the smectic layers therefore results in a uniformly polarized ferroelectric state with an optical axis at a tilt angle with respect to the smectic normal.

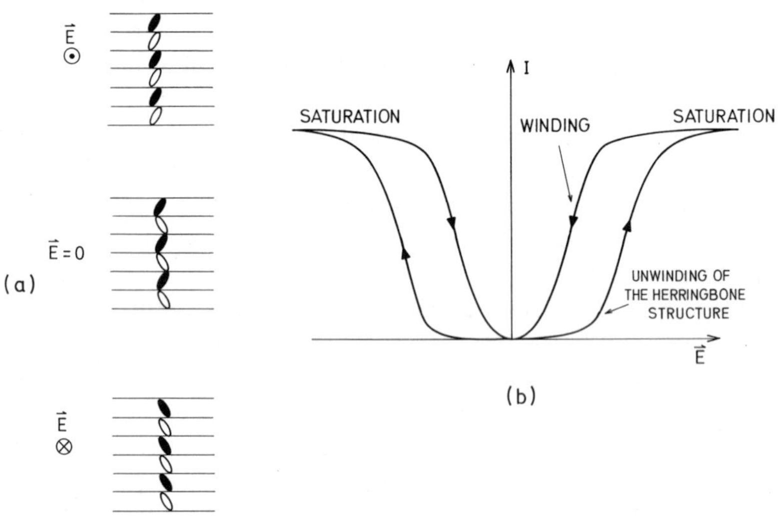

Fig.10.1.3. (a) The tristable switching in the antiferroelectric $smectic-C_A^*$ phase, confined in a bookshelf geometry between two closely spaced glass plates. (b) The optical transmission of such a cell shows three stable states, if the molecules at a surface can be switched. The tristable switching graph resembles the antiferroelectric D-E hysteresis loop of a solid antiferroelectric crystal.

Consequently, this state transmits some light due to the birefringence and the direction of the optical axis with respect to the polarizers. If this switching can also be induced at a surface, we have a thermodynamically stable state at least for some range of the applied voltages. By decreasing the voltage, this structure would "wind" again resulting in zero transmission. The same happens with the reversal of the field direction, with a difference that in this case the optical axis rotates via the surface of the tilt cone in the opposite direction. We have therefore three different states, which can be optically discriminated by a proper orientation of the polarizers and hence "the tristable switching".

10.2. Surface Stabilized Ferroelectric Liquid Crystal Displays

In a Surface Stabilized Ferroelectric Liquid Crystal Device (SSFLCD), the bulk helix of the ferroelectric $smectic-C^*$ phase is unwound by the surface forces and we obtain a more or less uniformly electrically polarized structure. Whereas we have in Chapter 4. discussed the implications of such a restricted geometry on the phase transitions between various smectic phases, we shall here analyze the same restricted geometry from the application point of view.

The driving force for the unwinding of the intrinsic helical modulation of the bulk smectic phase between two confining plates is the surface coupling between the substrate and the liquid crystalline molecules. In terms of spontaneous polarization $\vec{P}$, the surface coupling energy per unit area can be generally written as

$$W_s = -\gamma_1 \left(\frac{\vec{P}}{P} \cdot \vec{s} \right)^2 + \gamma_2 \left(\frac{\vec{P}}{P} \cdot \vec{s} \right) \tag{10.2.1}$$

Here $\vec{P}$ stands for the in-plane spontaneous polarization and $\vec{s}$ for the outward unit vector of the surface normal. The first term $(\gamma_1 \neq 0)$ does not depend on the sign of the direction of polarization with respect to the surface. It is of quadrupolar (non-polar) nature and occurs also in nematics and cholesterics. In ferroelectric liquid crystals it occurs if the surfaces are non-polar. On the other hand, the second term describes the specific polar surface interaction $(\gamma_2 \neq 0)$. It depends on the sign of the collective dipole moment (or the in-plane spontaneous polarization) and makes the two molecular director orientations at $\pm\theta$ non-equivalent. Such a term may arise for instance, from the electrostatic interaction of $\vec{P}$ with the electric field that is produced by surface charges or dipoles. For a discussion of these couplings, see, for instance Dijon et al., 1988, Handschy et al., 1983.

The surface coupling term Eq.10.2.1. is spatially homogeneous, which is incompatible with the intrinsic helical modulation of most ferroelectric liquid crystals. One can therefore conjecture, that the texture of a ferroelectric liquid crystal that is confined between two parallel plates would somehow reflect the competition between the helical modulation imposed by the bulk and the homogeneous structure imposed by the homogeneous surface. As a result of this competition, line defects arise in cells with a thickness that is comparable to the bulk helical period. These lines mediate the transition between the homogeneous part of the sample close to the surface and the helicaly modulated part far away from the surface. These, so-called dechiralization lines were first reported by Glogarova et al.(1983). In the following we shall briefly describe the possible structures of ferroelectric liquid crystals in a bookshelf geometry, the dechiralization lines and the response to an external electric field for the case of non-polar and polar anchoring

(i) Non-Polar Anchoring and Dechiralization Lines in Surface Stabilized Ferroelectric Liquid Crystals in Zero Electric Field.

M. Glogarova et al. (1983) introduced a simple model for the real structure of the $smectic-C^*$ phase confined between two closely spaced glass plates and proposed a mechanism for the unwinding of the helical $smectic-C^*$ structure by an electric field in such a geometry. Based on the topological model of Brunet and Williams (1978), they suggested that the sharp lines which are seen between crossed polarizes (Fig.10.2.1) correspond to linear defects - called dechiralization lines - which mediate the thin region of unwound $smectic-\overline{C}$ phase near the glass plates with the helical $smectic-C^*$ phase in the bulk of the sample. The coordinate axes and the molecular arrangement around the $\pm 2\pi$ dechiralization lines are shown in Fig.10.2.2. The molecules are represented by nails, the points of which represented those parts of the molecules, which are turned toward the observer. The disclination lines form pairs of $+2\pi$ and -2π twist dislocations which run parallel to the smectic planes and parallel to the glass surface. The two lines of a pair are located opposite to each other at a distance d from the glass surface and the distance between the lines in different smectic layers equals the helical pitch.

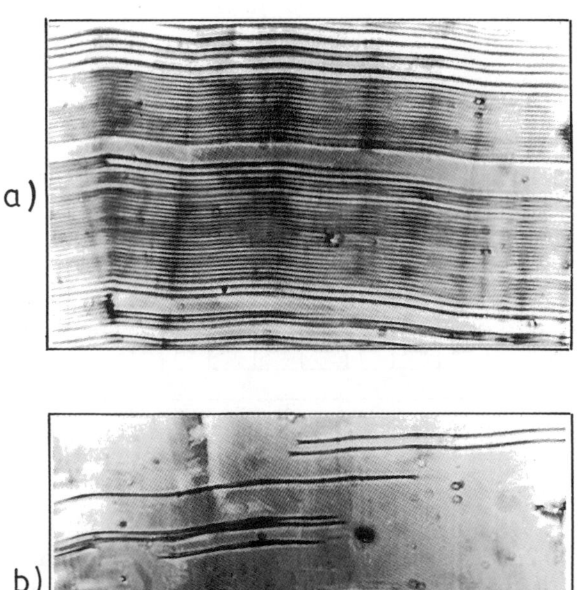

Fig.10.2.1.Texture of a $50\mu m$ thick sample of $smectic-C^*$ phase of DOBAMBC. a) $U=0V$. b) $E=3kV/cm$. Field of view $0.37\times0.25mm^2$. After Glogarova et al. (1983).

In a d.c. electric field the $\pm2\pi$ lines move either to the center of the cell or towards the two surfaces, depending on the polarity of the field. In the first case, $+2\pi$ and -2π twist disclination lines meet at the center of the cell and annihilate each other. In the second case, the disclinations move towards the surface, where for some value of the electric field, the molecules at the surface flip and the defect is annihilated. In both cases the annihilation of one pair of $\pm2\pi$ defects unwinds the helical structure by 2π . When all pairs annihilate, the helical structure is completely unwound.

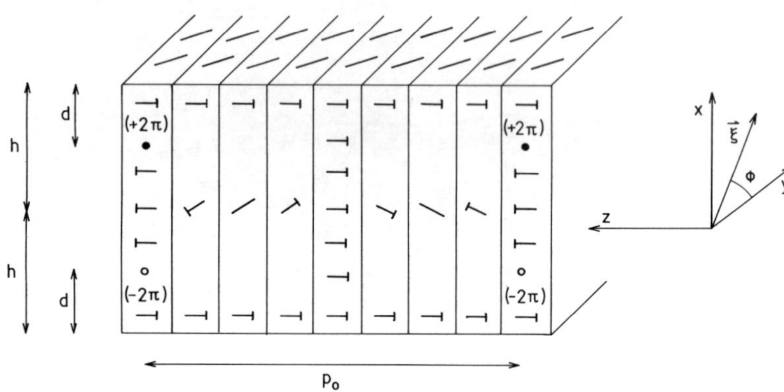

Fig.10.2.2.Schematic representation of molecular arrangement around the $\pm 2\pi$ dechiralization lines (dots indicated with $\pm 2\pi$). After Glogarova et al. (1983).

Following Glogarova et al. (1983), the emergence of disclination (or dechiralization) lines can be described by a simple Landau expansion of the free energy of a sample of thickness h and helical period $p_\circ$:

$$G = \int_0^h dx \int_{-p_\circ/2}^{+p_\circ/2} dz \left\{ \frac{K_1}{2}\left(\frac{\partial \Phi}{\partial x}\right)^2 + \frac{K_3}{2}\left(-\frac{\partial \Phi}{\partial z} + q_\circ\right)^2 - PE_z \cos\Phi \right\}$$

$$+ \int_{-p_\circ/2}^{+p_\circ/2} dz \left[W_s \sin^2 \Phi_\circ \right]_{x=+h} \qquad (10.2.2)$$

Here $\Phi(x,z)$ is the angle between the projection $\xi(x,z)$ of the molecular director onto the smectic layers, (see Fig.10.2.2) and $q_\circ = 2\pi / p_\circ$ is the wave-vector of the helix with a period $p_\circ$. The electric field $\vec{E} = (E,0,0)$ is here applied along the x-axis , i.e. in a direction perpendicular to the glass plates. K_1 and K_3 are the nematic-like elastic constants that describe the elastic deformation of the ξ director field in a smectic plane (i.e. splay-bend deformation) and along the smectic normal (i.e. twist-like deformation), respectively. A non-polar surface coupling has been considered, represented by the last term in Eq.10.2.2.

In the absence of the electric field, the variation of the free-energy, Eq.10.2.2.) with respect to the phase angle Φ yields the differential equation for the equilibrium phase of the director field

$$K_1 \frac{\partial^2 \Phi}{\partial x^2} + K_3 \frac{\partial^2 \Phi}{\partial z^2} = 0 \tag{10.2.3}$$

The boundary condition is

$$\Phi(x = \pm h) = \Phi_\circ \tag{10.2.4}$$

The solution of the differential equation 10.2.3. that corresponds to a periodical array of disclination lines with a period $p_\circ$ and is situated at a distance d from the upper and lower cell boundary plates is (Glogarova et al., 1983):

$$\Phi = \Phi_\circ + \sum_{n=-\infty}^{\infty} \sum_{k=-\infty}^{\infty} (-1)^k \left[arctg\left(\alpha \frac{x - 2hk - h + d}{z - np_\circ} \right) - arctg\left(\alpha \frac{x - 2hk + h - d}{z - np_\circ} \right) \right] \tag{10.2.5}$$

with $\alpha = \sqrt{K_3/K_1}$. Here, we have used the function $arctg(\alpha x/z)$ as an elementary solution of the Eq.10.2.3. It can be shown that the above solution describes both the perfect helical structure of the $smectic - C^*$ phase in the bulk and the system of disclination lines joining the unwound phase near the glass plates with the helical phase in the bulk.

The distance d of the dechiralization lines from the plates corresponds to a minimum of the free energy G with respect to d with fixed $\Phi_\circ$. For the simple case of $n = 0$ in Eq.10.2.5., one finds this distance from the equation

$$\sin\frac{\pi d}{h} = \frac{p_\circ}{4h\alpha} \tag{10.2.6}$$

For large thickness of the cell, this equation has two solutions. One of the solutions, that corresponds to a stable state is

$$d_1 = \frac{h}{\pi}\cdot\arcsin\frac{p_\circ}{4h\alpha} \tag{10.2.7}$$

For a large thickness, $\arcsin\beta \approx \beta$ and the distance of dechiralization lines from the surface is

$$d_1 \approx \frac{p_\circ}{4\pi\alpha} \tag{10.2.8}$$

In a thin sample, where the thickness of the sample is lower than the critical thickness $d_c < p_\circ/4\pi\alpha$, there are no dechiralization lines and the whole $smectic-C^*$ sample is unwound by surface anchoring.

(ii) Electric Field Effect on Dechiralization Lines in Surface Stabilized Ferroelectric Liquid Crystals for Non-Polar Anchoring

The above analysis of the free energy Eq.10.2.2. can be performed also in the presence of an external electric field. The variation of G with respect to Φ leads in this case to a Sine-Gordon equation in two dimensions. This means that soliton-like walls should appear in the system, because these are the solutions of the Sine-Gordon equation. Following the simple model of Glogarova et al. (1983), the dechiralization lines are displaced either towards the middle of the cell or towards the surfaces, when the DC electric field is introduced. The displacement $d = d(E)$ of the dichiralization lines with the applied field can be found from a minimization of G with respect to d as:

$$d(E) = \frac{h}{\pi} \cdot \arcsin\left[\pi^2 \sqrt{K_1 K_3} \Big/ 2p_\circ h \left(\frac{K_3}{2} q_\circ^2 - PE_z \cos\Phi_\circ\right)\right] \qquad (10.2.9)$$

The direction of the displacement depends on the direction of the electric field with respect to the polarization field $P(z)$, as shown in Fig.10.2.3.

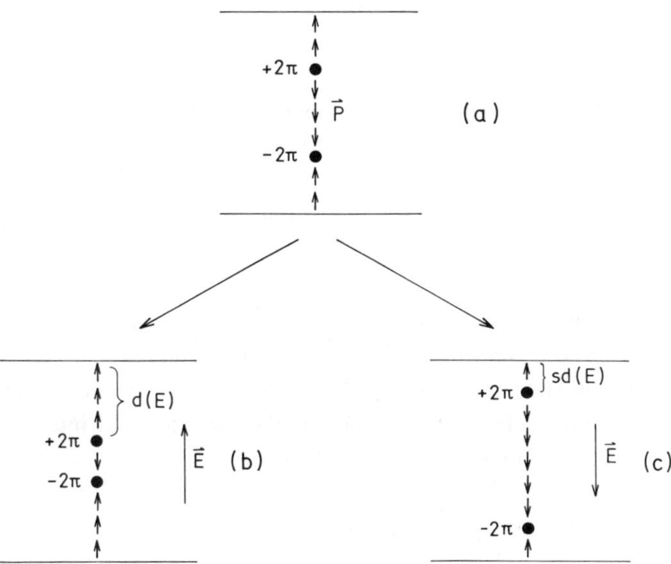

Fig.10.2.3.(a) Distribution of the polarization field across the cell for zero electric field. Depending on the direction of the electric field, the dechiralization lines move either towards the middle of the cell and annihilate (b) or towards the surface of the cell (c). In the latter case, switching is possible only if the molecules at the surface can be switched. If this is energetically costly, the cell is monostable.

This figure shows the polarization field across the cell and across a pair of two dechiralization lines. The polarization field in between the two defects is directed in the opposite direction than close to the surface. If, for example, the electric field is directed upwards (see Fig.10.2.3(b)), the direction of the polarization field in between the two dechiralization lines is energetically unfavorable and the two lines start to approach each other, because they both move towards the middle of the cell. When the field is high enough, these lines meet at the center of the cell, become unstable and anihilate each other at a critical electric field

$$E_c \approx \frac{\frac{1}{2}K_3 q_\circ^2 - \frac{\pi^2\sqrt{K_1K_3}}{2p_\circ h}}{|P|} \tag{10.2.10}$$

On the other hand, for the opposite direction of the external field (see Fig.10.2.3(c)), the direction of the polarization in between the two lines is favorable and the lines start to move apart, i.e. towards the surface. In this case, the distance $d(E)\to 0$ in the limit of very high fields and the line defects approach the surface.

In a thin sample, where the thickness of the sample is lower than the critical thickness $d_c < p_\circ/4\pi\alpha$, there are no dechiralization lines and the whole $smectic-C^*$ sample is unwound by surface anchoring. The critical field to change the orientation of the unwound sample as a whole is obtained from the balance of the electric field interaction energy and the maximum surface anchoring energy

$$2p_\circ hPE = 2p_\circ W_s \tag{10.2.11}$$

The critical electric field for switching the structure as a whole in this case depends only on the surface coupling energy and there is no dependence on the elastic constants

$$E_c \cong \frac{W_s}{hP} \tag{10.2.12}$$

In the limit of very weak surface coupling energy, the switching is without threshold and the switching time is inversely proportional to the applied electric field and spontaneous polarization:

$$\tau = \frac{\gamma}{PE} \tag{10.2.13}$$

Here, γ is rotational viscosity for the director motion on the cone of the tilt angle.

(iii) Polar Anchoring in Surface Stabilized Ferroelectric Liquid Crystals

The effect of polar surface anchoring on the director structure in thick cells of ferroelectric *smectic* $-C^*$ phase was considered by Handschy et al. (1983) and later by Pavel (1984) and others. Following Pavel's calculations, we shall consider the case of a planar sample where the *z*-axis is perpendicular to the layers and the *x*-axis is perpendicular to the glass plates, Fig.10.2.4.

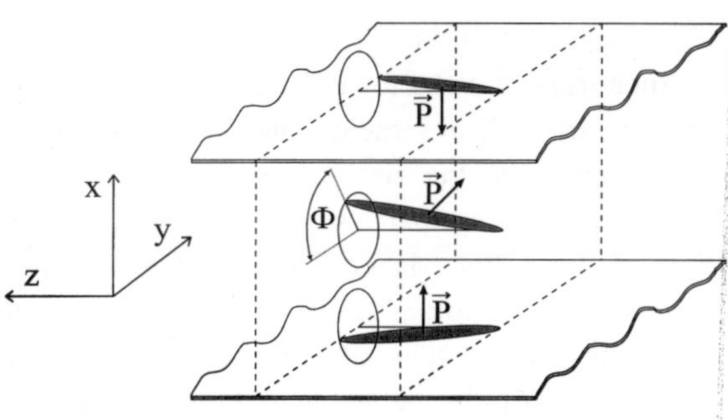

Fig.10.2.4. Polar anchoring of liquid crystalline molecules at the surfaces leads to the splayed state.

Let us assume that the glass plates force the liquid crystal molecules to be parallel to them. When the smectic layers are formed, there are just two orientations on the glass surface where the molecules can be parallel to the glass plates. These two director orientations on the surface make an angle $2\cdot\theta$ with each other and have opposite directions of dipole moments, as shown in Fig.10.2.4. We have therefore a splayed state, which has been considered in the limit of small fields in Chapter 4. In the constant tilt angle approximation (CAA) we can express the elastic free energy of this state as

$$G_{elastic} = \int_{-d/2}^{d/2} dx \int_{0}^{p_\circ} dz \left[\tfrac{1}{2} K_1 \left(\frac{\partial \Phi}{\partial x} \right)^2 + \tfrac{1}{2} K_3 \left(\frac{\partial \Phi}{\partial z} - q_\circ \right)^2 \right] \tag{10.2.14}$$

The corresponding Euler-Lagrange equation is the same as Eq.10.2.3:

$$K_1 \frac{\partial^2 \Phi}{\partial x^2} + K_3 \frac{\partial^2 \Phi}{\partial z^2} = 0 \tag{10.2.15}$$

The boundary conditions are now different because of opposite orientations of dipole moments on the two surfaces

The solutions of the Eq.10.2.15. are shown in Fig.10.2.5. for (a) non-polar anchoring and (b) for polar surface anchoring (J.Pavel, 1984). In order to join the helical structure in the middle of the sample and the unwound structure near the glass plates it is necessary to have a pair of $\pm 2\pi$ disclinations, i.e. pair of dechiralization lines. The spatial distribution of these lines is however different in the two cases. For quadrupolar surface coupling these lines face each other, whereas for polar surface anchoring, they alternate. When the sample thickness is reduced, the width of the helical part between the two dechiralization lines is also reduced. Finally, when the separation distance between the two dechiralization lines is reduced to zero, the sample becomes unwound.

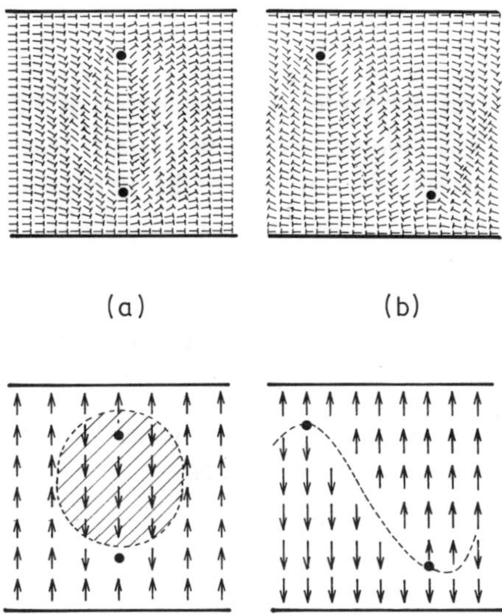

Fig.10.2.5. Structures, calculated from the Eq.10.2.15. for the ratio $K_3/K_1 = 0.35$, with (a) non-polar (quadrupolar) boundary conditions and (b) polar boundary conditions (after J.Pavel, 1984). (c) and (d) show the corresponding distribution of the electric dipole field. The dashed lines represent domain interfaces.

Let us now look into the structure of unwound samples with different anchoring at the sample surfaces and investigate the relative stabilities of the splayed and homogeneous states as a function of sample thickness. Let W_1 designate the anchoring energy for molecules on the surface oriented in such a way that their dipole moments are directed from the glass into the bulk of the sample and $W_2 > W_1$ for anchoring of molecules with their dipole moments pointing towards the sample wall. When the director orientations of the molecules at the two surfaces are opposite, the dipole moments at both surfaces point into the bulk of the sample and the free energy of this splayed state is

$$G_{splayed} = \tfrac{1}{2} \int_{-d/2}^{d/2} K_1 \left(\frac{\partial \Phi}{\partial x} \right)^2 dx + 2W_1 \tag{10.2.16}$$

When, however, the molecular orientations at the two surfaces are equal, the structure is homogeneous, $\Phi = const.$ At one surface the dipole moment points into the bulk and at the other into the wall, so that the free energy becomes

$$G_{\text{hom}} = \int K_1 \left(\frac{\partial \Phi}{\partial x} \right)^2 dx + W_1 + W_2 = W_1 + W_2 \tag{10.2.17}$$

After minimization of $G_{splayed}$ with respect to $\Phi(x)$, the free energies of the splayed and homogeneous states are

$$G_{splayed} = \tfrac{1}{2} K_1 \frac{\pi^2}{d} + 2W_1 \tag{10.2.18a}$$

$$G_{\text{hom}} = W_1 + W_2 \tag{10.2.18b}$$

We see that in the limit of large *d*, the splayed state is always energetically favorable with respect to the homogeneous state. By decreasing the distance between the glass plates, the splayed state, which is a state of elastic deformation, becomes energetically unfavorable. The transition from the splayed to the homogeneous state takes place at a critical distance

$$d_c = \frac{K_1 \pi^2}{2(W_2 - W_1)} \tag{10.2.19}$$

(iv) Effect of an Electric Field on Polar Anchoring in Surface Stabilized Ferroelectric Liquid Crystals

Let us now discuss the effect of an applied electric field on an helicaly unwound but transversely splayed planar sample with polar anchoring, following the approach of J.Pavel (1984). As mentioned before, in the absence of an electric field such a configuration is stable for $d > d_c$. When an external electric field is applied perpendicularly to the sample plates, the free energy becomes

$$G_{splayed} = \tfrac{1}{2} \int_{-d/2}^{d/2} K_1 \left(\frac{\partial \Phi}{\partial x} \right)^2 dx + 2W_1 + \int_{-d/2}^{d/2} PE \cos \Phi dx \qquad (10.2.20)$$

where $P(x)$ is the local spontaneous polarization. The variation of $G_{splayed}$ with respect to $\Phi(x)$ yields a one-dimensional Euler-Lagrange equation:

$$\frac{d^2\Phi}{dx^2} + \frac{PE}{K_1} \cdot \sin \Phi(x) = 0 \qquad (10.2.21)$$

This equation can be solved analytically using Jacobi's elliptic functions. The solution for the phase profile $\Phi(x)$ is a soliton wall

$$\sin \frac{\Phi(x)}{2} = sn(u, k) \qquad (10.2.22)$$

Here, $u = (x + d/2)\sqrt{(C + PE/K_1)/2}$ is the reduced coordinate and $k^2 = 2/(1 + CK_1/(PE))$ is the modulus of the elliptic functions. The constant of integration C is determined from

$$K(k) = d\sqrt{\frac{C + \dfrac{PE}{K_1}}{2}} \qquad (10.2.23)$$

where $K(k)$ is the complete elliptic integral of the first kind. The phase profile $\Phi(x)$ is shown in Fig.10.2.6. for different values of the electric field. In zero field, the phase increases linearly across the cell. For finite fields, the soliton wall is formed, which moves towards the surface when the field is increased.

The total free energy of the splayed state in the external electric field can be found by substituting Eq.10.2.22. into Eq.10.2.20. Similarly, we can obtain the total free energy of a homogeneous structure in the external electric field. It turns out that both energies depend on the sample thickness and the strength of the electric field, which determines the critical field for the switching (see Pavel, 1984).

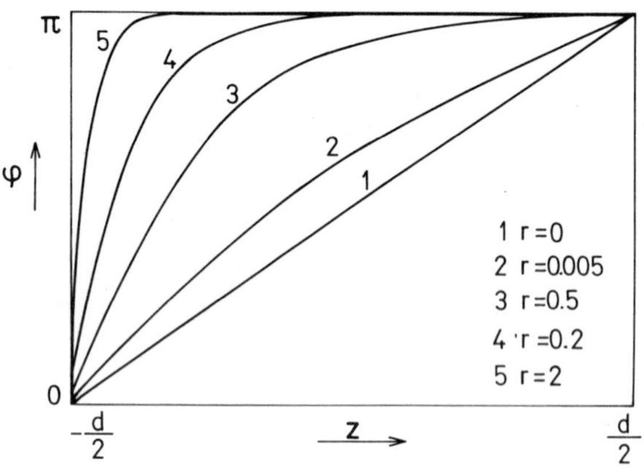

Fig.10.2.6. The phase profile of a splayed state in an external electric field. By increasing the field, the soliton wall moves towards the surface. The parameter $r = PE/K_1$. After J.Pavel (1984).

In should be stressed that we have presented here only simplified structures of ferroelectric $smectic-C^*$ phase in a planar restricted geometry. More detailed and extensive calculations and results can be found in the articles of Glogarova et al.(1983, 1984a-c), Bourdon et al(1982), Kondo et al.(1983), Hudak (1984), Kai et al. (1983), Pavel et al.(1991), Lejček et al. (1984, 1990), Nakagawa et al.(1986), Škarabot et al.(1998), Clark and Lagerwall (1984), Handschy et al.(1983), Handschy and Clark (1984), Maclennan et al.(1991), Cladis and Saarlos (1991) and Pikin(1991).

10.3. Ultrafast Electroclinic Effect

The electroclinic effect was one of the first linear electrooptic effects, reported by Garoff and Meyer (1977, 1979) soon after discovery of ferroelectric liquid crystals. In their experiment, Garoff and Meyer have measured the linear electrooptic response of DOBAMBC in homeotropic geometry, where the measuring electric field was applied in a direction perpendicular to the helix with a pair of transverse copper electrodes, as shown in Fig.10.3.1.

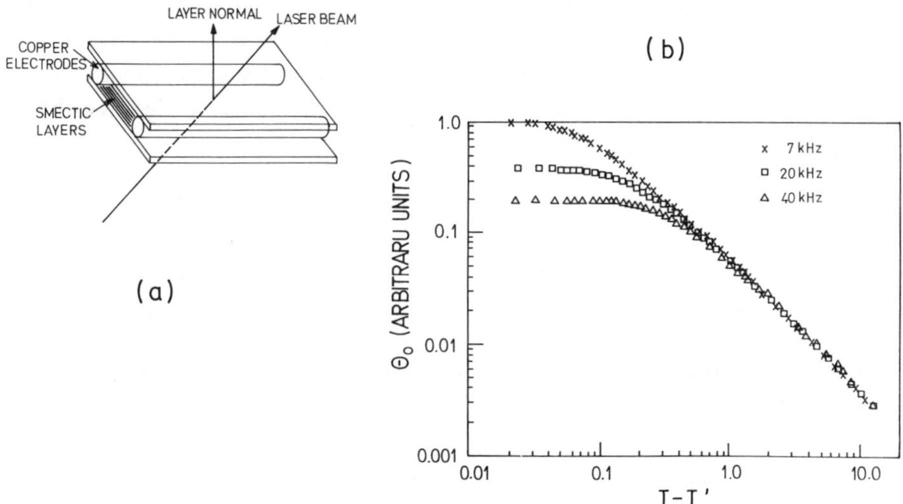

Fig.10.3.1. (a) The experimental geometry of Garoff and Meyer (1979), measuring the electroclinic effect in the *smectic* $-A$ phase of a ferroelectric liquid crystal DOBAMBC in homeotropic geometry. (b) The magnitude of the linear response increases as we approach the ferroelectric *smectic* $-C^*$ from above. After Garoff and Meyer, 1979.

In the $smectic-A$ phase they observed a small but very fast linear electrooptic signal, which was an evidence for an electric field induced tilt angle in the paraelectric $smectic-A$ phase. The frequency dispersion of this signal could be well described with a Debye-type overdamped relaxational collective motion of the director field.

This effect has been explained in Section 5.6. The external electric field $E_{\circ}e^{i\omega t}$ couples in the $smectic-A$ phase linearly to the polar soft mode at the wave-vector $\vec{q}=0$. Due to this coupling, a finite tilt of the molecules is induced in the $smectic-A$ phase

$$\delta\theta = \frac{\varepsilon C E_{\circ}}{a(T)+K_3 q_c^2}\cdot\frac{1}{1+i\omega\tau_\theta} \tag{10.3.1}$$

Here, the second term describes the Debye-type relaxational dynamics, ε is the dielectric constant of the paraelectric $smectic-A$ phase, C is the piezoelectric coefficient, $a(T)$ is the coefficient of the quadratic term in the free-energy expansion, K_3 is the torsional elastic constant and q_c is the critical wave-vector of the helix. τ_θ^{-1} is the corresponding relaxation rate of the soft mode, given by

$$\tau_\theta^{-1} = \frac{\alpha(T-T_c)}{\gamma}+\frac{K_3}{\gamma}q_c^2 \tag{10.3.2}$$

where γ is the rotational viscosity in the $smectic-A$ phase. From these two equations we can see the following:

(i) Far above the $smectic-A$ - $smectic-C^*$ phase transition, the coefficient $a(T)=\alpha(T-T_c)$ is very large. This means that the induced tilt is very small. It becomes larger, as we approach the phase transition from above. Typically, the magnitude of the induced tilt is less than one degree at a temperature of several degrees above the transition. In some compounds, a very large electroclinic effect can be obtained. An example of a very large electroclinic effect is shown in Fig.10.3.2. (after Bahr and Heppke, 1987).

(ii) Due to the smallness of the electroclinic tilt, the effect is very fast. Typically, the modulation frequencies for the electroclinic effect are between 100kHz and 1MHz. This is a consequence of the large thermodynamic restoring force, which drives the molecules back to equilibrium. Fig.10.3.3. shows the temperature dependence of maximum modulation frequency for the same compound as in Fig.10.3.2. (Bahr and Heppke, 1987).

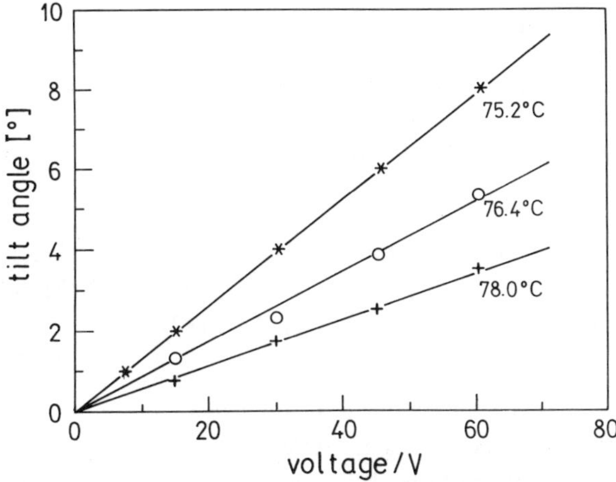

Fig.10.3.2. An extremely large electroclinic effect observed in 4-(3-methyl-2-chlorobutanoyloxy)-4'-heptyloxybiphenyl. The phase transition is around 73.5°C. After Bahr and Heppke, 1987.

The most convenient geometry for the observation of the electroclinic effect is the planar bookshelf geometry with a sample thickness of several micrometers. As shown in Fig.10.3.4., the sample is placed between ITO electrodes and the optical axis of the $smectic-A$ phase is aligned at an angle $\Omega_\circ$ with respect to the axis of crossed polarizers. In this case, the transmitted light intensity is

$$I(\Omega_\circ) = I_\circ \sin^2 2\Omega_\circ \cdot \sin^2 \frac{\pi \Delta n d}{\lambda} \tag{10.3.3}$$

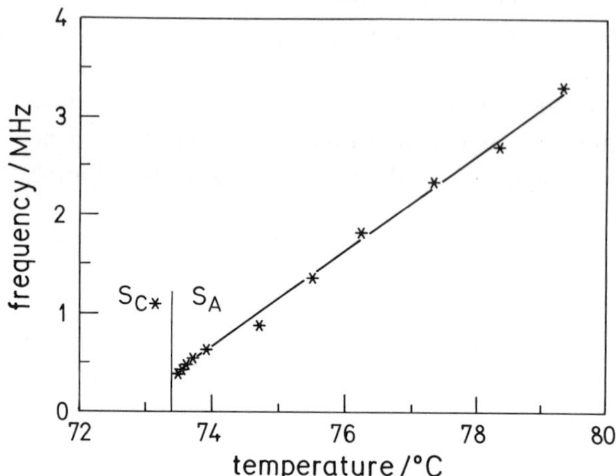

Fig.10.3.3. The temperature dependence of the relaxation frequency in the *smectic* – *A* phase of 4-(3-methyl-2-chlorobutanoyloxy)-4'-heptyloxybiphenyl (Bahr and Heppke 1987).

Here, $I_{\circ}$ is the maximum light intensity, Δn is the birefringence of the *smectic* – *A* phase, d is the thickness of the cell and λ is the vacuum wavelength of the light. When an electric field is applied, the molecules tilt collectively by an angle $\delta\theta$. This means that also the direction of the optical axis tilts, and, correspondingly, the angle $\Omega_{\circ}$ changes to $\Omega_{\circ} \pm \delta\theta$. As we can see from the Eq.10.3.3., this results in a change of the transmitted light intensity:

$$\delta I \approx \sin 4\Omega_{\circ} \cdot \sin^2 \frac{\pi \Delta n d}{\lambda} \cdot \delta\theta \tag{10.3.4}$$

The changes of intensity are largest when $\Omega_{\circ}$ is set to $22.5°$. Furthermore, due to interference nature of the transmitted light intensity, the thickness of the sample should be adjusted for a given wave-length and birefringence, so that the second (geometrical) term in the Eq.10.3.4. is maximum. More details and experimental results for various materials can be found in the work of Bahr and Heppke (1987, 1990), Lee and Patel (1989), Li et al.(1989, 1991), Etxebaria and Zubia (1991), Giebelman and Zugenmaier (1995), van Haaren and Rikken (1989), Levstik et al.(1992) and Qiu et al.(1988).

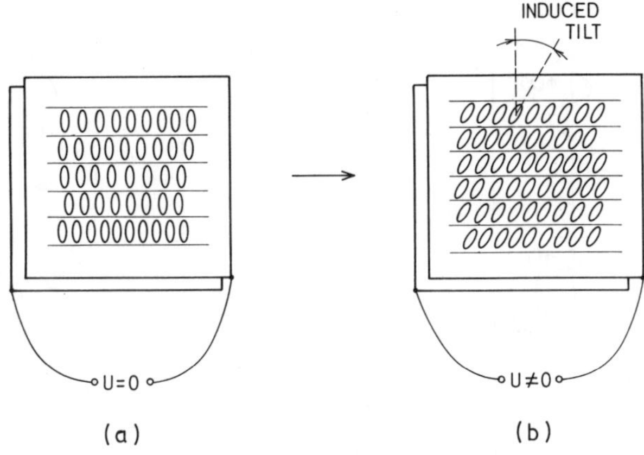

Fig.10.3.4. The electroclinic effect in a planar geometry. By applying an electric field between the ITO electrodes, the molecules flip collectively in the plane of the cell. This causes the modulation of light intensity.

10.4. Deformed Helix Mode Ferroelectric Displays

Deformed Helix Mode (DHM) ferroelectric displays are based on the linear electrooptic effect in the helicoidaly modulated $smectic-C^*$ phase. This type of display was first discussed by Ostrovski et al.(1980) and subsequently by Beresnev et al. (1989) and Funfschilling and Schadt (1989). Conceptually, it is based on the soliton-like deformation of the ferroelectric helix by an external electric field, as discussed in Chapters 3.4 and 5.6.

When an external DC electric field is applied in a direction perpendicular to the helical axis of the ferroelectric $smectic-C^*$ phase, an array of 2π soliton walls is induced, as shown in Fig.10.4.1.

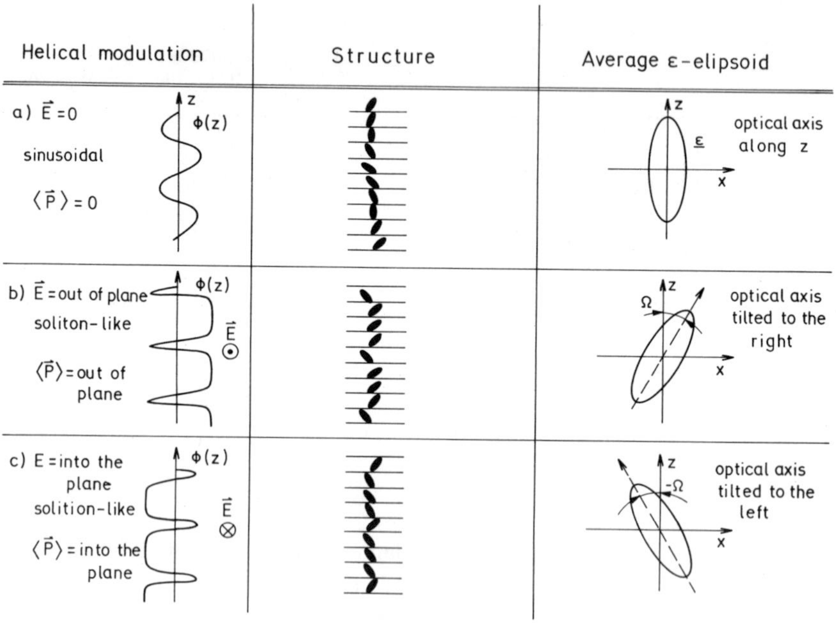

Helical modulation	Structure	Average ε-elipsoid
a) $\vec{E}=0$ sinusoidal $\langle\vec{P}\rangle=0$ z $\phi(z)$		z ε x optical axis along z
b) $\vec{E}$ = out of plane soliton-like $\langle\vec{P}\rangle$ = out of plane $\phi(z)$ $\vec{E}$ ⊙		Ω z x optical axis tilted to the right
c) E = into the plane solition-like $\langle\vec{P}\rangle$ = into the plane $\phi(z)$ $\vec{E}$ ⊗		z −Ω x optical axis tilted to the left

Fig.10.4.1. The principle of operation of the Deformed Helix Mode ferroelectric displays. The spatial average of the dielectric tensor for the optical frequency is important in this mode of operation.

When the electric field points out of the plane of the paper and is below the critical field for the unwinding of the helix, most of the molecules are aligned, let say, into the right direction. When we move along the smectic normal, we come to the region, where molecules turn by 2π at a very short distance. This is the 2π -soliton wall. By proceeding, we come to the next region of molecules, aligned into the right direction and so on. By reversing the direction of the field, we get a mirror image of this structure, as shown in Fig.10.4.1.

We have already seen in Chapter 5. that in the first order approximation to the wave-equation, the optical properties of helical birefringent phases are determined by the *spatial average of the dielectric tensor* $\langle \varepsilon(z) \rangle$. For zero electric field, the spatial averaging of a sinusoidally modulated helical structure yields the dielectric tensor with its principal axes along the z and x-axes, respectively. Optically the helical structure is equivalent to the $smectic-A$ phase with the optical axis along the normal to the smectic layers. When the field is applied, the principal axes of the spatially averaged dielectric tensor are not any more aligned along the smectic normal. Instead, the largest axis of the dielectric tensor is at some angle Ω with respect to the z-axis. This angle can be calculated from the soliton profiles of the phase profile in a deformed helix. For a reverse direction of the field, the direction of the axes of the space-averaged dielectric tensor is reversed. We have therefore two extreme spatial directions of the optical axes of the soliton-like distorted helix, which can be easily discriminated by crossed polarizers. It is also important, that this type of effect is continuous, so that a gray scale can be generated. The switching times of the DHF mode display is of the order of several $100\mu s$. Experimental details can be found in the publications of Ostrovski et al. (1980), Beresnev et al. (1989) and Funfschilling and Schadt (1989).

10.5. Chevrons in SSFLC

The $smectic-C^*$ phase is usually described as a stack of two-dimensional liquid layers where the average direction of the long molecular axis is tilted with respect to the layer normal. It has been also traditionally assumed that a homogeneously aligned $smectic-C^*$ sample between solid plates has a bookshelf structure with planar $smectic-C^*$ layers perpendicular to the plates.

An X-ray experiment by Rieker et al. (1987) showed that this is not generally the case. Instead, layer bending takes place inside surface stabilized ferroelectric liquid crystal (SSFLC) cells resulting in a "chevron" structure, shown in Fig.10.5.1.

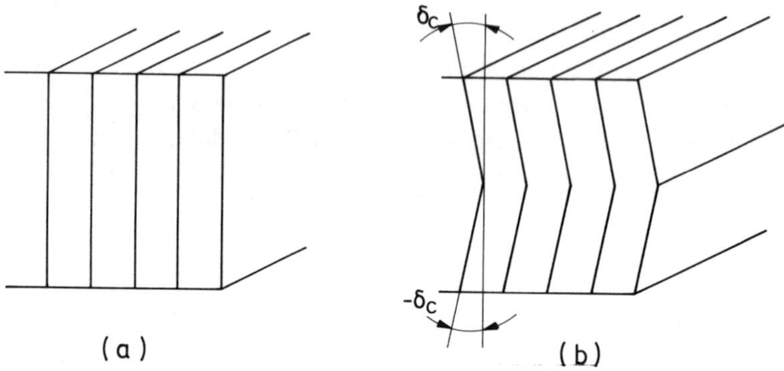

Fig.10.5.1. The traditional "bookshelf" geometry has been reconsidered after the experiment of Rieker et al. (1987). (a) Instead of planar layers, which are perpendicular to the walls of the cell, a chevron-like organization of smectic layers can be observed in most cases, as shown in (b).

The layers are planar but exhibit two distinct tilt angles $\pm\delta_C \neq 0$ with respect to the normal to the bounding plate. These two layer meet in the interior of the cell and make an abrupt reorientation at a planar interface, which is parallel to the bounding plates. This "chevron" structure also explains the zigzag walls that separate regions of uniform but different optical contrast in polarized transmission microscopy of SSFLC cells.

The first X-ray (synchrotron) experiment that revealed the chevron-like organization of smectic layers in confined smectics was performed on DOBAMBC by Rieker et al., 1987. They have used thin, $3\mu m$ cells with uniformly buffed Nylon and PVA as an aligning layer. The cells were first evaluated optically to ensure uniform alignment and suppression of the bulk ferroelectric liquid crystal director helix. The X-ray scattering geometry was chosen so that one could determine the direction of the smectic layers by performing angular scans and measuring the diffracted intensity. When the scattering wave vector matches the direction and magnitude of the smectic modulation, the X-rays are Bragg reflected and an increased intensity is observed, as shown in Fig.10.5.2.

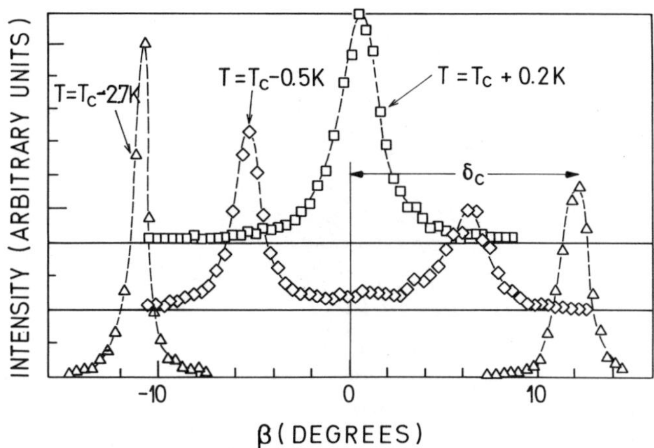

Fig.10.5.2. The angular dependence of the X-ray scattered intensity in the *smectic – A* and the unwound *smectic – C* phase for $\delta_A = 0$, after Rieker et al. (1987). β indicates the direction of the scattering wave-vector with respect to the normal to the plates. Note that in the *smectic – A* there is only one peak at $\beta = 0$ indicating planar bookshelf geometry. This peak splits into two asymmetrical peaks below T_c, clearly indicating chevron structure.

From the angular scans of the diffracted X-ray intensity, Rieker et al. could conclude that already in the *smectic – A* phase the layers may be tilted (because of shearing etc.) with respect to the bounding plate normal at an angle δ_A. From the intensity profiles in the ferroelectric phase, they could resolve two possible scenarios of chevron formation, which depend on the layer tilt δ_A in the *smectic – A* phase:

(i) ***Symmetric chevrons*** are formed when there is no tilting of the layers in the *smectic – A* phase, i.e. $\delta_A = 0$, as shown in Fig.10.5.3.

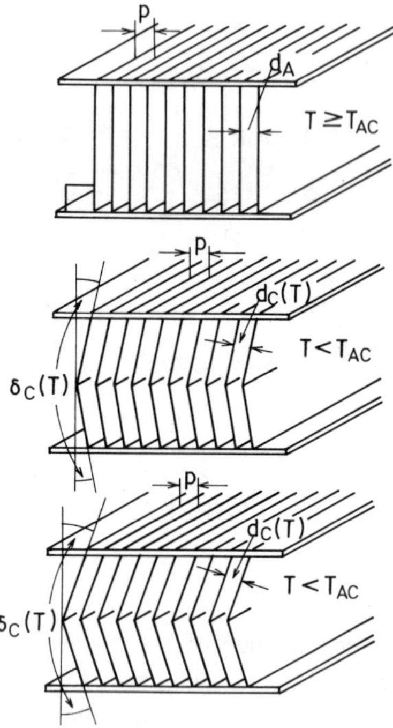

Fig.10.5.3. The formation of a symmetric chevron structure when cooling the *smectic* – *A* in a planar bookshelf geometry into the ferroelectric phase. In this case the layers are not tilted in the *smectic* – *A*, i.e. $\delta_A = 0$. After Rieker et al. (1987).

The temperature dependence of the layer tilt δ_C is shown in Fig.10.5.4. The two branches $+\delta_C$ and $-\delta_C$ are located symmetrically around $\delta = 0$. Deeply in the ferroelectric phase, the two curves saturate around $\delta_C = \pm 18^\circ$ and $\delta_C \to 0$ as one approaches the *smectic* – *A* phase. The two X-ray scattering peaks are here of equal intensity. The two curves $\delta_C(T)$ resemble the temperature dependence of the tilt angle $\theta(T)$. This allows for the conclusion that the above "symmetric chevron" structure is formed as a response to the shrinking of the *smectic* – *C* layers while anchored to the solid bounding plates.

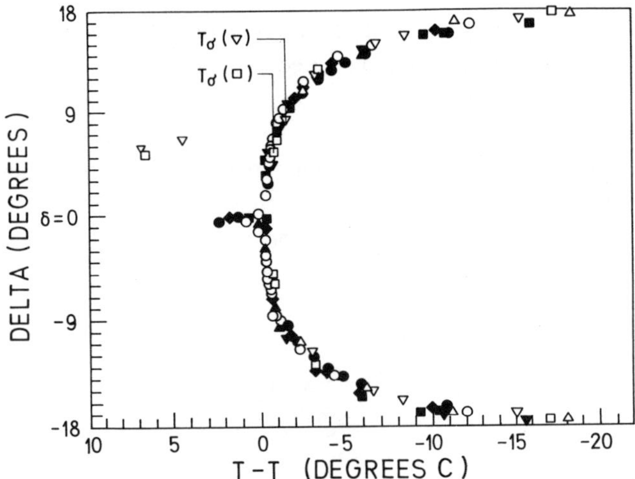

Fig.10.5.4. Layer tilt δ in a chevron structure, as a function of temperature. After Rieker et al. (1987)

The argument goes as follows: Once the layers have formed and are anchored, the anchored positions of the molecules are fixed and the molecules cannot move along the surface of the confining glass. This argument has been later independently proved by Cagnon and Durand (1993). Because the layer periodicity is fixed at the bounding surfaces, it cannot change by crossing the phase transition into the tilted phase. When the molecules tilt, the effective thickness of the smectic layers decreases. In order to meet the fixed smectic periodicity at the surface, the smectic layers have to tilt away from the normal to the glass plates.

The symmetric chevron is here the only kind of a tilted layer structure which preserves the anchoring condition at a surface. The $smectic-C$ chevron tilt angle δ_C is related to the $smectic-C$ layer spacing $d_C(T)$ via

$$\cos\delta_C(T) = \frac{d_C(T)}{d_A} \tag{10.5.1}$$

where d_A is the smectic layer spacing in the $smectic-A$ phase. It is equal to the layer spacing, imprinted at the surface.

As the layers can tilt in two directions at a given surface, there would be two kinds of symmetric chevrons, as indicated in Fig.10.5.5. As a result of this degeneracy, both types of symmetric chevrons would appear in the cell and the so-called zigzag walls in SSFLC cells separate regions of different directions of spontaneous layer tilt.

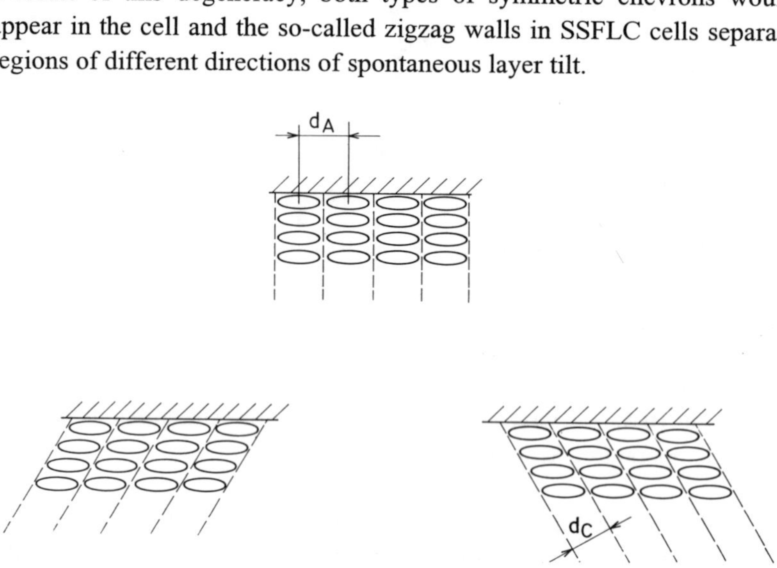

Fig.10.5.5. By cooling the $smectic-A$ phase into the tilted ferroelectric $smectic-C$ phase, the smectic layers shrink because of the tilt. Because the molecules and therefore also the smectic layers at the confining surfaces are positionaly fixed (cannot move), the smectic layers must tilt to accommodate this shrinking. This tilting can occur in two directions, hence two types of symmetric chevrons can be formed.

(ii) ***Asymmetric chevrons*** are formed when a finite layer tilt already exists in the $smectic-A$ phase. Here, the two tilted layer structures from upper and bottom plate do not meet at the center of the cell, but closer to one of the confining surfaces, as shown in Fig.10.5.6. The asymmetric

chevrons can be easily identified due to the asymmetric X-ray diffracted intensity profiles. The discovery of chevron structure in surface stabilized ferroelectric cells had a significant impact on the physics and application of devices based on surface stabilization. Because of technological importance, there is an enormous number of publications, describing various implications of chevron structures.

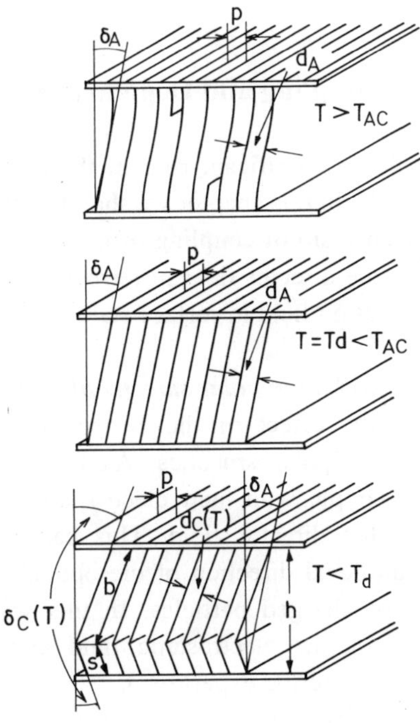

Fig.10.5.6. The formation of an asymmetric chevron. After Rieker et al.(1987).

Here we shall briefly review some of them. X-ray experiments and the chevron structure can be found in the publications of Findon et al.(1998), Gleeson and Morse (1996), Ouchi et al.(1988), Srajer et al.(1991) and Willis et al.(1992). Experiments on optical and switching properties of chevron cells are discussed in the papers of Hark et al.(1988), Hiji et al.(1988a-b), Itoh et al.(1992), Maclennan et al.(1990), Ouchi et al.(1987), Sako et al.(1997), Towler et al.(1991),Yamada et al.(1989), Yang and Chieu (1988) and Zhuang et al.(1990, 1991). Theoretical investigations are treated in the work of de Meyere and Dahl (1994), Kralj et al.(1994, 1998), Limat and Prost (1993), Vaupotič et al.(1996) and Yoneya et al.(1989). For a general review, the reader is advised to look at the recently published review of S.T.Lagerwall (1998).

10.6. Ion-Director Coupling and Depolarization Field in SSFLCD

The ionic impurities play a significant role in the physics and application of ferroelectric liquid crystals. In this chapter we shall briefly discuss two important phenomena: *(i)* The mechanism of coupling of the electrostatic (Coulomb) field of ionic impurities with the director field in ferroelectric liquid crystals and *(ii)* The problem of depolarization field in surface stabilized ferroelectric displays.

(i) Ion-director coupling in polar smectic phases.

Let us consider the effect of the Brownian motion of ions on the collective modes in polar smectics. As they drift through the liquid crystal, the ions couple with their electrostatic field to the surrounding polarization and the director field. This causes small but observable fluctuations of the local direction of the optical axis, which reflects the random motion of charged particles. In this way, the motion of ions becomes visible via the electrostatic field and can be observed, for example in a light scattering experiment.

For simplicity, we shall consider a non-chiral $smectic-C$ with the normal to the smectic layers in the z-direction and the projection of the director ξ on the smectic plane makes an angle Φ with respect to the x-axis. We consider the equilibrium concentration of the ionic impurities is c_i and the fluctuation of the local concentration is δc_i, as these ionic impurities follow a Brownian motion. The calculations for the ion-polarization coupling in helical $smectic-C^*$ phase are much more tedious, but can in principle be done following the same procedure. The total free energy density of the fluctuations in this system is:

$$g(\vec{r}) = g_d(\vec{r}) + g_c(\vec{r}) + g_e(\vec{r}) \tag{10.6.1}$$

Here $g_d(\vec{r})$ is the contribution of the director fluctuations to the total free energy, $g_c(\vec{r})$ results from the ionic impurities and $g_e(\vec{r})$ is the electrical contribution of the total charge density to the total free energy.

The director contribution to the free energy density is in the adiabatic approximation (Eq.2.4.11a)

$$g_d = g_A + \tfrac{1}{2}\tilde{a}\left(\xi_x^2 + \xi_y^2\right) + \tfrac{1}{4}b\left(\xi_x^2 + \xi_y^2\right)^2 + \tfrac{1}{2}K_3\left(\left(\frac{d\xi_x}{dz}\right)^2 + \left(\frac{d\xi_y}{dz}\right)^2\right) \tag{10.6.2}$$

Here, the Lifshitz term has been omitted, because it equals zero for the non-chiral system we are considering. $\delta\xi_x$ and $\delta\xi_y$ are the components of the order parameter fluctuations and $a = \alpha(T - T_c)$ with $\alpha > 0$ and T_c is the transition temperature. We shall consider only the temperature range below T_c.

The free energy density $g_c(\vec{r})$ of the impurity ions can be calculated from the chemical potential of the ionic impurities :

$$g_c(\vec{r}) = \tfrac{1}{2}k_B T\frac{(\delta c_i(\vec{r}))^2}{c_i} \tag{10.6.3}$$

Here k_B is the Boltzmann constant, T is the absolute temperature and $\delta c_i(\vec{r})$ is the fluctuation of the local impurity concentration c_i.

The electrical contribution to the free energy g_e originates from the electrostatic Coulomb interaction of both real charges $\rho_i(\vec{r})$ of the ionic impurities and fluctuation-induced polarization charges $\rho_P = -div\vec{P} = -\nabla\cdot\vec{P}(\vec{r},t)$ and is given by

$$g_e = \tfrac{1}{2}\rho(\vec{r})U(\vec{r}) \tag{10.6.4}$$

Here, $\rho(\vec{r}) = \rho_i(\vec{r}) + \rho_P(\vec{r}) = e_i\delta c_i(\vec{r}) - \vec{\nabla}\vec{P}$, and $U(\vec{r})$ is the local electric potential that satisfies the Poisson equation

$$\sum_{\alpha,\beta}\partial_\alpha\left(\varepsilon_{\alpha\beta}\partial_\beta U(\vec{r})\right) = -\frac{\rho(\vec{r})}{\varepsilon_\circ} \tag{10.6.5}$$

$\varepsilon_{\alpha\beta}$ is the dielectric tensor for high frequencies, $\varepsilon_\circ$ is the permittivity of free space and $\partial_\alpha \equiv \partial/\partial x_\alpha$. The spontaneous polarization is $P = -P_\circ\xi_y + P_\circ\xi_x$ and $P_\circ$ is the magnitude of the spontaneous polarization. Note that P_o is replaced by the fluctuation induced polarization $P_\circ = C\varepsilon$ in the $smectic - A$ phase. C is the flexoelectric coefficient and ε is dielectric constant of the $smectic - A$ phase.

Following the standard approach, we expand the fluctuations $\delta\xi_x$ and $\delta\xi_y$ of the order parameter in terms of plane waves

$$\delta\xi_x = \sum_{\vec{q}}\delta\xi_{x,q}\cdot e^{i\vec{q}\vec{r}} \tag{10.6.6a}$$

$$\delta\xi_y = \sum_{\vec{q}}\delta\xi_{y,q}\cdot e^{i\vec{q}\vec{r}} \tag{10.6.6b}$$

Here, $\vec{q} = \left(q_x \quad q_y \quad q_z\right)$ is the wave-vector of the fluctuation and $\delta\xi_{x,-\vec{q}} = \delta\xi^*_{x,q}$ to keep the physical variables real. Similarly, we introduce Fourier expansions for the concentration fluctuations

$$\delta c_i(\vec{q}) = \int\delta c_i(\vec{r})\cdot e^{i\vec{q}\vec{r}}\cdot d^3\vec{r} \tag{10.6.7}$$

and for the electric potential

$$U(\vec{q}) = \int U(\vec{r})\cdot e^{i\vec{q}\vec{r}}\cdot d^3\vec{r} \tag{10.6.8}$$

The electric field contribution to the total free energy is

$$G_e = \tfrac{1}{2}\sum_{\vec{q}} \frac{\rho(\vec{q})\rho(-\vec{q})}{\varepsilon\varepsilon_\circ q^2} \tag{10.6.9}$$

whereas the contribution from the concentration fluctuations is

$$G_c = \tfrac{1}{2}\sum_{\vec{q}} k_B T \frac{\delta c_i(\vec{q})\delta c_i(-\vec{q})}{c_i} \tag{10.6.10}$$

Together with the director field contribution, the total free energy of fluctuations can be represented as

$$G = \int_V g(\vec{r})d^3\vec{r} = \tfrac{1}{2}\sum_{\vec{q}} \delta\psi^+ \underline{D}(\vec{q})\delta\Psi_q \tag{10.6.11}$$

Here $\delta\Psi_q = [\delta\xi_{x,q} \quad \delta\xi_{y,q} \quad \delta c_i(q)]$ is the three-component wave-vector composed of director fluctuations and concentration fluctuations. The dynamical matrix $\underline{D}(\vec{q})$ is in this case

$$\underline{D}(\vec{q}) = \begin{bmatrix} A_q & -D_q & +iF_q \\ -D_q & B_q & -iE_q \\ -iF_q & +iE_q & C_q \end{bmatrix} \tag{10.6.12}$$

and describes the increase of the free energy when a coupled director-ionic fluctuation with a wave-vector $\vec{q}$ is created in a system. The matrix elements are

$$A_q = -a(T) + \tfrac{1}{2}K_3 q^2 + \frac{P_\circ^2}{2\varepsilon\varepsilon_\circ}\frac{q_y^2}{q^2} \tag{10.6.13a}$$

$$B_q = \tfrac{1}{2} K_3 q^2 + \frac{P_\circ^2}{2\varepsilon\varepsilon_\circ} \frac{q_x^2}{q^2} \tag{10.6.13b}$$

$$C_q = \frac{e_i^2}{2\varepsilon\varepsilon_\circ q^2} + \tfrac{1}{2} \frac{k_B T}{c_i} \tag{10.6.13c}$$

$$D_q = \frac{P_\circ^2}{2\varepsilon\varepsilon_\circ} \frac{q_x q_y}{q^2} \tag{10.6.13d}$$

$$E_q = \frac{e_i P_\circ}{2\varepsilon\varepsilon_\circ} \frac{q_x}{q^2} \tag{10.6.13e}$$

$$F_q = \frac{e_i P_\circ}{2\varepsilon\varepsilon_\circ} \frac{q_y}{q^2} \tag{10.6.13f}$$

The collective dynamics is calculated following the same procedure as in Chapter 2. We use the Landau-Khalatnikov equations for director fluctuations

$$\frac{\partial G}{\partial \xi_{-q}} = -\gamma \frac{\partial \delta \xi_q}{\partial t} \tag{10.6.14}$$

where γ is the rotational viscosity. For the dynamics of the ionic impurities we use the continuity equation

$$\frac{\partial \delta c_i(q)}{\partial t} = -\vec{\nabla} \cdot \vec{j}_i = -\vec{\nabla} \cdot \left(m_i c_i \vec{\nabla} \left(\frac{\partial G}{\partial \delta c_i(-q)} \right) \right) \tag{10.6.15}$$

which can be rewritten in the form of a generalized diffusion equation for charge:

$$\frac{\partial G}{\partial \delta c_i(-q)} = -\frac{1}{m_i c_i q^2}\frac{\partial \delta c_i(q)}{\partial t} \tag{10.6.16}$$

m_i is the mobility of the impurity ions and is connected to the ionic mobility μ_i in an external electric field by the equation $\mu_i = e_i \cdot m_i$ where e_i is the electric charge of an ion.

After considering the time dependence of the fluctuations as $\delta\xi_q = \delta\xi_{q,0} Exp(-t/\tau(q))$, the dynamics of the system reduces to a set of coupled linear equations that can be represented in a matrix form:

$$\begin{bmatrix} 2A_q - \gamma/\tau(q) & -2D_q & +2iF_q \\ -2D_q & 2B_q - \gamma/\tau(q) & -2iE_q \\ -2iF_q & +2iE_q & 2C_q - 1/\left(m_i c_i q^2 \tau(q)\right) \end{bmatrix} \cdot \begin{bmatrix} \delta\xi_{x,q} \\ \delta\xi_{y,q} \\ \delta c_{i,q} \end{bmatrix} = 0$$

(10.6.17)

The matrix elements are given by the Eq.10.6.13(a-f) and $\tau(q)$ is the unknown relaxation time for a coupled director-ion fluctuation with a wave-vector $\vec{q}$. The system of coupled equations (Eq.10.6.17) has a non-trivial solution if the corresponding determinant vanishes

$$\begin{vmatrix} 2A_q - \gamma/\tau(q) & -2D_q & +2iF_q \\ -2D_q & 2B_q - \gamma/\tau(q) & -2iE_q \\ -2iF_q & +2iE_q & 2C_q - 1/\left(m_i c_i q^2 \tau(q)\right) \end{vmatrix} = 0 \tag{10.6.18}$$

For a given wave-vector $\vec{q}$ we obtain three unknown relaxation rates, which are the solution of the above cubic equation for $\tau^{-1}(q)$. These solutions represent three distinct dispersion branches of the excitation spectrum of our system. We shall now consider three limiting situations, which will give physical insight into the nature of these solutions.

First, let us consider the case, where there is no spontaneous polarization in the system, i.e. $P_\circ = 0$. The determinant (Eq.10.6.18) which describes the collective excitations reduces in this case to (see coefficients given by Eq.10.6.13(a-f))

$$\begin{vmatrix} 2A_q - \gamma/\tau(q) & 0 & 0 \\ 0 & 2B_q - \gamma/\tau(q) & 0 \\ 0 & 0 & 2C_q - 1/\left(m_i c_i q^2 \tau(q)\right) \end{vmatrix} = 0 \tag{10.6.19}$$

The determinant is diagonal and it is straightforward to find the three relaxation rates. The first branch of excitations is the amplitudon (see also Eqs.2.4.15.)

$$\tau_{ampl}^{-1}(q) = 2\frac{\alpha(T_c - T)}{\gamma} + \frac{K_3}{\gamma} q^2 \tag{10.6.20}$$

The second branch of excitations is the gapless branch of phason modes

$$\tau_{phason}^{-1}(q) = \frac{K_3}{\gamma} q^2 \tag{10.6.21}$$

whereas the third branch of excitations describes the Brownian motion of ionic impurities

$$\tau_{ion}^{-1}(q) = k_B T m_i q^2 + \frac{m_i c_i e_i^2}{\varepsilon\varepsilon_\circ} \tag{10.6.22}$$

The fluctuation spectrum of the ionic impurities is therefore parabolic in $\vec{q}$ space and has a finite gap (the last term), which is due to the repulsive electrostatic interaction of like-charges. By assuming typical values of electric mobility, $\mu_i = 8.5\times10^{-11} m^2V^{-1}s^{-1}$ and concentration of $c_i \approx 10^{26}\ ions/m^3$, the room temperature relaxation rate is found to be $10-100Hz$ at the wave-vector that corresponds to visible light. One should however note, that ionic motion by itself is not visible, for example, in a light scattering experiment. It becomes visible, after we include in our calculations the coupling of ions with spontaneous polarization and consequently to the director field and the optical tensor.

Second, let us consider the case where there are no ions in the system, but we have the spontaneous polarization and the system is fluctuating. In this case, the determinant (Eq.10.6.18) is of dimension 2×2

$$\begin{vmatrix} 2A_q - \gamma/\tau(q) & -2D_q \\ -2D_q & 2B_q - \gamma/\tau(q) \end{vmatrix} = 0 \tag{10.6.23}$$

We have now two relaxation modes, which represent the coupled motion of phase and amplitude fluctuations. The coupling term D_q vanishes when either $q_x = 0$ or $q_y = 0$ and we obtain two polarization-renormalized relaxation branches. For $q_x = 0$ we obtain the amplitudon

$$\tau^{-1}_{ampl}(q) = \frac{2\alpha(T_c - T)}{\gamma} + \frac{K_3}{\gamma} q^2 + \frac{P_\circ^2}{\gamma\varepsilon\varepsilon_\circ} \tag{10.6.24}$$

which has an additional gap (the last term), whereas for $q_y = 0$ we obtain a phason branch of excitations, which has a finite gap due to the spontaneous polarization

$$\tau^{-1}_{hason}(q) = \frac{K_3}{\gamma} q^2 + \frac{P_\circ^2}{\gamma\varepsilon\varepsilon_\circ} \tag{10.6.25}$$

We have obtained the same result as in Section 4.4. (see Eq.4.4.16). Note that the gap in the originally gapless dispersion relation appears due to splay like fluctuations of the polarization field, which results in the appearance of fluctuation-induced charges. These interact via Coulomb interaction and result in a frequency gap $P_\circ^2/\gamma\varepsilon\varepsilon_\circ$.

Third, let us now consider the most general case, where there is a spontaneous polarization as well as fluctuating ions in the tilted smectic ferroelectric phase. Without loss of generality we shall choose the fluctuations with the wave-vector $\vec{q} = (q_x \quad 0 \quad 0)$ in the x-direction. The determinant (Eq.10.6.13) is now

$$\begin{vmatrix} 2A_q - \gamma/\tau(q) & 0 & 0 \\ 0 & 2B_q - \gamma/\tau(q) & -2iE_q \\ 0 & +2iE_q & 2C_q - 1/\left(m_i c_i q^2 \tau(q)\right) \end{vmatrix} = 0 \qquad (10.6.26)$$

and can be decomposed in three distinct branches of excitations. The first root of the determinant (Eq.10.6.26) gives us the polarization-renormalized amplitudon branch of excitations

$$\tau_{ampl}^{-1}(q) = \frac{2\alpha(T_c - T)}{\gamma} + K_3 q^2 + \frac{P_\circ^2}{\varepsilon\varepsilon_\circ} \qquad (10.6.27)$$

The two other branches of excitations are given by the roots of the 2×2 determinant

$$\begin{vmatrix} 2B_q - \gamma/\tau(q) & -2iE_q \\ +2iE_q & 2C_q - 1/\left(m_i c_i q^2 \tau(q)\right) \end{vmatrix} = 0 \qquad (10.6.28)$$

and represent coupled fluctuations of the phase of the director field with the ionic fluctuations. For weak coupling, the approximate solutions of the above determinant are

$$\tau_{phas}^{-1} = \frac{K_3}{\gamma} q^2 + \frac{P_\circ^2}{\gamma\varepsilon\varepsilon_\circ} + \Delta \qquad (10.6.29)$$

and

$$\tau_{ion}^{-1} = k_B T m_i q^2 + \frac{m_i c_i e_i^2}{\varepsilon\varepsilon_\circ} - \Delta \qquad (10.6.30)$$

where

$$\Delta = \frac{2m_i c_i e_i P_\circ q}{\frac{K_3}{\gamma} q^2 + \frac{P_\circ^2}{\gamma \varepsilon \varepsilon_\circ} - k_B T m_i q^2 - \frac{m_i c_i e_i^2}{\varepsilon \varepsilon_\circ}} \qquad (10.6.31)$$

is the splitting between the two branches. We see, that a coupling between the ionic and director fluctuations results in renormalization of both relaxation rates. Moreover, due to this coupling, the eigenmodes of the system are not any more pure director and ionic fluctuations, but mixed fluctuations. This means, that the ionic motion becomes visible via the electrostatic coupling with director modes and can be observed in a light scattering experiment or any other optical experiment that is sensitive to small fluctuations of the local index of refraction.

The intensities of these modes can be calculated from the equipartition theorem. This gives a Curie-Weiss law for the intensity of the amplitudon modes, a temperature independent intensity of the phason modes mode and a weakly temperature dependent intensity of the ionic modes. Similar calculations can be performed for the $smectic - A$ phase. It is interesting to note that the intensity of light scattered by ionic mode, which is coupled to the soft mode, is proportional to the inverse square of $(T - T_c)$.

(ii) Depolarization and ionic field in SSFLC

The principle of operation of SSFLC is based on the linear coupling between the local polarization $P(\vec{r})$ and the local electric field $\vec{E}(\vec{r})$. This local electric field in a SSFLC device can be expressed as a sum of external electric field $\vec{E}_{ext}$, the depolarization field E_{dep} of a uniformly polarized ferroelectric material and the ionic field $\vec{E}_{ion}$ that originates from the ionic impurities in the sample:

$$\vec{E} = \vec{E}_{ext} + \vec{E}_{dep} + \vec{E}_{ion} \tag{10.6.32}$$

It is important to note that for a given value of the external electric field $\vec{E}_{ext}$, which is determined by the voltage that is applied across the SSFLC cell, the system will tend to decrease the total electric field $E(\vec{r})$. This will lower the density of the electrostatic energy of the system, which is proportional to $\vec{E}\vec{D}$. The two other fields E_{dep} and E_{ion} will therefore tend to be directed against the external electric field.

The depolarization field originates from the uniform polarization of a ferroelectric material. If we consider a thin layer of a ferroelectric material, enclosed between two flat electrodes, as shown in Fig.10.6.1., the depolarization field originates from the surface charge density due to spontaneous polarization $P_\circ$

$$E_{dep} = \frac{\sigma}{\varepsilon_\circ} = \frac{P_\circ}{\varepsilon_\circ} \tag{10.6.33}$$

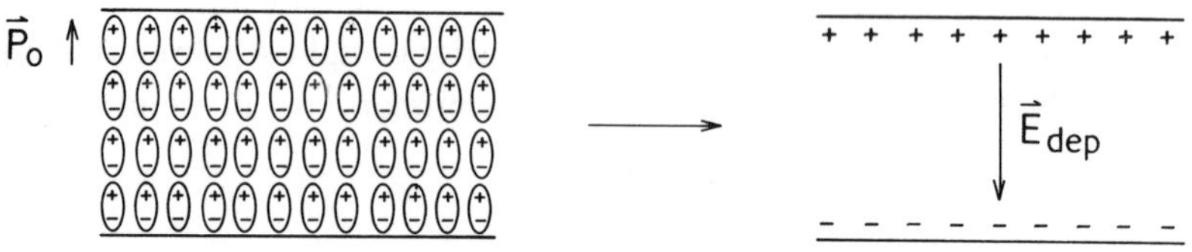

Fig.10.6.1. The depolarization field E_{dep} originates from the surface charge density due to a homogeneous spontaneous polarization. The direction of this field is opposite to the direction of the spontaneous polarization.

The ionic field is created because of the spatially inhomogeneous distribution of electric charges which are always present in liquid crystalline materials. They originate either from the impurities and dissociation of liquid crystalline material or are injected from the electrodes when a strong electric field is applied.

Let us assume that we have equal number of positive and negative ions in our system, so that we assure electroneutrality. If these ions are randomly distributed in the system, they do not give rise to any macroscopic electric field. However, when an external electric field is applied, the two ionic species drift in opposite directions, as they are driven by the electric force. The drift of ions results in accumulation of negative charges on one side and positive charges on the other side. Due to this spatial separation, they create their own macroscopic electric field, which is opposed to the direction of the external electric field, as shown in Fig.10.6.2.

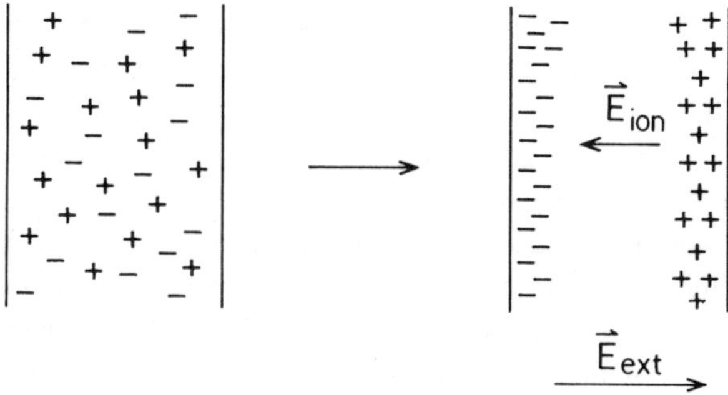

Fig.10.6.2. Under the influence of an external electric field, the ions separate and drift in opposite directions. As a result, they build the depolarization (or screening) field that is in the opposite direction with respect to external electric field.

The time that is required to build up the ionic field, depends on the mobility μ of the ions, distance d they have to travel and the strength of the electric field E

$$\tau_{ion} = \frac{d}{\mu E} \tag{10.6.34}$$

For $d = 1.5\mu m$, typical mobility $\mu = 10^{-10} m^2 s^{-1} V^{-1}$ and $U = 10V$ applied across the cell, the drift time for the ions is of the order of several milliseconds. Let us consider now what happens with the electric field and the electric potential in a liquid crystalline cell with ionic impurities, when we apply a switching electric voltage across the electrodes of the cell, as shown in Fig.10.6.3. When the voltage is switched on (Fig.10.6.3(a)), the potential rises linearly across the cell, which means that we have a uniform electric field across the electrodes. At this moment, the ions start to drift in opposite directions and build up a screening, depolarizing ionic field. This is reflected in a decrease of the slope of the electric potential versus distance, see Fig.10.6.3(b) and in the increase of the electric field across the aligning insulating layer. For a long enough time, the ions screen-out completely the internal field in the cell, where the electric potential is now constant, as shown in Fig.10.6.3(c).

When the electrodes are short-circuited, the electric potential of both electrodes becomes the same, as shown in Fig.10.6.3(d). As a result, a strong electric field is suddenly created due to the ionic field, which is still there. Note that this field is in the opposite direction with respect to the previously applied electric field. This reversed electric field due to a build-up of the ionic charges than slowly fades-away, as the ions drift thermally and the charge distribution becomes spatially uniform again. This means, that during the switching of the voltage across the SSFLC, a reverse ionic field can be created, if the ions have had enough time to migrate across the cell. This reverse electric field results in back-switching of the electric polarization and the SSFLC structure is monostable due to electrostatic reasons. These problems are more pronounced in ferroelectric liquid crystals with a high spontaneous polarization (see Dijon, 1992).

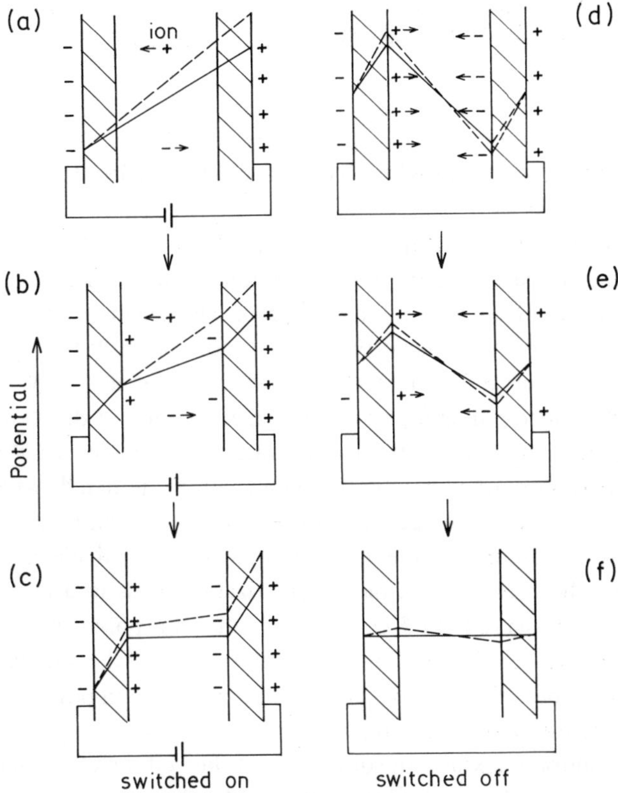

Fig.10.6.3. The time evolution of the electric potential in a liquid crystalline cell with ionic impurities. The hatched region is the dielectric aligning layer. The vertical axis denotes the electric potential and the horizontal axis is along the cell thickness. The switch-on process is shown on the left, the switch-off process on the right side. After Mada and Osajima (1986).

Different aspects of the problem of ionic impurities in liquid crystals can be found in the work of Mada and Osajima (1986), Sugimura et al. (1989), Valenti et al. (1990), Barbero and Durand (1990) and others. The specific problems of the switching of ferroelectric thin layers in SSFLC geometry in the presence of mobile ionic impurities can be found in the work of Yang et al.(1989), Saxena et al. (1990), Zou et al. (1991), Maximus et al.(1991) and others.

10.7. Towards Complexity: Volume Stabilization by Polymers

The concept of "volume stabilization" of the molecular ordering in ferroelectric liquid crystals is based on the polymerization of a small amount of polymer precursor that is added to the liquid crystal. The technique was first used by Mariani *et al.*(1986) for smectic phases and more recently by Hikmet *et al.* for the nematic liquid crystal (Hikmet 1990, 1991, Hikmet and de Witz, 1991). The amount of the added polymer precursor is small enough so that it does not affect significantly the degree of molecular ordering or alignment of the liquid crystal. The polymerization of the polymer precursor is initiated when the liquid crystalline film acquires the desired orientation, which can be determined either by boundary conditions or the interaction with external electric or magnetic fields. Under these conditions, the polymerization leads to a highly anisotropic texture. It was observed (Pirš *et al.*, 1992) that the polymer forms a network through the entire liquid crystal and "volume stabilizes" the structure. It was also observed that a part of the polymer phase separates upon polymerization and is expelled from the bulk of the liquid crystal to the bounding surfaces where it forms an anisotropic texture. In some parts of the cell the phase separation leads to a spontaneous nucleation and formation of droplets that grow from the surface into the bulk , eventually join the other surface and thus act as spacers.

As a result of polymerization and phase separation, one thus obtains a polymer network that is built into the liquid crystalline structure and eventually binds the two confining surfaces. This has very important technological implications, as the polymer structure significantly hinders the flow of the liquid crystal under external mechanical stress and thus preserves the director configuration of smectic liquid crystalline phases that are well known for their high sensitivity to external forces and mechanical shock.

The volume stabilized ferroelectric and antiferroelectric liquid crystalline cells are prepared essentially in the same way as conventional surface stabilized devices. ITO covered glass plates are spin- or dip-coated with a Nylon orienting layer that is uniformly rubbed in one direction. The thickness of the cells is of the order of 1-3μm. A small amount (0.5 to 3wt%) of monomeric acrylate (Desolite D044 or similar) is added in the isotropic phase of liquid crystals. The cells are filled with a liquid crystal-monomer solution and then slowly cooled down to the $smectic-C^*$ phase. The helical period of most commercial ferroelectric liquid crystal mixtures is around 10μm so that a cell thickness of 1-3μm is sufficient to surface-unwind the helix. Together with homogeneous boundary conditions this usually results in a homogeneous chevron molecular ordering in the ferroelectric phase. Using strong (80V), slowly alternating (≈3Hz) electric field pulses, a well-known electric field induced stripe texture (Clark and Rieker, 1988, Shao *et al.*, 1991) can be obtained in pure as well as polymer-doped samples. The angle between the directions of extinction of the two neighboring stripe-like domains is several degrees. This results in high contrast, very good bistability and very high angle between the optical axis of the two bistable states. As the liquid crystal-monomer solution is filled in the isotropic phase of the liquid crystal, a part of the polymer precursor phase-separates, when the solution is cooled down to the smectic phase. This phase separation is optically observed as the appearance of small (<1 μm) microdroplets, that condense at the two surfaces.

The polymerization of the polymer precursor is induced in the ferroelectric phase using an UV light source for a few minutes. As a result of polymerization, polymer phase separates mainly in the vicinity of the domain walls. Defects in the structure thus act as polymerization centers, which nucleate the growth of the phase-separated polymer. This observation can be also proved by heating the ferroelectric liquid crystal into the isotropic phase and then cooling it down to the ferroelectric $smectic-C^*$ phase. The structure that reappears in the $smectic-C^*$ phase is identical to the one observed before heating-up the cell. This indicates, that the polymer structure is indeed formed in the disordered regions close to domain walls.

The surface of the confining glass plates can be conveniently analyzed using electron microscopy and Atomic Force Microscopy (AFM). The cells are disassembled and flushed with solvent, thus removing the liquid crystal and hopefully preserving the polymer network. The quasi-ordered polymer network can be observed by electron microscopy as shown in Fig.10.7.1. One can clearly see that the preferential direction of the ordering of the network globally matches the direction of the smectic layer normal and thus also the direction of domain walls. The separation between the massive polymer structure is of the order of a few microns which is consistent with the observed width of the stripe-like

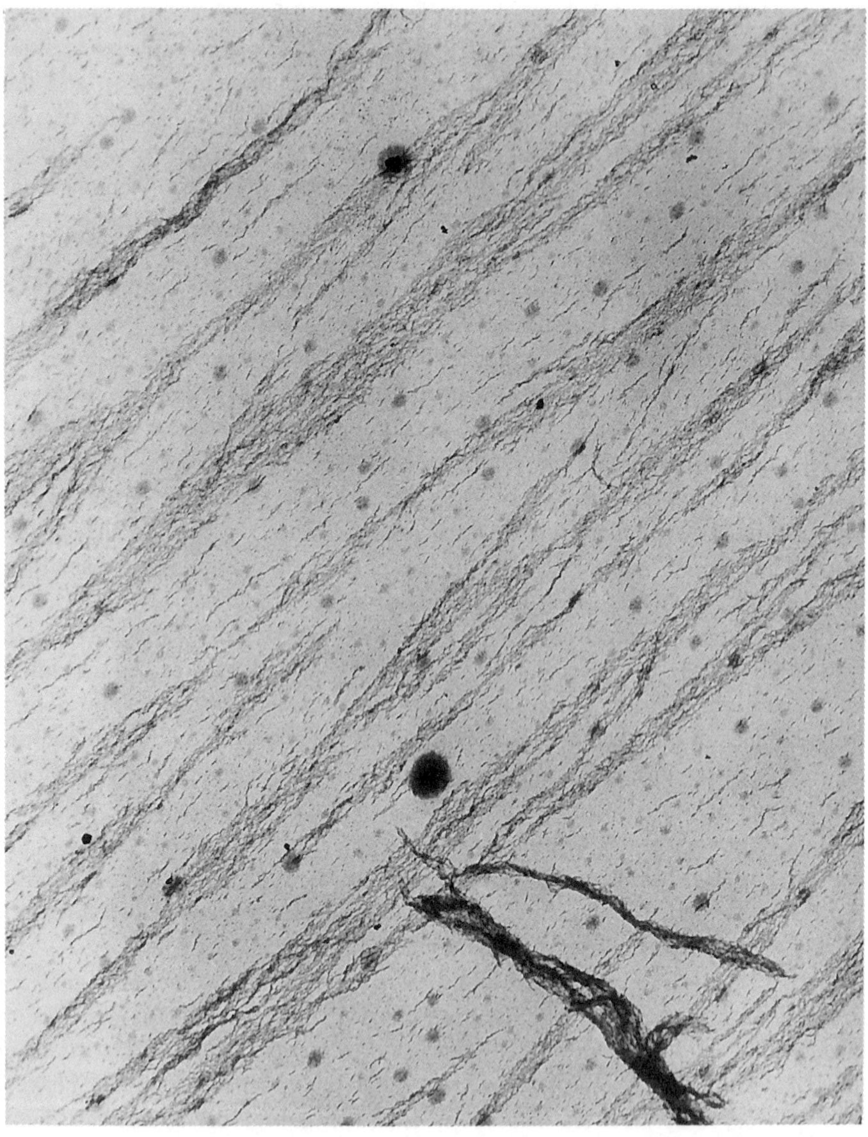

Fig.10.7.1. TEM micrograph of a the ordered polymer texture formed during UV activated polymerization process in the polymer-ferroelectric liquid crystal mixture.

ferroelectric domains. One can notice in Fig.10.7.1. the presence of microdroplets and a very fine structure between them. This fine structure can be further analyzed with the Atomic Force Microscopy to reveal the structural details on the nanometer scale.

The results of the AFM study of surfaces of disassembled volume stabilized liquid crystalline cells are shown for different length-scales in Fig.10.7.2., 10.7.3., and 10.7.4., respectively.

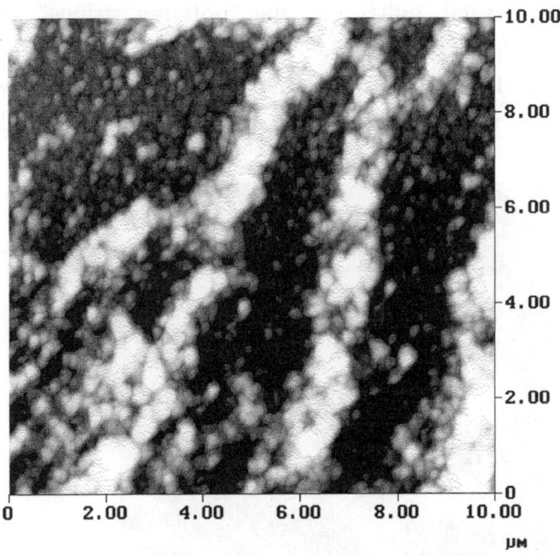

Fig.10.7.2. AFM image of the polymer texture formed on the surface of confining glass plates. Height between dark and white regions is 200nm.

Figure 10.7.2. shows a $10x10\mu m^2$ region that clearly reveals local preferential direction of the polymer network. The width of the "polymer stripes" is of the order of a micron and the height is of the order of 100nm. This indicates that the observed polymer stripes are the remnants of polymer that was phase separated in the region of the domain walls. Figure 10.7.3. shows the enlarged part of the surface in Fig.10.7.2. One can see a fine-grained structure that is present in between the polymer strips. The width of the grains is of the order of 100nm, whereas the height is of the order of 20nm. The grains are elongated in the direction of the smectic layer normal. Fig.10.7.4. shows an enlarged part of the surface, shown in Fig.10.7.3. Here, one can see more clearly the anisotropic shape of the "polymer hills" and the dimensions are in agreement with the objects, shown in the previous figure. We have to stress, however, that on this length- and height-scale, the AFM image changes with time of scanning, indicating that the surface is perturbed on the nanometer scale even in the low-force regime. This indicates that the polymer surface is rather soft and not very compact. From the above measurements we can estimate the distribution and the amount of the polymer material that is phase separated from the ferroelectric phase. A rough calculation shows that almost all of the polymer is expelled from the bulk of liquid crystal, either into the "polymer stripes", microdroplets or surface-condensed "polymer hills". Approximately 30% of the material is condensed at the surface, 30% in the disordered regions in between the stripe-like domains and the rest is phase separated in the form of microdroplets.

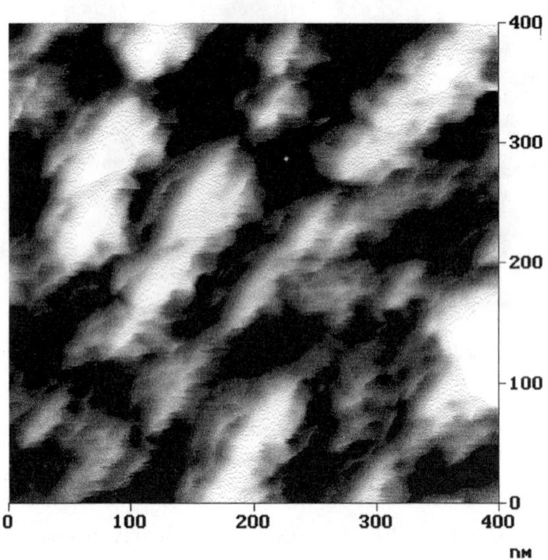

Fig.10.7.3. Enlarged AFM image of the polymer texture shown in Fig.10.7.2.

The polymer phase separation, which is induced in the smectic phases leads to the formation of three different structures on the nanometer scale. Due to the presence of wall defects and disordered regions in smectic phases, a strip-like polymer network is phase-separated, extending between the two confining surfaces. In addition, a large amount of the polymer "condenses" at the confining surfaces, where it forms "hill-like" structures on the 10 nm scale. In some parts of the sample, spontaneous growth of phase-separated microdroplets eventually leads to a formation of large droplets, binding the two surfaces firmly together. Whereas the appearance of these large microdroplets is important from the technological point of view (Pirš *et al.*, 1992), the appearance of a surface-condensed polymer may be important as it creates a random surface. Here, the extent of the randomness of the confining surfaces could be externally controlled by the amount of the polymer that phase separates at the surface.

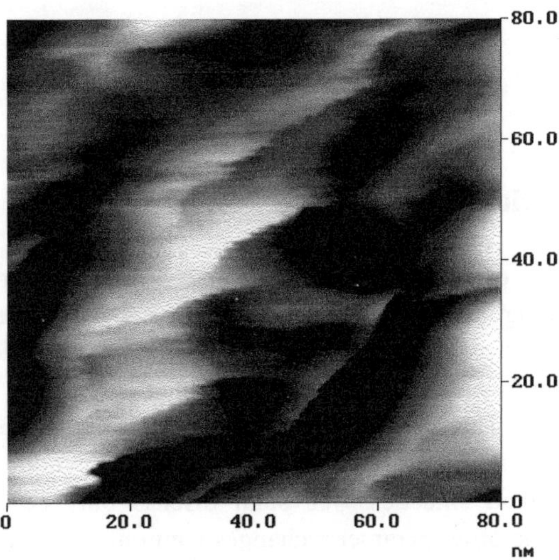

Fig.10.7.4. Enlarged AFM image of the polymer texture shown in Fig.10.7.3.

Appendix A

Landau Theory of Second Order Phase Transitions

A1. Introduction

The Landau theory of phase transitions is based on the concept of an order parameter describing the change in the system on going through the phase transition. The order parameter η for the transition in question is a physical quantity which equals zero above the transition and is non-zero below the phase transition point T_c. This can be for example the electric polarization $\vec{P}$ in ferroelectric materials, the magnetization $\vec{M}$ in ferromagnetic materials or the tilt angle θ in tilted smectics. We shall discuss only second order phase transitions where the order parameter changes continuously in the vicinity of the phase transition. Furthermore we shall consider the case where the phase of lower symmetry is spatially inhomogeneous in one direction only, i.e. $\eta = \eta(z)$.

The basic idea of the Landau theory is that the non-equilibrium thermodynamic potential $\Phi(p,T)$ of a system is also a function of the order parameter η, $\Phi = \Phi(p,T,\eta)$. Here, p is the pressure and T is the temperature of a system. This potential is in most cases the free energy G

$$G = U - T \cdot S \tag{A1.1}$$

where U is the internal energy and S is the entropy of the system. The minimum of the internal energy corresponds to an ordered state, and the maximum of the entropy to a disordered state. For a given temperature T, the system will therefore settle in a minimum of the free energy G. At low temperatures, $T \to 0$, the internal energy will be dominant and the system would normally be in an ordered state. On the other hand, at high temperatures, $T \to \infty$, the entropy term will prevail and the system will be in a disordered state.

A2. Equilibrium properties of the Landau theory

For simplicity, let us assume that the order parameter η is a one-component vector. We shall also assume that the order parameter is a function of the spatial coordinate z. This will allow us to consider spatially inhomogeneous fluctuations of the order parameter as well as inhomogeneously ordered phases. Because we are discussing second order phase transition, the order parameter $\eta(T)$ changes continuously across the second order transition and therefore obtains arbitrarily small values in the vicinity of the transition temperature T_c. We assume that the nonequilibrium free energy density $g = dG/dV$ can be expanded near T_c in terms of the order parameter η and its derivative $d\eta/dz = \nabla\eta$

$$g(\eta,\nabla\eta) = g_\circ + a(T)\cdot\eta^2 + b\eta^4 + d(\nabla\eta)^2 - h\cdot\eta \qquad \text{(A2.1)}$$

Here, $g_\circ$ is the free-energy density of the high-symmetry phase $\eta = 0$ which occurs for $T > T_c$. We also assume that the coefficient of the quadratic term is linearly temperature dependent, $a = \alpha(T - T_c)$, whereas the rest of the coefficients is constant. We also assume that there the symmetry $\eta \leftrightarrow -\eta$ forbids the odd powers of the order parameter in the expansion (A2.1). The fourth term $d(\nabla\eta)^2$ represents the increase of the free energy density due to fluctuations of the order parameter. The last term describes the coupling of the order parameter η to an external field h, which is the ***conjugate field to the order parameter***. In ferroelectric liquid crystals, the conjugate field is for example an external electric field that couples linearly to the polarization.

Let us consider the role of the signs of individual coefficients in the expression A2.1. This can be done by plotting the free energy density as a function of η and $\nabla\eta$. It is straightforward to see that the coefficient of the fourth order b should be positive because of the stability of the free energy. If this coefficient is for some reason zero or negative, higher-order positive terms have to be taken into account, to keep the system stable.

(i) For positive values of the coefficients $a > 0$, $b > 0$ and $c > 0$, the function $g(\eta,\nabla\eta)$ has for $h = 0$ a minimum at $\eta = 0$ and $\nabla\eta \equiv 0$, as shown in Fig.A2.1. In equilibrium, the system would settle at the bottom of this nearly parabolic pot. This would correspond to the disordered, spatially homogeneous phase with $\eta = 0$ and $\nabla\eta \equiv 0$.

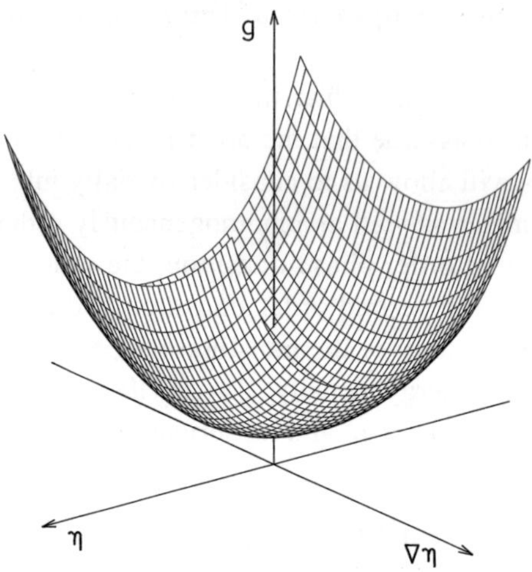

Fig.A2.1. For $a > 0$, $b > 0$ and $c > 0$, the minimum of the free-energy density (Eq.A2.1.) for $h = 0$ corresponds to a spatially homogeneous ($\nabla\eta \equiv 0$), disordered phase ($\eta = 0$).

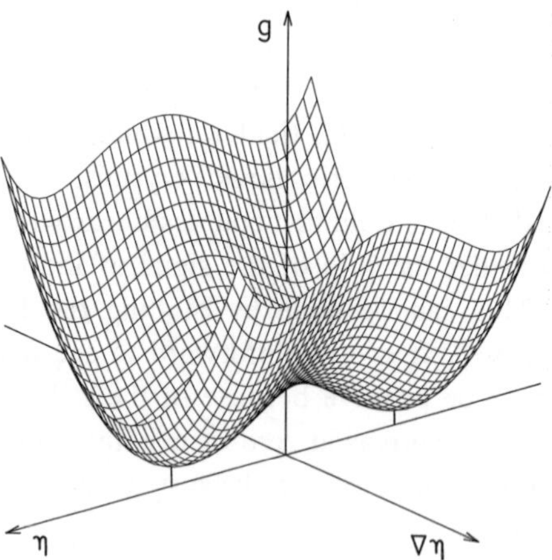

Fig.A.2.2. For $a < 0$, $b > 0$ and $c > 0$, the minimum of the free-energy density (Eq.A2.1.) corresponds for $h = 0$ to a spatially homogeneous ($\nabla\eta \equiv 0$), ordered phase ($\eta \neq 0$).

(ii) In the case when the coefficient of the quadratic term becomes negative, $a < 0$, the free energy density has for $h = 0$ a minimum at $\eta \neq 0$ and $\nabla\eta \equiv 0$ and the system is in the ordered homogeneous phase, as shown in Fig.A2.2. The temperature, where the coefficient a changes sign is therefore the phase transition temperature between the disordered $(\eta = 0)$ and ordered $(\eta \neq 0)$ phase. The evolution of the free-energy density as a function of temperature is shown in Fig.A2.3.

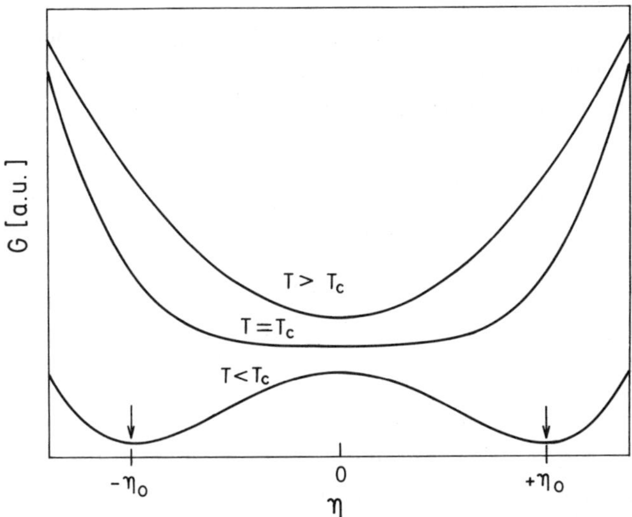

Fig.A2.3. The evolution of the free-energy density (Eq.A2.1) in the vicinity of the second order phase transition temperature T_c.

The inhomogeneous term $d \cdot (\nabla\eta)^2$ has a special role in the free-energy expansion (Eq.A2.1). For positive values of the coefficient $d > 0$, this term damps-out spatial variations of the order parameter. This can be easily understood, because any spatial variation of the order parameter $\eta(z)$ leads to a finite derivative $|\nabla\eta| \neq 0$, which always increases the free-energy density for positive $d > 0$. This means, that the equilibrium of the system corresponds to $\nabla\eta \equiv 0$ and the system is spatially homogeneous. A very interesting situation arises when this term for some reason tends to zero and eventually changes sign at some value of an external parameter. For example, this happens in the case of the Lifshitz point in ferroelectric liquid crystals, when the magnetic field reaches a certain value. Then, for negative values, $d < 0$, the system is unstable with respect to any spatial variation of the order parameter and one has to add a positive term $f\cdot(\nabla^2\eta)^2$ to stabilize the free-energy density. Consequently, this will result in a finite spatial variation of the order parameter even in the equilibrium state and the system will be spatially inhomogeneous. As we have mentioned, the special point where the $d\cdot(\nabla\eta)^2$ term changes sign and at the same time $a(T)\cdot\eta^2$ changes sign, is the so-called Lifshitz point. It was first introduced by Hornreich, Luban and Shtrikman (1975) and was discussed in Chapter 3.

Let us further consider the phase transition in the case where $d>0$ and the conjugate field is zero, $h=0$. The free energy of the system is obtained by the integration of the free-energy density over the volume V

$$G=\int_V g(\vec{r})\cdot d^3\vec{r} \tag{A2.2}$$

which gives

$$G=G_\circ+V\cdot\left(a\cdot\eta^2+b\cdot\eta^4\right) \tag{A2.3}$$

In the thermodynamical equilibrium the system settles in the minimum of the free-energy, which corresponds to an equilibrium value $\eta_\circ$ of the order parameter. The conditions for this minimum are

$$\left(\frac{\partial G}{\partial\eta}\right)_{\eta_\circ}=0 \qquad \Rightarrow \qquad \eta_\circ\left(a+2b\eta_\circ^2\right)=0 \tag{A2.4a}$$

$$\left(\frac{\partial^2 G}{\partial\eta^2}\right)_{\eta_\circ}>0 \qquad \Rightarrow \qquad a+6b\eta_\circ^2>0 \tag{A2.4b}$$

Remembering that $a(T)$ is positive above T_c, we see from (A2.4a and b) that the solution $\eta_\circ=0$ is stable for $T>T_c$:

$$T>T_c \quad : \qquad \eta_\circ(T)=0 \tag{A2.5}$$

whereas below T_c a finite order parameter appears:

$$T < T_c \quad : \qquad \eta_\circ(T) = \pm\sqrt{-\frac{a}{2b}} \approx \pm\sqrt{(T_c - T)} \tag{A2.6}$$

The form of free energy (Eq.A2.3) in the order parameter space is shown schematically in Fig.A2.3. for different temperatures. Whereas above T_c the minimum of the free energy is at $\eta_\circ = 0$, this minimum splits into two symmetric minima $\eta_\circ = \pm\sqrt{-a/2b}$ below the phase transition temperature, while $\eta_\circ = 0$ corresponds to a maximum. At the phase transition the harmonic term in the free energy expansion equals zero and $G(\eta)$ has a very broad minimum, determined by the fourth order term in the expansion (Eq.A2.3).

The temperature dependence of the equilibrium order parameter, as obtained from the simple free-energy expansion (Eq.A2.3) is shown in Fig.A2.4.

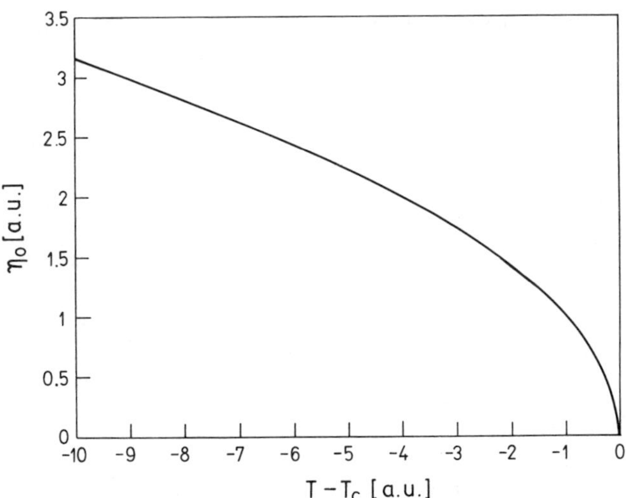

Fig.A2.4. The temperature dependence of the equilibrium order parameter $\eta_\circ$ shows a characteristic square root dependence when only the second and fourth order terms in the free-energy density expansion are considered.

So far we have neglected the $-h\cdot\eta$ term which describes the linear coupling of the order parameter to the conjugate field. Let us now consider the role of this term in the limit of small fields, $h=\delta h \rightarrow 0$. In this case the conjugate field can be considered as a small perturbation which infinitesimally changes the equilibrium value of the order parameter

$$\delta h \rightarrow 0 \; : \qquad \eta = \eta_\circ + \delta\eta \tag{A2.7}$$

Here we define the susceptibility of the system as the linear response of the system $\delta\eta$ to a small external field δh, in the limit of small field

$$\chi = \left(\frac{\delta\eta}{\delta h}\right)_{\delta h \rightarrow 0} \tag{A2.8}$$

The susceptibility of the system in the vicinity of the phase transition temperature can be calculated by minimizing the free-energy (Eq.A2.1) and writing $\eta = \eta_\circ + \delta\eta$, where $\eta_\circ$ is the equilibrium value of the order parameter, given by the Eqs.A2.5 and A2.6. The linear response of the system is then given by

$$\delta\eta = \frac{\delta h}{2a + 12b\eta_\circ^2} \tag{A2.9}$$

which yields

$$T > T_c \; : \qquad \chi(T) = \frac{1}{4a(T)} \approx (T - T_c)^{-1} \tag{A2.10a}$$

$$T < T_c \; : \qquad \chi(T) = -\frac{1}{2a(T)} \approx (T_c - T)^{-1} \tag{A2.10b}$$

From the above expressions one can see that the linear susceptibility of a system diverges as one approaches the phase transition temperature from both sides, as shown in Fig.A2.5. At the phase transition temperature the system is thus unusually "soft" and even a small external perturbation results in large response. Formally this can be explained by the fact that the free-energy minimum at the phase transition temperature is very broad because the harmonic term equals zero and only the quartic term is driving the system back to the equilibrium.

As a result of the enhanced susceptibility of the system near T_c, thermally excited fluctuations of the order parameter increase dramatically, as they are produced with a relatively little cost in the energy of the system. This means that very near T_c, the gradient term $d \cdot (\nabla\eta)^2$ cannot be considered as a small quantity because of increase of the susceptibility of the system near the phase transition.

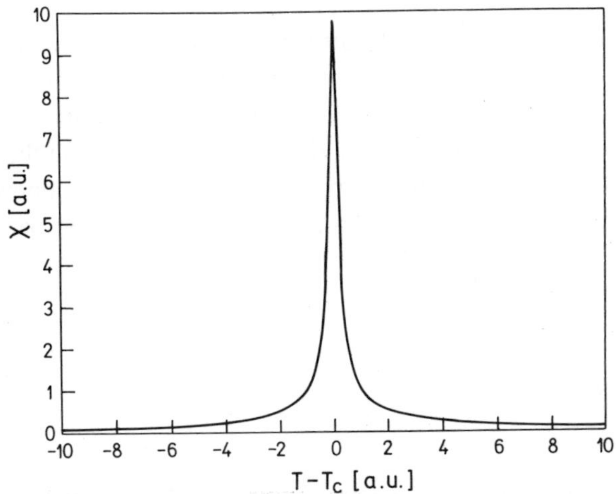

Fig.A2.5. Divergence of the linear susceptibility of a system near the second order phase transition.

This leads to the well known breakdown of the Landau theory in the very vicinity of the phase transition. The range of its validity can be estimated by the so-called Landau-Ginzburg criterion (see, for example Chaikin and Lubensky, 1995), which takes into account thermal fluctuations of the order parameter near T_c. Beyond this criterion, one has to consider truly critical dynamics, which can be described within the framework of the renormalization group theory (see Chaikin and Lubensky).

A3. Critical Slowing Down and Correlation of Collective Excitations near the Second Order Phase Transition

One of the most interesting aspects of second order phase transitions is the existence of a collective mode with a wave-vector $\vec{q}_c$, the relaxation frequency of which critically slows down as we approach the phase transition from above

$$T \to T_c \ : \qquad \frac{1}{\tau_{q_c}} \to 0 \tag{A3.1}$$

The existence of this so-called soft-mode follows directly from the Landau thermodynamic potential. Although it is one of the most remarkable features of phase transitions, it was Cochran (1959) who pointed out the significance of its appearance, although it was earlier discussed by several authors. In order to discuss the phenomena of critical slowing down within the framework of Landau theory, one has to consider the time-fluctuations of the order parameter $\delta\eta(\vec{r},t)$ around its equilibrium value $\bar{\eta}_{\circ}(\vec{r})$

$$\eta(\vec{r},t) = \eta_{\circ}(\vec{r}) + \delta\eta(\vec{r},t) \tag{A3.2}$$

Let us first consider the dynamics of thermally excited order parameter fluctuations above T_c, where the equilibrium value of the order parameter equals zero, $\eta_{\circ} = 0$. The instantaneous nonequilibrium free-energy density is obtained by inserting the Eq.A3.2 into Eq.A2.1. We shall not consider, for simplicity, the coupling term between the order parameter and the conjugate field in the free-energy expansion (A2.1.) and keep only the harmonic terms in the expansion:

$$g(\eta(\vec{r},t), \nabla\eta(\vec{r},t)) = g_{\circ} + a(T)\cdot\delta\eta^2(\vec{r},t) + d\cdot(\nabla\delta\eta(\vec{r},t))^2 \tag{A3.3}$$

The free-energy is now a functional

$$G = \int_V g(\delta\eta, \nabla\delta\eta)\cdot d^3\vec{r} \tag{A3.4}$$

because it depends also on the spatial form of the fluctuations $\delta\eta(\vec{r},t)$.

According to (Eq.A3.3), any small fluctuation of the order parameter $\delta\eta(\vec{r},t)$ increases the free-energy of the system, which is then driven back to equilibrium. We assume that this thermodynamic force, which drives the system back to equilibrium, is proportional to the first functional derivative of the free-energy, $\Delta G/\Delta\delta\eta$:

$$\frac{\Delta G}{\Delta\delta\eta} = \frac{\partial g}{\partial(\delta\eta)} - \frac{d}{dz}\left(\frac{\partial g}{\partial\left(\frac{d(\delta\eta)}{dz}\right)}\right) \tag{A3.5}$$

Here we have chosen for simplicity only the fluctuations $\delta\eta(z,t)$ which propagate in the z-direction. It is straightforward to generalize the calculations for the other spatial directions as well. Furthermore we assume that the regression of the fluctuation $\delta\eta(z,t)$ towards zero is proportional to the first functional derivative (Eq.A3.5):

$$\frac{d(\delta\eta)}{dt} = -\Gamma\cdot\frac{\Delta G}{\Delta\delta\eta} \tag{A3.6}$$

The above equation of motion was first used by Landau and Khalatnikov (1954). Here the so-called kinetic coefficient Γ is introduced, which is assumed to be temperature independent in the vicinity of the phase transition. In liquid crystals, this coefficient is the inverse viscosity. We should also mention that we have neglected inertial terms, which are not important for the collective dynamics of liquid crystals. In our case, the equation of motion for the order parameter fluctuations is deduced from the Eq.A3.3 and Eq.A3.5:

$$\frac{d(\delta\eta)}{dt} = -2\Gamma\left(a(T)\cdot\delta\eta(z,t) - d\cdot\frac{d^2(\delta\eta(z,t))}{dz^2}\right) \tag{A3.7}$$

The solution of the above Landau-Khalatnikov equation is an overdamped plane wave fluctuation, which represents collective excitations of the system:

$$\delta\eta(z,t) = \delta\eta_\circ\cdot\cos(qz+\Phi_q)\cdot e^{-\frac{t}{\tau(q)}} \tag{A3.8}$$

After inserting the above expression into the Eq.A3.7, we obtain the dispersion relation for the order parameter fluctuations in the disordered phase

$$\frac{1}{\tau(q,T)} = 2\Gamma(a(T) + d\cdot q^2) \tag{A3.9}$$

The dispersion relation for the order parameter excitations is thus parabolic and centered at $q=0$, as shown in Fig.A3.1. The minimum of the dispersion relation equals $2\Gamma a(T)$ and goes to zero linearly as we approach T_c from above

$$T \to T_c: \qquad \tau^{-1}(q=0) \approx a(T) \approx (T-T_c) \to 0 \tag{A3.10}$$

This is the so-called critical correlation frequency, which equals to the overdamped soft mode of the transition which condenses at T_c. In our case it appears in the form of a spatially homogeneous, overdamped wave which relaxes more and more slowly as we approach T_c, until it freezes out because there is no thermodynamic restoring force to drive it back to zero. This softening is a direct consequence of the "flattening" of the thermodynamic potential in the order parameter space (see Fig.A2.3) due to the vanishing of the coefficient of the harmonic term.

The soft mode can appear in very different forms in different systems. It is often underdamped in solid ferroelectrics but always overdamped in liquid crystals. Whereas it appears in the form of a spatially uniform wave at a phase transition into a homogeneous phase, it appears in the form of a plane-wave with a wave-vector $\vec{q}_c \neq 0$ for a phase transition into an inhomogeneous phase. For example, at the *isotropic* to *nematic* transition, it appears as a spatially homogeneous orientational wave (Blinc et al., 1974b; de Gennes, 1968a,b). At the *nematic* to $smectic-A$ transition it appears in the form of a density wave with a wave-vector corresponding to the thickness of a smectic layer. In chiral smectics, the soft mode appears in the form of a helical orientational wave at the $smectic-A$ to $smectic-C^*$ transition (see Blinc and Žekš, 1978). Here the order parameter has two components and the soft mode of the $smectic-A$ phase is doubly degenerate. This degeneracy is lifted in the $smectic-C^*$ phase as the soft mode splits into phason and amplitudon branch. In all these cases, the frozen-in soft mode breaks the symmetry of the high temperature phase. For example, it breaks the continuous translational symmetry of the *nematic* phase and the continuous rotational symmetry of the $smectic-A$ phase. This continuous symmetry breaking is very characteristic of liquid crystals and results in a zero frequency, symmetry restoring Goldstone mode below T_c. In the ferroelectric liquid crystals, this is the so-called "phason" mode at a critical wave-vector (Blinc and Žekš, 1978).

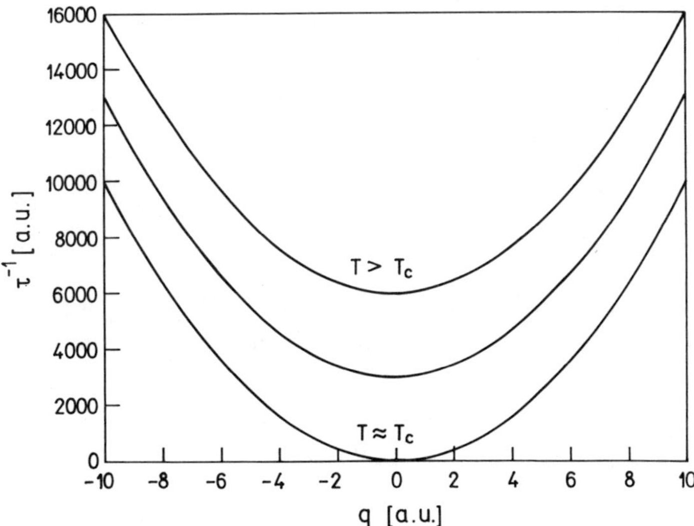

Fig.A3.1. Dispersion relations for the order parameter excitations near T_c.

Critical slowing down near the second order phase transition is accompanied by another very interesting fundamental feature, which is reflected in the enhanced space-correlation of the order parameter excitations. This phenomenon was first considered by Ornstein and Zernike in 1916.

The amount of correlation between the order parameter at various places in a system can be described by the static pair correlation function $G(\vec{r},\vec{r}\,')$, which is defined as the thermal average of order parameter correlation at different point in space

$$G(\vec{r},\vec{r}\,')=\langle\eta(\vec{r})\cdot\eta(\vec{r}\,')\rangle \tag{A3.11}$$

We shall take again a simple free-energy density expansion (Eq.A3.3.) and will consider the space-correlation of fluctuations above T_c where the equilibrium order parameter is zero. A more extensive and rigorous treatment of this problem can be seen in the paper of Kadanoff et al. (1967). The order parameter is expanded in a Fourier series

$$\delta\eta(\vec{r})=\sum_{\vec{q}}\delta\eta_q\cdot e^{i\vec{q}\vec{r}} \tag{A3.12}$$

where

$$\delta\eta_{-q}=\delta\eta_q^* \tag{A3.13}$$

to keep the order parameter fluctuation a real quantity. Here, $\delta\eta_q^*$ is the complex conjugate of $\delta\eta_q$. The corresponding nonequilibrium free-energy density is

$$g=g_\circ+a(T)\sum_{q,q'}\delta\eta_q\cdot\delta\eta_{q'}\cdot e^{i(q+q')z}-d\sum_{q,q'}\delta\eta_q\cdot\delta\eta_{q'}\cdot q\cdot q'\cdot e^{i(q+q')z} \tag{A3.14}$$

After integration, the increase in the free energy due to the order parameter fluctuations is

$$\frac{\Delta G}{V} = \frac{G - G_{\circ}}{V} = a(T)\sum_{q} \left|\delta\eta_{q}\right|^{2}\left(1 + \xi^{2} q^{2}\right) \tag{A3.15}$$

The order fluctuations therefore contribute independently to the free energy density of the system. This is a characteristic for normal modes of any system and leads to the application of the equipartition theorem, where a $\frac{1}{2}k_{B}T$ of energy corresponds to each degree of freedom. We obtain the mean-square amplitude of the normal fluctuation mode with the wave-vector q

$$\left\langle \left|\delta\eta_{q}\right|^{2} \right\rangle = \frac{k_{B}T}{2a(T)\cdot\left(1 + \xi^{2} q^{2}\right)} \tag{A3.16}$$

which diverges for homogeneous $q = 0$ fluctuations at T_{c}. On the contrary, for finite wave-vectors, the amplitudes of the normal modes are finite and therefore behave non-critically at T_{c}.

In the above expression, ξ is a characteristic length of a system

$$\xi^{2} = \frac{d}{a(T)} \approx \left(T - T_{c}\right)^{-1} \tag{A3.17}$$

which is also called the coherence length. We shall see later on, that this length is a measure of the amount of correlation of the order parameter at different points in space. The correlation length diverges as we approach the phase transition from above. This expresses the fact that larger and larger regions of the disordered phase are coherently ordered as we approach the phase transition.

We can evaluate the pair-correlation function $G(\vec{r},\vec{r}')$ at points $\vec{r}$ and $\vec{r}'$ by using Fourier transforms (Eq.A3.12)

$$G(\vec{r},\vec{r}')=\left\langle \delta\eta(\vec{r})\delta\eta(\vec{r}+\vec{R})\right\rangle = \left\langle \iint d\vec{q}d\vec{q}'\delta\eta_q \delta\eta_{q'} e^{i(\vec{q}+\vec{q}')\vec{r}} e^{i\vec{q}\vec{R}} \right\rangle =$$

$$= \int d\vec{q}\left\langle \left|\delta\eta_q\right|^2\right\rangle \cdot e^{i\vec{q}\vec{R}} \approx \left\langle \left|\delta\eta_q\right|^2\right\rangle \cdot \frac{e^{-|\vec{R}|/\xi}}{|\vec{R}|} \qquad \text{(A3.18)}$$

For temperatures above T_c, the pair correlation function decreases rapidly, because the correlation length is finite

$$T > T_c \quad : \qquad \xi .... \text{finite} \Rightarrow G(\vec{r},\vec{r}') \approx e^{-R/\xi} \qquad \text{(A3.19a)}$$

whereas at T_c long range order appears in the system, because the correlation length diverges

$$T = T_c \quad : \qquad \xi \to \infty \Rightarrow G(\vec{r},\vec{r}') \approx \frac{1}{R} \qquad \text{(A3.19b)}$$

The above results illustrates that there is considerable ordering within the distance of the order of correlation length of the system. On approaching the phase transition the volume of these ordered regions diverges until the new ordering is fully developed at T_c.

These conclusions lead to a very interesting consequence for the dynamics near second order phase transition. As the correlation length increases on approaching T_c, it will necessarily become much longer than the characteristic length (unit cell, for example) of the system. As a consequence, the behavior of the system in the vicinity of the phase transition will not depend on the nature of microscopic interactions which drive the transition, but will exhibit a universal behavior, that is determined by the dimensionality of the system and the dimensionality of the order parameter. This universal behavior is expected to be common to certain classes of thermodynamic systems which are grouped into the so-called universality classes. These exhibit similar critical behavior (see Chaikin and Lubensky, 1995).

As a final remark, we should notice a close connection between the existence of gapless excitations and the magnitude of the correlation length in a system. Let us consider again the dispersion relation (Eq.A3.9), as derived from the free-energy given by Eq.A3.3:

$$\tau^{-1}(q,T) = 2\Gamma\left(a(T) + d \cdot q^2\right) = 2\Gamma d\left(\frac{a(T)}{d} + q^2\right) = 2\Gamma d\left(\frac{1}{\xi^2} + q^2\right) \qquad \text{(A3.20)}$$

This dispersion is shown in Fig.A3.2. for two different correlation lengths, i.e. $\xi \to \infty$ in Fig.A3.2(a) and finite ξ in Fig.A3.2(b). The inverse square of the correlation length is proportional to the gap of the dispersion relation for a given branch of excitations. For infinite correlation lengths, the dispersion is gapless and the excitations are "massless" because there is a zero frequency Goldstone mode. On the other hand, for finite correlation length, the modes are "massive" and the dispersion relation has a finite gap. One should here note the analogy to similar phenomena in particle physics.

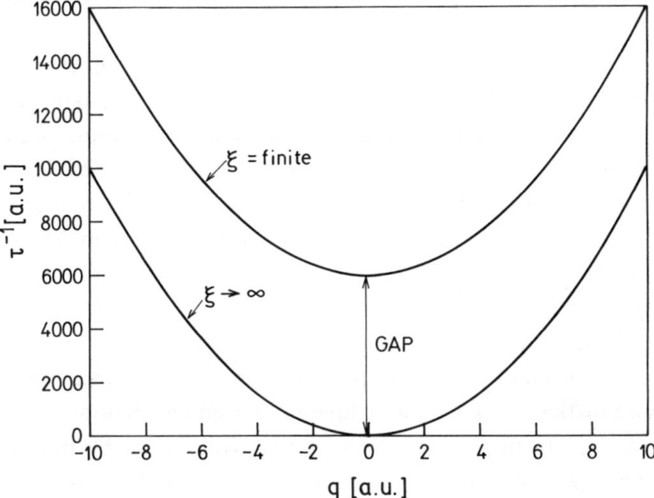

Fig.A3.2. (a) "Massless" modes are connected to infinite correlation length and the existence of a zero frequency, symmetry restoring Goldstone mode. (b) "Massive" modes are characteristic of finite correlation length and the dispersion relation shows a finite frequency gap.

The fact that a zero frequency mode is a sign of an infinite space-correlation length thus simply reflects the fact that different parts of the system are correlated on a time scale that is infinitely long.

Appendix B

Survey of Different Experimental Methods

B1. Controlling the Temperature with a Milli-Kelvin Resolution

In liquid crystalline research, it is often necessary to control precisely the temperature of the liquid crystalline material. This is in particular necessary in the studies of phase transitions between various liquid crystalline phases, where sometimes a temperature resolution and stability of the order of $1mK$ are required. As an example, the critical phenomena close to the $smectic-A$ - $smectic-C^*$ are observable in a temperature interval less than $1K$ around the transition. This means that one has to assure that the temperature gradients over the measured volume of the sample and temperature stability are at least two order of magnitude smaller, i.e. of the order of $10mK$ or less.

It is not difficult to obtain a milli-Kelvin stability of a sample. However, it is much more difficult to keep the temperature gradients below this value. One should therefore try to minimize the measuring volume to the minimum value required by the experimental signal to noise ratio. This reduction is in particular simple in optical experiment, where one usually focuses the measuring beam to a relatively small spot of $10\mu m$ diameter, where the temperature gradients are usually negligible.

One of the most important reasons for poor temperature stability is usually poor or insufficient thermal isolation of the sample. In a laboratory, a typical source of large temperature variations is the flow of the air through laboratory, as well as air turbulences induced by persons walking close to the measuring place. For a typical single-stage temperature controlled experiment, the resulting fluctuations of the sample temperature are of the order of $0.1K$ If one wants to achieve milli-Kelvin stability of the experiment, one should plan to set-up the experiment in a quiet and well isolated place.

Second, it is practically impossible to achieve milli-Kelvin stability using a single-stage temperature control. One can achieve a temperature stability of the order of $10mK$ using a single stage temperature control. The stability can be improved by using a double-stage temperature controlled set-up, which is schematically shown in Fig.B1.1. In this kind of set-up, the first temperature controlled stage is imbedded in a second temperature controlled stage. This second temperature controlled stage (i.e. the outer stage) provides a very stable temperature of the environment for the inner stage, which is imbedded in it. In this way, we significantly reduce the temperature fluctuations of the surroundings of the inner stage, and a stability of the order of $1mK$ can be easily achieved.

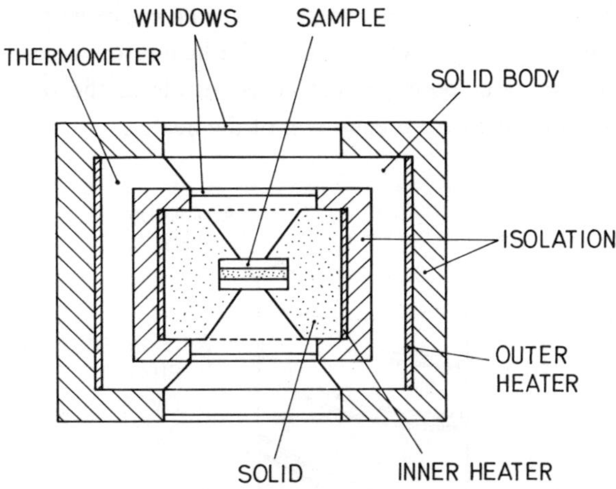

Fig.B1.1. The schematic view of a double-stage temperature controlled set-up. Two independent heating stages are used. The inner stage is imbedded in the outer stage, in order to reduce temperature fluctuations of the surroundings, as seen by the inner stage.

Third most important aspect of good temperature control is good design of the heater. Here, very good thermal coupling between the temperature sensor and the heating wire is most important. A typical heater is shown in Fig.B1.2. We use a piece of aluminum of cylindrical shape as a heater element. At a curved surface of this cylinder, small groves are machined in a spiral pattern, as indicated in Fig.B1.2. These groves are made just large enough that a heating wire can be placed and fixed in these grooves using a high-stability epoxy, as for example Torr-Seal epoxy (Varian). In this way, we achieve a good thermal contact with a heating block and a good mechanical stability of the heater. Electrical short-circuit of the heating wire is prevented by anodizing the aluminum heater after the groves have been machined. This makes a thin, electrically insulating surface layer that prevents electrical short-circuit of the heating wire. Second, we use the so-called manganin-wire, which can be purchased with a thin inorganic electrically insulating layer. This wire can be purchased in different diameters to accommodate the maximum heating current. Typically, we use $0.3mm$ diameter wire that can sustain several amperes of electric current when in good thermal contact with the heater. It should be stressed that the entire length of the heating wire should be in good thermal contact with the heater, or we can have thermal overheating in places with poor contact. This results in the destruction of the isolating layer and the electrical breakdown of the heater.

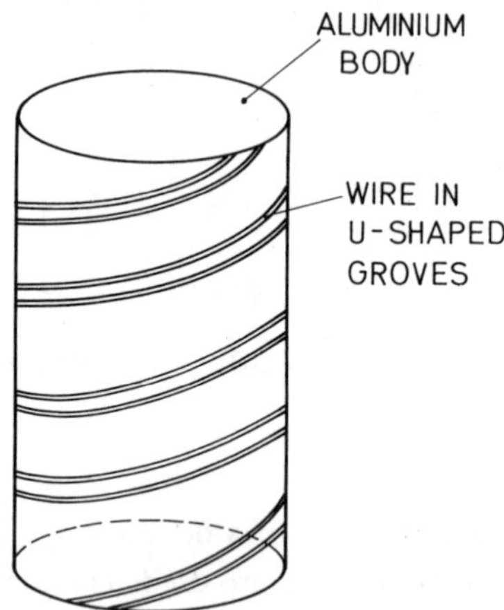

Fig.B1.2. The heater is made by winding a heating wire around the cylindrically-shaped heater block. The wire should be in good thermal contact with the body and at the same time electrically isolated from the aluminum body.

A good thermal contact between the heating wire and the temperature sensor is achieved by placing the sensor as close to the heating wire, as possible. This is shown in the cross-section in Fig.B1.3. Close proximity of the sensor relative to the source of the heat results in negligible time-delay between the heat input and the temperature sensing. This facilitates the tuning of the electronic unit and in general improves time response and temporal stability of the system. If the temperature-sensing element is placed away from the heat-source, a time-delay is introduced in the system. This time delay is due to the finite time, necessary for the heat diffusion from the heating wire towards the thermometer. This time delay causes troubles to the stability of electronics and is the third most important fact that should be considered in a design of the set-up.

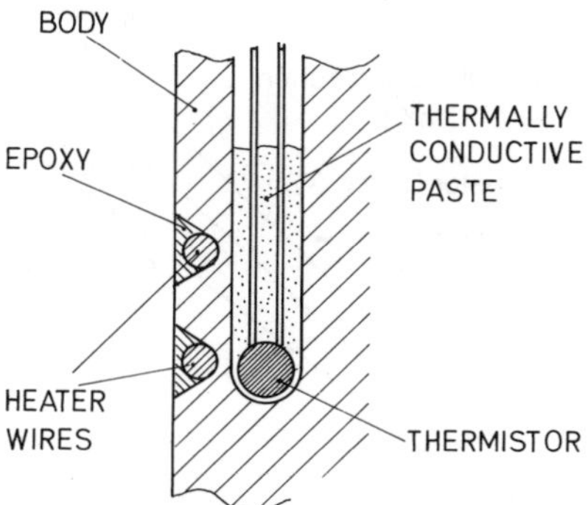

Fig.B1.3. The sensor should be positioned as close to the heater as possible.

One of the results of placing the temperature sensing element close to the heater is that the temperature, measured by this thermometer is not necessarily equal to the temperature of the sample. For this reason, we use a second temperature-sensing element, which is very close to the sample. The read-out of this thermometer is used to determine the sample temperature, whereas the other thermometer is used only for the temperature control.

One of the important aspects of good temperature control is the choice of the temperature sensing element. Here, there is a wide choice of different possible thermometers and they are briefly discussed in the Appendix C. In most cases, platinum resistors and semiconductor thermistors are used. Thermistors have nearly an order of magnitude larger temperature coefficient of the resistance, which makes them very suitable for high-resolution temperature control. Thermistors also show very small magneto-resistance, which makes them very suitable for room-temperature experiments in high magnetic fields. Capacitive sensors (Lake Shore) also have a very small magnetic field dependence and can be used in magnetic fields as well. Fig.B1.4. shows a comparison of magnetic-field-induced change in the temperature due to magneto-resistance in Pt-100 and YSI-44004 thermistor (Seppen 1987).

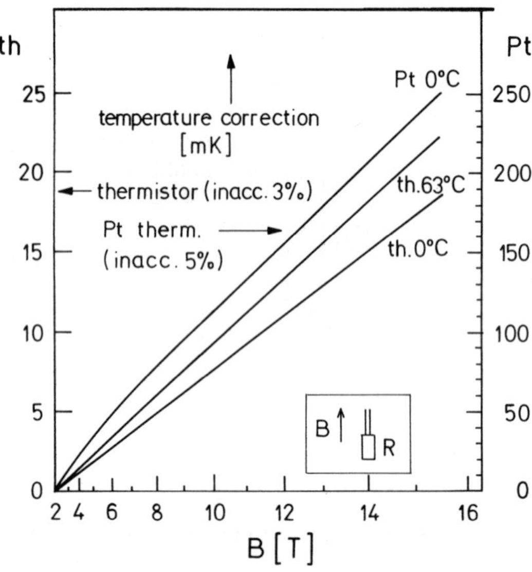

Fig.B1.4. Correction in temperature due to the magnetic field. The field is given in quadratic scale. It is applied along the axis of the sensors.

Finally, let us discuss the last component of a temperature-controlled set-up, i.e. the controller itself. There are several sources of high-quality temperature controllers (Oxford Instruments, Lake Shore, Instec) that can be just plugged-into the experiment. We shall here however briefly describe a controller that can be assembled from a standard equipment, that can be found in a physics laboratory. The schematic drawing of the temperature controller is shown in Fig.B1.5.

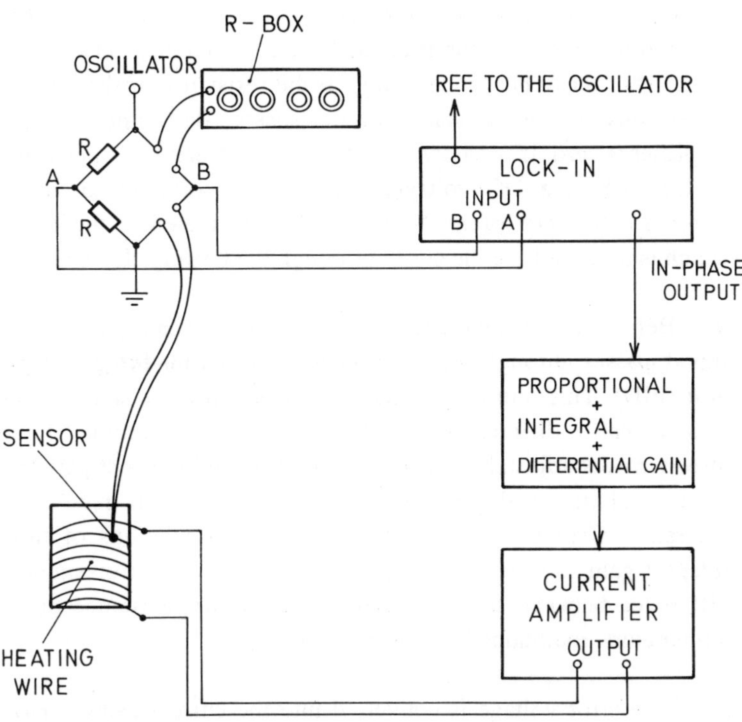

Fig.B1.5. Block diagram of a temperature controller, built from standard laboratory equipment.

It consists of:

(i) ***The Wheatstone bridge***, where we have in one branch of the bridge a resistance-box and the resistive sensor, which is placed in the heating cell

and is connected with two BNC cables the metal box that contains the bridge. The bridge is connected to a signal generator ($100Hz - 1kHz$, $\approx 500mV_{pp}$).

(ii) The error-signal of the bridge (points A and B) is connected to ***a differential input of a lock-in amplifier***. The in-phase output of the lock-in amplifier gives the amplified DC voltage difference between the points A and B of the Wheatstone bridge. If the resistance of the resistance-box is set equal to the resistance of the sensor, the output of the lock-in amplifier is zero. If these two resistance are different, we get at the output of the lock-in a DC voltage, which is proportional to the potential difference between point A and B. This is therefore the error signal, which is used to drive the current through the heating element.

(iii) Before converting the error voltage into the heating current, the error signal passes through an additional ***Proportional-Integral-Differential unit*** (PID). This unit gives a sum of : original error signal + integrated error signal + differentiated error signal. The magnitude of individual three terms is adjustable and can be used to find the best performance of the controller. Usually only proportional and integral gains are needed, whereas differential gain is likely to cause oscillations of the system. Integral gain is needed to minimize the long-term error signal (i.e. the difference between the temperature, set by the resistance box and the temperature, maintained by the system).

(iv) The error voltage is converted into the current using a ***current or voltage amplifier***, which is able to deliver several amperes into a typically $10 - 20\Omega$ heater (Kepco). Usually, an electric output power of several tens of Watts is needed to keep a typical heater at a temperature of $\approx 100^\circ C$.

A simple and inexpensive temperature controller is shown in the wiring diagrams in Appendix C.

B2. Optical Polarization Modulation Techniques: Measurements of Birefringence, Optical Rotation and Circular Dichroism

In optical polarization modulation techniques, the polarization of light is modulated by optical properties of matter, which is exposed to external electric, magnetic or stress fields. This polarization modulation allows us not only to modulate, but also to analyze the polarization state of light, as it enters the optical polarization set-up. Various polarization modulation elements are available, such as the Pockel cell, Faraday rotator and the Photoelastic modulator (for a review, see Drake 1986). These polarization modulation techniques have been widely used in ellipsometry (Jasperson et al., 1969, 1973), circular dichroism (Southerland et al., 1975; Cheng et al., 1975; Chabay et al., 1975; Gale et al., 1972; Gafni et al., 1972, Shatwell et al., 1974, Hipps et al., 1978). Due to its wide aperture and versatility, we shall here describe the optical polarization techniques, based on Photoelastic modulators (PEM), first introduced by Kemp (1969).

A photoelastic modulator is an optical element, whose optical retardation is controlled or varied in time by external mechanical stress, as shown in Fig.B2.1.

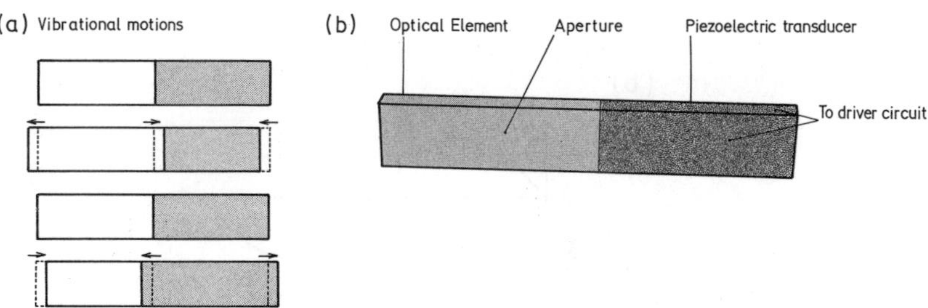

Fig.B2.1. Vibrational motion of the quartz plate and the modulator optical assembly (courtesy Hinds Instruments).

Typically, the modulator is made from a large crystal of quartz, which is cut into a predefined direction. If an external electric field is applied in a suitable crystal direction, an elastic mechanical stress is introduced due to the inverse piezoelectric effect. Since mechanical stress changes the optical index of crystal, an optical retardation is introduced by the application of an external field. In a photoelastic modulator (PEM), the applied electric field usually varies at the lowest mechanical resonance frequency of a given quartz crystal. This results in well-defined, time-dependent optical retardation of the PEM, which is oscillating at the excited resonance frequency of the PEM. The peak magnitude of the retardation can be controlled by the magnitude of the vibration of the PEM. Let us see on a simple example, what happens when a linearly polarized, monochromatic light is sent through a PEM and the input polarization is at 45° with respect to the two eigen-axis of the PEM, as shown in Fig.B2.2. The magnitude of mechanical vibration of the PEM is set to a half-wave peak retardation.

(a)

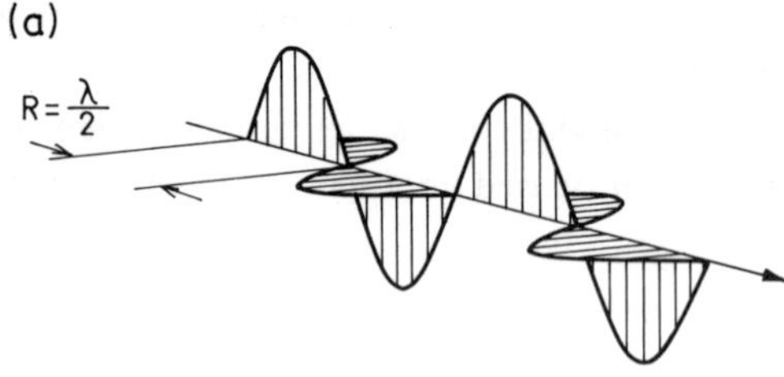

(b)

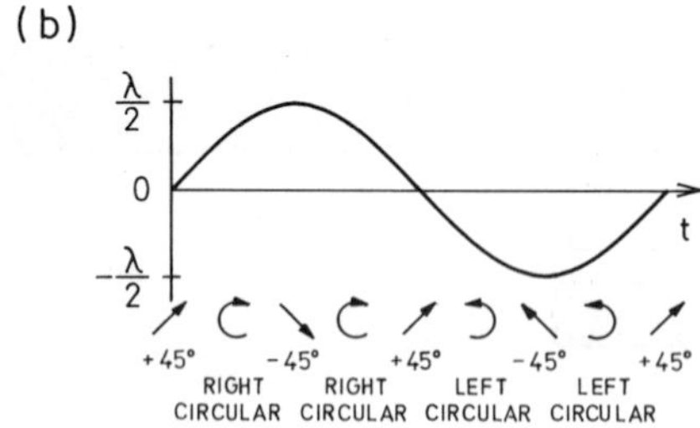

Fig.B2.2. (a) After passing through the PEM, the two spatially orthogonal optical eigenwaves are phase shifted with respect to each other. The maximum phase shift is 180° for a half-wave peak retardation. (b) Schematic view of the resulting polarization as a function of time.

At $t = 0$, mechanical stress is zero and there is no retardation in the PEM. The output polarization is the same as the input one, i.e. at 45° with respect to the axis of the PEM. Due to increased stress, the retardation increases with time and the resulting output polarization changes from linear via right elliptic to right circular for the retardation of one quarter of the wavelength. Upon further increasing retardation, the output polarization changes again from right-circular through elliptic to a linear polarization for one-half of the wavelength retardation. The direction of the polarization is however now rotated by 90° with respect to the input polarization. At this moment the stress is at a maximum value and starts to decrease with time. This leads to inverse sequence of output polarizations till we reach initial polarization for zero stress. By increasing time, the stress becomes negative, which results in a similar sequence of output polarization with a different handedness, which is now opposite to the one for positive stress.

This simple modulation of the polarization of light can be used to analyze the polarization property of light, as will be shown in a simple example. Let us take for simplicity a light, composed of some circularly-polarized component as well as of some linearly polarized component, as illustrated in Fig.B2.3.

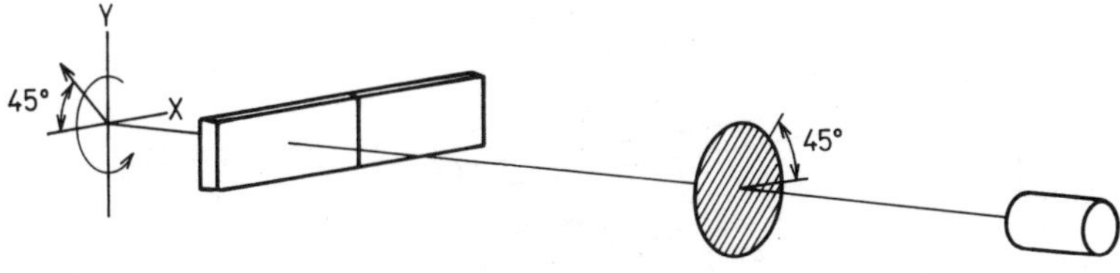

Fig.B2.3. A simple illustration where PEM is used to analyze the polarization properties of incident light.

We use a PEM, an analyzer with the polarization axis at 45° with respect to the two axis of the PEM and a photodetector. The maximum retardation of PEM is set to one quarter-wave. A net circularly polarized component of input light will be totally transmitted through the analyzer when the PEM adds additional quarter-wave retardation and will be totally attenuated when PEM subtracts one quarter-wave retardation. This means that for this input polarization, the detector will see a light intensity, which is oscillating at the modulator frequency f . This can be detected with a suitable lock-in technique, where the f signal of the lock-in gives the intensity of the circularly polarized input light. For the same quarter-wave retardation and a linear input polarization, the detector will detect a signal at a doubled frequency $2f$. Again, this can be detected with a lock-in amplifier.

There are several standard schemes for measuring the birefringence, optical rotation and circular dichroism. In the following, we shall briefly review some of them.

(i) High resolution measurements of birefringence using a photoelastic modulator.

A simple set-up for the measurements of birefringence is shown in Fig.B2.4.

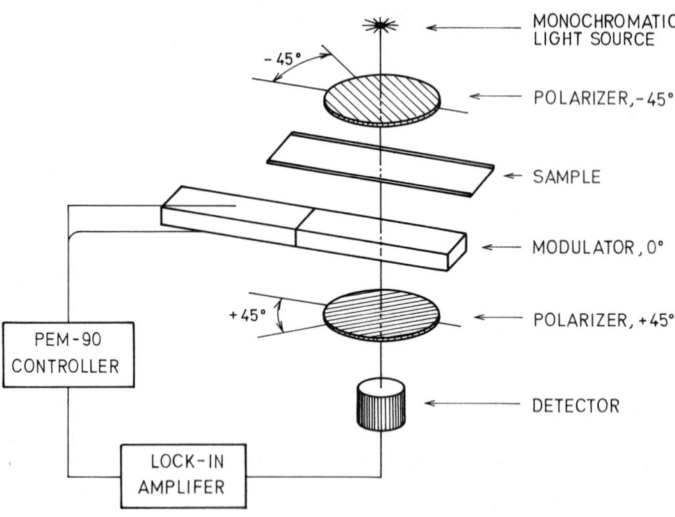

Fig.B2.4. A simple set-up for the measurement of optical birefringence (courtesy Hinds Instr.).

The PEM is operated at quarter-wave peak retardation and is set slightly off-axis, to reduce the problems with stray interference from multiple reflections. The $1 \cdot f$ signal of the lock-in is proportional to the birefringence of the sample. The sample must be rotated until a maximum signal is obtained, which indicates the birefringent axis. For small values of the retardation, the retardation can be determined by comparing the lock-in signal from the sample to a signal, obtained from standard retarder, such as $1/30 \cdot \lambda$ retarder. In this case, the birefringence of the sample B_{sample} is

$$B_{sample} = B_{s\tan dard} \cdot \frac{V_{1f(sample)}}{V_{1f(s\tan dard)}} \tag{B2.1}$$

An advanced set-up for the measurement of the birefringence is shown in Fig.B2.5. A straightforward analysis leads to the light intensity I at the detector, (Kemp, 1969)

$$I = 1 - \cos(B) \cdot \cos(A) + \sin(B) \cdot \sin(A) \tag{B2.2}$$

Here, B is the optical retardation of the sample and A is the time-oscillating retardation of the PEM,

$$A = A_{\circ} \cdot \cos(\Omega t) \tag{B2.3}$$

A Fourier series expansion of Eq.B2.2. yields the following expression

$$\begin{aligned} I = 1 - \cos(B) \cdot J_{\circ}(A_{\circ}) + \\ + 2 \cdot \sin(B) \cdot J_1(A_{\circ}) \cdot \cos(\Omega t) + \\ + 2 \cdot \cos(B) \cdot J_2(A_{\circ}) \cdot \cos(2\Omega t) \end{aligned} \tag{B2.4}$$

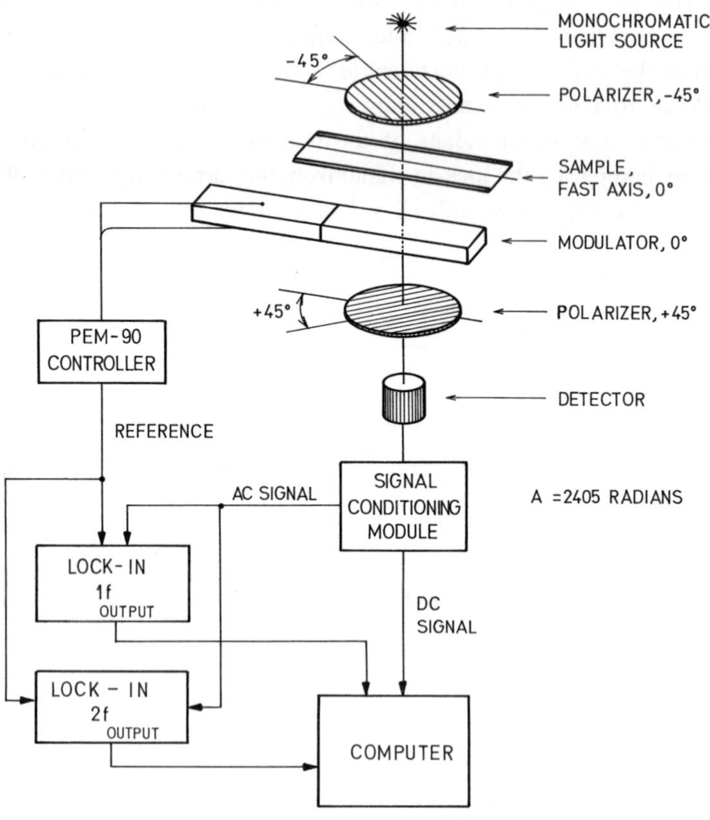

Fig.B2.5. An advanced set-up for the high-resolution measurements of birefringence.

Here, the term in the first line represents the transmitted DC light intensity, the term in the second line the intensity of light, which is oscillating at a frequency Ω of PEM modulator and the last term represents the light intensity, oscillating at a doubled frequency 2Ω of the PEM. $J_o(A_o)$, $J_1(A_o)$ and $J_2(A_o)$ are the values of the Bessel functions at a constant amplitude of the retardation A_o, which can be adjusted by the PEM controller. The retardation B of the sample can be in this set-up measured in two ways:

(a) Using a single lock-in and a DC voltage meter. We adjust the amplitude $A_{\circ}$ of the modulation such that $J_{\circ}(A_{\circ})=0$. This occurs for $A_{\circ}=2.405$ radians (approximately $0.383\cdot\lambda$). In this case, the DC term in the Eq.B2.4. does not depend on the retardation of the sample, and it depends only of the intensity of the external light, which has been set to 1 in the above equations. The DC signal can be therefore used to "normalize" the $1\cdot f$ signal and the retardation of the sample is

$$B=\arcsin\left(\frac{V_{1\Omega,RMS}}{V_{DC}}\cdot\frac{1}{\sqrt{2}\cdot J_1(A_{\circ})}\right) \qquad \text{(B2.5)}$$

Here, $V_{1\Omega,RMS}$ is the RMS output of the lock-in amplifier and the $\sqrt{2}$ is due to the relation between the peak and RMS amplitudes of sinusoidal signals.

(b) Using two independent lock-in amplifiers, which measure the signal at a fundamental and the doubled frequency of the light intensity. In this case, the retardation of the sample is

$$B=arctg\left(\frac{V_{1\Omega,RMS}}{V_{2\Omega,RMS}}\cdot\frac{J_2(A_{\circ})}{J_1(A_{\circ})}\right) \qquad \text{(B2.6)}$$

In both cases, one needs to know the length of the path L through the sample in order to determine the birefringence from the relation

$$\Delta n=\frac{\lambda_{\circ}}{2\pi L}\cdot B \qquad \text{(B2.7)}$$

Here, $\lambda_{\circ}$ is the wavelength of the light in vacuum.

Using the set-up shown in Fig.B2.5., one can achieve typical resolution of $\delta B\approx 0.01\deg$ phase difference between the two eigenwaves. For good temporal stability in long-term, high resolution experiments, it is ***advised to control the temperature of the head of the PEM to better than*** $0.1K$. Experiences show, that in the very high resolution regime, the amplitude of the PEM retardation A_o drifts significantly with temperature resulting in significant experimental error. A somewhat lower resolution in retardation can be achieved by measuring the birefringence using a Pockel cell as a polarization modulator. An example of such a set-up (Seppen 1987; Maret et al., 1983) is shown in Fig.B2.6.

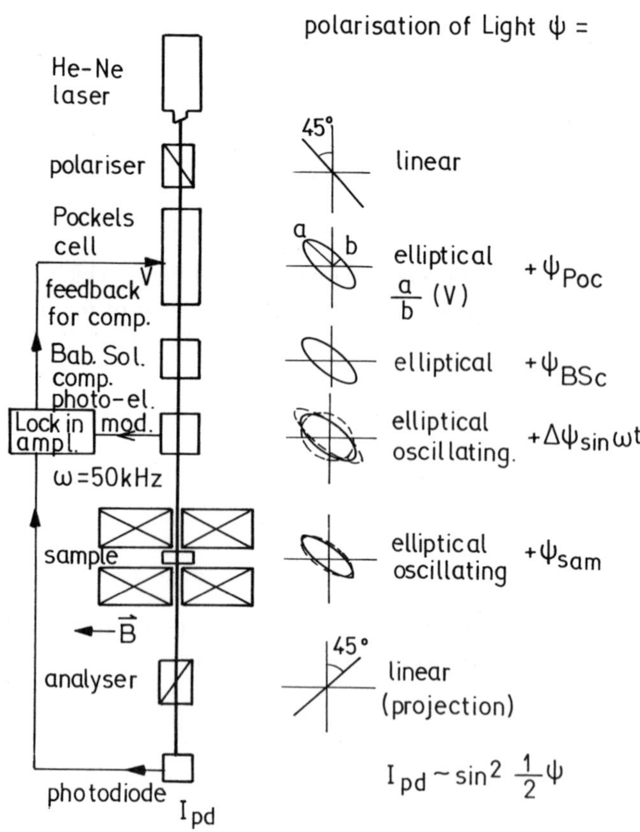

Fig.B.2.6. Set-up for the measurements of birefringence using Pockels cell (Seppen 1987). This set-up was used in high magnetic fields up to 13T.

***(ii)* High resolution measurements of the optical rotation using PEM**

In chiral phases, the optical rotation of the phase gives valuable information on the structure and phase transitions in a system. It is therefore important to measure accurately the rotation of polarization of light, as it travels through the sample. This rotation can be measured with the PEM technique to better than $0.01\deg$ resolution using a set-up, shown in Fig.B2.7.

The analysis of the light intensity signal, as detected by the photodiode shows that it consists of several harmonics of the fundamental frequency of the PEM. Most accurately, the angle Ψ of rotation of polarization of light can be determined from the measurements of the first and second harmonics, using lock-in technique. The angle of rotation is in this case

$$\Psi = \frac{1}{2} \cdot arctg\left(\frac{V_{1\Omega,RMS}}{V_{2\Omega,RMS}} \cdot \frac{J_2(A_\circ)}{J_1(A_\circ)} \right) \tag{B2.8}$$

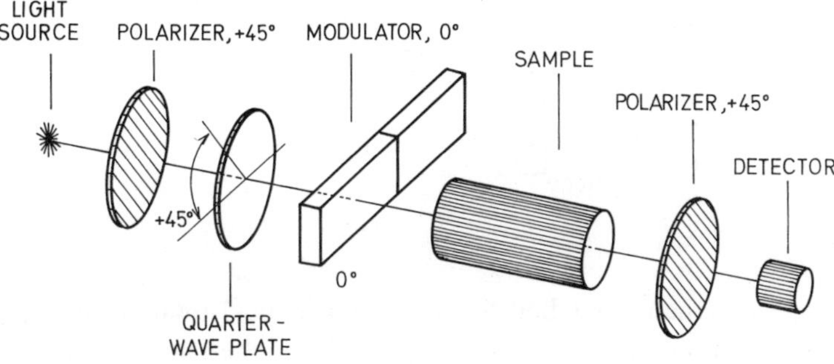

Fig.B2.7. Set-up for the measurement of the optical rotation.

An alternative set-up can be used, which has a single lock-in amplifier and an additional Faraday modulator, which is driven by the signal from the lock-in output, as shown in Fig.B2.8. The Faraday rotator compensates the optical rotation of the sample, so that the light intensity, detected by the photodiode is minimal. In this way, we assure the linearity of the measurement due to the linearity of the Faraday rotator. The measuring range of this method is therefore limited by the maximum angle of rotation of the Faraday rotator.

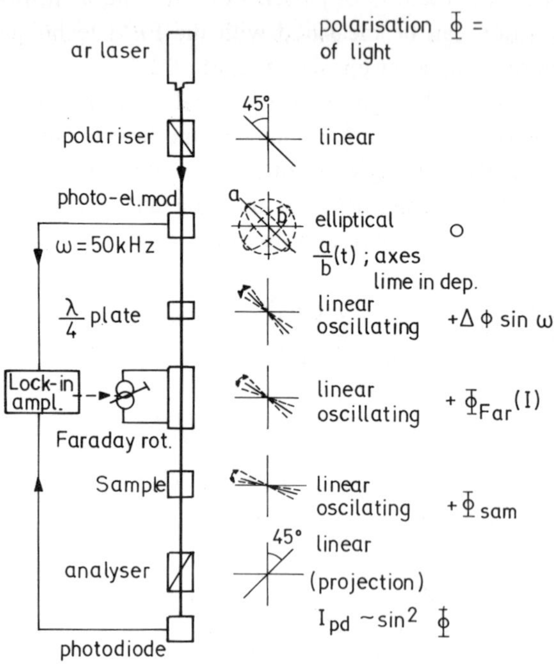

Fig.B2.8. Alternative method for measuring optical rotation, using Faraday rotator to compensate the polarization rotation due to the sample (Seppen 1987).

(iii) Measuring circular dichroism with PEM

Circular dichroism is the differential absorption between left and right circular polarized light components. This phenomena occurs naturally with chiral compounds and is therefore connected to the electronic transitions in molecules. However, it should be pointed that phenomena which is similar to molecular circular dichroism can be expected in chiral, helicoidally modulated liquid crystal phases with a given handedness. In this phases, selective Bragg reflection is observed for a certain sign of circular polarization of the input light. This means that the reflective coefficient for a given handedness of circularly polarized light can be significantly different from that for the opposite one. If one measures only the transmission of light, it is therefore likely to interpret the selective reflection as circular dichroism of the sample, which is related to the absorption. The basic set-up for the circular dichroism measurements is shown in Fig.B2.9. The modulation amplitude $A_{\circ}$ of the PEM is set to quarter-wave. This means that we have during each cycle of the PEM at a certain moment a left-handed circular wave, which is followed after one-half of the period by the right-handed circular wave. The absorption coefficient (or reflection coefficient in the case of selective Bragg reflection) therefore changes during each cycle of the PEM, reaching a maximum value only once during each cycle. This means that the circular dichroism CD of the sample is proportional to the intensity of detected light, which oscillates at a fundamental frequency of the PEM,

$$CD \approx \frac{V_{1\Omega,RMS}}{V_{DC}} \tag{B2.9}$$

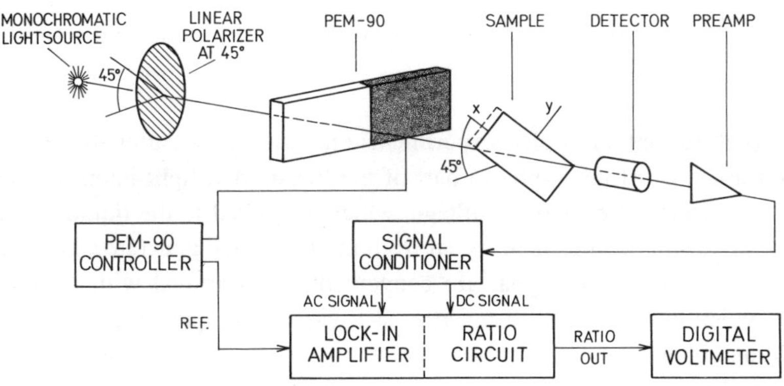

Fig.B2.9. The set-up for the measurement of circular dichroism.

B3. Measurements of Linear Electrooptic Response

The principles of the linear electrooptic response were discussed in Section 5.5. Here we shall briefly describe basic experimental set-up for these kind of measurements, which is shown in Fig.B3.1. It consists of a laser source, polarizer, two lenses sample with electrodes, analyzer, photodetector with preamplifier, lock-in detector and a computer.

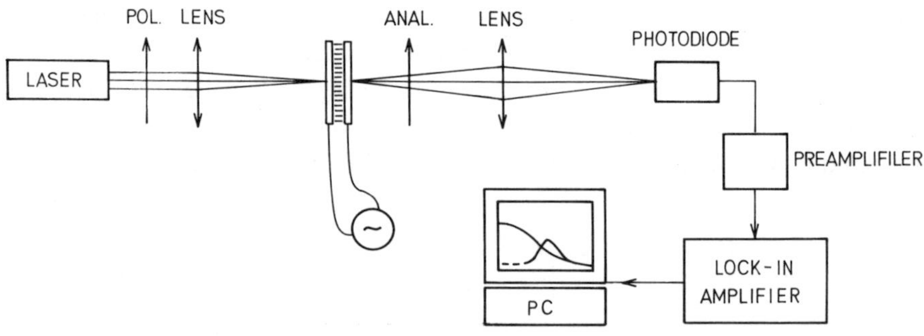

Fig.B3.1. Basic set-up for the measurements of linear electrooptical response.

In the linear electrooptic response experiments one usually measures the real (in-phase) and imaginary (out-of-phase) part of the transmitted light intensity, with respect to the reference excitation voltage, which is applied to the liquid crystal. This is usually a small, sinusoidal signal from the frequency generator. Both real and imaginary parts of the signal are conveniently determined with a lock-in amplifier, which in most cases also serves as a variable frequency generator for the excitation voltage.

The linear electrooptic response measurements are usually performed at a set of different frequencies, which allows us to determine the relaxation rates of the eigenmodes in liquid crystal, that contribute to the linear response. As shown in Sections 5.5. and 6.1., this can be determined either from the maximum of the imaginary part of measured signal (see Fig. 5.5.5) or from the Cole-Cole plot (see Fig.6.2.4.). It is also important to note that the set-up with a lock-in amplifier allows for a straightforward measurements of the nonlinear response, because most of the lock-in amplifiers are capable of measuring several higher harmonics with respect to the frequency of the excitation voltage.

The critical part of the set-up shown in Fig.B3.1. is a photodiode with a preamplifier. These two elements should not only provide efficient conversion of light into electric voltage, but should also be phase-linear in the entire frequency range. Any nonlinearity of the phase is reflected in the shape of the measured frequency spectrum of the linear response signal and can be misleading in the analysis. There are not many commercial photodiode preamplifiers that would meet the requirement of having a rather high gain and being phase-linear up to 1MHz. A simple photodiode preamplifier, which is phase-linear from DC to 1MHz and has appropriate gain for the liquid crystal experiments is shown in the Appendix C.

B4. Quasielastic Light Scattering Experiments

The theoretical aspects of quasielastic light scattering in liquid crystals were discussed in Section 5.4. Here we shall describe the experimental aspects of a quasielastic light scattering experiment.

The set-up for a light scattering experiment is shown in Fig.B4.1. It consists of a laser, the optics for polarizing and focusing the input beam, the optics for analyzing and detecting the scattered photons and the autocorrelator. The laser is one of the critical components of the set-up. First, it should provide at least $10-20mW$ of power to assure that the scattered light intensity is considerable. Typically, a $100mW$ laser is a good choice for such an experiment. Second, the laser should provide a "clean" output, with no $50Hz$ modulation of the beam intensity. The best choice for light scattering experiments are frequency-doubled Nd-YAG lasers that give a very clean beam with a wavelength of $532nm$ and several $100mW$ of output optical power. The polarization rotator and the polarizer serve to prepare either π or σ polarized incident light, which is slightly focused at the sample using a long focal length lens. Typically, the light is focused to a spot with several $10\mu m$ diameter. The sample is mounted into a temperature-controlled oven that is put on a rotating stage, and provides an easy selection of the incident angle. The stage will be described further on. The scattered light is collected with a collecting optics, that is mounted on an arm, attached to the second rotation stage that is co-axial with

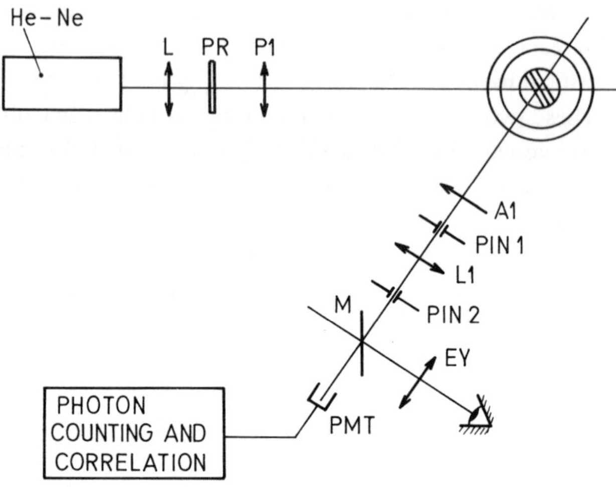

Fig.B4.1. Set-up for the light scattering experiment. L-lens with f=100mm; PR-polarization rotator; P1-polarizer; A1-analyzer; PIN1-pinhoče; L1-lens; PIN2-pinhole; M-metallic mirror with a $\approx 200\mu m$ pinhole; EY-microscope eyepiece; PMT-photomultiplier for single photon detection.

the first rotation stage. This second rotation stage thus provides a selection of the angle of the scattered light. In combination with a first rotation stage we have therefore a possibility to independently determine the direction and the magnitude of the scattering wave vector. The collecting optics consists of an analyzer that determines the polarization of the scattered photons. The angle of the scattered light is determined with a pinhole, which is typically a $1mm$ hole in a thin piece of metal that is placed at a distance of $200mm$ from the sample. Several other pinholes can be used to reduce the collection of the stray light. Lens L1 is used to project the image of the illuminated part of the sample onto a metal mirror that is mounted at 45°. A microscope eyepiece is placed close to this mirror, so that the image of the illuminated part of the sample is clearly visible. The lens L1 and the eyepiece EY therefore form a telescope, which allows visual inspection of the illuminated area of the sample. At the same time, one clearly sees the position of the scattered light beam on the metal mirror. This mirror has a small pinhole, which is etched in the metal film. By moving the lens L1 we can adjust the position of the scattered light so that it enters the pinhole and is detected by the photomultiplier PMT. A careful observation of the illuminated spot shows that we have a diffraction pattern due to the diffraction of scattered light on a pinhole PIN2. The bright middle part of this diffraction pattern represents a single coherence area of the scattered light. Here the phase difference of the scattered light is less than one wavelength. By selecting a larger pinhole PIN2, this coherence area shrinks and vice-versa. It is important that the size of the pinhole, which is etched in the metal mirror is less than one coherence area. Then, we are in the so-called single coherence area regime of light scattering.

The photomultiplier PMT serves as a single photon detector and is one of the critical parts of a light-scattering set-up. The most critical are here the after-pulses that are created in the photomultiplier after a photon has been detected and converted into an electric pulse. Good photomultipliers for light scattering set-up are selected for minimum afterpulsing, short response time, as well as low dark-current. In some applications, where high-speed is necessary, one can use a combination of two equal photomultiplier tubes. The scattered light is split into two beams of equal intensities and detected independently by the two photomultipliers. The two sequences of pulses are cross-correlated in a cross-correlator. This procedure reduces the effect of afterpulsing, because these are different for the two tubes and therefore not correlated. As a result, one obtains fast autocorrelation of the scattered light intensity with minimized effects of the photomultiplier afterpulsing. Finally, the photoelectron pulses from the photomultiplier are autocorrelated by an autocorrelator.

An example of the autocorrelation function from an ALV-5000 autocorelator that has logarithmic time scale is shown in Fig.B.4.2. It allows for the identification of relaxation processes that are distributed from several MHz to several Hz. This means that we can measure over nearly six orders of magnitudes of time in a single experimental run. Here, we should comment on the short-time value of the autocorrelation function, which is an indication of the homodyne or heterodyne regime of detection. As we have already mentioned, in the light scattering experiment we usually mix the inelastically scattered light with a reference light. This leads to the so-called mixing or heterodyning, which is very important for the interpretation of the autocorrelation functions (see for example Berne and Pecora 1976). There are two "clean" regimes of the detection: *(i)* In the homodyne regime ***only the inelastically scattered light*** is

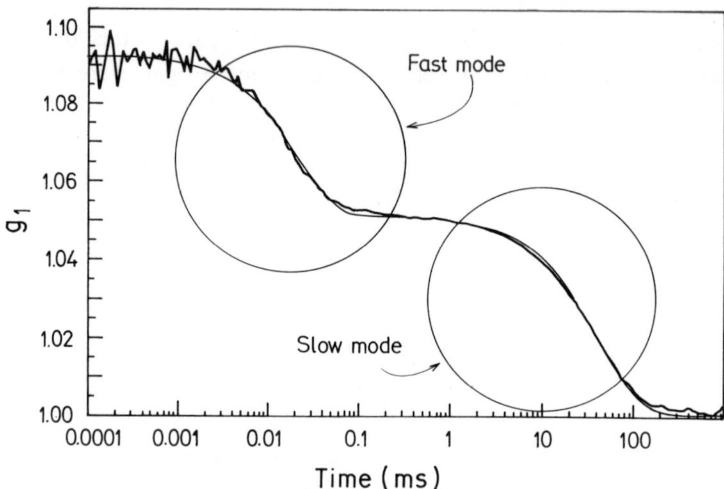

Fig.B4.2. An example of the normalized autocorrelation function of the scattered light intensity, as detected by the ALV-5000 autocorrelator. One can clearly see two single exponential relaxations.

detected. The normalized autocorrelation function of the scattered light intensity reaches a value of 2 in the short-time limit and approaches value of 1 in the long time limit. For a single relaxation light scattering process, the autocorrelation function decays twice faster with time than the corresponding relaxation itself. *(ii)* Usually, the inelastically scattered light is mixed coherently with a much stronger reference beam. In reality, this is done by collecting the light, which is elastically scattered on defects or impurities in a sample and mixing it with the inelastically scattered light. The magnitude of the autocorrelation function of the scattered light intensity is now for short times smaller than 2. Actually, we can control this value by adding more of the elastically scattered light, which generally reduces this value. It is very important that in this heterodyne regime, the short-time value of the autocorrelation function is smaller than approximately 1.1. This is an indication, that we are in the pure heterodyne regime. In this case, a single exponential relaxation process in the sample will be observed as a single exponential form of the autocorrelation function of the scattered light. The two relaxation times are then identical. If, however, the short-time value of the detected autocorrelation function of the scattered light is somewhere in between, let say $g_1 \approx 1.5$, the experiment is in the mixed regime. Here, the relaxation time of the scattering process cannot be determined unambiguously from the data. In some experiments, it is nearly impossible to bring the experiment into the pure heterodyne regime, because the elastic scattering on imperfections is for example, small. In this case it is possible to bring some light from the laser directly to the PMT and mix it coherently with the scattered light. This can be done by using optical fibers, but the set-up is very sensitive to mechanical perturbations and changes in temperature. One should note that in this case we have actually an interferometer, where two coherent beams are mixed onto the photomultiplier.

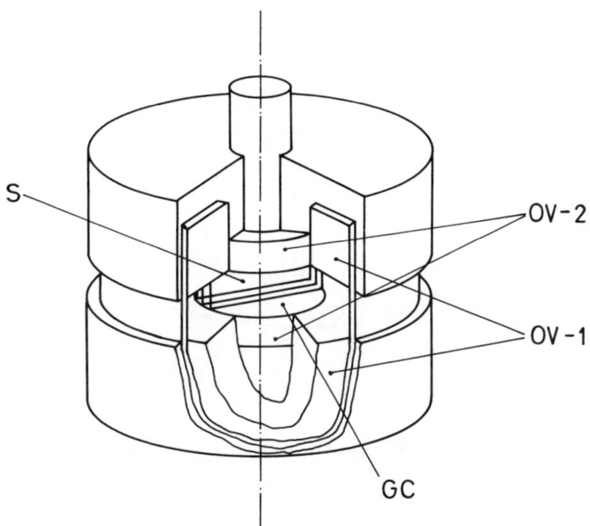

Fig.B4.3. Temperature-controlled stage for the quasielastic light scattering set-up. OV1, OV2-ovens; S-sample; GC-glass half-cylinders.

The sample, which is used in a light scattering experiment, is mounted in a suitable temperature-controlled oven, which is mounted on a rotation stage, to allow selection of the angle of the incident light. An example of such a stage is shown in Fig.B4.3. The stage is a double temperature controlled stage with two independent heaters that form the outer and the inner oven, respectively. The sample is mounted in the middle of the cell and is placed between two equal half-cylinders, made of glass, as shown in Fig.B4.4. The space between the sample and the glass cylinders is filled with transparent thermoplastic materials that softens at approximately $120^{\circ}C$. In this way, the sample and the two half-cylinders form a transparent cylinder with a diameter of $\approx 60mm$. This simple geometry with nearly no refraction at the interfaces facilitates the calculations of the scattering wave-vector and also extends the range of accessible wave-vectors in the experiment. The sample is mounted by simply melting the thermoplastic material and pressing both glass pieces against the sample surface.

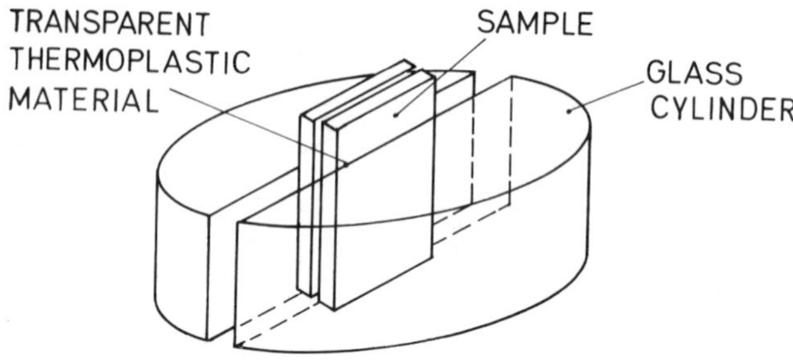

Fig.B4.4. The sample for the light scattering experiments is put in between two equal half cylinders which are made of glass.

One of the most important steps in a light scattering experiment is the calculation of the scattering wave-vector. The scattering experiments in birefringent media are here particularly demanding, because the scattering wave-vector depends on *(i)* the angle of incidence and the scattering angle, *(ii)* the polarization of the incident and scattered beam. Of course one should know the birefringence and the orientation of the optical axis.

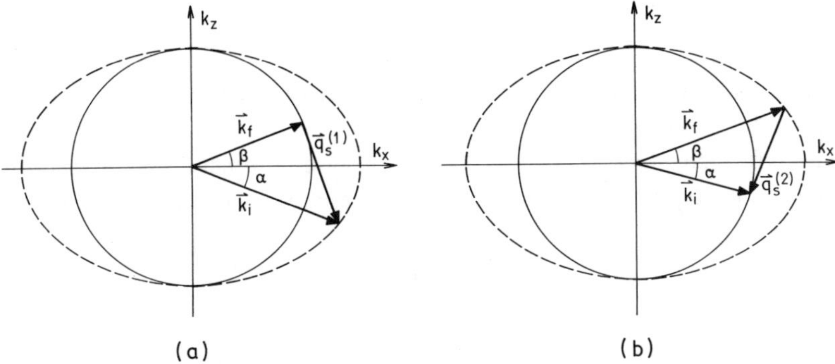

Fig.B4.5. In birefringent medium, the scattering wave-vector depends on the polarization properties of the incident and scattered beam. (a) Extraordinary incident beam is scattered into the ordinary beam. (b) The ordinary incident beam is scattered into the extraordinary beam. Note the difference in the two scattering wave-vectors.

Figures B4.5.(a) and (b) illustrate the importance of polarizations of the incident and scattered beam for the scattering wave-vector. In both cases the incident and the scattering angles are equal, the polarizations are however reverted in case (b) with respect to (a). One can clearly see that the two scattering wave-vectors $\vec{q}_s^{(1)}$ and $\vec{q}_s^{(2)}$ are quite different due to the optical anisotropy of the liquid crystal.

Finally, let us discuss the calculation of the scattering wave-vector in the birefringent medium. In this calculation, we shall consider a homeotropically aligned liquid crystal between two glass half-cylinders, as shown in Fig.B4.6. The input beam is ordinary polarized and the scattered beam is extraordinary polarized. It is straightforward to do similar calculations for other geometries and other polarizations as well.

Let us denote the incident angle by α and the scattering angle (or detector angle) by β. The difference between these two angles is denoted by $\delta = \beta - \alpha$. Note that there is no refraction at the glass-air interface. The index of refraction of glass is n_g , the ordinary index of refraction of liquid crystal is n_o , whereas the extraordinary index of liquid crystal depends on the direction of light propagation in liquid crystal

$$\frac{1}{n_e^2(\alpha')} = \frac{\sin^2\alpha'}{n_e^2} + \frac{\cos^2\alpha'}{n_o^2} \qquad \text{(B4.1.)}$$

Here, n_e is the maximum value of the extraordinary index of refraction of liquid crystal and α' is the angle of propagation of extraordinary wave in liquid crystal. After considering the refraction of both ordinary and extraordinary waves at glass-liquid crystal interface, the components of the scattering wave-vector are

$$q_x = \frac{4\pi}{\lambda_o} \cdot n_g \cdot \cos\left(\beta + \frac{\delta}{2}\right) \cdot \sin\frac{\delta}{2} \qquad \text{(B4.2a)}$$

$$q_z = \frac{2\pi}{\lambda_o} \cdot n_o \cdot \left[\sqrt{1 - \frac{n_g^2}{n_e^2} \cdot \sin^2\alpha} - \sqrt{1 - \frac{n_g^2}{n_o^2} \cdot \sin^2\beta}\right] \qquad \text{(B4.2b)}$$

Here, λ_o is the vacuum wavelength of laser light that is used in the experiment.

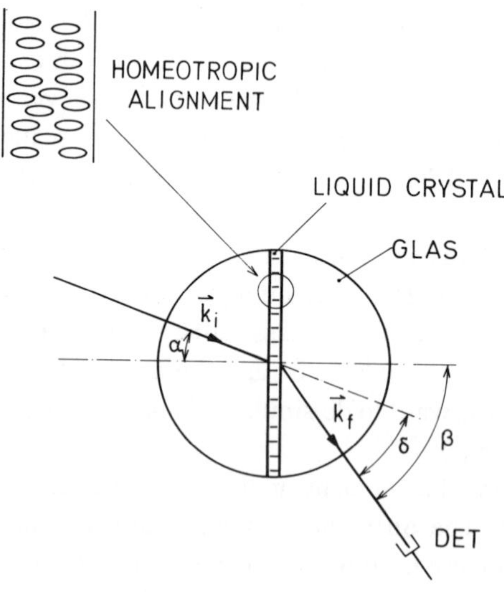

Fig.B4.6. The calculation of the scattering wave-vector for a simple homeotropic geometry.

B5. Experiments in High Magnetic Fields

The theoretical aspects of experiments with liquid crystals in an external magnetic field were discussed in Chapter 3., Symmetry breaking by external fields. In this Section, we shall briefly review some of the experimentally important aspects of these experiments.

High magnetic fields are generated in superconducting, resistive (Bitter) and combined (hybrid) magnets. There are several installations in the world, where access to high magnetic fields is possible. These are for example the High Field Magnet Laboratory (HFML) of the University of Nijmegen, The Netherlands, where most of the experiments, presented in this book, were performed, the Max-Planck Institute, Grenoble, France and the National Laboratory for High Magnetic Fields, Tallahassee, Florida State University, USA. In these installations, steady magnetic fields up to 40T can be generated for several hours. As an example, we shall describe the set-up for the quasielastic light scattering experiment in high magnetic fields. The light scattering set-up was planned for the 60 mm bore, resistive Bitter magnet at the HFML in Nijmegen, which is shown in Fig.B5.1.

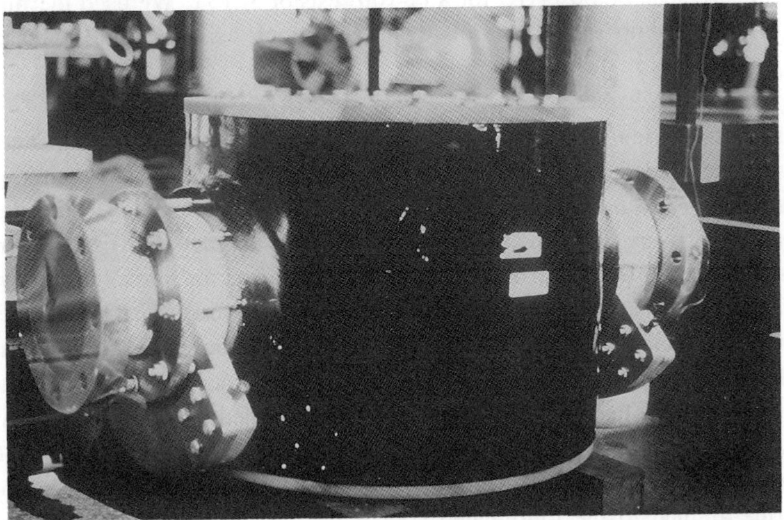

Fig.B5.1. 60 mm bore Bitter magnet at the HFML of the Katholieke Universiteit Nijmegen, The Netherlands. The highest magnetic field in this magnet is 15T.

Because this is an optical experiment in an external magnetic field, the following precautions had to be taken into account when constructing the set-up:

(i) Laser source and photomultipliers are sensitive to magnetic fields. For this reason, the laser should be placed not too close to the magnet, whereas the photomultiplier should be placed far away from the magnet.

(ii) High magnetic field induces large rotation of polarization of light, when light is traveling along the magnetic field. This Faraday rotation is especially large in condensed matter, like glass.

(iii) Magnetic field gradients exert large forces on the ferromagnetic and even diamagnetic materials, which are put into the magnet bore. We therefore used nonmagnetic metals like alumina or brass for the mechanical construction of the set-up.

(iv) Resistive temperature sensors are quite sensitive to external magnetic fields due to magnetoresistance. Temperature shifts of the order of half a degree can be observed for a Pt-100 resistor at 15T. We used thermistors for the temperature control, which are much less sensitive to magnetic fields.

The schematic view of the set-up is shown in Fig.B5.2. and B5.3. He-Ne laser light was expanded via L_1 and L_2 to a 6 mm diameter beam and then slightly focused to $\approx 100\mu m$ spot in the liquid crystalline sample (S), placed in a thermostated oven (OV) in the center of a Bitter magnet (BM). A system of adjustable miniature Al mirrors AM_1 and AM_2 was used for fine alignment of the beam. The polarization of the incoming beam was selected by the polarizer POL, whereas the incidence angle was defined by the rotation of the sample around the direction of the field. The scattered light in the quasielastic light scattering experiment was collected with an optical system formed by an adjustable mirror AM_3, pinhole (PIN1), lens (L_3), analyzer (AN), tilted mirror with a pinhole (M_4) and an eyepiece (EY). The collecting optics was mounted on a rotatable arm and was fine aligned to the illuminated spot with the mirror AM_3. The scattering plane of the experiment was thus perpendicular to the direction of the magnetic field.

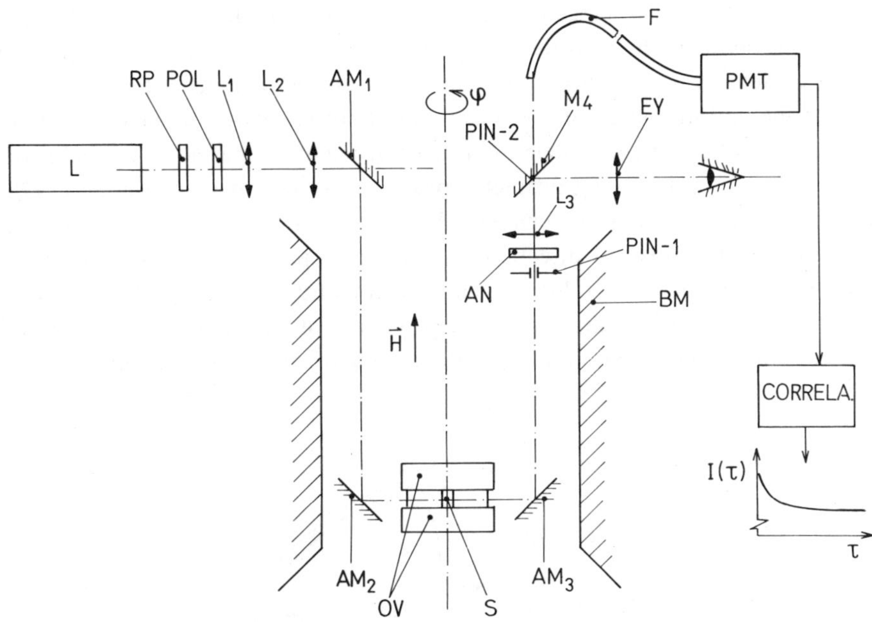

Fig.B5.2. The set-up for light scattering and linear response measurements in high magnetic fields. Definitions: L, laser; RP, polarization rotator; Pol, polarizer; L_1, L_2, L_3 lenses; AM_1, AM_2, AM_3 adjustable mirrors; OV, oven; S, sample; PIN-1, PIN-2, pinholes; AN, analyzer; M_4 mirror; EY, eyepiece; F, fibre; PMT, photomultiplier; BM, Bitter magnet.

The illuminated spot was observed with the lens L_3 and the eyepiece EY and was fine aligned to the $\approx 100 \mu m$ pinhole in the tilted mirror by translating the lens L_3. After inserting the pinhole PIN1, the scattered light from less then one coherence area was transmitted through the pinhole in the tilted mirror and captured into a multimode fibre (F). The collected light was led to the photomultiplier (PMT), placed far away from the magnet and connected to the photon counting unit and the digital clipped autocorrelator.

The same set-up was also used for the measurements of the linear electrooptic response and helical period. In the measurements of the electrooptic response, the light intensity, transmitted through the sample placed between crossed polarizers was detected. The polarization of the light was set at 45 deg to the scattering plane, whereas the angle between the sample normal and the light direction was set to the half-intensity point of the conoscopic figure. The probing electric field was applied in the direction of the magnetic field, and the magnitude and the phase of the transmitted light intensity was detected with a lock-in amplifier. The magnetic field dependence of the period of the helix was determined from the Bragg diffracted peaks. As shown by Garoff et al. (1978), the Bragg diffraction peak can be observed as a strong, static depolarized scattering in the forward direction in a homeotropic sample with sufficiently long period of the helix. This depolarized, forward scattered peak was indeed observed by recording the angular dependence of the transmitted light with ordinary incident and extraordinary outgoing polarizations or vice versa.

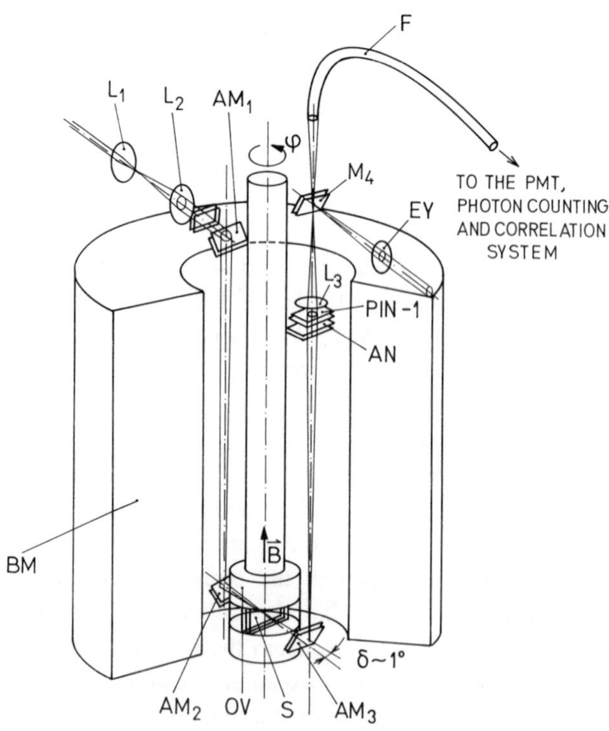

Fig.B5.3. The quasielastic light scattering set-up, shown in perspective.

Appendix C

Nuts and Bolts Approach

C1. Liquid Crystal Experimentalist's Collection of Nuts and Bolts

1. Temperature sensors:

(i) ***Platinum resistive thermometers.*** In these sensors, a thin wire or layer of platinum is used to measure the temperature via the temperature dependence of the resistance. Most popular are the Pt-100 resistors (Degussa, Sensycone), which have a resistance of 100Ω at zero degrees Celsius. The resistance increases linearly with increasing temperature, with a change of 0.39Ω per degree, or approximately 0.39%.

$^{\circ}C$	**0**	**1**	**2**	**3**	**4**	**5**	**6**	**7**	**8**	**9**
0	100.00	100.39	100.78	101.17	101.56	101.95	102.34	102.73	103.12	103.51
10	103.90	104.29	104.68	105.07	105.46	105.85	106.24	106.63	107.02	107.40
20	107.79	108.18	108.57	108.96	109.35	109.73	110.12	110.51	110.90	111.28
30	111.67	112.06	112.45	112.83	113.22	113.61	113.99	114.38	114.77	115.15
40	115.54	115.93	116.31	116.70	117.08	117.47	117.85	118.24	118.62	119.01
50	119.40	119.78	120.16	120.55	120.93	121.32	121.70	122.09	122.47	122.86
60	123.24	123.62	124.01	124.39	124.77	125.16	125.54	125.92	126.31	126.69
70	127.07	127.45	127.84	128.22	128.60	128.98	129.37	129.75	130.13	130.51
80	130.89	131.27	131.66	132.04	132.42	132.80	133.18	133.56	133.94	134.32
90	134.70	135.08	135.46	135.84	136.22	136.60	136.98	137.36	137.74	138.12
100	138.50	138.88	139.26	139.64	140.02	140.39	140.77	141.15	141.53	141.91
110	142.29	142.66	143.04	143.42	143.80	144.17	144.55	144.93	145.31	145.68
120	146.06	146.44	146.81	147.19	147.57	147.94	148.32	148.70	149.07	149.45
130	149.82	150.20	150.57	150.95	151.33	151.70	152.08	152.45	152.83	153.20
140	153.58	153.95	154.32	154.70	155.07	155.45	155.82	156.19	156.57	156.94

150	157.31	157.69	158.06	158.43	158.81	159.18	159.55	159.93	160.30	160.67
160	161.04	161.42	161.79	162.16	162.53	162.90	163.27	163.65	164.02	164.39
170	164.76	165.13	165.50	165.87	166.24	166.61	166.98	167.35	167.72	168.09
180	168.46	168.83	169.20	169.57	169.94	170.31	170.68	171.05	171.42	171.79
190	172.16	172.53	172.90	173.26	173.63	174.00	174.37	174.74	175.10	175.47
200	175.84	176.21	176.57	176.94	177.31	177.68	178.04	178.41	178.78	179.14

Table C1.1. Reference table for a Pt-100 resistor. The resistance is given in Ohms.

These resistors can be supplied in various shapes and dimensions as small as $1.5 \times 3mm$ (Sensycone, H 2104). They are stable, do not drift with time and can be used up to $400^{\circ} C$. However, they have rather large magneto-resistance and it is not advised to use them in magnetic-field experiments. Unfortunately, the relative change of the resistance of these resistors with temperature is relatively small and it is therefore difficult to obtain milli-Kelvin resolution with Pt-100 resistors. Table C1.1. shows the reference table for Pt-100 platinum resistance element.

(ii) ***Thermistors***: in contrast to metallic resistors, the resistance of a thermistor decreases nonlinearly with increasing temperature. The relative temperature change of the thermistor's resistance is typically an order of magnitude larger than for metallic resistors, see Table C1.2.

$^{\circ}C$	**0**	**1**	**2**	**3**	**4**	**5**	**6**	**7**	**8**	**9**
0	7355	6989	6644	6319	6011	5719	5444	5183	4937	4703
10	4482	4273	4074	3886	3708	3539	3378	3226	3081	2944
20	2814	2690	2572	2460	2354	2252	2156	2064	1977	1894
30	1815	1739	1667	1599	1533	1471	1412	1355	1301	1249
40	1200	1152	1107	1064	1023	983.8	946.2	910.2	875.3	842.8
50	811.3	781.1	752.2	724.5	697.9	672.5	648.1	624.8	602.4	580.9
60	560.3	540.5	521.5	503.3	485.8	469.0	452.9	437.4	422.5	408.2
70	394.5	381.2	368.5	356.2	344.5	333.1	322.3	311.8	301.7	292.0
80	282.7	273.7	265.0	256.7	248.6	240.9	233.4	226.2	219.3	212.6
90	206.1	199.9	193.9	188.1	182.5	177.1	171.9	166.9	162.0	157.3
100	152.8	148.4	144.2	140.1	136.1	132.3	128.6	125.0	121.6	118.2

110	115.0	111.8	108.8	105.8	103.0	100.2	97.6	95.0	92.5	90.0
120	87.7	85.4	83.2	81.1	79.0	77.0	75.0	73.1	71.3	69.5
130	67.8	66.1	64.4	62.9	61.3	59.8	58.4	57.0	55.6	54.3
140	53.0	51.7	50.5	49.3	48.2	47.0	45.9	44.9	43.8	42.8

Table C1.2. Reference table for YSI-44004 thermistor. The resistance is given in Ohms.

2. Temperature controllers:

A simple and inexpensive temperature controller is shown in the wiring diagrams in Figs.C1.4 and C1.5. The first part represents the Wheatstone bridge detector and a low-pass filter and the second part represents the PID controller and the current driving unit. The controller easily obtains a stability of several $10mK$ in a single-stage heating unit.

3. Phase linear photodiode preamplifier

The preamplifier uses BPX65 photodiode and a wideband operational amplifier OPA643 (Burr-Brown). OPA 643 amplifier has 1.5 GHz gain-bandwidth and has negligible phase shift up to 1MHz at a gain of 10^2.

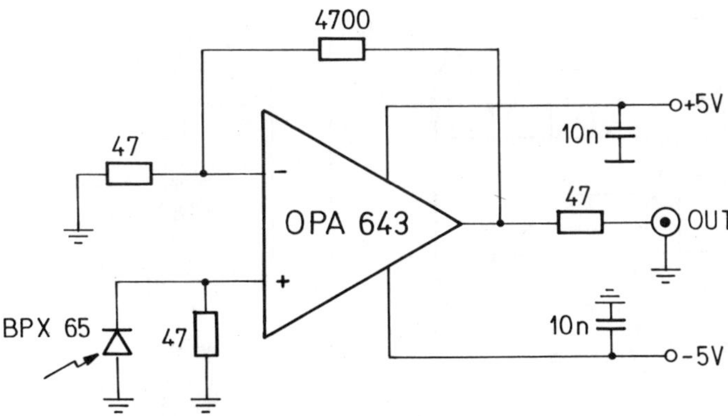

Fig.C1.3. The electronic circuit of the phase-linear photodiode preamplifier, which is phase linear up to 1MHz at a gain $10^2\ V/A$.

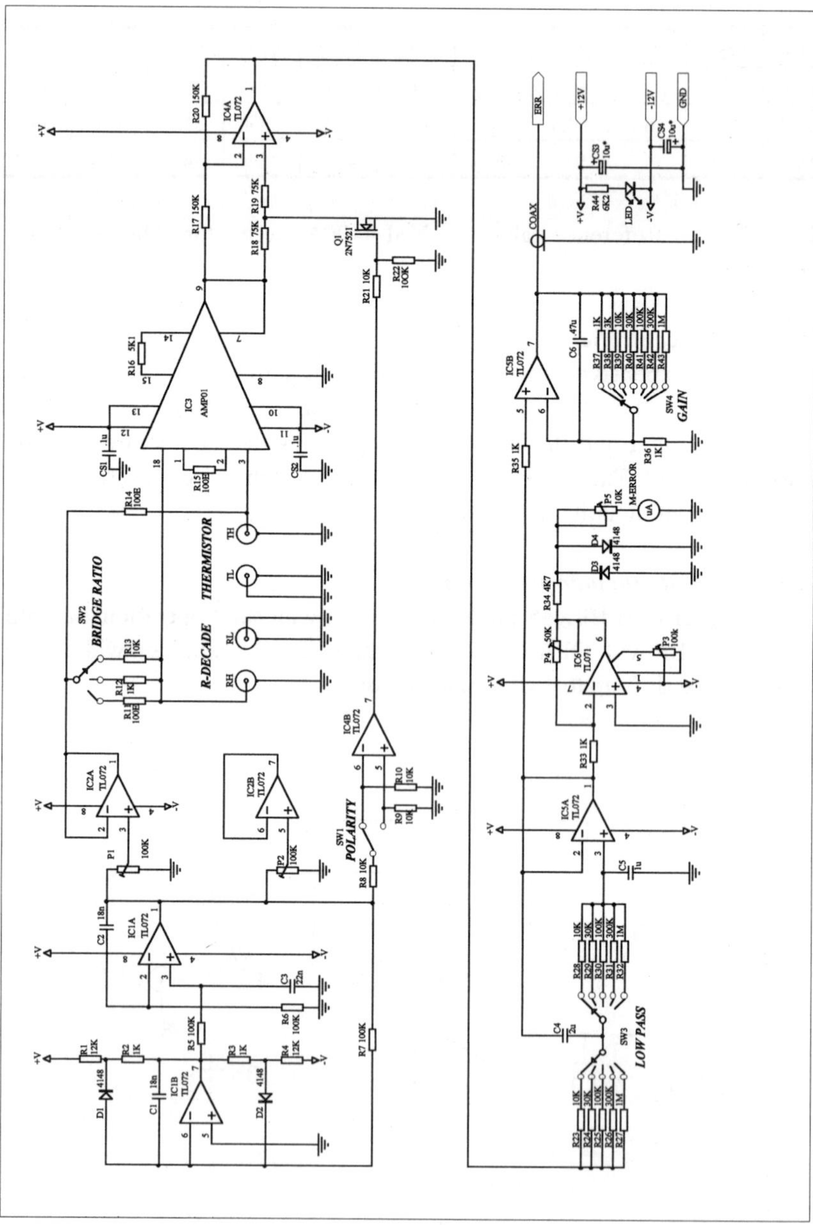

Fig.C1.4. The Wheatstone bridge with variable ratio and the low-pass filter of the input stage of the controller.

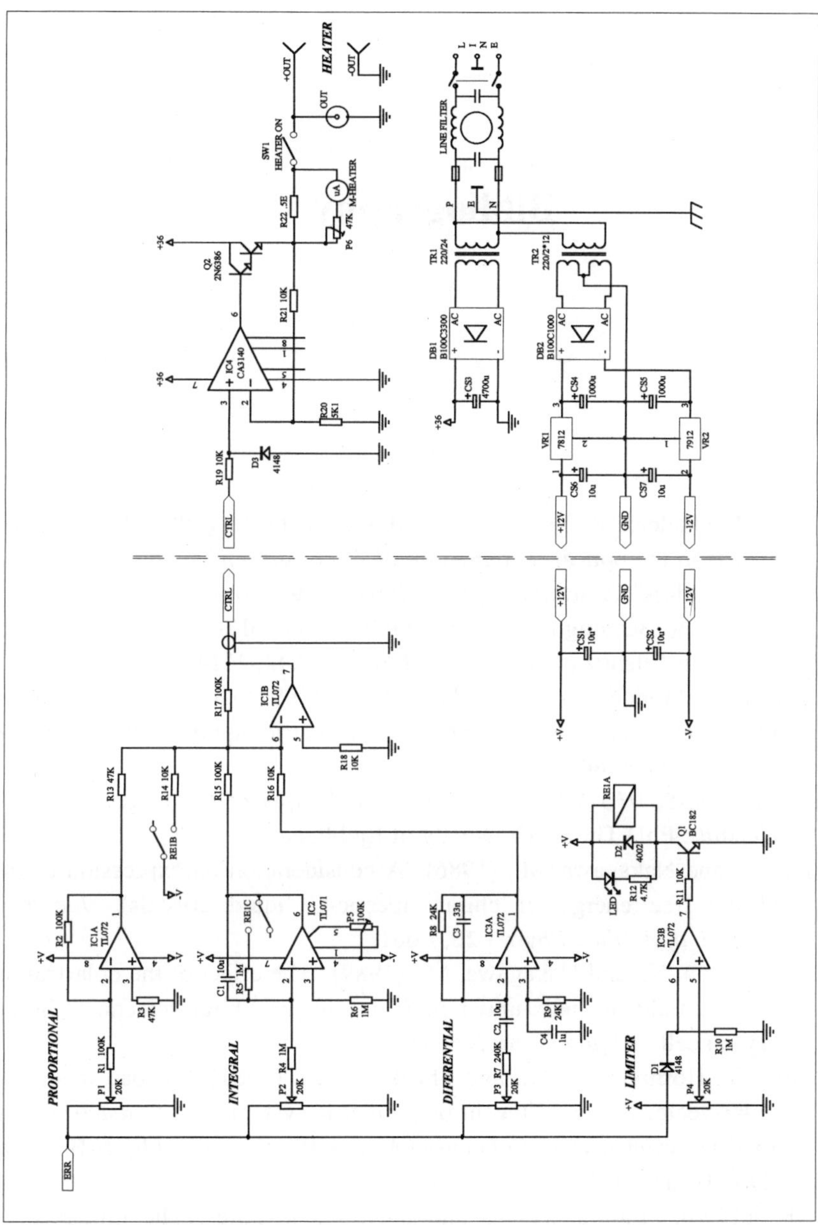

Fig.C1.5. The PID unit with the current driver and system power supply.

Bibliography

Abbate, G., Maddalena, P., Marucci, L., and Santamato, E., (1991) "Polarization effects in the optical reorientation of freely suspended smectic-C liquid crystal films", *Journal of Applied Physics* **69**, 1269-1274.

Abdulhalim, I., and Benguigui, L., 1990, "Optics of chiral smectic liquid crystals near their Lifshitz point", *Physical Review* **A42**, 2114.

Abdulhalim, I., Moddel, G., and Clark, N.A., (1992) "Director-polarization reorientation via solitary waves in ferroelectric liquid crystals", *Applied Physics Letters* **60**, 551-553.

Abramič, D., 1995, *^2H NMR Study of Hexatic Phases of Ferroelectric Liquid Crystals*, PhD Thesis, University of Ljubljana.

Akahane, T., and Nakagawa, M., (1986) "A consideration on expression of the elastic free energy of chiral smectic-C liquid crystals", *Japanese Journal of Applied Physics* **25**, L661.

Akahane, T., Itoh, N., and Nakagawa, M., (1989) "The effect of the polarization electric field on helix-unwinding in a planar chiral smectic C liquid crystal cell", *Liquid Crystals* **5**, 1107-1113.

Alexander, S., Hornreich, R.M., and Shtrikman, S., (1981) "Landau theory of polar-vector ordering in centro-symmetric systems", in *Symmetries and Broken Symmetries in Condensed Matter Physics*, edited by N.Boccara, IDSET-Paris, 1981.

Aliev, F., (1996), "Liquid crystals and polymers in pores: The influence of confinement on dynamic and interfacial properties", Chapter 17. in *Liquid Crystals in Complex Geometries*, edited by G.P.Crawford and S.Žumer, Taylor&Francis.

Allender, D.W., and Doane, J.W., (1978) "Biaxial order parameter s in liquid crystals: Their meaning and determination with NQR", *Physical Review* **A 17**, 1177.

Allender, D.W., Henderson, G.L., and Johnson, D.L., (1981) "Landau theory of wall-induced phase nucleation and pretransitional birefringence at the isotropic-nematic transition", *Physical Review* **A 24**, 1086-1089.

Allia, P., Galatola, P., Oldano, C., Rajteri, M., and Trossi, L., (1994), "Form birefringence in helical liquid crystals", Journal de Physique II France **4**, 333.

Amador, S.M., and Pershan, P.S., (1990), "Light scattering and ellipsometry studies of two-dimensional smectic-C to smectic-A transition in thin liquid crystal films", *Physical Review* **A 41**, 4326.

Anderson, P.W., (1981), "Some general thoughts about broken symmetry", in *Symmetries and Broken Symmetries in Condensed Matter Physics,* edited by N.Boccara, Idset-Paris.

Asahina, S., Sorai, M., Fukuda, A., Takezoe, H., Furukawa, K., Terashima, K., Suzuki, Y., and Kawamura, I., (1993) "Heat capacity and phase transitions of the antiferroelectric liquid crystals MHPOBC and MHPOCBC", *Abstracts* 4th International Conference on Ferroelectric Liquid Crystals, Tokyo, P-35, 147.

Ayton, G., and Patey, G.N., (1996) "Ferroelectric order in model discotic nematic liquid crystals", *Physical Review Letters* **76**, 239.

Bahr, Ch. and Heppke, G., (1987a) "Optical and dielectric investigations of the electroclinic effect exhibited by a ferroelectric liquid crystal with high spontaneous polarization", *Liquid Crystals* **2**, 825-831.

Bahr, Ch., Heppke, G., and Sharma, N.K., (1987b) "Dielectric studies of the SmC*-SmA transition of a ferroelectric liquid crystal with high spontaneous polarization", *Ferroelectrics* **76**, 151.

Bahr, Ch., and Heppke, G., (1990) "Influence of electric field on a first-order smectic-A-ferroelectric-smectic-C* liquid crystal phase transition: a field induced critical point", *Physical Review* **A 41**, 4335-4342.

Bahr, Ch., and Heppke, G., (1991) "Critical exponents of the electric-field-induced smectic-A-ferroelectric smectic-C* liquid crystal critical point", *Physical Review* **A 44**, 3669.

Bahr, Ch., and Fliegner, D., (1992a) "Behavior of a first order smectic-A-smectic-C transition in free-standing liquid crystal films", *Physical Review* **A 46**, 7657-7663.

Bahr, Ch., Heppke, G., and Wuthe, K., (1992b) "Dielectric, viscosity and pitch measurements of a ferroelectric liquid crystal exhibiting the phase sequence SmA-SmC", *Liquid Crystals* **12**, 997-1003.

Bahr, Ch., and Fliegner, D., (1993a) "Ferroelectric-antiferroelectric phase transition in a two-molecular-layer free-standing liquid-crystal film", *Physical Review Letters* **70**, 1842-1845.

Bahr, Ch., and Fliegner, D., (1993b) "Free-standing films of ferroelectric and antiferroelectric liquid crystals", *Ferroelectrics* **147**, 1-11.

Barberi, R., Giocondo, M., Sayko, G.V., Zvedin, A.K., (1994) "AFM experimental observation and fractal characterization of an SiO coated plate for nematic bistable anchoring", *Journal of Physics: Condensed Matter* **6**, A275.

Barrat, A., Silberzan, P., Bourdieu, L., and Chatenay, D., (1992) "How are the wetting properties of silanated surfaces affected by their structure? An AFM study", *Europhysics Letters* **20**, 633.

Bellini, T., Clark, N.A., and Schaefer, D.W., (1995) "Dynamic light scattering study of nematic and smectic-A liquid crystal ordering in silica aerogel", *Physical Review Letters* **74**, 2740-2743.

Belyakov, V.A., and Dmitrenko, V.E., (1989) *Optics of Chiral Liquid Crystals*, Soviet Scientific Reviews Section **A**, Vol.**13**, Part 1, Harwood Academic Publishers GmbH.

Benguigui, C., (1982) "Dielectric relaxation in a liquid crystal with helicoidal dipole ordering", *Journal de Physique (Paris)* **43**, 915.

Benguigui, L., (1984) "Dielectric properties and dipole ordering in liquid crystals", *Ferroelectrics* **58**, 269.

Benguigui, L., and Jacobs, A.E., (1988) "A model of a reentrant phase transition of a smectic-C* liquid crystal under magnetic field", *Ferroelectrics* **84**, 379.

Benguigui, L., and Jacobs, A.E., (1994) "Reentrant smectric-C and smectic C* phases in liquid crystals under an electric field", *Physical Review* **E 49**, 4221.

Benguigui, L., and Martinoty, P., (1997) "Characterizing the nature of the smectic-A-smectic-C and smectic-A-smectic-C* transition", *Journal de Physique II France* **7**, 225.

Beresnev, L.A., Chigrinov, V.G., Dergachev, D.I., Poshidaev, E.P., Funfschilling, J., and Schadt, M., (1989) "Deformed helix ferroelectric liquid crystal display: A new electrooptic mode in ferroelectric chiral smectic C liquid crystal", *Liquid Crystals* **5**, 1171.

Berne, B.J., and Pecora, R., (1976) *Dynamic Light Scattering,* John Wiley&Sons, Inc., New York.

Berreman, D.W., (1973a), "Twisted smectic-C phase: Unique optical properties", *Molecular Crystals and Liquid Crystals* **22**, 175.

Berreman, D.W., (1973b), "Optics in smoothly varying anisotropic planar structures: Applications to liquid crystal twist cells", *Journal of the Optical Society of America* **63**, 1374.

Binder, K., and Hohenberg, P.C., (1972) "Phase transitions and static spin correlations in Ising models with free surface", *Physical Review* **B 6**, 3461-3487.

Binnig, G., Rohrer, H., Gerber, Ch., and Weibel, E., (1982a) "Surface studies by scanning tunneling microscopy", *Physical Review Letters* **49**, 57.

Binnig, G., Rohrer, H., Gerber, Ch., and Weibel, E., (1982b) "Tunneling through a controllable vacuum gap", *Applied Physics Letters* **40**, 178.

Binnig, G., Quate, C.F., and Gerber, Ch., (1986) "Atomic Force Microscope", *Physical Review Letters* **56**, 930.

Biradar, A.M., Wrobel, S., and Haase, W., (1989) "Dielectric relaxation in the smectic-A and smectic-C* phases of a ferroelectric liquid crystal", *Physical Review* **A 39**, 2693-2702.

Birgeneau, R.J., Garland, C.W., Kortan, A.R., Litster, J.D., Meichle, M., Ocko, B.M., Rosenblatt, C., Yu, L.J., and Goodby, J., (1983) "Smectic-A-smectic-C transition: Mean field or critical", *Physical Review* **A 27**, 1251.

Bittel, H., Hellmiss, G., and Unruh, H.G., (1965) "Polarisationsschwankungen eines ferroelektrikums in der umbegung eines umwandlungspunktes zweiter art (Curie-punkt)", *Zeitschrift fur Physik* **184**, 1.

Blinc, R., and Žekš, B., (1974a) *Soft Modes in Ferroelectrics and Antiferroelectrics*, North Holland Publishing Company-Amsterdam, Oxford.

Blinc, R., Lugomer, S., and Žekš, B., (1974b) "Soft-mode dynamics in nematic liquid crystals", *Physical Review* **A 9**, 2214.

Blinc, R., Burgar, M., Luzar, M., and Pirš, J., (1974c) "Anisotropy of self-diffusion in the smectic-A and smectic-C phases, *Physical Review Letters* **33**, 1192.

Blinc, R., (1975) "Soft mode dynamics of smectic-C type ferroelectric liquid crystals", *Physica Status Solidi* **B 70**, K29.

Blinc, R., (1976) "Soft mode dynamics in ferroelectric liquid crystals", *Ferroelectrics* **14**, 603.

Blinc, R, and Žekš, B., (1978) "Dynamics of helicoidal ferroelectric smectic-C* liquid crystals", *Physical Review* **A 18**, 740-745.

Blinc, R., Seliger, J., Vilfan, M., and Žagar, V., (1979) " Temperature dependence of orientational ordering in the smectic-H and smectic-VI phases of TBBA", *Journal of Chemical Physics* **70**, 778.

Blinc, R., Vilfan, M., and Seliger, J., (1983) "^{14}N NQR and orientational ordering in smectic liquid crystals", *Bulletin of Magnetic Resonance* **5**, 51.

Blinc, R., Dolinšek, J., Luzar, M., and Seliger, J., (1988) "^{13}C NMR and ^{14}N NQR in ferroelectric liquid crystals: Polar versus quadrupolar order", *Liquid Crystals* **3**, 663.

Blinc, R., Muševič, I., Žekš, B., and Seppen, A., (1991a) "Ferroelectric liquid crystals in a static magnetic field", *Physica Scripta* **T35**, 38.

Blinc, R., (1991b) "Solid and liquid crystalline ferroelectrics and antiferroelectrics", *Condensed Matter News* **1**, 17.

Blinov, L.M., (1997), "Electric field effects in liquid crystals", chapter in *Handbook of Liquid Crystal Research*, editors Collings, P.J. and Patel, J.S., Oxford University Press.

Bloembergen, N., (1965), *Nonlinear Optics*, W.A.Benjamin, NY, Amsterdam.

Bludman, S.A., and Klein, A., (1963) "Broken symmetries and massless particles", *Physical Review* **131**, 2364.

Boccara, N., (1981), *Symmetries and Broken Symmetries,* editor, IDSET-Paris.

Born, M., and Wolf, E., *Principles of Optics*, Pergamon Press 1959.

Bouchiat, H., and Ocio, M., (1988) "Experimental investigation of 1/f magnetic fluctuations in spin glasses", *Comments on Condensed Matter Physics* **14**, 163.

Bourdon, L., Sommeria, J., and Kleman, M., (1982) "Sur l'existence de lignes singulieres dans les domaines focaux en phases SmC et SmC*", *Journal de Physique* **43**, 77-90.

Brand, H.R., and Pleiner, H., (1991), *Molecular Crystals and Liquid Crystals Letters* **8**, 11.

Brandow, S.L., Harrison, J.A., DiLella, D.P., Colton, R.J., Pfeiffer, S., and Shashidar, R., (1993) "STM study of the interfacial order in a ferroelectric liquid crystal", *Liquid Crystals* **13**, 163.

Brown, M.E., Chaney, J.D., Santasiero, B.D., and Hollingsworth, M.D., (1996), *Chemical Matterials* **8**, 1588.

Bruce, A.D., Cowley, R.A., and Murray, A.F., (1978a) "The theory of structurally incommensurate systems: II. Commensurate-incommensurate phase transitions", *Journal of Physics C: Solid State Physics* **11**, 3591.

Bruce, A.D., and Cowley, R.A., (1978b) "The theory of structurally incommensurate systems: III. The fluctuation spectrum of incommensurate phases", *Journal of Physics C: Solid State Physics* **11**, 3609.

Bruinsma, R., and Prost, J., (1994) "Fluctuation forces and Devil's staircase of ferroelectric smectic-C*'s", *Journal de Physique II France* **4**, 1209.

Brunet, M., and Williams, C., (1978) "Defauts dans les smectiques chiraux", *Annales de Physique* **3**, 237.

Byrd, P.F., and Friedman, M.D., (1954), *Handbook of elliptic integrals for engineers and physicists*, Springer Verlag.

Cagnon, M., and Durand, G., (1993) "Positional anchoring of smectic liquid crystals", *Physical Review Letters* **70**, 2742.

Callen, H.B., and Welton, Th.A., (1951) "Irreversibility and generalized noise", *Physical Review* **83**, 34.

Candel, V., and Galerne, Y., (1993a) "Anchoring strength onto simple surface edge dislocations in an induced smectic-*O* film", *Physical Review Letters* **70**, 4083-4086.

Candel, V., and Galerne, Y., (1993b) "Anomalous anisotropy of the elastic constants in the induced smectic-O films", *Liquid Crystals* **15**, 541.

Carlsson, T., Žekš, B., Filipič, C., Levstik, A., and Blinc, R., (1988) "Thermodynamic model of ferroelectric chiral smectic C* liquid crystal", *Molecular Crystals and Liquid Crystals* **163**, 11-71.

Carlsson, T., and Žekš, B., (1989) "The concept of rotational viscosities of smectic-C and chiral smectic-C* liquid crystals", *Liquid Crystals* **5**, 359.

Carlsson, T., Žekš, B., Filipič, C., and Levstik, A., (1990) "Theoretical model of the frequency and temperature dependence of the complex dielectric constant of a ferroelectric liquid crystals near the SmC*-SmA phase transition", *Physical Review* **A 42**, 877.

Cava, R.J., Patel, J.S., Collen, K.R., Goodby, J.W., and Rietman, E.A., (1987) "Thin cell dielectric response of a ferroelectric liquid crystal", *Physical Review* **A 35**, 4378-4388.

Chaikin, P.M., and Lubensky, T.C., (1995), *Principles of Condensed Matter Physics*, Cambridge University Press, Cambridge, England.

Chandani, A.D.L., Gorecka, E., Ouchi, Y., Takezoe, H., and Fukuda, A., (1989) "Antiferroelectric chiral smectic phases responsible for the tristable switching in MHPOBC", *Japanese Journal of Applied Physics* **28**, L1265-L1268.

Chandani, D.D.L., Ouchi, Y., Takezoe, H., Fukuda, A., and Terashima, K., (1990) "Novel phases exhibiting tristable switching", *Japanese Journal of Applied Physics* **28**, L1261.

Chen, W., Ouchi, Y., Moses, T., Shen, R., and Yang, K.H., (1992) "Surface electroclinic effect on the layer structure of a ferroelectric liquid crystal", *Physical Review Letters* **68**, 1547-1550.

Chen, W., Martinez-Miranda, L.J., Hsiung, H., and Shen, Y.R., (1989) "Orientational wetting behavior of a liquid-crystal homologues series", *Physical Review Letters* **62**, 1860.

Cheng, J.C., Naffie, L.A., and Stephens, P.J., (1975) "Polarization scrambling using a photoelastic modulator: Application to circular dichroism measurements", *Journal of the Optical Society of America* **65**, 1031.

Chiang, S., (1992) "Molecular imaging by STM", Chapter 9. in *Scanning Tunneling Microscopy I*, Springer Series in Surface Science, edited by Guntherodt, H.J., Wiesendanger, R.

Cladis, P.E., Brand, H.R., and Finn, P.L., (1983) "Soliton switch in chiral smectic liquid crystals", *Physical Review* **A 28**, 512.

Cladis, P.E., and van Saarloos, W., (1991), "Some nonlinear problems in anisotropic systems", Chapter 4. in *Solitons in Liquid Crystals*, edited by L.Lam and J.Prost, Springer Verlag

Cladis, P.E., and Brand, H.R., (1993) "Electrooptic response of smectic-O and smectic-O*", *Liquid Crystals* **14**, 1327.

Clark, N.A. and Lagerwall, S.T., (1980) "Submicrosecond bistable electro-optic switching in liquid crystals", *Applied Physics Letters* **36**, 899-901.

Clark, N.A., and Lagerwall, S.T., (1984) "Surface-stabilized ferroelectric liquid crystal electro-optics: New multistate structures and devices", *Ferroelectrics* **59**, 25-67.

Clark, N.A., and Rieker, T.P., (1988) "Smectic-C "chevron", a planar liquid-crystalline deffect: Implications for the surface stabilized ferroelectric liquid crystal geometry", *Physical Review* **A 37**, 1053-1057.

Clark, N.A. and Lagerwall, S.T., (1991), "Introduction to ferroelectric liquid crystals", Chapter 1. in *Ferroelectric Liquid Crystals*, edited by G.W.Taylor, Gordon and Breach Science Publishers.

Clark, N.A., and K.M.Johnson, (1992) "Applications of liquid crystals in optical computing", in *Liquid Crystals: Applications and uses, Vol.1.*, editor B.Bahadur, World Scientific.

Clark, N.A., Bellini, T., Malzbender, R.M., Thomas, B.N., Rappaport, A.G., Muzny, C.D., Schaefer, D.W., and Hrubesh, L., (1993) "X-ray scattering study of smectic ordering in a silica aerogel", *Physical Review Letters* **71**, 3505.

Cluzeau, P., Nguyen, H.T., Destrade, C., Isaert, N., Barois, P., and Babeau, A., (1995) "New chiral tolane series with antiferroelectric properties", *Molecular Crystals Liquid Crystals* **260**, 69.

Coates, D., (1992) "Smectic A LCD's", in *Liquid Crystals: Applications and uses, Vol.1.*, editor B.Bahadur, World Scientific.

Cochran, W., (1959) "Crystal stability and the theory of ferroelectricity", *Physical Review Letters* **3**, 412.

Cognard, J., (1982) "Alignment of nematic liquid crystals and their mixtures", *Molecular Crystals and Liquid Crystals*, Supplement 1.

Collings, P.J., and Hird, M., 1997, *Introduction to Liquid Crystals-Chemistry and Physics*, Taylor and Francis.

Conradi, M., Kityk, A.V., Škarabot, M., Blinc, R., and Muševič, I., (1999), "Electric field induced birefringence in a ferroelectric liquid crystal", *Liquid Crystals* **26**, 1179.

Cowley, R.A., and Bruce, A.D., (1978) "The theory of structurally incommensurate systems: I. Disordered-incommensurate phase transitions", *Journal of Physics C: Solid State Physics* **11** 3577.

Crandall, K.A., Tripathi, S., and Rosenblatt, Ch., (1992) "Surface-mediated electroclinic effect in a chiral nematic liquid crystal", *Physical Review* **A 46**, R715-718.

Crawford, G.P., Yang, D.-K, Žumer, S., Finotello, D., and Doane, J.W., (1991) "Ordering and self diffusion in the first molecular layer at a liquid crystal-polymer interface", *Physical Review Letters* **66**, 723.

Crawford, G.P., and Žumer, S., (1996), "Historical perspective of liquid crystals confined to curved geometries: From freely suspended droplets to flat panel displays", Chapter 1. in *Liquid Crystals in Complex Geometries*, Taylor and Francis,

Cummins, H.Z., and Pike, E.R., (1977) *Photon Correlation Spectroscopy and Velocimetry*, Nato Advanced Study Institute Series, Plenum Press, New York and London.

Čepič, M., (1993), *Antiferroelectric Liquid Crystals*, PhD thesis, University of Ljubljana.

Čepič, M., and Žekš, B., (1995) "Influence of competing interlayer interactions on the structure of the SmC_α^* phase", *Molecular Crystals and Liquid Crystals* **263**, 61-67.

Čepič, M., and Žekš, B., (1996) "Dynamics of chiral antiferroelectric liquid crystal materials exhibiting the SmC_α^* phase", *Liquid Crystals* **20**, 29-34.

Čepič, M., Žekš, B., (1997) "Electrostatic interlayer interactions in tilted polar smectic liquid crystals", *Molecular Crystals and Liquid Crystals* **301**, 221.

Čepič, M., Žekš, B., and Mavri, J., (1999),"Structures of possible phases of achiral polar smectic liquid crystals formed by bow shaped molecules", *Molecular Crystals and Liquid Crystals* **328**, 47.

de Gennes, P.G., (1968a) "Calcul de la distorsion d'une structure cholesterique par un champ magnetique", *Solid State Communications* **6**, 163.

de Gennes, P.G., (1968b) "Fluctuations d'orientation et diffusion Rayleigh dans un cristal nematique", *Comptes Rendus Academie des Sciences Paris*, **266**, B15-17.

de Gennes, P.G., (1969a) "Phenomenology of short-range-order effects in the isotropic phase of nematic materials", *Physics Letters* **30A**, 454.

de Gennes, P.G., (1969b) "Conjectures sur l'etat smectique", *Journal de Physique* **C4**, C4-65.

de Gennes, P.G., (1971) "Short range order effects in the isotropic phase of nematics and cholesterics", *Molecular Crystals Liquid Crystals* **12**, 193.

de Gennes, P.G., (1974), *The Physics of Liquid Crystals*, Oxford University Press.

Delaye, M., and Keller, P., (1976) "Critical angular fluctuations of molecules above a second order smectic-A to smectic-C phase transition", *Physical Review Letters* **37**, 1065.

Delaye, M., (1979) "Coherence length and angular susceptibility divergences above a smectic-A to smectic-C phase transition observed by Rayleigh scattering", *Journal de Physique* Colloque **C3**, C3-350

deMeyere, A., and Dahl, I., (1994) "Modelling the molecular distribution in chevron FLCD's", *Liquid Crystals* **17**, 397.

Demikhov, E., Hoffmann, U., and Stegemeyer, H., (1994) "Surface ordering and destruction of the first order phase transition smectic-A-smectic-C* in free standing smectic films", *Journal de Physique II France* **4**, 1865-1874.

Demikhov, E.I., Hoffmann, E., Stegemeyer, H., Pikin, S.A., and Strigazzi, A., (1995a) "Modulated structures of the smectic-C* phase in free-standing films with high spontaneous polarization", *Physical Review* **E 51**, 5954-5961.

Demikhov, E.I., (1995b) "Surface and bulk periodic structures in smectic-C films with one free boundary", *Physical Review* **E 51**, R12-R15.

de Vries, H.P., (1951), *Acta Crystallographica* **4**, 219.

Dianoux, A.J., Volino, F., (1979) "Molecular orientational order in TBBA: Results of a consistent analysis of neutron, NQR and NMR data", *Journal de Physique* **40**, 181.

Dierker, S.B., and Pindak, R., (1987) "Dynamics of thin tilted hexatic liquid crystal films", *Physical Review Letters* **59**, 1002-1005.

Dijon, J., Ebel.Ch., and Mulatier, L., (1988) "The role of polar and dispersive parts of surface tension on the bistability of ferroelectric liquid crystal cells", *Ferroelectrics* **85**, 47.

Dijon, J., (1992) "Ferroelectric LCD's", in *Liquid Crystals: Applications and uses, Vol.1.*, editor B.Bahadur, World Scientific.

Dmitrienko, V.E., and Belyakov, V.A., (1980*), Žurnal Eksperimentalnoj i Teoretičeskoj Fiziki* **78**, 4.

Doane, J.W., (1979) "NMR of liquid crystals", in *Magnetic Resonance of Phase Transitions*, ed. F.J.Ovens et al., New York: Academic Press.

Drevenšek, I., Muševič, I., and Čopič, M., (1990) "Dispersion of the SmC* order parameter fluctuations in the SmA and SmC* phases of 4-(2'-methylbutyl)phenyl 4'-n-octylbiphenyl-4-carboxylate", *Physical Review* **E 41**, 923-928.

Drevenšek, I., Muševič, I., and Čopič, M., (1991) "Orientational fluctuations in ferroelectric liquid crystals in static external electric fields", *Molecular Crystals and Liquid Crystals* **207**, 199-203.

Durand, G., Leger, L., Rondelez, F., and Veyssie, M., (1969) "Magnetically induced cholesteric-to-nematic phase transition in liquid crystals", *Physical Review Letters* **22**, 227.

Durand, G., and Martinot-Lagarde, Ph., (1980) "Physical properties of ferroelectric liquid crystals", *Ferroelectrics* **24**, 89.

Dumrongrattana, S., and Huang, C.C., (1986a) "Polarization and tilt angle measurements near the smectic-A-chiral-smectic-C transition of DOBAMBC", *Physical Review Letters* **56**, 464.

Dumrongrattana, S., Huang, C.C., Nounesis, G., Lien, S.C., and Viner, J.M., (1986b) "Tilt angle, polarization, and heat capacity measurements near the SmA-SmC* phase transitions of DOBAMBC", *Physical Review* **A 34**, 5010.

Dumrongrattana, S., and Huang, C.C., (1986c) "Temperature dependence of electrical critical field for one SmC* material", *Journal de Physique* **47**, 2117-2120.

Dzyaloshinskii, I.E., (1964) "Theory of helical structures in antiferromagnets I. Nonmetals", *Soviet Physics JETP* **19**, 960.

Elston, S.J., and Sambles, J.R., (1992) "Observation of dielectric biaxiality in a ferroelectric liquid crystal by the excitation of optic modes", *Molecular Crystals and Liquid Crystals* **220**, 99-104.

Ema, K., Yao, H., Kawamura, I., Chan, T., and Garland, C.W., (1993) "High resolution calorimetric study of the antiferroelectric liquid crystals MHPOBC and MHPOCBC", *Physical Review* **E 47**, 1203.

Ema, K., Watanabe, J., and Yao, H., (1994) "Heat capacity of antiferroelectric liquid crystal", *Ferroelectrics* **156**, 173.

Ema, K., Watanabe, J., Takagi, A., and Yao, H., (1995) "Critical behavior of heat capacity at the $smectic-C_{\alpha}^{*}$-$smectic-A$ transition of the antiferroelectric liquid crystal MHPOBC", *Physical Review* **E 52**, 1216-1221.

Ema, K., Takagi, A., and Yao, H., (1996a) "Gaussian tricritical behavior of heat capacity at the smectic-A-smectic-C liquid crystal transition", *Physical Review* **E 53**, R3036.

Ema, K., Ogawa, M., Takagi, A., and Yao, H., (1996b) "Crossover from XY critical to tricritical behavior of heat capacity at the smectic-A-chiral-smectic-C liquid crystal", *Physical Review* **E 54**, R25.

Ema, K., Yao, H., Fukuda, A., Takanishi, Y., Takezoe, H., (1996c) "Landau critical behavior of heat capacity at the $smectic-A$-$smectic-C_{\alpha}^{*}$

transition of the fantiferroelectric liquid crystal MHPOCBC", *Physical Review* **E 54**, 4450-4453.

Ema, K., Takagi, A., and Yao, H., (1997) "Gaussian tricritical behavior of the heat capacity at the $smectic-A-smectic-C^*$ liquid crystal transition in a racemic mixture of MHPOBC", *Physical Review* **E 55**, 508-515.

Ema, K., and Yao, H., (1998) "Crossover from XY critical to tricritical behavior of heat capacity at the smectic-A-chiral-smectic-C liquid crystal transition", *Physical Review* **E 57**, 6677.

Erdmann, J.H., Žumer, S., and Doane, J.W., (1990) "Configuration transition in a nematic liquid crystal confined to small spherical cavity", *Physical Review Letters* **64**, 1907.

Etxebarria, J., and Zubia, J., (1991) "Electroclinic effect in a liquid crystal with chiral nematic and smectic-A phases", *Physical Review* **A 44**, 6626.

Ezcurra, A., Perez Jubindo, M.A., de la Fuente, M.R., Etxebarria, J., Remon, A., and Tello, M.J., (1989) "Dielectric and optical studies near the chiral smectic C-smectic A transition of (S)-2-hydroxy-4-decyloxy benzylidene - 4'-amino-2p9-methylbutylcinnamate", *Liquid Crystals* **4**, 125.

Fan, C., Kramer, L., and Stephen, M.J., (1970) "Fluctuations and light scattering in cholesteric liquid crystals", *Physical Review* **A 2**, 2482.

Filipič, C., Levstik, A., Levstik, I., Blinc, R., and Žekš, B., (1987) "Spontaneous polarization of the helicoidal ferroelectric smectic-C* liquid crystals", *Ferroelectrics* **73**, 295.

Filipič, C., Carlsson, T., Levstik, A., Žekš, B., Blinc, R., Gouda, F., Lagerwall, S.T., and Skarp, K., (1988) "Dielectric properties near the SmC*-SmA phase transition of some ferroelectric liquid crystalline systems with a very large spontaneous polarization", *Physical Review* **A38**, 5833.

Findon, A., Gleeson, H.F., and Lydon, J., (1998) "An unusual structure in a chiral smectic-C liquid crystal device", *Liquid Crystals* **25**, 631.

Florin, E.L., Radmacher, M., Fleck, B., and Gaub, H., (1994) "Atomic force microscope with magnetic force modulation", *Review of Scientific Instruments* **65**, 639.

Forster, D., Lubensky, T.C., Martin, P.C., Swift, J., and Pershan, P.S., (1971) "Hydrodynamics of liquid crystals", *Physical Review Letters* **26**, 1016-1019.

Forster, D., (1975) *Hydrodynamic Fluctuations, Broken Symmetry and Correlation Functions*, Frontiers in Physics, W.A.Benjamin, Inc.

Foster, J.S., and Frommer, J.E., (1988) "Imaging of liquid crystals using a tunneling microscope", *Nature* **333**, 542.

Frank, F.C., (1958) "On the theory of liquid crystals", *Discussions of Faraday Society* **25**, 19.

Frisken, B.J., and Cannell, D.S., (1992) "Critical dynamics in the presence of a silica gel", *Physical Review Letters* **69**, 632-635.

Fukuda, A., Takanishi, Y., Isozaki, T., Ishikawa, K., and Takezoe, H., (1994) "Antiferroelectric chiral smectic liquid crystals", *Journal of Material Chemistry* **4**, 997.

Funfschilling, J., and Schadt, M., (1989) "Fast responding and highly multiplexible distorted helix ferroelectric liquid-crystal displays", *Journal of Applied Physics* **66**, 3877.

Funfschilling, J., and Schadt, M., (1991) "New short pitch bistable ferroelectric liquid crystal displays", *Japanese Journal of Applied Physics* **30**, 741.

Gale, R., McCaffery, A.J., and Shatwell, R., (1972), *Chemical Physics Letters* **17**, 416.

Galerne, Y., Martinand, J.L., Durand, G., and Veyssie, M., (1972) "Quasielastic light scattering in a smectic-C liquid crystal", *Physical Review Letters* **29**, 562-564.

Galerne, Y., (1981) "Interferometric measurements at a smectic-A-smectic-C-phase transition", *Physical Review* **A 24**, 2284

Galerne, Y., and Liebert, L., (1989) "The antiferroelectric smectic-*O* liquid crystal phase", Communication O-27, Second International Conference on Ferroelectric Liquid Crystals, Goteborg, Sweden.

Galerne, Y., and Liebert, L., (1990) "Smectic-*O* films", *Physical Review Letters* **64**, 906-909.

Galerne, Y., and Liebert, L., (1991) "Antiferroelectric chiral smectic-*O** films", *Physical Review Letters* **66**, 2891-2894.

Galerne, Y., (1992) "Optical analysis of the surface smectic-*O* films", *Europhysics Letters* **18**, 511-516.

Galerne, Y., (1996) "Self-screening of the ferroelectric polarization in smectic-C* in the bookshelf geometry", *Applied Physics Letters* **69**, 34.

Garoff, S., (1977a), *Ferroelectric Liquid Crystals*, PhD Thesis, Harward University, Cambridge, Massachusetts.

Garoff, S., and Meyer, R.B., (1977b) "Electroclinic effect at the A-C phase change in a chiral smectic liquid crystal", *Physical Review Letters* **38**, 848-851.

Garoff, S., Meyer, R.B., and Barakat, R., (1978a) "Kinematic and dynamic light scattering from the periodic structure of a chiral smectic C liquid crystal", *Journal of the Optical Society of America* **68**, 1217.

Garoff, S., and Meyer, R.B., (1978b) "Reply to: Behavior of electric susceptibility and electroclinic coefficient near the chiral smectic-A-C* transition", *Physical Review* **A 18**, 2739.

Garrof, S. and Meyer, R.B., (1979) "Electroclinic effect at the A-C phase change in a chiral smectic liquid crystal", *Physical Review* **A 19**, 338-347.

Geer, R., Huang, C.C., Pindak, R., and Goodby, J.W., (1989) "Heat-capacity anomaly from four-layer liquid crystal film", *Physical Review Letters* **63**, 540-543.

Geer, R., Stoebe, T., Pitchford, T., and Huang, C.C., (1991) "An AC calorimeter for measuring heat capacity of free-standing liquid-crystal films", *Review of Scientific Instruments* **62**, 415-421.

Giebelmann, F., and Zugenmaier, P., (1995) "Mean-field coefficients and the electroclinic effect of a ferroelectric liquid crystal", *Physical Review* **E 52**, 1762.

Gisse, P., Lorman, L., Pavel, J., and Nguyen, H.T., (1996) "On the existence of two ferrielectric phases in antiferroelectric liquid crystals", *Ferroelectrics* **178**, 297.

Gleeson, H.F., and Morse, A.S., (1996) "Microsecond studies of layer motion in a ferroelectric liquid crystal device", *Liquid Crystals* **21**, 755.

Glogarova, M., Lejček, L., Pavel, J., Janovec, V., and Fousek, J., (1983) "The influence of an external electric field on the structure of chiral SmC* liquid crystal", *Molecular Crystals and Liquid Crystals* **91**, 309-325.

Glogarova, M., and Pavel, J., (1984a) "The structure of chiral SmC* liquid crystal in planar samples and its change in an electric field", *Journal de Physique* **45**, 143-149.

Glogarova, M., and Pavel, J., (1984b) "The behavior of thin samples of ferroelectric liquid crystal", *Molecular Crystals and Liquid Crystals* **114**, 249-257.

Glogarova, M., Fousek, J., Lejček, L., and Pavel, J., (1984c) "The structure of ferroelectric liquid crystals in planar geometry and its response to electric fields", *Ferroelectrics* **58**, 161-178.

Glogarova, M., Sverenyak, H., Nguyen, H.T., and Destrade, Ch, (1993) "Dielectric and electrooptical study of a new antiferroelectric liquid crystal with the thiobenzoate group", *Ferroelectrics* **147**, pp.43-52.

Glogarova, M., Sverenyak, H., Holakovsky, J., Nguyen, H.T., and Destrade, C., (1995) "The thickness mode contribution to the permitivity of ferroelectric liquid crystals", *Molecular Crystals and Liquid Crystals* **263**, 245-254.

Goldstone, J., Salam, A., and Weinberg, S., (1962) "Broken symmetries", *Physical Review* **127**, 965.

Golemme, A., Žumer, S., Allender, D.W., and Doane, J.W., (1988) " Continuous nematic-isotropic transition in submicron-size liquid crystal droplets" , *Physical Review Letters* **61**, 2937.

Gorecka, E., Chandani, A.D.L., Ouchi, Y., Takezoe, H., and Fukuda, A., (1990) "Molecular orientational structures in ferroelectric, ferrielectric and antiferroelectric smectic liquid crystal phases as studied by conoscope observations", *Japanese Journal of Applied Physics* **29**, 131.

Gouda, F.M., (1992a), *Dielectric relaxation spectroscopy of chiral smectic liquid crystals*, PhD Thesis, Chalmers University of Technology, Goteborg, Sweden.

Gouda, F., Kuczynski, W., Lagerwall, S.T., Matuszczyk, M., Matuszczyk, T., and Skarp, K., (1992b) "Determination of the dielectric biaxiality in a chiral SmC* phase", *Physical Review* **A 46**, 951-958.

Groh, B., and Dietrich, S., (1994) "Long-ranged orientational order in dipolar fluids", *Physical Review Letters* **72**, 2422.

Guntherodt, H.J., Wiesendanger, R., editors, (1992) *Scanning Tunneling Microscopy I*, Springer Series in Surface Science.

Hall, A.W., Hollingshurst, J., and Goodby, J.W., 1997, "Chiral and achiral calamitic liquid crystals for display applications", Chapter 2. in *Handbook of liquid crystals*, edts.P.J.Collings and J.S.Patel, Oxford University Press.

Handschy, M.A., Clark, N.A., and Lagerwall, S.T., (1983) "Field-induced first order orientation transitions in ferroelectric liquid crystals", *Physical Review Letters* **51**, 471-474.

Handschy, M.A. and Clark, N.A., (1984) "Structures and responses of ferroelectric liquid crystals in the surface stabilized geometry", *Ferroelectrics* **59**, 69-116.

Hara, M., Iwakabe, Y., Tochigi, K., Sasabe, H., Garito, A.F., and Yamada, A., (1990) "Anchoring structure of smectic liquid crystal layers on MoS_2 observed by STM", *Nature* **344**, 228.

Hark, S.K., Hull, V.J., and Wysocki, J.J., (1988) "Switching response of surface stabilized ferroelectric liquid crystal in full-wave plate geometry", *Ferroelectrics* **85**, 551.

Hartshorne, N.H., and Stuart, A., (1964), *Practical Optical Crystallography*, Edward Arnold Publishers, London.

Heinekamp, S., Pelcovits, R.A., Fontes, E., Chen, E.Yi, Pindak, R., and Meyer, R.B., (1984) "Smectic-C* to smectic-A transition in variable-thickness liquid-crystal films: Order parameter measurements and theory", *Physical Review Letters* **52**, 1017-1020.

Heppke, G., Kleineberg, P., and Lötzsch, D., (1993) "Synthesis and liquid-crystalline properties of a homologous series of antiferroelectric liquid crystals and their respective racemates", *Liquid Crystals* **14**, 67-71.

Heppke, G., Kruerke, D., Loehning, C., Loetzsch, D., Rauch, S., and Sharma, N.K., (1997), poster P70, presented at "Freiburger Arbeitstagung Flussige Kristalle 1997, Freiburg.

Hide, F., Clark, N.A., Nito, K., Yasuda, A., and Walba, D.M., (1995) "Dynamic polarized infrared spectroscopy of electric field-induced molecular reorientation in a chiral smectic-A liquid crystal", *Physical Review Letters* **75**, 2344.

Hiji, N., Ouchi, Y., Takezoe, H., and Fukuda, A., (1988a) "Determination of chevron direction and sign of the boat-shaped disclination in SSFLC", *Japanese Journal of Applied Physics* **27**, L1.

Hiji, N., Ouchi, Y., Takezoe, H., and Fukuda, A., (1988b) "Structures of twisted states in ferroelectric liquid crystals studied by microspectrophotometry and numerical calculations", *Japanese Journal of Applied Physics* **27**, 8.

Hikmet, R.A.M., (1990) "Electrically induced light scattering from anisotropic gels", *Journal of Applied Physics* **68**, 4406-4412.

Hikmet, R.A.M., (1991) "Anisotropic gels and plasticized networks by liquid crystalline molecules", *Liquid Crystals* **9**, 405-416.

Hikmet, R.A.M., and de Witz, C., (1991) "Gel layers for inducing adjustable pretilt angles in liquid crystalline system", *Journal of Applied Physics* **70**, 1265-1269.

Hiller, A.C., and Bard, A.J., (1997) "AC-mode AFM imaging in air and solutions with thermally driven bimetallic cantilever probe", *Review of Scientific Instruments* **68**, 2082.

Hinshaw, G.A.Jr., Petschek, R.G., and Pelcovits, R.A., (1988) "Modulated phases in thin ferroelectric liquid-crystal films", *Physical Review Letters* **60**, 1864-1867.

Hiraoka, K., Taguchi, A., Ouchi, Y., Takezoe, H., and Fukuda, A., (1990) "Observation of three subphases in $smectic-C^*$ phase of MHPOBC by dielectric measurements", *Japanese Journal of Applied Physics* **29**, L103-L106.

Hiraoka, K., Takezoe, H., and Fukuda, A., (1993) "Dielectric relaxation modes in the antiferroelectric $smectic-C_A^*$ phase", *Ferroelectrics* **147**, pp.13-25.

Hiraoka, K., Uematsu, Y., Takezoe, H., and Fukuda, A., (1996) "Molecular fluctuations in smectic phases possesing antiferroelectric ordering", *Japanese Journal of Applied Physics* **35**, 6157.

Hohenberg, P.C., and Halperin, B.I., (1977) "Theory of dynamic critical phenomena", *Reviews of Modern Physics* **49**, 435.

Hornreich, R.M., Luban , M., and Shtrikman, S., (1975) "Critical behavior at the onset of k-space instability on the λ line", *Physical Review Letters* **35**, 1678.

Hornreich, R.M., and Shtrikman, S., (1977) "Critical behavior of the smectic-A to C phase transition in a magnetic field", *Physics Letters* **A 36**, 39.

Hsiung, H., Rasing, Th., and Shen, Y.R(1986) "Wall-induced orientational order of a liquid crystal in the isotropic phase-an evanescent-wave-ellipsometry study", *Physical Review Letters* **57**, 3065-3068.

Huang, C.C., and Viner, J.M., (1982) "Nature of the smectic-A-smectic-C phase transition in liquid crystals", *Physical Review* **A 25**, 3385.

Huang, C.C., and Dumrongrattana, S., (1986) "Generalized mean-field model for the smectic-A-chiral smectic-C phase transition", *Physical Review* **A34**, 5020.

Huang, C.C., Dumrongrattana, S., Nounesis, G., Stofko, J.J., and Arimili, P.A., (1987a) "Heat-capacity, tilt-angle, and polarization measurements near the smectic-A-chiral-smectic-C transition of one liquid crystal compound", *Physical Review* **A35**, 1460.

Huang, C.C., (1987b) "Nature of the smectic-A chiral smectic-C phase transition", *Molecular Crystals and Liquid Crystals* **144**, 1-16.

Huang, H.W., (1971) "Hydrodynamics of liquid crystals", *Physical Review Letters* **26**,1525.

Hudak, O., (1981a) "Double sine-Gordon equation: a stable 2π-kink and commensurate-incommensurate phase transitions", *Physics Letters* **A 82**, 95.

Hudak, O., (1981b) "The double sine-Gordon equation: On the nature of internal oscillations of the 2π-kink, *Physics Letters* **A 86**, 208.

Hudak, O., (1983), "Modulated smectic-C* uniform smectic-C phase transition in an electric field", *Journal de Physique* **44**, 57-66.

Hudak, O., (1984) "The smectic-C* structure disclinations in an applied electric field", *Ferroelectrics* **58**, 135-138

Indenbom, V.L., Pikin, S.A., and Loginov, E.B., (1976), *Kristalografija* **21**, 1093. See the translation: "Phase transitions and ferroelectric structures in liquid crystals", *Soviet Physics Crystallography* **21**, 632(1976).

Isozaki, T., Ishikawa, K., Takezoe, H., and Fukuda, A., (1993a) "Subphases in antiferroelectric liquid crystalline binary mixtures with infinite helical pitch or zero spontaneous polarization", *Ferroelectrics* **147**, 121.

Isozaki, T., Fujikawa, T., Takezoe, H., Fukuda, A., Hagiwara, T., Suzuki, Y., and Kawamura, I., (1993b) "Devil's staircase formed by competing interactions stabilizing the ferroelectric smectic-C* phase and the antiferroelectric phase in liquid crystalline binary mixtures", *Physical Review* **B 48**, 13439.

Israelachvili, J., 1995, *Intermolecular and surface forces*, Academic Press.

Israeloff, N.E., (1996) "Dielectric polarization noise through the glass transition", *Physical Review* **B 18**, R11913.

Itoh, N., Kido, M., Tagawa, A., Koden, M., Miyoshi, S., and Wada, T., (1992) "Effect of surface pretilt angle on optical properties of SSFLC", *Japanese Journal of Applied Physics* **31**, L1089.

Iwakabe, Y., Hara, M., Kondo, K., Tochigi, K., Mukoh, A., Garito, A.F., Sasabe, H., and Yamada, A., (1990) "Two types of anchoring structure in smectic liquid crystal molecules", *Japanese Journal of Applied Physics* **29**, L2243.

Jacobs, A.E., and Walker, M.B., (1980), "Phenomenological theory of charge-density-wave state in trigonal-prismatic, transition-metal dichalcogenides", *Physical Review* **E 21**, 4132.

Jacobs, A.E., and Benguigui, L., (1989) "Landau theory of the helicoidal C* phase in smectic liquid crystals: Re-entrance of the smectic-C* phase and order of the smectic-C-smectic-C* transition", *Physical Review* **A 39**, 3622.

Jähnig, F., and H.Schmidt, (1971), *Annals of Physics* **71**, 129.

Johno, M., Itoh, K., Lee, J., Ouchi, Y., Takezoe, H., Fukuda, A., and Kitazume, T., (1990) "Temporal and spatial behaviour of the field-induced transition between the antiferroelectric and ferroelectric phases in chiral smectics", *Japanese Journal of Applied Physics* **29**, L107-L110.

Johnson, J.B., (1928), *Physical Review* **32**, 97.

Kadanoff, L.P., Goetze, W., Hamblen, D., Hecht, R., Lewis, E.A.S., Palciauskas, V.V., Rayl, M., and Swift, J., (1967) "Static phenomena near critical points: Theory and experiments", *Reviews of Modern Physics* **39**, 395.

Kaganov, M.I., and Omelyanchuk, A.N., (1971) "Phenomenological theory of phase transitions in thin ferromagnetic films", *Zhournal Eksperimentalnoj i Teoreticheskoj Fiziki* **61**, 1679-1685. See also translation in *Soviet Physics, JETP* **34**, 895(1972).

Kai, S., Nakagawa, M., Narushige, Y., and Imasaki, M., (1983) "Anomalous behavior of helicoidal pitch in surface induced transition from winding to unwinding chiral smectic C", *Japanese Journal of Applied Physics* **22**, L488-L490.

Kemp, J.C., (1969) "Piezo-optical birefringence modulators: New use for a long-known effect", *Journal of the Optical Society of America* **59**, 950.

Killingbeck, J., and Cole, G.H.A., 1971, *Mathematical Techniques and Phzsical Applications*, Academic Press.

Kim, K.H., Takanishi, Y., Ishikawa, K., Takezoe, H., and Fukuda, A., (1994) "Phase transitions and conformational changes in an antiferroelectric liquid crystal MHPOBC", *Liquid Crystals* **16**, 185-202.

Kim, K.H., Ishikawa, K., Takezoe, H., and Fukuda, A., (1995) "Orientation of alkyl chains and hindered rotation of carbonyl groups in the smectic-C* phase of antiferroelectric liquid crystals studied by polarized FTIR spectroscopy", *Physical Review* **E 51**, 2166.

Kimura, Y., Hayakawa, R., Okabe, N., and Suzuki, Y., (1996) "Nonlinear dielectric relaxation spectroscopy of the antiferroelectric liquid crystal TFMHPOBC", *Physical Review* **E 53**, pp.6080-6084.

Koda, T., and Kimura, H., (1996) "Head-tail assymetry of a molecule as a microscopic origin of the successive phase transitions in chiral smectics", *Journal of the Physical Society of Japan* **65**, 2880.

Kondo, K., Takezoe, H., Fukuda, A., and Kuze, E., (1982) "Temperature sensitive helical pitches and wall anchoring effects in homogeneous monodomains of ferroelectric SmC* liquid crystal nOBAMBC (n=6-10)", *Japanese Journal of Applied Physics* **21**, 224-229.

Kondo, K., Takezoe, H., Fukuda, A., Kuze, E., Flatischler, K., and Skarp, K., (1983) "Surface induced helix-unwinding process in thin homogeneous ferroelectric smectic cells of DOBAMBC", *Japanese Journal of Applied Physics* **22**, L294-L296.

Konovalov, D., and Sprunt, S., (1998) "Light scattering study of the $smectic-C_{\alpha}^{*}$ phase of a chiral liquid crystal", *Physical Review* **E 58**, 6869.

Kralj, S., and Sluckin, T.J., (1994) "Landau-de Gennes theory of the chevron structure in a smectic-A liquid crystal", *Physical Review* **E 50**, 2940-2951.

Kralj, S., Pirš, J., Žumer, S., and Petkovšek, R., (1998) "Influence of the inhomogeneous surface pretilt on zig-zag deffects", *Journal of Applied Physics* **84**, 4761.

Kraus, I., Pieranski, P., Demikhov, E., Stegemeyer, H., and Goodby, J., (1993) "Destruction of a first-order smectic-A-smectic-C* phase transition by dimensional crossover in free-standing films", *Physical Review* **E 48**, 1916-1923.

Kremer, F., Vallerien, S.U., Kapitza, H., Zentel, R., and Fischer, E.W., (1990) "Constant molecular rotation at the smectic-A to smectic-C* transition in ferroelectric liquid crystal", *Physical Review* **A 42**, 3667.

Krishna Prasad, S., Khened, S.M., Raja, V.N., Chandrasekhar, S., and Shivkumar, B., (1993) "Dielectric studies in the vicinity of the A-C* transition", *Ferroelectrics* **138**, 37-49.

Kubo, R., (1957) "Statistical-mechanical theory of irreversible processes", *Journal of the Physical Society of Japan* **12**, 579.

Kuczynski, W., Stryla, B., Hoffman, J., and Malecki, J., (1990) "Investigation of the smectic C*-smectic A phase transition in electric field", *Molecular Crystals and Liquid Crystals* **192**, 301-305.

Kutnjak, Z., Levstik, A., and Žekš, B., (1990) "Degeneracy of rotational viscosities at the chiral SmC*-SmA transition in DOBAMBC", *Liquid Crystals* **8**, 589-592.

Kutnjak, Z., (1991), *Dielectric properties and the rotational viscosity in ferroelectric liquid crystals*, MsC Thesis, University of Ljubljana.

Kutnjak-Urbanc, B., (1990), *Spontaneous polarization in ferroelectric liquid crystals,* MsC Thesis, University of Ljubljana.

Kutnjak-Urbanc, B., and Žekš, B., (1992) "The phase transition from the SmC* to the SmC phase induced by an external magnetic field", in *Phase Transitions in Liquid Crystals*, Edts.S.Martellucci and A.N.Chester, Plenum Press, New York.

Kutnjak-Urbanc, B., (1993a) *Ferroelectric liquid crystals in an external magnetic and electric field*, Ph.D. Thesis, University of Ljubljana

Kutnjak-Urbanc, B., and Žekš, B., (1993b) "Theoretical investigations of the behavior of ferroelectric liquid crystals in a magnetic or in a high-frequency electric field", *Physical Review* **E 48**, 455.

Kutnjak-Urbanc, B., Žekš, B., and Rovšek, B., (1993c) "The influence of finite dimensions on the static ordering of the SmC* phase in an electric field", *Liquid Crystals* **14**, 999.

Kutnjak-Urbanc, B., and Žekš, B., (1995a) "Phase excitation spectrum of ferroelectric liquid crystals in an external static electric field", *Physical Review* **E 52**, 3892-3903.

Kutnjak-Urbanc, B., and Žekš, B., (1995b) "Microscopic origin of spontaneous polarization in ferroelectric smectic-C* liquid crystals", *Liquid Crystals* **18**, 483.

Lagerwall, S.T., (1984) "Ferroelectric liquid crystals", *Molecular Crystals and Liquid Crystals* **114**, 151.

Lagerwall, S.T., (1996) "Can liquids be macroscopically polar?", *Journal of Physics: Condensed Matter* **8**, 9143.

Lagerwall, S.T., (1998) "Ferroelectric Liquid Crystals", in *Handbook of Liquid Crystals*, edited by D.Demus, J.Goodby, G.W.Gray, H.W.Spiess, V.Vill, Wiley-VCH.

Lallane, J.R., Buchert, J., Destrade, C., Nguyen, H.T., and Marcerou, J.P., (1989) "Slowing down of molecular rotation at the smectic-A-smectic-C* transition in liquid crystals", *Physical Review Letters* **26**, 3046.

Landau, L.D., and Khalatnikov, I.M., (1954) "On the anomalous absorption of sound near the second order phase transitions", *Doklady Akademii Nauk SSSR* **96**, 469.

Landau, L.D., and Lifshitz E.M., (1958) *Quantum Mechanics*, 2nd edition, Oxford, Pergamon.

Lange, R.V., (1995), "Goldstone theorem in nonrelaticvistic theories", *Physical Review Letters* **14**, 3.

Lange, R.V., (1966), "Nonrelativistic theorem analogous to the Goldstone theorem", *Physical Review* **146**, 301.

Leadbetter, A.J., Richardson, R.M., and Frost, J.C., (1979) "The dynamics of the crystal S_E, S_C and S_A phases of IBPBAC", *Journal de Physique Colloque* **C3-40**, 125.

Lee, Eung Sang, Saito, Y., and Uchida, T., (1993) "Detailed morphology of rubbed alignment layers and surface anchoring of liquid crystals", *Japanese Journal of Applied Physics* **32**, L1822.

Lee, Sin-Doo and Patel, J., (1989) "Temperature and field dependence of the switching behaviour of induced molecular tilt near the smectic A-C* transition", *Applied Physics Letters* **55**, 122.

Lee, Sin-Doo, and Patel, J.S., (1990) "Symmetry breaking effect of interfacial interactions on electro-optical properties of liquid crystals", *Physical Review Letters* **65**, 56-59.

Lejček, L., (1984) "The disclination model of field induced transition between twisted and uniform structure in thin planar SmC* samples", *Ferroelectrics* **58**, 139-148.

Lejček, L., and Pirkl, S., (1990) "Memory effect in thin samples of chiral smectic-C liquid crystals", *Liquid Crystals* **8**, 871.

Leung, On Man, and Goh, M.C., (1992) "Orientational ordering of polymers by AFM tip-surface interaction", *Science* **255**, 64.

Levelut, A.M., (1976) "Etude de l'ordre local lie a la rotation des molecules dans la phase smectique B", *Journal de Physique* **C3-37**, 51.

Levelut, A.M., Germain, C., Keller, P., Liebert, L., and Billard, J., (1983) "Two new mesophases in a chiral compound", *Journal de Physique* **44**, 623.

Levstik, A., Prelovšek, P., Filipič, C., and Žekš, B., (1982) "Dielectric constant in the incommensurate phase of Rb_3ZnCl_4", *Physical Review* **B 25**, 3419.

Levstik, A., Carlsson, T., Filipič, C., Levstik, I., and Žekš, B., (1987) "Goldstone mode and soft mode at the smectic-A-smectic-C* phase transition studied by dielectric relaxation", *Physical Review* **A 35**, 3527-3534.

Levstik, A., Carlsson, T., Filipič, C., and Žekš, B., (1988) "Dielectric relaxation and the determination of the SmA-SmC* phase transition in DOBAMBC", *Molecular Crystals and Liquid Crystals* **154**, 259.

Levstik, A., Kutnjak, Z., Filipič, C., Levstik, I., Bregar, Z., Žekš, B., and Carlsson, T., (1990) "Dielectric method for determining the rotational viscosity in thick samples of ferroelectric chiral SmC* liquid crystals", *Physical Review* **A 42**, 22042210.

Levstik, A., Kutnjak, Z, Žekš, B., Dumrongrattana, S., and Huang, C.C., (1991a) "Dielectric method for determining the electric critical field in ferroelectric liquid crystals", *Journal de Physique II (Paris)* **1**, 797.

Levstik, A., Kutnjak, Z., Filipič, C., Levstik, I., Žekš, B., and Carlsson, T., (1991b) "A dielectric method for determining the rotational viscosity in ferroelectric liquid crystals", *Ferroelectrics* **113**, 207-217.

Levstik, A., Kutnjak, Z., Filipič, C., and Žekš, B., (1992) "Rotational viscosities in ferroelectric liquid crystal SCE9", *Ferroelectrics* **126**, 139.

Li, J.-F., Rosenblatt, C, Li, and Zili, (1995) "Tilt susceptibility at an antiferroelectric smectic-C_A*-smectic-A phase transition", *Physical Review* **E 51**, 462.

Li, Zili and Rosenblatt, Ch., (1989) "Origin of the anomalous electroclinic susceptibility exponent in a ferroelectric liquid crystal", *Physical Review* **A 39**, 1594.

Li, Zili, Akins, R.B., DiLisi, G.A., Rosenblatt, Ch., and Petschek, R.G., (1991) "Anomaly in the dynamic behaviour of the electroclinic effect below the nematic-smectic-A phase transition", *Physical Review* **A 43**, 852.

Lim, Khoon-Cheng, and Ho, John T., (1978) "Critical suppression of birefringence in a smectic-A liquid crystal near the smectic-C phase", *Physical Review Letters* **40**, 1576.

Limat, L., and Prost, J., (1993) "A model for the chevron structure obtained by cooling a smectic-A liquid crystal in a cell of finite thickness", *Liquid Crystals* **13**, 101-113.

Link, D.R., Natale, G., Shao, R., Maclennan, J.E., Clark, N.A., Körblova, E., and Walba, D., (1997), "Spontaneous formation of macroscopic chiral domains in a fluid smectic phase of achiral molecules", *Science* **278**, 1924.

Link, D.R., Natale, G., Clark, N.A., Maclennan, J.E., Walsh, M., Keast, S.S., and Neubert, M.E., (1999), "Antoclinic smectic-C surfaces on smectic-A freely suspended liquid-crystal film", *Physical Review Letters* **82**, 2508.

Litster, J.D., and T.W.Stinson, (1970) "Critical slowing of fluctuations in a nematic liquid crystal", *Journal of Applied Physics* **41**, 996.

Litster, J.D., Stinson, T.W., and Clark, N.A., (1971)"Observation of coupling between orientational order fluctuations and hydrodynamic shear modes in a liquid crystal", *Bulletin of the American Physical Society* **16**, 519.

Litster, J.D., (1983) "Scattering spectroscopy of liquid crystals", in *Light Scattering near Phase Transitions*, edited by H.Z.Cummins and A.P.Levanyuk (North-Holland Publishing Company).

Lo, W.S., Pelcovits, R.A., Pindak, R., and Srajer, G., (1990) "Dynamical behavior of thin ferroelectric liquid crystal films in an ac electric fields", *Physical Review* **A 42**, 3630-3633.

Lorman, V.L., Bulbitch, A.A., and Toledano, P., (1994) "Theory of reorientational transitions in ferrielectric liquid crystals", *Physical Review* **E 49**, 1367-1374.

Lorman, V.L., (1995) "Antiferroelectric and ferrielectric structures induced by multilayer ordering in chiral smectics", *Molecular Crystals and Liquid Crystals* **262**, 437-453.

Lorman, V.L., (1996) "Ferrielectric smectic phases: Liquid crystal structure and macroscopic fluctuations", *Liquid Crystals* **20**, 267.

Lu, Min-Hua, Crandall, K.A., and Rosenblatt, Ch., (1992) "Polarization-induced renormalization of the B_1 elastic modulus in a ferroelectric liquid crystal", *Physical Review Letters* **68**, 3575-3578.

Luban, M., Mukamel, D., and Shtrikman, S., (1974) "Transition from the cholesteric storage mode to the nematic phase in critical restricted geometries", *Physical Review* **A 10**, 360-367.

Luzar, M., Rutar, J., Seliger, J., Burgar, M., and Blinc, R., (1980) "^{13}C NMR study of the ferroelectric liquid crystal DOBAMBC", *Ferroelectrics* **24**, 215.

Luzar, M., Rutar, V., Seliger, J., and Blinc, R., (1984) "^{13}C NMR in ferroelectric smectic liquid crystals", *Ferroelectrics* **58**, 115.

Ma, Shang-Keng, (1976), *Modern Theory of Critical Phenomena*, W.A.Benjamin, Inc., Reading Massachusetts.

Mach, P., Pindak, R., Levelut, A.M., Barois, P., Nguyen, H.T., Huang, C.C., and Furenlid, L., (1998) "Structural characterization of various chiral Smectic-C phases by resonant X-ray scattering", *Physical Review Letters* **81**, 1015.

Macdonald, R., Kentischer, F., Warnick P, and Heppke, G., (1998) "Antiferroelectricity and chiral order in new liquid crystals of nonchiral molecules studied by optical second harmonic generation", *Physical Review Letters* **81**, 4408.

Maclennan, J.E., Clark, N.A., Handschy, M.A., and Meadows, M.R., (1990) "Director orientation in chevron SSFLC cells", *Liquid Crystals* **7**, 753.

Maclennan, J.E., Clark, N.A., and Handschy, M.A., (1991), "Solitary waves in ferroelectric liquid crystals", Chapter 5. in *Solitons in Liquid Crystals*, edited by Lam, L., and Prost, J., Springer-Verlag.

Maclennan, J.E., Sohling, U., Clark, N.A., and Seul, M., (1994) "Textures in hexatic films of nonchiral liquid crystals: Symmetry breaking and modulated phases", *Physical Review* **E 49**, 3207-3224.

Mahajan, M.P., Taylor, P.L., and Rosenblatt, C., (1997), "Magnetic levitation of liquid crystals", *Liquid Crystals* **23**, 547-550.

Maret, G., and Weill, G., (1983), *Biopolymers* **22**, 2727.

Mariani, P., Samori, B., Angeloni, A.S., and Ferruti, P., (1986) "Polymerization of bisacrylic monomers within a liquid crystalline smectic-B solvent", *Liquid Crystals* **1**, 327-336.

Martin, P.C., Parodi, O., and Pershan, P.S., (1972) "Unified hydrodynamic theory for crystals, liquid crystals and normal fluids", *Physical Review* **A 6**, 2401-2420.

Martinot-Lagarde, Ph., (1976) "Observation of ferroelectrical monodomains in the chiral smectic-C* liquid crystal", *Journal de Physique Colloque* **37**, C3-129.

Martinot-Lagarde, Ph., (1977) "Direct electrical measurement of the permanent polarization of a ferroelectric chiral smectic-C liquid crystal", *Journal de Physique Lettres* **38**, L-17.

Martinot-Lagarde, Ph., and Durand, G., (1980) "Flexoelectric effects in ferroelectric liquid crystals?", *Journal de Physique Lettres* **41**, L-43.

Martinot-Lagarde, Ph., and Durand, G., (1981a) "Dielectric relaxation in a ferroelectric liquid crystal", *Journal de Physique (Paris)* **42**, 269.

Martinot-Lagarde, Ph., Duke, R., and Durand, G., (1981b), "Temperature dependence of tilt, pitch and polarization in ferroelectric liquid crystals", *Molecular Crystals and Liquid Crystals* **75**, 249-286.

Maruyama, N., (1980) "Critical behavior in ferroelectric liquid crystal", *Journal of the Physical Society of Japan* **49**, Supplement B, 175.

Mauger, A., Zribi, G., Mills, D.L., and Toner, J., (1984) "Substrate-induced orientational order in the isotropic phase of liquid crystals", *Physical Review Letters* **53**, 2485-2488.

McMillan, W.M., (1973) "Simple molecular theory of the smectic-C phase", *Physical Review* **A8**, 1921.

Meier, P., Ohmes, E., and Kothe, G., (1986) "Multipulse dynamic NMR of phospholipid membranes", *Journal of Chemical Physics* **85**, 3598.

Meister, R., Dumoulin, H., Halle, M.A., and Pieranski, P., (1996a) "The anchoring of a cholesteric liquid crystal at the free surface", *Journal de Physique II France* **6**, 827.

Meister, R., Dumoulin, H., Halle, M.A., and Pieranski, P., (1996b) "Structure of the cholesteric focal conic domains at the free surface", *Physical Review* **E 54**, 3771.

Mertelj, A., (1998), *Dynamics of a liquid crystal in silica aerogel near the phase transition into the isotropic phase*, PhD Thesis, University of Ljubljana.

Meyer, R.B., (1968) "Effects of electric and magnetic fields on the structure of cholesteric liquid crystals", *Applied Physics Letters* **12**, 281.

Meyer, R.B., (1969) "Piezoelectric effects in liquid crystals", *Physical Review Letters* **22**, 918.

Meyer, R.B., Liebert, L., Strzelecki, L., and Keller, P., (1975) "Ferroelectric liquid crystals", *Journal de Physique Lettres* **36**, L69-71.

Meyer, R.B., (1977) "Ferroelectric liquid crystals: A review", *Molecular Crystals Liquid Crystals* **40**, 33.

Michel, L., (1980), "Symmetry defects and broken symmetry. Configurations. Hidden Symmetry", *Reviews of Modern Physics* **52**, 617.

Michelson, A., (1977a) "Physical realization of a Lifshitz point in liquid crystal", *Physical Review Letters* **39**, 464-467.

Michelson, A., Benguigui, L., and Cabib, D., (1977b) "Phenomenological theory of the polarized helicoidal smectic-C* phase", *Physical Review* **A 16**, 394.

Michelson, A., (1977c) "Phase diagrams near the Lifshitz point. I. Uniaxial magnetization", *Physical Review* **B 16**, 577.

Michelson, A., (1977d) "Phase diagrams near the Lifshitz point: II. Systems with cylindrical, hexagonal and rhombohedral symmetry having an easy plane of magnetization", *Physical Review* **B 16**, 585.

Michelson, A., and Cabib, D., (1977e) "Electric field induced tricritical point in chiral polarized liquid crystals", *Journal de Physique Lettres* **38**, L-321.

Michelson, A., Cabib, D., and Benguigui, L., (1977f) "Symmetry changes and dipole orderings in the smectic-A to smectic-C phase transitions of second order", *Journal de Physique* **38**, 961.

Miyachi, K., Matsushima, J., Takanishi, Y., Ishikawa, K., Takezoe, H., and Fukuda, A., (1995) "Spontaneous polarization parallel to the tilt plane in the antiferroelectric chiral phase of liquid crystals observed by polarized IR spectroscopy", *Physical Review* **E 52**, R2153.

Miyachi, K., Takanishi, Y., Ishikawa, K., Takezoe, H., and Fukuda, A., (1997) "Relaxation time of molecular rotation about its long axis in the smectic-A phase of ferroelectric and antiferroelectric liquid crystals as observed by picosecond optical Kerr effect", *Physical Review* **E 55**, 1632.

Miyano, K., (1979) "Wall-induced pretransitional birefringence: a new tool to study boundary aligning forces in liquid crystals", *Physical Review Letters* **43**, 51-54.

Moriyama, T., Takanishi, Y., Ishikawa, K., Takezoe, H., and Fukuda, A., 1995, "Binary mass diffusion in smectic C and CA phases as observed by forced Rayleigh scattering", *Liquid Crystals* **18**, 639.

Moses, T., Ouchi, Y., Chen, W., and Shen, Y.R., (1993) "Pretransitional surface phenomena in ferroelectric liquid crystals", *Molecular Crystals and Liquid Crystals* **225**, 55-65.

Muševič, I., Žekš, B., Blinc, R., Rasing, Th. and Wyder, P., (1982) "Phase diagram of a ferroelectric liquid crystal near the Lifshitz point", *Physical Review Letters* **48**, 192.

Muševič, I., Žekš, B., Blinc, R., Rasing, Th., and Wyder, P., (1983) "Dielectric study of the modulated SmC*-uniform SmC transition in a magnetic field", *Physica Status Solidi* **B 119**, 727-733.

Muševič, I., Žekš, B., Blinc, R., Jansen, L., Seppen, A., and Wyder, P., (1984) "Temperature dependence of the pitch of the helix in a chiral ferroelectric smectic liquid crystal", *Ferroelectrics* **58**, 71.

Muševič, I., Blinc, R., Žekš, B., Filipič, C., Čopič, M., Seppen, A., Wyder, P., and Levanyuk, A., (1988) " Observation of phason dispersion in a feroelectric liquid crystal by light scattering", *Physical Review Letters* **60**, 1530.

Muševič, I., Drevenšek, I., Blinc, R., Kumar, S., and Doane, J.W., (1989) "Temperature dependence of molecular tilt angle and rotatory power in a ferroelectric liquid crystal with polarization sign reversal", *Molecular Crystals and Liquid Crystals* **172**, 217-222.

Muševič, I., Žekš, B., Blinc, R.,Wierenga, H.A., and Rasing, Th., (1992) "Phason dispersion and magnetic-field induced band gap in a ferroelectric liquid crystal", *Physical Review Letters* **68**, 1850.

Muševič, I., (1993a), *Elementary excitations and broken symmetries in a ferroelectric liquid crystal,* PhD Thesis, University of Ljubljana.

Muševič, I., Žekš, B., Blinc, R., and Rasing, Th., (1993b) "Magnetic field induced biaxiality in an antiferroelectric liquid crystal", *Physical Review* **E 47**, 1094-1100.

Muševič, I., Blinc, R., Žekš, B., Čopič, M., Wittebrood, M.M., Rasing, Th., Orihara, H., and Ishibashi, Y., (1993c) "Gapless phason in an antiferroelectric liquid crystal", *Physical Review Letters* **71**, 1180.

Muševič, I., Žekš, B., Blinc, R., and Rasing, Th., (1994) "Phasons and broken symmetries", *Physical Review* **B 49**, 9299-9311.

Muševič, I., Škarabot, M., Blinc, R., Schranz, W., and Dolinar, P., (1996a) "Birefringence in smectic and hexatic chiral phases of CE-8", *Liquid Crystals* **20**, 771.

Muševič, I., Rastegar, A., Čepič, M., Žekš, B., Čopič, M., Moro, D., and Heppke, G., (1996b) "Observation of an optic-like phase mode in an antiferroelectric liquid crystal", *Physical Review Letters* **77**, 1769-1772.

Muševič, I., Kityk, A., Škarabot, M. and Blinc, R., (1997) "Polarization Noise in a Ferroelectric Liquid Crystal", *Physical Review Letters* **79**, 1062.

Muševič, I., Škarabot, M., Kityk, A.V., Blinc, R., Moro, D., and Heppke, G., (1997) "Birefringence and optical rotation in chiral ferroelectric and antiferroelectric liquid crystals", *Ferroelectrics* **203**, 133.

Muzny, C.D., and Clark, N.A., (1992) "Direct observation of the Brownian motion of a liquid-crystal topological deffect", *Physical Review Letters* **68**, 804-807.

Nakai, T., Miyajima, S., Takanishi, Y., Yoshida, S., and Fukuda, A., (1999a),"High resolution ^{13}C NMR study of an antiferroelectric liquid crystal: Verification of the bent chain structure" *Journal of Physical Chemistry* **B103**, 406.

Nakai, T., Fujimora, H., Kuwahari, D., Miyajima, S., (1999b),"Complete assignment of 13C NMR spectra and determination of orientational order parameter for antiferroelectric liquid-crystalline MHPOBC" *Journal of Physical Chemistry* **B103**, 417.

Nakagawa, M., and Akahane, T., (1986a) "Effect of polarization field on elastic deformation in a 180° twisted sample of chiral smectic C liquid crystal", *Journal of the Physical Society of Japan* **55**, 1516-1522.

Nakagawa, M., and Akahane, T., (1986b) "A study of homogeneous-twist transition in surface stabilized ferroelectric liquid crystal", *Journal of the Physical Society of Japan* **55**, 4492-4499.

Nakagawa, M., (1993a) "On the phase transition of SmC_A^* liquid crystals", *Journal of the Physical Society of Japan* **62**, 2260-2269.

Nakagawa, M., (1993b) "On the dynamics of ferro- and antiferroelectric liquid crystals", *Liquid Crystals* **14**, 1763-1784.

Nguyen, H.T., Rouillon, J.C., Cluzeau, P., Sigaud, G., Destrade, C., and Isaert, N., (1994) "The chiral thiobenzoate series with antiferroelectric mesophases", *Liquid Crystals* **17**, 571.

Niori, T., Sekine, T, Watanabe, J., Furukawa, T., and Takezoe, H., (1996),"Distinct ferroelectric smectic liquid crystals consisting of banana shaped achiral molecules" *Journal of Material Chemistry* **6**, 1231.

Nyquist, H., (1928) "Thermal agitation of electric charge in conductors", *Physical Review* **32**, 110.

Obayashi, K., Orihara, H., and Ishibashi, Y., (1995) "Nonlinear dielectric spectroscopy of the Goldstone mode in an antiferroelectric liquid crystal", *Journal of the Physical Society of Japan* **64**, 3188-3191.

O'Brien, J.P., Moses, T., Chen, W., Freysz, E., Ouchi, Y., and Shen, Y.R., (1993) "Reorientation dynamics of ferroelectric liquid crystal molecules near the smectic-A-smectic-C* transition", *Physical Review* **E 47**, R2269-R2272.

Ocio, M., Bouchiat, H., and Monod, P., (1985) "Observation of 1/f magnetic fluctuations in a spin glass", *Jornal de Physique Lettres* **46**, L-647.

Okabe, N., Suzuki, Y., Kawamura, I., Isozaki, T., Takezoe, H., and Fukuda, A., (1992) "Reentrant antiferroelectric phase in MHPBC", *Japanese Journal of Applied Physics* **31**, L793.

Okano, K., (1986) "Electrostatic contribution to the distortion free-energy density of ferroelectric liquid crystal", *Japanese Journal of Applied Physics* **25**, L846-847.

Oldano, C., Miraldi, E., and Taverna Valabrega, P., (1983) "Dispersion Relation for Propagation of Light in Cholesteric Liquid Crystals", *Physical Review* **A 27**, 3291-3299.

Oldano, C., (1984) "Existence of a Critical Tilt Angle for the Optical Properties of Chiral Smectic Liquid Crystals", *Physical Review Letters* **53**, 2413-2416.

Oldano, C., Allia, P., and Trossi, L., (1985) "Optical Properties of Anisotropic Periodic Helical Structures", *Journal de Physique* **46**, 573-582.

Orihara, H., amd Ishibashi, Y., (1990) "A phenomenological theory of the antiferroelectric phase transition in smectic liquid crystals", *Japanese Journal of Applied Physics* **29**, L115-118.

Orihara, H., and Ishibashi, Y., (1993a) "A phenomenological theory of nonlinear dielectric response in a ferroelectric liquid crystal", *Journal of the Physical Society of Japan* **62**, 489-496.

Orihara, H., Fukase, A., Izumi, S., and Ishibashi, Y., (1993b) "Nonliear response of a ferroelectric liquid crystal", *Ferroelectrics* **147**, pp.411-418.

Orihara, H., Fukase, A., and Ishibashi, Y., (1995a) "Nonlinear dielectric spectroscopy of the Goldstone mode in a ferroelectric liquid crystal", *Journal of the Physical Society of Japan* **64**, 976-980.

Orihara, H., and Ishibashi, Y., (1995b) "Electro-optic effect and third order nonlinear dielectric response in antiferroelectric liquid crystal", *Journal of the Physical Society of Japan* **64**, 3775-3786.

Orihara, H., and Ishibashi, Y., (1995c) "A phenomenological theory of non-linear dielectric response", *Journal of the Physical Society of Japan* **64**, 99-105.

Orsay Liquid Crystal Group, (1969) "Dynamics of fluctuations in nematic liquid crystals", *Journal of Chemical Physics* **51**, 816-822.

Osipov, M.A., and Pikin, S.A., (1995) "Dipolar and quadrupolar ordering in ferroelectric liquid crystals", *Journal de Physique II France* **5**, 1223.

Ostrovski, B.I., Rabinovich, A.Z., Sonin, A.S., Strukov, B.A., and Taraskin, S.A., (1978) "Ferroelectric properties of smectic liquid crystals", *Ferroelectrics* **20**, 189-191.

O'Shea, S.J., Welland, M.E., and Pethica, J.B., (1994) "AFM of local compliance at solid-liquid interfaces", *Chemical Physics Letters* **223**, 336.

Ouchi, Y., Takezoe, H., and Fukuda, A., (1987) "Switching process in FLC: Disclination dynamics of the surface stabilized state", *Japanese Journal of Applied Physics* **26**, 1.

Ouchi, Y., Lee, Ji, Takezoe, H., Fukuda, A., Kondo, K., Kitamura, T., and Mukoh, A., (1988) "Smectic-C* chevron layer structure by X-ray diffraction", *Japanese Journal of Applied Physics* **27**, L725.

Overney, R.M., Meyer, E., Frommer, J., Guntherodt, H.J., Decher, G., Reibel, J., and Sohling, U., (1993) "A comparative AFM study of liquid crystal films: Transferred freely suspended vs LB. Morphology, lattice and manipulation", *Langmuir* **9**, 341.

Palierne, J.F., (1986) "Elastic-like contribution of electric origin to the distortion free-energy of nematics", *Physical Review Letters* **56**, 1160-1163.

Panarin, Y.P., Kalmykov, Y.P., MacLughadha, S.T., Xu, H., and Vij, J.K., (1994a) "Dielectric response of surface stabilized ferroelectric liquid crystal cells", *Physical Review* **E 50**, 4763-4772.

Panarin, Y.P., Xu, Huang, MacLughadha, S.T. and Vij, J., (1994b) "Dielectric response of ferroelectric liquid crystal cells", *Japanese Journal of Applied Physics* **33**, 2648-2650.

Panarin, Yu.P., Mac Lughadha, S.T., and Vij, J.K., (1995) "In-plane anchoring energy in ferroelectric liquid crystals: Evidence for its existence and measurement", *Physical Review **E** * **52**, R17.

Panarin, Y., Rosenblatt, Ch., and Aliev, F.M., (1998) "Appearance of ferrielectric phases in a confined liquid crystal investigated by photon correlation spectroscopy", *Physical Review Letters* **81**, 2699.

Pang, J., Muzny, Ch.D., and Clark, N.A., (1992) "String deffects in freely suspended liquid-crystal films", *Physical Review Letters* **69**, 2783-2786.

Pargelis, A.N., Finn, P., Goodby, J.W., Panizza, P., Yurke, B., and Cladis, P.E., (1992) "Defect dynamics and coarsening dynamics in smectic-C films", *Physical Review* **A 46**, 7765-7776.

Parks, D.C., N.A.Clark, D.M.Walba and P.D.Beale, (1993) "STM of coexisting 2D crystalline and 1D stacking-disordered phases at the chiral-liquid-crystal-graphite interface", *Physical Review Letters* **70**, 607.

Parodi, O., (1975) "Light Propagation Along the Helical Axis in Chiral Smectics C*", *Journal de Physique Coloque* **C1, 36**, C1-325.

Parsons, D., and Hayes, C.F., (1974) "Fluctuations of a cholesteric liquid crystal in a static magnetic field", *Physical Review* **A 9**, 2652.

Pausak, S., Pines, A., and Waugh, J.S., (1973) "Carbon-13 chemical shielding tensors in single crystal durene", *Journal of Chemical Physics* **59**, 591.

Pavel, J., (1984) "Behaviour of thin planar smectic-C* samples in an electric field", *Journal de Physique* **45**, 137.

Pavel, J., and Glogarova, M., (1987) "Dielectric permitivity of ferroelectric liquid crystals influenced by a biasing electric field", *Ferroelectrics* **76**, 221-232.

Pavel, J., and Glogarova, M., (1991a) "The effect of biasing electric field on relaxations in FLC investigated by the dielectric and optical methods", *Ferroelectrics* **121**, 45.

Pavel, J., and Glogarova, M., (1991b) "A new type of layer structure defects in chiral smectics", *Liquid Crystals* **9**, 87.

Pelzl, G., Diele, S., and Weissflog, W., (1999), "Banana-shaped compounds-A new field of liquid crystals", *Advanced Materials* **11**, 707.

Peterson, M.A., (1983) "Light propagation and light scattering in cholesteric liquid crystals", *Physical Review* **A 27**, 520.

Philip, J., Lalanne, J.-R., Marcerou, J.-P., and Sigaud, G., (1994) "Comparison between the tilted SmC^*_α and SmC^*_γ phases of MHPOBC studied by optical techniques", *Journal de Physique II France* **4**, 2149-2159.

Philip, J., Lallane, J.R., Marcerou, J.P., and Sigaud, G., (1995) "Optical investigations in the various phases of an antiferroelectric liquid crystal", *Physical Review* **E 52**, 1846-1856.

Photinos, D.J., and Samulski, E.T., (1995) "On the origins of spontaneous polarization in tilted smectic liquid crystals", *Science* **270**, 783.

Pikin, S.A., and Indenbom, V.L.,(1978), *Uspekhi Fizičeskih Nauk* **125**, 251.

Pikin, S.A., and Yoshino, K., (1981) "Influence of boundaries on the phase transition point in ferroelectric smectics", *Japanese Journal of Applied Physics* **20**, L557.

Pikin, S.A., (1991a), *Structural transformations in liquid crystals*, Gordon and Breach Science Publishers.

Pikin, S.A., (1991b) "Physical properties of ferroelectric liquid crystalline thin films", *Ferroelectrics* **117**, 197.

Pikin, S.A., (1995) "Short pitch modes approach to the problem of antiferroelectricity in liquid crystals", *Molecular Crystals and Liquid Crystals* **262**, 425-435.

Pindak, R., Young, C.Y., Meyer, R.B., and Clark, N.A., (1980), "Macroscopic orientation patterns in smectic-C films", *Physical Review Letters* **45**, 1193.

Pines, A., Gibby, M.G., and Waugh, J.S., (1972) "Proton-enhanced nuclear induction spectroscopy. ^{13}C chemical shielding anisotropy in some organic solids", *Chemical Physics Letters* **15**, 373.

Pines, A., and Chang, J.J., (1974) "Study of isotropic-nematic-solid transition in a liquid crystal by ^{13}C-^{1}H double resonance", *Physical Review* **A 10**, 946,

Pines, A., Chang, J.J., and Griffin, R.G., (1974b) "Carbon-13 NMR in solid ammonium tartrate", *Journal of Chemical Physics* **61**, 1021.

Pleiner, H., (1984) "Broken symmetries and hydrodynamics of hexatic-B and smectic-F and I liquid crystal phases", *Molecular Crystals and Liquid Crystals* **114**, 103.

Pleiner, H., Brand, H.R., (1994) "Dielectric degrees of freedom in chiral smectic-C* liquid crystals", *Molecular Crystals and Liquid Crystals* **257**, 289.

Pirš, J., Blinc, R., Marin, B., Pirš, S., Muševič, I., Žumer, S., and Doane, J.W., (1992), poster presented at 14th ILCC Conference in Pisa, Italy.

Povše, T., Muševič, I., Žekš, B., and Blinc, R., (1993) "Phase transitions in ferroelectric liquid crystals in a restricted geometry", *Liquid Crystals* **14**, 1587-1598.

Prelovšek, P., (1982) "Tricritical point in the incommensurate systems", *Journal of Physics C: Solid State Physics* **15**, L269.

Prost, J., and Bruinsma, R., (1993) "Fluctuation forces and devil's staircase of ferro-antiferroelectric smectics C*", *Ferroelectrics* **148**, 25-29.

Qiu, Ruozi, John, T.Ho, and Hark, S.K., (1988) "Electroclinic effect above the smectic-A-smectic-C* transition", *Physical Review* **A 38**, 1653-1655

Rapini, A., and Papoular, M., (1969) "Distorsion d'une lamelle nematique sous champ magnetique conditions d'ancrage aux parois", *Journal de Physique Colloque* **30**, C4-54-56.

Rapini, A., (1972) "Instabilites magnetiques d'un smectique C", *Journal de Physique* **33**, 237.

Rastegar, A., (1996) *Light cattering in the Ferroelectric and Antiferroelectric Liquid Crystals in Confined Geometries,* PhD Thesis, University of Ljubljana.

Rastegar, A., Muševič, I., and Čopič, M., (1996a) "Finite size effects in the dynamic behaviour of smectic C* phase studied by light scattering", *Ferroelectrics* **181**, 219-226.

Rastegar, A., Muševič, I., and Čopič, M., (1996b) "Dynamic light scattering near the smectic A- smectic C* phase transition in thin planar cells", *Molecular Crystals and Liquid Crystals* **282**, 27-34.

Rauch, S., (1996), *Elektrooptische Eigenschaften antiferroelektrischer Flussigkristalle*, Diplomarbeit Technische Universitat Berlin.

Raya, V.N., Krishna Prased, S., Khened, S.M., and Shankar Rao, D.S., (1991) "Experimental studies in the vicinity of the C*-I* transition", *Ferroelectrics* **121**, 343.

Reed, L., Stoebe, T., Huang, C.C., (1995) "Critical fluctuations near the smectic-A-smectic-C transition of a partially perfluorinated compound", *Physical Review* **E 52**, R2157.

Rieker, T.P., Clark, N.A., Smith, G.S., Parmar, D.S., Sirota, E.B., and Safinya, C.R., (1987) "Chevron local layer structure in surface stabilized ferroelectric smectic-C cells", *Physical Review Letters* **59**, 2658.

Rosenblatt, Ch., Pindak, R., Clark, N.A., and Meyer, R.B., (1979), "Freely suspended ferroelectric liquid crystal films: absolute measurements of polarization, elastic constants, and viscosities", *Physical Review Letters* **42**, 1220.

Rosenblatt, Ch., Meyer, R.B., Pindak, R., and Clark, N.A., (1980) "Temperature dependence of ferroelectric liquid crystal thin films: A classical X-Y system", *Physical Review* **A 21**, 140-147.

Rovšek, B., and Žekš, B., (1995) "The SmA-SmC* phase transition in a planar cell with a polar surface coupling", *Molecular Crystals and Liquid Crystals* **263**, 49.

Roy, A., and Madhusudana, N.V., (1996) "A simple model for phase transitions in antiferroelectric liquid crystals", *Europhysics Letters* **36**, 221.

Roy, A., Madhusudana, N.V., Toledano, P., and Figueiredo., Neto, A.M (1999) "Longitudinal spontaneous polarization and longitudinal electroclinic effect in achiral smectic phases with bent-shaped molecules", *Physical Review Letters* **82**, 1466.

Rozanski, S.A., and Kuczynski, W., (1984) "The influence of an electric field on the pitch in chiral smectic C* in the vicinity of the SmC*-SmA transition", *Chemical Physics Letters* **105**, 104.

Safinya, C.R., Kaplan, M., Als-Nielsen, J., Birgenau, J., Davidov, D., Litster, J.D., Johnson, D.L., and Neubert, M.E., (1980) "High-resolution X-ray study of a smectic-A-smectic-C phase transition", *Physical Review* **B 21**, 4149.

Sako, T., Kimura, Y., Hayakawa, R., Okabe, N., and Suzuki, Y., (1996) "Dielectric and electrooptic response of antiferroelectric liquid crystals in the smectic-A phase", *Japanese Journal of Applied Physics* **35**, 679.

Sako, T., Itoh, N., Sakaigawa, A., and Koden, M., (1997) "Switching behaviour of SSFLC induced by pulse voltages", *Applied Physics Letters* **71**, 461.

Schaub, B., and Mukamel, D., (1983) "Phase diagrams of systems exhibiting disorder-incommensurate transitions", *Journal of Physics C: Solid State Physics* **16**, L225.

Schneider, T., and Meier, P.F., (1973), "Phase transitions, symmetry breaking and asymptotic properties of the Goldstone-mode frequency", *Physica* **67**, 521.

Schwartz, D.K., Steinberg, S., Israelachvili, J., and Zasadzinski, J.A.N., (1992) "Growth of self-assembled monolayer by fractal agregation", *Physical Review Letters* **69**, 3354.

Sekine, T., Niori, T., Watanabe, J., Furukawa, T., Choi, S.W., and Takezoe, H., (1997a),"Spontaneous helix formation in smectic liquid crystals comprising achiral molecules", *Journal of Material Chemistry* **7**, 1307.

Sekine, T., Niori, T., Sone, M., Watanabe, J., Choi, S.W., Takanishi, Y., and Takezoe, H., (1997b),"Origin of helix in achiral banana-shaped molecular systems", *Japanese Journal of Applied Physics* **36**, 183.

Seliger, J., Osredkar, R., Žagar, V., and Blinc, R., (1977) "Orientational order in biaxial liquid crystals" *Physical Review Letters* **38**, 411.

Seliger, J., Žagar, V., and Blinc, R., (1978) "^{14}N NQR study of orientational ordering in the smectic phases of achiral TBBA and chiral TBACA" *Physical Review* **A 17**, 1149.

Seppen, A., (1987), *Phase Transitions and Fluctuations of biaxial liquid crystals in magnetic fields*, Ph.D. Thesis, University of Nijmegen, The Netherlands.

Seppen, A., Muševič, I., Maret, G., Žekš, B., Wyder, P., and Blinc, R., (1988) "Temperature dependence of the rotatory power in the SmC* of DOBAMBC", *Journal de Physique France* **49**, 1569-1573.

Shao, R.F., Willis, P.C., and Clark, N.A., (1991) "The field induced stripe texture in surface stabilized ferroelectric liquid crystalline cells", *Ferroelectrics* **121**, 127-136.

Sheng, P., (1976) "Phase transition in surface-aligned nematic films", *Physical Review Letters* **37**, 1059-1062.

Shigeno, M., Mizutani, W., Suginoya, M., Ohmi, M., Kajimura, K., and Ono, M., (1990) "Observation of liquid crystals on graphite by STM", *Japanese Journal of Apliede Physics* **29**, L119.

Shubnikov, A.V., and Koptsik, V.A., (1974), *Symmetry in Science and Art*, Plenum Press.

Sidebottom, D.L., Bergman, R., Börjesson, and L.M.Torell, (1992) "Observation of scaling behavior in the liquid-glass transition range from dynamic light scattering in poly(propylene glycol)", *Physical Review Letters* **68**, 3587-3590.

Simmons, J., (1963) "Generalized formula for the electric tunnel effect between similar electrodes separated by a thin insulating film", *Journal of Applied Physics* **34**, 1793.

Skarp, K., (1988) "Rotational viscosities in ferroelectric smectic liquid crystals", *Ferroelectrics* **84**, 119.

Sluckin, T.J., and Poniewierski, A., (1985) "Novel surface phase transition in nematic liquid crystals: Wetting and the Kosterlitz-Thouless transition", *Physical Review Letters* **55**, 2907-2910.

Sluckin, T.J., and Poniewierski, A., (1990) "Wetting and capillary condensation in liquid crystal systems", *Molecular Crystals and Liquid Crystals* **179**, 349-364.

Smith, D.P.E., Bryant, A., Quate, C.F., Rabe, J.P., Gerber, Ch., and Swalen, J.D., (1987) "Imaging of a lipd bilayer at molecular resolution by STM", *Proceedings of the National Academy of Sciences USA* **84**, 969.

Smith, D.P.E., Horber, H., Gerber, Ch., and Binnig, G., (1989) "Smectic liquid crystal monolayers on graphite observed by STM", *Science* **245**, 43.

Spector, M.S., Sprunt, S., and Litster, J.D., (1993) "Novel dynamical mode in a tilted smectic liquid-crystal film", *Physical Review* **E 47**, 1101-1107.

Sprunt, S., and Litster, J.D., (1987) "Light-scattering study of bond orientational order in a tilted hexatic liquid crystal film", *Physical Review Letters* **59**, 2682-2685.

Srajer, G., Pindak, R., and Patel, J., (1991) "Electric field induced layer reorientation in FLC: An X-ray study", *Physical Review* **A 43**, 5744.

Stanarius, R., (1998), "Diamagnetic properties of nematic liquid crystals", Section 2.3. in *Handbook of Liquid Crystals*, edts.Demus, Goodby, Gray, Spiess and Vill, Wiley-VCH.

Stephen, M.J., 1970, "Hydrodynamics of liquid crystals", Physical Review **A 2**, 1558-1562.

Stoebe, T., Huang, C.C., and Goodby, J.W., (1992a) "Simultaneous measurement of heat capacity and in-plane density of thin free-standing liquid-crystal films", *Physical Review Letters* **68**, 2944-2947.

Stoebe, T., Geer, R., Huang, C.C., and Goodby, J.W., (1992b) "Layer-by-layer surface ordering near a continuous transition in free-standing liquid crystal films", *Physical Review Letters* **69**, 2090-2093.

Stoebe, T., Mach, P., and Huang, C.C., (1994) "Surface tension of free-standing liquid crystal films", *Physical Review* **E 62**, R3587-3590.

Sun, H., Orihara, H., and Ishibashi, Y., (1991) "Observation of the phase mode in an antiferroelectric liquid crystal by photon correlation technique", *Journal of the Physical Society of Japan*, **60**, 4175.

Sun, H., Orihara, H., and Ishibashi, Y., (1993a) "Relaxation modes and visco-elastic properties in an antiferroelectric liquid crystal (R)-MHPOBC observed by photon-correlation spectroscopy", *Journal of the Physical Society of Japan* **62**, 2066. See also Erratum, *Journal of the Physical Society of Japan* **62**, 3766.

Sun, H., Orihara, H., and Ishibashi, Y., (1993b) "A phenomenological theory of ferroelectric and antiferroelectric liquid crystals based on a discrete model", *Journal of the Physical Society of Japan* **62**, 2706-2718.

Sutherland, B., (1973) "Some exact results for one-dimensional model of solids", *Physical Review* **A 8**, 2514.

Suzuki, A., Orihara, H., Ishibashi, Y., Yamada, Y., Yamamoto, N., Mori, K., Nakamura, K., Suzuki, Y., Hagiwara, T., Kawamura, I., and Fukui, M., (1990) "Layer structure of antiferroelectric liquid crystal MHPOBC studied by X-ray diffraction", *Japanese Journal of Applied Physics* **29**, L336.

Swanson, B.D., Stragier, H., Tweet, D.J., and Sorensen, L.B., (1989) "Layer-by-layer surface freezing of freely suspended liquid crystal films", *Physical Review Letters* **62**, 909-912.

Škarabot, M., (1997), *Phase transitions of ferroelectric and antiferroelectric liquid crystals in confined geometry,* Ph.D. Thesis, University of Ljubljana.

Škarabot, M., Čepič, M., Žekš, B., Blinc, R., Heppke, G., Kityk, A.V., and Muševič, I., (1998a) "Birefringence and tilt angle in the antiferroelectric, ferroelectric and intermediate phases of chiral smectic liquid crystals", *Physical Review* **E 58**, 575.

Škarabot, M., Muševič, I., Blinc, R., (1998b) "Solitonlike dynamics in submicron ferroelectric liquid crystal cells", *Physical Review* **E 57**, 6725.

Škarabot, M., Kralj, S., Blinc, R., Muševič, I., (1999a), "The smectic-A-smectic-C* transition in submicron confined geometry", *Liquid Crystals* **26**, 723-729.

Škarabot, M., Kočevar, K., Blinc, R., Heppke, G., and Muševič, I., (1999b), "Critical behavior of birefringence in the smectic-A phase of chiral smectic liquid crystals", *Physical Review* **E59**, R1323.

Takanishi, Y., Ouchi, Y., Takezoe, H., and Fukuda, A., (1989) "Chevron layer structure in the smectic A phase of 8CB", *Japanese Journal of Applied Physics* **28**, L487.

Takanishi, Y., Hiraoka, K., Agrawal, V.K., Takezoe, H., Fukuda, A., and Matsushita, M., (1991) "Stability of antiferoelectricity and causes for its appearance in SmC_α and SmC_A phases of a chiral smectic liquid crystal MHPOBC", *Japanese Journal of Applied Physics* **30**, 2023.

Takanishi, Y., Takezoe, H., Johno, M., Yui, T., and Fukuda, A., (1993) "Tristable switching in $smectic-O$ of MHTAC and its miscibility with $smectic-C_A^*$ of antiferroelectric chiral smectic liquid crystal", *Japanese Journal of Applied Physics* **32**, 4605-4610.

Takanishi, Y., Ikeda, A., Takezoe, H., and Fukuda, A., (1995) "Higher smectic-layer order parameters in liquid crystals determined by X-ray diffraction and the effect of antiferroelectricity", *Physical Review* **E 51**, 400.

Takezoe, H., Furuhata, K., Nakagiri, T., Fukuda, A., and Kuze, E., (1978) "Birefringence in the Sm A phase and the disappearance of helicoidal structure in the Sm C* phase caused by an electric field in DOBAMBC, *Japanese Journal of Applied Physics* **17**, 1219.

Takezoe, H., Kondo, K., Miyasato, K., Abe, S., Tsuchiya, T., Fukuda, A., and Kuze, E., (1984) "On the methods of determining material constants in ferroelectric smectic C* liquid crystals", *Ferroelectrics* **58**, 55-70.

Takezoe, H., Chandani, A.D.L., Furukawa, K., and Kishi, A., (1989) Communication O26, Second International Conference on Ferroelectric Liquid Crystals, Goteborg, Sweden.

Tarczon, J.C., and Miyano, K., (1980) "Surface-induced ordering of a liquid crystal in the isotropic phase", *Journal of Chemical Physics* **73**, 1994-1998.

Taupin, D., Guyon, E., and Pieranski, P., (1978), "Propagation de la lumiere dans les milieux anisotropes periodiques", *Journal de Physique* **39**, 406.

Taylor, G.W., (1991), editor, *Ferroelectric Liquid Crystals*, Gordon and Breach Science Publishers.

Tersoff, J., and Hamann, D.R., (1985) "Theory of STM", *Physical Review* **B 31**, 805.

Toledano, P., Neto, A.M.Figueiredo, (1997) "In search of a chiral smectic liquid crystal state with a macroscopic spontaneous feroelectric polarization", *Physical Review Letters* **79**, 4405.

Towel, M.J., Anderson, M.H., Jones, J.C., Raynes, E.P., (1991) "Optical study of thin layers of smectic C materials", *Ferroelectrics* **121**, 137.

Tripathi, Sanjay, Lu, Min-Hua, Terentjev, E.M., Petschek, R.G., and Rosenblatt, Ch., (1991) "Surface-induced polarization at a chiral-nematic-substrate interface", *Physical Review Letters* **67**, 3400-3403.

Tredgold, R.H., (1990) "Ferroelectric liquid crystals do not have to be chiral", *Journal of Physics D: Applied Physics* **23**, 119-120.

Tournilhac, F., Blinov, L.M., Simon, J., and Yablonsky, S.V., (1992) "Ferroelectric Liquid Crystals from Achiral Molecules", *Nature* **359**, 621.

Urbach, W.Z., and Billard, J., (1972) "Sur le phases smectiques du type C", *Comptes Rendus Academie de Sciences (Paris), Serie* **B 274**, 1287.

Urbanc, B., (1990), *Spontaneous polarization in ferroelectric liquid crystals*, MsC Thesis, University of Ljubljana.

Urbanc, B., Žekš, B., and Carlsson, T., (1991) "Nonlinear effects in the dielectric response of ferroelectric liquid crystals", *Ferroelectrics* **113**, 219.

Vallerien, S.U., Kremer, F., Kapitza, H., Zentel, R., and Frank, W., (1989) "Field dependent soft and Goldstone mode in a ferroelectric liquid crystal as studied by dielectric spectroscopy", *Physics Letters* **A 138**, 219-222.

van der Ziel, A.M.A., (1959) *Fluctuation Phenomena in Semi-Conductors*, Butterworths Publications, Ltd, London ; In USA, Academic Press Inc., Publishers, New York.

van der Hart, (1976) "Characterization of the methylene 13C chemical shift tensor in the normal alkane n-$C_{20}H_{42}$", *Journal of Chemical Physics* **64**, 830.

van Haaren, J.A.M.M., and Rikken, G.L.J.A., (1989) "Electric field and thickness effect on the electroclinic temporal behavior in a chiral smectic-A liquid crystal", *Physical Review* **A 40**, 5476.

van Hove, L., (1954) "Correlations in space and time and Born approximation scattering in systems of interacting particles", *Physical Review* **95**, 249.

van Sprang, H.A., (1983) "Surface-induced order in a layer of nematic crystal", *Journal de Physique* **44**, 421-426.

van Winkle, D.H., and Clark, N.A., (1984) "Direct measurement of orientational correlations: Observation of Landau-Peierls divergence in a freely suspended tilted smectic film", *Physical Review Letters* **53**, 1157-1160.

van Winkle, D.H., and Clark, N.A., (1988) "Direct measurements of orientation correlations in a two-dimensional liquid crystal system", *Physical Review* **A 38**, 1573-1589.

Vaupotič, N., Kralj, S., Čopič, M., and Sluckin, T.J., (1996) "Landau-de-Gennes theory of the chevron structure in a smectic liquid crystal", *Physical Review* **E 54**, 3783.

Vilfan, I., Vilfan, M., and Žumer, S., (1989) "Orientational order in bipolar nematic microdroplets close to the phase transition", *Physical Review* **A 40**, 4724.

Vilfan, Marija, Blinc, R., Seliger, J., and Žagar, V., (1980a) "Molecular orientational ordering in the smectic-C* and H* phases of chiral TABCA" *Ferroelectrics* **24**, 211.

Vilfan, Marija, Seliger, J., Žagar, V., and Blinc, R., (1980b) "^{14}N NQR and ^{1}H NMR study of the smectic-B phase of IBPBAC", Physics Letters **79 A**, 186.

Vilfan, Marija, and Vrbančič-Kopač, N., (1996) " NMR of liquid crystals with embedded polymer network", in *Liquid Crystals in Complex Geometries*, editer by S.Žumer and G.P.Crawford, Taylor and FRancis 1996.

Walba, D.M., Stevens, F., Parks, D.C., Clark, N.A., and Wand M.D., (1995) "Near atomic resolution imaging of ferroelectric liquid crystal molecules on graphite by STM", *Science* **267**, 1144.

Wang, X.Y., and Taylor, P.L., (1996) "Devil's staircase, critical thickness and propagating fingers in antiferroelectric liquid crystals", *Physical Review Letters* **76**, 640.

Wang, X.Y., Kyu, T., Rudin, A.M., and Taylor, P.L., (1998) "Freedericksz transition in antiferroelectric liquid crystals and cooperative motion of smectic layers", *Physical Review* **E 58**, 5919.

Wang, Z.H., Sun, Z.M., and Feng, D., (1991) "Effects of high-frequency a.c. electric field on the helical structure in a ferroelectric liquid crystal", *Europhysics Letters* **14**, 785.

Whittaker, E.T., and Watson, G.N., (1935), *Modern Analysis*, Cambridge Universty Press, Cambridge, England.

Wiesendanger, R., and Guntherodt, H.J., editors, (1992) *Scanning Tunneling Microscopy II*, Springer Series in Surface Science.

Willis, P.C., Clark, N.A., and Safinya, C.R., (1992) "Molecular director and layer response of chevron SSFLC to low electric field", *Liquid Crystals* **11**, 581.

Wise, R.A., Smith, D.H., and Doane, J.W., (1973) "Nuclear magnetic resonance in the smectic phase", *Physical Review* **A 7**, 1366.

Wrobel, S., Biradar, A.M., and Haase, W., 1989, "Soft mode behaviour in the SmA and SmC* phases as studied by dielectric spectroscopy", *Ferroelectrics* **100**, 271-279.

Wu, B.-G., and Doane, J.W., (1987) "Deuterium NMR and magneto-aligned samples of the chiral smectic-C* phase", *Journal of Magnetic Resonance* **75**, 39-49.

Wu, X.I., Goldburg, W.I., Liu, M.X., and Xue, J.Z., (1992) "Slow dynamics of isotropic-nematic phase transition in silica gels", *Physical Review Letters* **69**, 470-473.

Xue, Jiuzhi, Clark, N.A., and Meadows, M.R., (1988) "Surface orientation transitions in surface stabilized ferroelectric liquid crystals", *Applied Physics Letters* **53**, 2397.

Xue, Jiuzhi, and Clark, N.A., (1990) "Surface electroclinic effect in chiral smectic-A liquid crystal", *Physical Review Letters* **64**, 307-310.

Yamada, Y., Yamamoto, N., Inoue, T., Orihara, H., and Ishibashi, Y., (1989) "Investigations of switching behaviour in a FLC aligned on obliquely deposited SiO films", *Japanese Journal of Applied Physics* **28**, 50.

Yamashita, M., Kimura, H., and Nakano, H., (1981) "On the stability of cholesteric phase exposed to magnetic field", *Progress in Theoretical Physics* **65**, 1504.

Yamashita, M., (1983) "Reentrant phase transition of chiral smectic-C* liquid crystal in electric field", *Journal of the Physical Society of Japan* **52**, 3735.

Yamashita, M., (1984) "Vortex-street structure of surface induced SmC*-SmC transition in ferroelectric liquid crystal", *Ferroelectrics* **58**, 149.

Yamashita, M., (1985), "Multisolitons and soliton lattices in sine-Gordon system with variable amplitude", *Progress in Theoretical Physics* **74**, 622.

Yamashita, M., (1988,)"Phase transitions of ferroelectric liquid crystals in a magnetic field", *Journal of the Physical Society of Japan* **57**, 2051.

Yamashita, M., and Kimura, H., (1988) "Phase transitions and solitons in liquid crystals", *Molecular Crystals and Liquid Crystals* **165**, 213-232.

Yamashita, M., (1992) "Solitons and phase transitions in ferroelectric smectics", in *Solitons in Liquid crystals*, edited by Lam, L., and Prost, J.,, Springer-Verlag. See also the references therein.

Yamashita, M., (1991) "On the first order phase transitions of the antiferroelectric liquid crystals", *Journal of The Physical Society of Japan 60*, 219-226.

Yamashita, M., and Miyazima, S., (1993) "Successive phase transitions ferro-ferri and antiferroelectric smectics", *Ferroelectrics* **148**, 1.

Yamashita, M., (1996a) "Mechanism to stabilize the ferrielectric phase in ferroelectric smectics", *Journal of the Physical Society of Japan* **65**, 2904.

Yamashita, M., (1996b) "Long range interactions and the sense of molecular long axis in ferroelectric smectics", *Journal of the Physical Society of Japan* **65**, 134.

Yamashita, M., (1997) "Mechanism to stabilize the ferrielectric phase in ferroelectric smectics-the order of the phase sequence", *Journal of the Physical Society of Japan* **66**, 130.

Yang, K.H., and Chieu, T.C., (1988) "Bipolar switching of layer-tilted state in FLC", *Ferroelectrics* **85**, 153.

Yang, Y.B., Bang, T., Mochizuki, A., and Kobayashi, S., (1991) "The surface anchoring dependence of the dielectric constant of an FLC material showing electroclinic effect", *Ferroelectrics* **121**, 113-125.

Yang, F., Bradberry, G.W., and Sambles, J.R., (1994) "Optical confirmation of the extended mean-field theory for a smectic-C*-smectic-A transition", *Physical Review* **E 50**, 2834.

Yokoyama, H., (1988) "Nematic-isotropic transition in bounded thin films", *Journal of Chemical Society, Faraday Transactions* **84**, 1023-1040.

Yoneya, M., Kondo, K., Odamura, M., and Kitamura, T., (1989) "A new operation mode of a ferroelectric liquid crystal cell with an asymmetrical surface condition", *Japanese Journal of Applied Physics* **28**, 2247.

Yoshida, S., Jin, B., Takanishi, Y., Tokumaru, K., Ishikawa, K., Takezoe, H., Fukuda, A., Kusumoto, T., Nakai, T., and Miyajima, S., 1999, "A bent and assymetrically hindered chiral alkyl chain in smectic-A phase of antiferroelectric liquid crystal as observed by ^{2}H-NMR", *Journal of the Physical Society of Japan* **68**, 9.

Yoshizawa, A., Nishiyama, I., Kikuzaki, H., and Ise, N., (1992) "C-13 NMR and X-ray investigations of the phase transitions in an antiferroelectric liquid crystal", *Japanese Journal of Applied Physics* **31**, L860.

Young, C.Y., Pindak, R., Clark, N.A., and Meyer, R.B., (1978) "Light-scattering study of two-dimensional molecular orientation fluctuations in a freely suspended ferroelectric liquid crystal film", *Physical Review Letters* **40**, 773-776.

Zalar, B., Gregorovič, A., Simsič, M., Zidanšek, A., and Blinc, R., (1998) "Anisotropy of the critical magnetic field in a ferroelectric liquid crystal", *Physical Review Letters* **80**, 4458.

Ziman, J.M., (1962), *Electrons and Phonons*, Oxford University Press, London.

Zhuang, Z., Clark, N.A., and Maclennan, J.E., (1990) "Optical symmetry of ferroelectric liquid crystal cells", *Japanese Journal of Applied Physics* **29**, L2239.

Zhuang, Z., Clark, N.A., and Maclennan, J.E., (1991) "Visible polarized light transmission spectroscopy of the electro-optic switching of SSFLC cells", *Liquid Crystals* **10**, 409.

Žekš, B., Levstik, A., and Blinc, R., (1979) "On the dynamics of helicoidal ferroelectric SmC* liquid crystals", *Journal de Physique* **40**, C3-409-412.

Žekš, B., (1984) "Landau free-energy expansion for chiral ferroelectric smectic liquid crystal", *Molecular Crystals Liquid Crystals* **114**, 259.

Žekš, B., Carlsson, T., Filipič, C., and Urbanc, B., (1988) "Thermodynamic model of ferroelectric liquid crystals and its microscopic basis", *Ferroelectrics* **84**, 3-14.

Žekš, B., Blinc, R., and Čepič, M., (1991) "Landau-Khalatnikov dynamics of helicoidal antiferroelectric liquid crystals", *Ferroelectrics* **122**, 221-228.

Žekš, B., and Blinc, R., (1992) "Dielectric properties of smectic-C* liquid crystals", *Molecular Crystals and Liquid Crystals* **220**, 63.

Žekš, B., and Čepič, M., (1993) "A phenomenological model of antiferroelectric liquid crystals", *Liquid Crystals* **14**, 445-451.

Subject Index

C

D

E

F

G

H

I

J

L

M

N

O

P

Q

R

S

T

U

V

W

X

Z